AF348957

Barry G. Adams

# Algebraic Approach to Simple Quantum Systems

## With Applications to Perturbation Theory

Springer-Verlag Berlin Heidelberg GmbH

Prof. Barry G. Adams
Department of Mathematics and Computer Science
Laurentian University
Ramsey Lake Road
Sudbury, Ontario P3E 2C6
Canada

**Additional material to this book can be downloaded from http://extras.springer.com.**

ISBN 978-3-540-57801-7      ISBN 978-3-642-57933-2 (eBook)
DOI 10.1007/978-3-642-57933-2

Cip data applied for



Typesetting: Camera-ready by author
SPIN:10129034      51/3020-5 4 3 2 1 0 - Printed on acid-free paper

To my mother and father

# Preface

This book provides an introduction to the use of algebraic methods and symbolic computation for simple quantum systems with applications to large order perturbation theory. It is the first book to integrate Lie algebras, algebraic perturbation theory and symbolic computation in a form suitable for students and researchers in theoretical and computational chemistry and is conveniently divided into two parts.

The first part, Chapters 1 to 6, provides a pedagogical introduction to the important Lie algebras so(3), so(2,1), so(4) and so(4,2) needed for the study of simple quantum systems such as the D-dimensional hydrogen atom and harmonic oscillator. This material is suitable for advanced undergraduate and beginning graduate students. Of particular importance is the use of so(2,1) in Chapter 4 as a spectrum generating algebra for several important systems such as the non-relativistic hydrogen atom and the relativistic Klein-Gordon and Dirac equations. This approach provides an interesting and important alternative to the usual textbook approach using series solutions of differential equations.

The second part, Chapters 7 to 10, provides an introduction to large order perturbation theory, using the so(4,2) algebraic approach developed in the first part, to large order perturbation theory for the Stark and Zeeman perturbations in hydrogenic systems and central field perturbations arising from the charmonium, harmonium, screened Coulomb and Yukawa potentials. The emphasis here is on the symbolic computation of the energy and wavefunction corrections using the Maple computer algebra system. Many results previously available only in research journals are presented here along with some new results and applications.

The general concepts of Lie algebras and their matrix representations are introduced in Chapter 1 using only the familiar concepts of vector spaces, linear transformations and matrices with examples from angular momentum theory. In Chapter 2 commutation relations involving position, momentum and angular momentum operators $r$, $p$ and $L$ are derived in a systematic way. They are needed to obtain the coordinate realizations of the Lie algebras so(3), so(2,1), so(4) and so(4,2) which are of primary importance in algebraic perturbation theory.

A brief review of so(3) is presented in Chapter 3. This is just the familiar angular momentum theory from an algebraic viewpoint. Since angular momentum theory is covered in considerable detail in many specialized books and textbooks on quantum mechanics using both the algebraic and differential equations approaches, we consider only the basic ideas which illustrate, in a familiar context, the algebraic approach used in subsequent chapters to study more complicated Lie algebras. The important concept of a vector operator with respect to so(3) is also introduced here and the matrix representation of a vector operator is derived. These results are useful later in the derivation of matrix representations of so(3,1) and so(4).

In Chapter 4 the matrix representations of so(2,1) are derived following the same approach used in the previous chapter for so(3). The general theory is the same in either case. However, there are fundamental differences when we restrict our representations to the unitary irreducible representations (unirreps) needed for the applications of so(2,1) as a spectrum generating algebra for a class of radial Schrödinger equations. In fact all the unirreps of so(3) are finite-dimensional whereas those of so(2,1) are infinite-dimensional.

Next, realizations of the so(2,1) generators in the coordinate representation are derived in a form suitable for expressing the radial eigenvalue problems for various simple quantum systems such as the hydrogen atom. We use these realizations, the matrix representations of so(2,1) and a simple scaling transformation to show that the radial Schrödinger equation for the D-dimensional hydrogen atom and harmonic oscillator can be expressed as eigenvalue problems for one of the so(2,1) generators. Since the classification of the eigenvalue spectra of this generator are known from the matrix representation theory, the formulas for the energy levels and wavefunctions are directly obtained, thus providing a purely algebraic alternative to the usual differential equation approach.

We also consider other systems such as the Klein-Gordon equation and the relativistic Dirac equation. Finally we show how to separate the Schrödinger equation for the 3-dimensional hydrogen atom in parabolic coordinates and we obtain a parabolic realization of the so(2,1) generators. This realization will be useful in the application of perturbation theory to the Stark effect in Chapter 8.

To proceed to higher Lie algebras containing both so(3) and so(2,1) as subalgebras it is necessary to consider the representation theory and realizations of so(4), the Lie algebra of the 4-dimensional rotation group. In Chapter 5 so(4) and so(3,1) are introduced by closing out the commutation relations of so(3) with a vector operator. Then the general representation theory is developed. Again when we specialize to unirreps it is found that all the unirreps of so(4) are finite-dimensional whereas those of so(3,1) are infinite-dimensional.

In the so(4) case, in which we are primarily interested, an important realization of this vector operator is provided by the quantum mechanical version of the classical Laplace-Runge-Lenz vector first used by Pauli in his purely algebraic group theoretical treatment of the hydrogen atom. This energy-dependent realization of so(4) is not suitable for merging with our so(2,1) realizations so we apply a scaling transformation to obtain an energy-independent realization referred to as the scaled hydrogenic realization of so(4). It follows that the scaled hydrogenic basis functions introduced in Chapter 4 provide a matrix representation of both so(2,1) and our scaled hydrogenic representation of so(4).

Chapter 6 concludes the first part of the book. Here we finally merge together the Lie algebras so(3), so(2,1) and so(4) into a bigger Lie algebra by closing out the commutation relations with three additional vector operators. The result is a Lie algebra with 15 generators called so(4,2). The scaled hydrogenic realization of these generators is then obtained. The important result is that the complete set of scaled hydrogenic wavefunctions form a basis for a single infinite dimensional unirrep of so(4,2). This means that we can easily calculate the matrix elements of any operator expressible in terms of the so(4,2) generators. In particular the various perturbations considered in the second part of the book are easily expressed in terms of the so(4,2) generators and their matrix elements are simple finite sums.

In Chapter 7 Rayleigh-Schrödinger perturbation theory (RSPT) is presented in a general context for a nondegenerate reference state in preparation for several detailed examples of hydrogenic perturbation theory in the next chapters. First we develop the conventional RSPT formalism in terms of iterative formulas for the wavefunction and energy corrections.

Our interest is in hydrogenic perturbation theory for which the unperturbed Schrödinger equation is defined by a hydrogenic hamiltonian. In this important case it is well known that conventional perturbation theory is incomplete in the sense that the unperturbed radial hydrogenic eigenfunctions for bound states do not form a complete set for the expansion of a bound state solution to the perturbed Schrödinger equation. The continuum states must also be taken into account in the perturbation expansions. This problem begins in second order and higher orders become unmanageable due to the multiple integrations over the continuum state contributions. Even if these contributions were negligible there remain slowly converging infinite sums over the discrete set of unperturbed bound states. We briefly discuss the Dalgarno and Lewis method for avoiding the continuum states by directly solving the inhomogeneous differential equations. However this method can not be easily extended in a systematic way to higher orders.

The most promising methods for avoiding the continuum state contributions are those which do not require the direct calculation of the wavefunction (see Chapter 10) and the algebraic perturbation theory based on the Lie algebra so(4,2) which relies on a simple scaling transformation to replace the unperturbed hamiltonian by a new unperturbed eigenvalue problem which has only discrete states. Here the matrix representations of the most important perturbations are not expressed as infinite sums over this complete set of discrete states but as finite sums with a small number of terms. This perturbation theory is referred to as modified algebraic RSPT and the formalism is developed in Chapter 7.

In Chapter 8 the algebraic RSPT for the Stark effect is developed to large order for the ground state using symbolic computation and the Maple computer algebra system. The Stark effect is a special case in the sense that the conventional perturbation theory is also applicable to high order using the separation in parabolic coordinates. We consider the symbolic computation for the perturbation of a general parabolic state in the two-dimensional and three-dimensional cases, thus extending the original work of Alliluev and Malkin and of Silverstone. This is an excellent example of the application of symbolic methods to perturbation theory.

In Chapter 9 the Zeeman effect is considered in detail. Unlike the Stark effect, which is reducible to a one-dimensional problem by separation in parabolic coordinates, the Zeeman effect is inherently a two-dimensional problem. In this chapter we show how the modified algebraic RSPT of Chapter 7 can be used to obtain the energy and wavefunction corrections in rational form to high order using Maple. We also develop the modified algebraic RSPT for degenerate states in the special case of a doubly degenerate state by adapting the general case of conventional degenerate RSPT and applying it to the $3s - 3d_0$ sublevel which is the first case where degenerate RSPT is needed for the Zeeman effect. Finally we show how the method of moments, which does not require the wavefunction, can also be used to obtain the ground state energy corrections in rational form.

The modified algebraic RSPT is then applied to several spherically symmetric perturbations of a hydrogenic system in Chapter 10. We consider the power potentials of the form $r^d$ for the charmonium ($d = 1$) and harmonium ($d = 2$) cases. Next we consider alternative approaches to obtaining the energy series that do not require the wavefunction but are easily adaptable to symbolic computation. The first method is a power series or difference equation approach which is applied to the Yukawa potential and the second method is the HVHF method which is based on the Hypervirial and Hellman-Feynman theorems. This method is applied to the charmonium and harmonium potentials and to the screened Coulomb and Yukawa potentials.

Several appendices containing tables of symbolic results for various energy series and the solutions to all exercises are also included.

To conclude, I would like to thank Professors Josef Paldus and Jiří Čížek for many fruitful and stimulating discussions over the the past several years. This book, which had its beginnings in the articles [CI77a] and [AD88b], is a direct result of my collaboration with them.

# Contents


**1 General Discussion of Lie Algebras**      **1**
  1.1 Introduction . . . . . . . . . . . . . . . . . . . . . . . 1
  1.2 Definition of a Lie Algebra . . . . . . . . . . . . . . . . 1
      1.2.1 Example: the general linear Lie algebra . . . . . . . . . . 3
  1.3 Representations of a Lie Algebra . . . . . . . . . . . 4
      1.3.1 Examples of matrix representations . . . . . . . . . . 4
  1.4 Realizations of a Lie Algebra . . . . . . . . . . . . . . 5
      1.4.1 Examples of realizations . . . . . . . . . . 7
  1.5 Irreducible Representations . . . . . . . . . . . . . . 10
      1.5.1 Example: a reducible representation . . . . . . . . . . 11
  1.6 Exercises . . . . . . . . . . . . . . . . . . . . . . . 12

**2 Commutator Gymnastics**      **17**
  2.1 Introduction . . . . . . . . . . . . . . . . . . . . . . . 17
  2.2 General Commutator Identities . . . . . . . . . . . . . 17
  2.3 Commutators Involving r and p . . . . . . . . . . . . . 19
  2.4 Commutators Involving L . . . . . . . . . . . . . . . . 21
  2.5 Exercises . . . . . . . . . . . . . . . . . . . . . . . 23

**3 Angular Momentum Theory and so(3)**      **25**
  3.1 Introduction . . . . . . . . . . . . . . . . . . . . . . . 25
  3.2 Ladder Operators for so(3) . . . . . . . . . . . . . . . 26
  3.3 General Eigenvalue Spectrum of $J_3$ . . . . . . . . . . 28
  3.4 Unirreps of so(3) . . . . . . . . . . . . . . . . . . . 28
  3.5 so(3) Scalar Operators . . . . . . . . . . . . . . . . 30
  3.6 so(3) Vector Operators . . . . . . . . . . . . . . . . 30
  3.7 Selection Rules for a Vector Operator . . . . . . . . . . 31
  3.8 Matrix Elements of a Vector Operator . . . . . . . . . . 33
  3.9 Wigner-Eckart Theorem . . . . . . . . . . . . . . . . 37
  3.10 Exercises . . . . . . . . . . . . . . . . . . . . . . . 39

**4 Representations and Realizations of so(2,1)**      **41**
  4.1 Introduction . . . . . . . . . . . . . . . . . . . . . . . 41
  4.2 Ladder Operators for so(2,1) . . . . . . . . . . . . . . 42
  4.3 Unirreps of so(2,1) . . . . . . . . . . . . . . . . . . 44

4.4   Realizations of so(2,1) .......................... 45
4.5   3-dimensional Hydrogenic Atom .................. 50
4.6   3-dimensional Harmonic Oscillator ............... 53
4.7   Generalization to D Dimensions ................. 54
4.8   D-dimensional Hydrogenic Atom ................. 56
4.9   D-dimensional Harmonic Oscillator ............... 56
4.10  Hermiticity of so(2,1) Generators ............... 58
4.11  Explicit $T_3$ Eigenfunctions ..................... 60
4.12  D-dimensional Relativistic Systems .............. 66
       4.12.1  Klein-Gordon equation .................. 66
       4.12.2  Dirac-Coulomb equation ................ 68
4.13  General Radial Equations ..................... 71
4.14  Parabolic Hydrogenic Realization of so(2,1) ........ 72
       4.14.1  Parabolic coordinate system ............. 72
       4.14.2  Separation in parabolic coordinates ....... 73
       4.14.3  Algebraic solution: method 1 ............ 76
       4.14.4  Algebraic solution: method 2 ............ 77
4.15  Exercises ............................... 78

**5  Representations and Realizations of so(4)    81**
5.1   Introduction ............................. 81
5.2   Representations of so(4) ...................... 83
5.3   so(4) as a Coupling Problem ................... 89
5.4   Laplace-Runge-Lenz Vector ................... 94
       5.4.1  Classical case ...................... 94
       5.4.2  Quantum mechanical case .............. 95
5.5   Hydrogenic Realizations of so(4) ................ 97
       5.5.1  Pauli's hydrogenic realization of so(4) ...... 98
       5.5.2  Scaled hydrogenic realization of so(4) ...... 100
5.6   Exercises ............................... 102

**6  Scaled Hydrogenic Realization of so(4,2)    105**
6.1   Introduction ............................. 105
6.2   Scaled Hydrogenic Realization of so(4,1) ........... 106
6.3   Scaled Hydrogenic Realization of so(4,2) ........... 108
6.4   Matrix Representation of so(4,2) ................ 111
6.5   Hydrogenic Tower of States ................... 114
6.6   Exercises ............................... 116

**7  Lie Algebraic Perturbation Theory    119**
7.1   Introduction ............................. 119
7.2   Conventional RSPT ........................ 120
       7.2.1  Conventional iteration formula ........... 122
       7.2.2  Conventional low order formulas .......... 122
       7.2.3  Conventional high-order iteration scheme ..... 123

      7.2.4   Example: continuum difficulties for Stark effect . . . . . 124
      7.2.5   Dalgarno and Lewis method for Stark effect . . . . . . . 125
  7.3  Scaled Hydrogenic Hamiltonian . . . . . . . . . . . . . . . . 127
  7.4  Modified Algebraic RSPT . . . . . . . . . . . . . . . . . . . 129
      7.4.1   Modified iteration formula . . . . . . . . . . . . . . . . 131
      7.4.2   Symmetric energy formula . . . . . . . . . . . . . . . . 131
      7.4.3   Modified low order formulas . . . . . . . . . . . . . . . 132
      7.4.4   Modified high order iteration scheme . . . . . . . . . . 132
      7.4.5   Example: ground state Stark effect . . . . . . . . . . 134
  7.5  Exercises . . . . . . . . . . . . . . . . . . . . . . . . . . . . 136

**8 Symbolic Calculation of the Stark Effect**           **137**
  8.1  Introduction . . . . . . . . . . . . . . . . . . . . . . . . . . . 137
  8.2  Symbolic Calculations for the Ground State . . . . . . . . . . 138
      8.2.1   Modified high order formalism . . . . . . . . . . . . . . 138
      8.2.2   Renormalized matrix elements . . . . . . . . . . . . . . 141
      8.2.3   Program *gsstark* for $E^{(j)}$ and $\Psi^{(j)}$ . . . . . . . . . . 144
      8.2.4   Program for symmetric energy formula . . . . . . . . . 149
  8.3  Summing the Ground State Series . . . . . . . . . . . . . . . 152
      8.3.1   Direct summation of the series . . . . . . . . . . . . . 152
      8.3.2   Summation using Padé approximants . . . . . . . . . . 153
  8.4  Stark Effect in Parabolic Coordinates . . . . . . . . . . . . . 158
      8.4.1   Separation and perturbation formalism . . . . . . . . . 158
      8.4.2   Maple program for $f^{(0)}$ to $f^{(4)}$ . . . . . . . . . . . 162
      8.4.3   Unravelling the energy expansion . . . . . . . . . . . 164
      8.4.4   Higher orders: general case . . . . . . . . . . . . . . . 166
      8.4.5   Higher orders: special cases . . . . . . . . . . . . . . 172
  8.5  2-Dim Stark Effect in Parabolic Coordinates . . . . . . . . . 174
  8.6  Exercises . . . . . . . . . . . . . . . . . . . . . . . . . . . . 177

**9 Symbolic Calculation of the Zeeman Effect**         **179**
  9.1  Introduction . . . . . . . . . . . . . . . . . . . . . . . . . . . 179
  9.2  Low Order Calculations for the Ground State . . . . . . . . . 180
  9.3  Symbolic Calculations for the Ground State . . . . . . . . . . 182
      9.3.1   Modified high order formalism . . . . . . . . . . . . . . 182
      9.3.2   Renormalized matrix elements . . . . . . . . . . . . . . 184
      9.3.3   Program *gszeeman* for $E^{(j)}$ and $\Psi^{(j)}$ . . . . . . . . 186
      9.3.4   Program for symmetric formula . . . . . . . . . . . . . 189
      9.3.5   Summation of the ground state series . . . . . . . . . . 191
  9.4  Degenerate RSPT for the Zeeman Effect . . . . . . . . . . . 191
      9.4.1   Doubly-degenerate modified RSPT . . . . . . . . . . . 192
      9.4.2   The $3s-3d_0$ degenerate level . . . . . . . . . . . . . 195
      9.4.3   Procedure *zee3s3d* for the $3s$ and $3d_0$ states . . . . . . . 197
      9.4.4   Energy corrections for the $3s$ and $3d_0$ states . . . . . . . 204
  9.5  Method of Moments . . . . . . . . . . . . . . . . . . . . . . 206

9.6    Exercises . . . . . . . . . . . . . . . . . . . . . . . . . . 209

**10 Spherically Symmetric Systems**                                    **211**
  10.1  Introduction . . . . . . . . . . . . . . . . . . . . . . . 211
  10.2  RSPT for Charmonium and Harmonium . . . . . . . . . . 212
       10.2.1  Low order calculations . . . . . . . . . . . . . . 213
       10.2.2  Symbolic calculations . . . . . . . . . . . . . . . 214
       10.2.3  Maple programs for charmonium and harmonium . . . . 216
  10.3  RSPT for Screened Coulomb Potential . . . . . . . . . . 226
  10.4  Power Series Method . . . . . . . . . . . . . . . . . . . 228
  10.5  HVHF Perturbation Method . . . . . . . . . . . . . . . . 233
       10.5.1  Hypervirial and Hellmann-Feynman theorems . . . . . 234
       10.5.2  HVHF method for hydrogenic systems . . . . . . . . 234
  10.6  HVHF Method for Power Potentials . . . . . . . . . . . . 236
  10.7  HVHF Method: Screened Coulomb Case . . . . . . . . . . 240
  10.8  Exercises . . . . . . . . . . . . . . . . . . . . . . . . 243

**A  The Levi-Civita Symbol**                                           **245**
  A.1  Basic Properties . . . . . . . . . . . . . . . . . . . . . 245
  A.2  Important Identities . . . . . . . . . . . . . . . . . . . 245
  A.3  Exercises . . . . . . . . . . . . . . . . . . . . . . . . 246

**B  Lie Groups and Lie Algebras**                                      **247**
  B.1  Some Definitions . . . . . . . . . . . . . . . . . . . . . 247
  B.2  One-parameter Subgroups . . . . . . . . . . . . . . . . . 248
  B.3  Rotation Group SO(2) . . . . . . . . . . . . . . . . . . . 250
  B.4  Rotation Group SO(3) . . . . . . . . . . . . . . . . . . . 250
  B.5  Special Unitary Group SU(2) . . . . . . . . . . . . . . . 252
  B.6  Mapping from SU(2) to SO(3) . . . . . . . . . . . . . . . 253
  B.7  $SO(n)$ and its Lie Algebra $so(n)$ . . . . . . . . . . . 253
       B.7.1  Special case: so(3) . . . . . . . . . . . . . . . . 255
       B.7.2  Special case: so(4) . . . . . . . . . . . . . . . . 256
  B.8  $SO(p,q)$ and its Lie Algebra $so(p,q)$ . . . . . . . . . 256
       B.8.1  Special case: so(2,1) . . . . . . . . . . . . . . . 258
       B.8.2  Special case: so(4,1) . . . . . . . . . . . . . . . 259
       B.8.3  Special case: so(4,2) . . . . . . . . . . . . . . . 259
  B.9  Exercises . . . . . . . . . . . . . . . . . . . . . . . . 260

**C  The Tilting Transformation**                                       **261**
  C.1  Transformation of Functions . . . . . . . . . . . . . . . 261
  C.2  Transformation of Operators . . . . . . . . . . . . . . . 262
  C.3  Transformation of the Radial Equation . . . . . . . . . . 262
       C.3.1  General case . . . . . . . . . . . . . . . . . . . 262
       C.3.2  Special case of hydrogenic atom . . . . . . . . . . 263
  C.4  Tilted Laplace-Runge-Lenz Vector . . . . . . . . . . . . 264

C.5   Exercises . . . . . . . . . . . . . . . . . . . . . . . . 264

**D   Perturbation Matrix Elements**                                        **265**
   D.1   Basic Formulas for Matrix Elements . . . . . . . . . . . . 265
         D.1.1   3-dimensional case  . . . . . . . . . . . . . . . 265
         D.1.2   D-dimensional case . . . . . . . . . . . . . . . 266
   D.2   Symbolic Calculation of Tables  . . . . . . . . . . . . . 267
   D.3   Special Values of Matrix Elements . . . . . . . . . . . . 274
         D.3.1   Matrix elements of $R$ for $n = 1,\ldots,5$ . . . . . . . . . . 274
         D.3.2   Matrix elements of $W = RZ$ for $n = 1,\ldots,5,\, m = 0$ . . . 275

**E   Tables of Stark Effect Energy Corrections**                           **281**
   E.1   Ground State Energy to Order 100 . . . . . . . . . . . . 281
   E.2   Symbolic Energy Corrections for Parabolic States . . . . . . . . 288

**F   Tables of Zeeman Effect Energy Corrections**                          **295**
   F.1   Ground State Energy to Order 60 . . . . . . . . . . . . . 295

**G   Tables of Charmonium Energy Corrections**                             **313**
   G.1   Ground State Energy to Order 100  . . . . . . . . . . . . 313
   G.2   Symbolic Energy Corrections to Order 14 . . . . . . . . . . 326

**H   Tables of Harmonium Energy Corrections**                              **331**
   H.1   Ground State Energy to Order 100  . . . . . . . . . . . . 331
   H.2   Symbolic Energy Corrections to Order 10 . . . . . . . . . . 349

**I   Tables of Screened Coulomb Energy Corrections**                       **355**
   I.1   Ground State Energy to Order 100 . . . . . . . . . . . . 355
   I.2   Ground State Yukawa Energy for D Dimensions . . . . . . . . 368
   I.3   Energy Series for Screened Coulomb Potential . . . . . . . . 371
   I.4   Energy Series for Yukawa Potential . . . . . . . . . . . . 375

**J   Solutions to Exercises**                                              **379**
   J.1    Solutions to Chapter 1 Exercises . . . . . . . . . . . . 379
   J.2    Solutions to Chapter 2 Exercises . . . . . . . . . . . . 385
   J.3    Solutions to Chapter 3 Exercises . . . . . . . . . . . . 389
   J.4    Solutions to Chapter 4 Exercises . . . . . . . . . . . . 393
   J.5    Solutions to Chapter 5 Exercises . . . . . . . . . . . . 399
   J.6    Solutions to Chapter 6 Exercises . . . . . . . . . . . . 409
   J.7    Solutions to Chapter 7 Exercises . . . . . . . . . . . . 415
   J.8    Solutions to Chapter 8 Exercises . . . . . . . . . . . . 424
   J.9    Solutions to Chapter 9 Exercises . . . . . . . . . . . . 424
   J.10  Solutions to Chapter 10 Exercises . . . . . . . . . . . . 426
   J.11  Solutions to Appendix A Exercises . . . . . . . . . . . . 427
   J.12  Solutions to Appendix B Exercises . . . . . . . . . . . . 428
   J.13  Solutions to Appendix C Exercises . . . . . . . . . . . . 432

**Bibliography**                                                   **435**

**Index of Symbols Used**                                          **443**

**Index**                                                          **447**

# Chapter 1

# General Discussion of Lie Algebras

## 1.1  Introduction

In this chapter a discussion of the general aspects of Lie algebras is presented using only the familiar concepts of vector spaces, linear transformations and operators, matrices and commutators. Our treatment is brief in order to provide only the necessary background for an understanding of the applications in later chapters. The relationship between a Lie algebra and its corresponding Lie group is discussed in Appendix B.

We first define an algebra in terms of a vector space with a bilinear multiplication and then define a Lie algebra as a particular kind of nonassociative, noncommutative algebra. Next the concepts of a representation, matrix representation, irreducible representation, and realization of a Lie algebra are discussed since they are important for our applications in later chapters of Lie algebras to the evaluation of matrix elements of quantum mechanical operators.

## 1.2  Definition of a Lie Algebra

First we define a linear algebra over a field $\mathcal{F}$ as a vector space $\mathcal{A}$ over $\mathcal{F}$ having a bilinear law of composition (multiplication) denoted by $A \circ B$ which has the following properties, for all $\alpha \in \mathcal{F}$ and $A, B, C \in \mathcal{A}$.

$$
\begin{aligned}
&(1) \quad && A \circ B \in \mathcal{A}, \\
&(2a) \quad && A \circ (B + C) = A \circ B + A \circ C, \\
&(2b) \quad && (A + B) \circ C = A \circ C + B \circ C, \\
&(3) \quad && \alpha(A \circ B) = (\alpha A) \circ B = A \circ (\alpha B).
\end{aligned}
$$

The product of $A$ and $B$ is often denoted, using juxtaposition, by $AB$. We shall always assume that $\mathcal{F}$ is either the real or complex number field.

The first rule guarantees the closure of the algebra under the multiplication. The distributive rules (2) and rule (3) combine this multiplication with the multiplication by the scalars of the underlying vector space $\mathcal{A}$ and imply the usual bilinearity rule

$$\left(\sum_j \alpha_j A_j\right) \circ \left(\sum_k \beta_k B_k\right) = \sum_{j,k} \alpha_j \beta_k (A_j \circ B_k), \qquad (1.1)$$

for all scalars $\alpha_j, \beta_k \in \mathcal{F}$ and elements $A_j, B_k \in \mathcal{A}$.

Different types of algebras can be defined by imposing further rules. For example a commutative algebra would have the additional rule that the multiplication be commutative:

$$(4')\quad A \circ B = B \circ A.$$

In this case rule $(2b)$ follows from rule $(2a)$ and would be redundant. An associative algebra would have the additional rule that the multiplication be associative:

$$(4'')\quad A \circ (B \circ C) = (A \circ B) \circ C.$$

A commutative associative algebra would include both rules $(4')$ and $(4'')$. A standard example of an algebra that is associative but not commutative is the $n^2$-dimensional algebra of $n \times n$ matrices over the real or complex number fields with respect to the usual operations of scalar multiplication, matrix addition and matrix multiplication.

In general a Lie algebra $\mathcal{L}$ is neither associative nor commutative. The Lie multiplication is usually denoted by $[A, B]$ rather than $A \circ B$ and is also called the bracket of $A$ and $B$. If we use $\mathcal{L}$ to denote both the Lie algebra and the underlying vector space then the Lie multiplication has the following properties, for all $\alpha \in \mathcal{F}$ and $A, B, C \in \mathcal{L}$.

$$
\begin{aligned}
&(1)\quad [A, B] \in \mathcal{L}, \\
&(2)\quad [A, B + C] = [A, B] + [A, C], \\
&(3)\quad \alpha[A, B] = [\alpha A, B] = [A, \alpha B], \\
&(4)\quad [A, B] = -[B, A], \\
&(5)\quad [A, [B, C]] + [B, [C, A]] + [C, [A, B]] = 0.
\end{aligned}
$$

Thus a Lie algebra is an algebra with the additional rules (4) and (5). Rule (4) shows that the Lie multiplication is anticommutative. The left distributive rule $[A + B, C] = [A, C] + [B, C]$ follows from rules (2) and (4). Rule (5) shows that a Lie algebra is not associative. It is called the Jacobi identity and replaces rule $(4'')$ of an associative algebra. Rules (2) and (3) imply the bilinearity rule analogous to (1.1)

$$\left[\sum_j \alpha_j A_j, \sum_k \beta_k B_k\right] = \sum_{j,k} \alpha_j \beta_k [A_j, B_k]. \qquad (1.2)$$

Lie algebras can often be obtained from an associative algebra $\mathcal{A}$ of matrices, linear transformations or linear operators on some vector space, by defining the Lie multiplication of two elements $A, B \in \mathcal{A}$ as the commutator

$$[A, B] = AB - BA, \qquad (1.3)$$

in terms of the associative multiplication (denoted here by juxtaposition). The importance of the commutator is that it satisfies the defining rules (2) to (5) of a Lie algebra (see Exercise 1.2). Thus to obtain a Lie algebra using the commutator it is only necessary to verify the closure rule (1) in any particular case.

Since a Lie algebra has an underlying vector space structure a basis for the vector space is also said to be a basis for the Lie algebra. If $\dim(\mathcal{L}) = n$ and $\{E_j : j = 1,\ldots,n\}$ is a basis for $\mathcal{L}$ the closure property (1) implies that the Lie products of the basis vectors have the form

$$[E_j, E_k] = \sum_\ell c_{jk\ell} E_\ell. \tag{1.4}$$

The constants $c_{jk\ell}$ are called structure constants of the Lie algebra. Because of the bilinearity rule (1.2), the Lie product of any two elements of the Lie algebra is completely specified by (1.4): if $A = \sum_j \alpha_j E_j$ and $B = \sum_k \beta_k E_k$ then $[A, B] = C$, where $C = \sum_\ell \gamma_\ell E_\ell$ and $\gamma_\ell = \sum_{jk} \alpha_j \beta_k c_{jk\ell}$.

We should also mention that to each Lie algebra there corresponds one or more Lie groups. In fact each basis vector of a Lie algebra is an infinitesimal generator of a one-parameter subgroup of some Lie group. The study of Lie groups is quite complex, involving both topology and differentiable manifolds, and will not be undertaken here. A brief discussion of the relationship between Lie algebras and Lie groups is given in Appendix B.

### 1.2.1  Example: the general linear Lie algebra

The set of all linear transformations (operators) $A : \mathcal{V} \to \mathcal{V}$ on a vector space $\mathcal{V}$ is also a vector space and an associative algebra with respect to the usual scalar multiplication and addition and composition of linear transformations. If the commutator (1.3) is used as the Lie multiplication, a Lie algebra denoted by $\mathrm{gl}(\mathcal{V})$ is obtained. This algebra is called the general linear Lie algebra. In a sense to be made precise in the next section, all finite dimensional Lie algebras are isomorphic to a subalgebra of some $\mathrm{gl}(\mathcal{V})$.

If an $n$-dimensional basis is chosen for $\mathcal{V}$ then each linear transformation is represented by an $n \times n$ matrix $A$ with matrix elements $(A)_{jk} \in \mathcal{F}$. This set of matrices is an $n^2$-dimensional vector space and also an associative algebra with respect to the usual matrix addition and matrix product. The standard basis for this algebra is $\{E_{jk} : 1 \leq j, k \leq n\}$ where $E_{jk}$ is the matrix with 1 in position $(j, k)$ and 0 elsewhere. Matrix element $(r, s)$ of $E_{jk}$ can be expressed as

$$(E_{jk})_{rs} = \delta_{jr}\delta_{ks}, \tag{1.5}$$

where $\delta_{jk}$ denotes the Kronecker delta symbol equal to 1 if $j = k$ and 0 otherwise. Every $n \times n$ matrix can be expressed as the linear combination $A = \sum_{jk} (A)_{jk} E_{jk}$. This algebra can be made into a Lie algebra called $\mathrm{gl}(n, \mathcal{F})$ using the commutator (1.3) as the Lie multiplication. The closure property (1) is expressed by the defining commutation relations (see Exercise 1.5)

$$[E_{jk}, E_{\ell m}] = \delta_{k\ell} E_{jm} - \delta_{mj} E_{\ell k}. \tag{1.6}$$

## 1.3 Representations of a Lie Algebra

The important concept of a representation of a Lie algebra can be defined in terms of a homomorphism of the underlying vector spaces of two Lie algebras which also preserves the Lie product. Thus, if $\mathcal{L}$ and $\mathcal{L}'$ are two Lie algebras then a linear transformation $\boldsymbol{T} : \mathcal{L} \to \mathcal{L}'$ is a homomorphism if

$$\boldsymbol{T}(\alpha A + \beta B) = \alpha \boldsymbol{T}(A) + \beta \boldsymbol{T}(B), \tag{1.7}$$

$$\boldsymbol{T}([A, B]) = [\boldsymbol{T}(A), \boldsymbol{T}(B)], \tag{1.8}$$

for all $\alpha, \beta \in \mathcal{F}$ and $A, B \in \mathcal{L}$. A representation of $\mathcal{L}$ can now be defined as a homomorphism $\boldsymbol{T} : \mathcal{L} \to \mathrm{gl}(\mathcal{V})$. If $\boldsymbol{T}$ is 1 to 1 it is called an isomorphism and the representation is said to be faithful. The vector space $\mathcal{V}$ on which the linear operators $\boldsymbol{T}(A)$ act is referred to as the representation space.

We are primarily interested in matrix representations of finite dimensional Lie algebras. If we choose a basis for $\mathcal{V}$ then we can define an $n \times n$ matrix representation of $\mathcal{L}$ as a homomorphism $\boldsymbol{T} : \mathcal{L} \to \mathrm{gl}(n, \mathcal{F})$ and we say that each element $A$ of the Lie algebra is represented by the matrix $\boldsymbol{T}(A)$. There is no loss of generality here since it can be shown that every finite dimensional Lie algebra is isomorphic to a subalgebra of some $\mathrm{gl}(\mathcal{V})$.

Another way to introduce matrix representations is to start with the defining commutation relations (1.4) of a Lie algebra and consider the basis vectors $E_j$ as operators acting on some $N$-dimensional vector space $\mathcal{V}$. If $\{|j\rangle : j = 1, \ldots, N\}$ is a basis for $\mathcal{V}$ then $E_\ell |k\rangle = \sum_j |j\rangle \langle j|E_\ell|k\rangle$, $j, k = 1, \ldots, N$ and $\ell = 1, \ldots, n$ where $\langle j|E_\ell|k\rangle$ denotes matrix element $(j, k)$ of $E_\ell$. These matrices also satisfy the defining commutation relations (1.4) and we say that they are a basis for a matrix representation of the Lie algebra.

### 1.3.1 Examples of matrix representations

Consider the real Lie algebra, denoted by $\mathrm{so}(3)$ or $\mathrm{su}(2)$, having basis vectors $E_j$, $j = 1, 2, 3$ and the defining commutation relations

$$[E_1, E_2] = E_3, \quad [E_2, E_3] = E_1, \quad [E_3, E_1] = E_2, \tag{1.9}$$

which can be compactly expressed as $[E_j, E_k] = \sum_\ell \epsilon_{jk\ell} E_\ell$ using the fully anti-symmetric Levi-Civita symbol $\epsilon_{jk\ell}$ (see Appendix A). A two-dimensional matrix representation is given by the three matrices

$$X_1 = \begin{pmatrix} 0 & -\dfrac{i}{2} \\ -\dfrac{i}{2} & 0 \end{pmatrix}, X_2 = \begin{pmatrix} 0 & -\dfrac{1}{2} \\ \dfrac{1}{2} & 0 \end{pmatrix}, X_3 = \begin{pmatrix} -\dfrac{i}{2} & 0 \\ 0 & \dfrac{i}{2} \end{pmatrix}, \tag{1.10}$$

and a three-dimensional one is given by the three matrices

$$Y_1 = \begin{pmatrix} 0 & 0 & 0 \\ 0 & 0 & -1 \\ 0 & 1 & 0 \end{pmatrix}, Y_2 = \begin{pmatrix} 0 & 0 & 1 \\ 0 & 0 & 0 \\ -1 & 0 & 0 \end{pmatrix}, Y_3 = \begin{pmatrix} 0 & -1 & 0 \\ 1 & 0 & 0 \\ 0 & 0 & 0 \end{pmatrix}, \tag{1.11}$$

under the isomorphisms $T_1(E_j) = X_j$ and $T_2(E_j) = Y_j$. This shows that there can be many matrix representations of a given Lie algebra.

Even though some of the matrix elements in (1.10) are complex numbers, the reference to the Lie algebra as a real one is to the underlying vector space structure and means that every element has the form $X = \sum_j \alpha_j X_j$ for $\alpha_j \in \mathbb{R}$, the real number field. Alternatively, we could consider the complex Lie algebra by letting the coefficients $\alpha_j$ be complex numbers.

As a second example consider the real Lie algebra, denoted by so(2,1), having defining commutation relations

$$[Z_1, Z_2] = -Z_3, \quad [Z_2, Z_3] = Z_1, \quad [Z_3, Z_1] = Z_2. \tag{1.12}$$

A three-dimensional matrix representation of this Lie algebra is given by the matrices

$$Z_1 = \begin{pmatrix} 0 & 0 & 0 \\ 0 & 0 & -1 \\ 0 & -1 & 0 \end{pmatrix}, Z_2 = \begin{pmatrix} 0 & 0 & -1 \\ 0 & 0 & 0 \\ -1 & 0 & 0 \end{pmatrix}, Z_3 = \begin{pmatrix} 0 & 1 & 0 \\ -1 & 0 & 0 \\ 0 & 0 & 0 \end{pmatrix}. \tag{1.13}$$

It is important to note that the real Lie algebras so(3) and so(2,1) are not isomorphic since the transformation $Z_1 = iE_1$, $Z_2 = iE_2$ and $Z_3 = E_3$ which transforms the defining commutation relations (1.12) into (1.9) is a complex linear transformation, not a real one. As complex Lie algebras they are isomorphic. For an example of a representation of the 8-dimensional Lie algebra su(3) see the end of chapter exercises.

## 1.4    Realizations of a Lie Algebra

For our purposes a realization of a Lie algebra is a homomorphism which associates a concrete set of operators with each abstract basis vector of the Lie algebra. It is more of a physical concept than a mathematical one and is of great importance in the application of Lie algebras.

In quantum mechanics such operators are often differential operators expressed in terms of the position and momentum operators, $q$ and $p$, and acting on some Hilbert space of quantum mechanical states. Alternatively, the realization may also be expressed in terms of matrix operators as is the case for the spin operators which are naturally constructed in terms of the Pauli spin matrices. We shall also refer to such a set of concrete operators and states as a realization of the Lie algebra. There are two ways to approach this connection between operators and abstract basis vectors.

One approach is to begin with an abstract Lie algebra and try to find a suitable set of operators which satisfy the same commutation relations as the abstract basis vectors of the Lie algebra. Alternatively, the more common approach is to start with a set of physically meaningful operators $\mathcal{A} = \{A_j\}$ and evaluate their commutators. If $[A_j, A_k] = \sum_\ell c_{jk\ell} A_\ell$ for each pair of operators in $\mathcal{A}$ then $\mathcal{A}$ is closed under commutation and is a realization of some Lie algebra. Otherwise we can attempt to extend the original set by including in it any new operators arising as commutators and not expressible as linear combinations of the operators in $\mathcal{A}$. More precisely, let $\mathcal{C}^{(1)} = \{C_\ell^{(1)}\}$ be the set of operators such that each $C_\ell^{(1)} = [A_j, A_k]$ for some pair of indices $j$, $k$ and such that $C_\ell^{(1)}$ is not expressible in the form $\sum_k \alpha_k A_k$. Now consider the commutators of all operators in the extended set $\mathcal{A}^{(1)} = \mathcal{C}^{(1)} \cup \mathcal{A}$. We can iteratively continue this process to obtain $\mathcal{A}^{(2)}$, $\mathcal{A}^{(3)}$, $\cdots$, $\mathcal{A}^{(k)}$. It may happen after a finite number of steps $k$ that a closed set of operators is obtained and we have a basis for some Lie algebra.

Another variation of this approach is to merge together the realizations of two Lie algebras to obtain a realization of a larger Lie algebra. Thus if the sets $\mathcal{A} = \{A_j\}$ and $\mathcal{B} = \{B_j\}$ are realizations of the basis vectors of two Lie algebras we can ask if their union $\mathcal{A} \cup \mathcal{B}$ is a basis of a larger Lie algebra. This will be the case only if the commutators of the form $[A_j, B_k]$ can be expressed as linear combinations of the operators in $\mathcal{A} \cup \mathcal{B}$. Otherwise we must extend $\mathcal{A} \cup \mathcal{B}$ to include the set $\mathcal{C}^{(1)}$ of operators $C_\ell^{(1)} = [A_j, B_k]$ which are not expressible in this form. Continuing in this fashion we may or may not arrive after a finite number of steps at a set $\mathcal{A} \cup \mathcal{B} \cup \mathcal{C}^{(1)} \cup \cdots \cup \mathcal{C}^{(k)}$ of operators which form the basis of a Lie algebra.

This merging of two Lie algebras will be done several times in subsequent chapters. For example the Lie algebra so(4) will be obtained by merging two so(3) algebras. In this case no additional operators are needed to "close out" the commutation relations so so(4) is generated by 6 operators. Then these 6 generators of so(4) will be merged with the 3 generators of so(2,1). In the process 6 new operators are necessary to "close out" the commutation relations and obtain the 15 generators of the Lie algebra so(4,2).

A realization of a Lie algebra in terms of operators is usually accompanied by a realization of the underlying vector space on which the operators act and we are naturally led to consider matrix representations of the operators using a concrete basis for this vector space and hence to matrix representations of the Lie algebra. We should emphasize however that there are many advantages to working with the abstract basis of the Lie algebra. In fact we shall see that the matrix representations can be obtained without using a particular realization of the basis or the underlying vector space. This is one of the main advantages of Lie algebras. Moreover if we have a set of operators and we can show that they form a realization of the basis vectors of a Lie algebra whose representations are known then we can use these known results to evaluate the matrix elements of our operators. This is the case for all of the Lie algebras considered in this book.

### 1.4.1 Examples of realizations

**Orbital angular momentum and so(3)**

The three components $L_j$, $j = 1, 2, 3$, of the orbital angular momentum $\boldsymbol{L} = \boldsymbol{r} \times \boldsymbol{p}$, where $\boldsymbol{r} = (x_1, x_2, x_3)$ is the position vector in $\mathbb{R}^3$ and $\boldsymbol{p} = -i\nabla$ is the momentum vector[1], provide a realization of the basis vectors of the Lie algebra so(3) since they satisfy the commutation relations

$$[L_1, L_2] = iL_3, \quad [L_2, L_3] = iL_1, \quad [L_3, L_1] = iL_2, \tag{1.14}$$

which can be written together as

$$[L_j, L_k] = i \sum_\ell \epsilon_{jk\ell} L_\ell. \tag{1.15}$$

These commutation relations are equivalent to those in (1.9) if we define $E_j = -iL_j$, $j = 1, 2, 3$.

In quantum mechanics it is more usual to consider so(3) using the components of the angular momentum rather than the basis vectors $E_j$ even though the $L_j$ are not basis vectors of the real so(3) Lie algebra (they are hermitian but the commutator of two hermitian operators is skew hermitian not hermitian). Physically they generate infinitesimal rotations of the real Lie group SO(3) considered as a transformation group on $\mathbb{R}^3$. In quantum mechanics we are mainly interested in unitary representations of Lie groups and these correspond to representations of the Lie algebra for which the infinitesimal generators are hermitian (see Appendix B).

In Cartesian coordinate space the $L_j$ are differential operators and they act on some suitable vector space of functions $\psi(\boldsymbol{r})$. Alternatively a spherical coordinate system can be used and a realization of the basis vectors can be chosen as the set of spherical harmonic functions $\{Y_{\ell m}(\theta, \phi) : \ell = 0, 1, \cdots, \infty, -\ell \leq m \leq \ell\}$. From this point of view the $Y_{\ell m}$ are simultaneous eigenfunctions of $L^2 = L_1^2 + L_2^2 + L_3^2$ and $L_3$ satisfying the differential equations $L^2 Y_{\ell m} = \ell(\ell + 1)Y_{\ell m}(\theta, \phi)$ and $L_3 Y_{\ell m}(\theta, \phi) = m Y_{\ell m}(\theta, \phi)$. Thus the study of Lie algebras, their representations and their realizations, is closely connected with the solution of various classes of differential equations.

We can also consider SO(3) as the three parameter $3 \times 3$ matrix group of rotations in $\mathbb{R}^3$ (see Appendix B). This gives rise to a natural realization of the so(3) commutation relations in terms of the infinitesimal generators in (1.11) or, defining $L_j = iY_j$, $j = 1, 2, 3$, in terms of the hermitian matrices

$$L_1 = \begin{pmatrix} 0 & 0 & 0 \\ 0 & 0 & -i \\ 0 & i & 0 \end{pmatrix}, \quad L_2 = \begin{pmatrix} 0 & 0 & i \\ 0 & 0 & 0 \\ -i & 0 & 0 \end{pmatrix}, \quad L_3 = \begin{pmatrix} 0 & -i & 0 \\ i & 0 & 0 \\ 0 & 0 & 0 \end{pmatrix}. \tag{1.16}$$

---

[1] we use atomic units throughout this book with $\hbar = 1$

It is shown in Appendix B that each matrix $L_j$ generates a one-parameter unitary subgroup of SO(3). For example, if $\alpha \in \mathbb{R}$ then the matrices $U(\alpha) = e^{-i\alpha L_3} = e^{\alpha Y_3}$ are a one-parameter subgroup of SO(3) of rotations about the $x_3$ axis in $\mathbb{R}^3$.

Another useful set of commutation relations equivalent to (1.15) can be obtained using the so-called ladder operators (also called raising and lowering operators)

$$L_+ = L_1 + iL_2, \tag{1.17}$$
$$L_- = L_1 - iL_2, \tag{1.18}$$

instead of $L_1$ and $L_2$. Then the commutation relations among $L_+$, $L_-$ and $L_3$ are given by

$$[L_+, L_-] = 2L_3, \tag{1.19}$$
$$[L_3, L_+] = L_+, \tag{1.20}$$
$$[L_3, L_-] = -L_-. \tag{1.21}$$

This form of the so(3) commutation relations and similar ones for so(2,1) will be used in later chapters in the construction of the irreducible representations of so(3) and so(2,1). They can often arise naturally from physical considerations as shown in the isospin and quasispin examples below.

### Spin and su(2)

Consider the real Lie algebra su(2), which corresponds to the real Lie group SU(2) of $2 \times 2$ unitary unimodular matrices and also has the defining commutation relations (1.15), or equivalently (1.9) (see Appendix B). The natural realization of su(2), which is also a matrix representation is given by the components of the one-electron spin vector $\boldsymbol{S} = \frac{1}{2}\boldsymbol{\sigma}$ defined in terms of the Pauli spin matrices

$$\sigma_1 = \begin{pmatrix} 0 & 1 \\ 1 & 0 \end{pmatrix}, \; \sigma_2 = \begin{pmatrix} 0 & -i \\ i & 0 \end{pmatrix}, \; \sigma_3 = \begin{pmatrix} 1 & 0 \\ 0 & -1 \end{pmatrix}. \tag{1.22}$$

The spin components $S_i$ also satisfy the su(2) commutation relations (1.15) and are related to the matrices in (1.10) by $X_j = -\frac{i}{2}\sigma_j$. The components of the total one-electron angular momentum $\boldsymbol{J} = \boldsymbol{L} + \boldsymbol{S}$ also satisfy these commutation relations and provide another realization of su(2).

### Isospin and elementary particles

In nuclear physics we might consider neutrons and protons to be different nucleon states. A state vector consisting of several nucleons could then be described in terms of fermion annihilation and creation operators. An over-simplified picture, ignoring the space and spin dependence of the states, is to

assume that there are only two states for a nucleon: it is either a proton or a
neutron. Then we can define $a_p^+$ to be a proton creation operator and $a_p$ to
be a proton annihilation operator. Similarly, define $a_n^+$ and $a_n$ for neutrons.
These operators satisfy the characteristic fermion anti-commutation relations

$$
\begin{aligned}
\{a_p, a_p^+\} &= \{a_n, a_n^+\} = 1, \\
\{a_p, a_p\} &= \{a_n, a_n\} = \{a_p^+, a_p^+\} = \{a_n^+, a_n^+\} = 0, \\
\{a_p^+, a_n\} &= \{a_n^+, a_p\} = \{a_n, a_p\} = \{a_n^+, a_p^+\} = 0,
\end{aligned}
\tag{1.23}
$$

where $\{A, B\} \equiv AB + BA$ defines the anticommutator of two operators. We
are interested in products of pairs of these operators which describe excitations
preserving the number of nucleons. There are four such operators:

$$
E_{12} = a_p^+ a_n, \quad E_{21} = a_n^+ a_p, \quad E_{11} = a_p^+ a_p, \quad E_{22} = a_n^+ a_n.
\tag{1.24}
$$

The first operator, $E_{12}$, describes the annihilation of a neutron followed
by the creation of a proton. We can think of this as a state change in which
a neutron is changed into a proton. Similarly, $E_{21}$ describes the opposite
situation. The last two operators are called number operators: $E_{11}$ gives the
number of protons (0 or 1 in our simple case) and similarly $E_{22}$ gives the
number of neutrons. Since none of the operators in (1.24) changes the number
of nucleons it follows that the operator

$$
N = a_p^+ a_p + a_n^+ a_n,
\tag{1.25}
$$

gives the total number of nucleons and commutes with each of the four operators in (1.24). This can be verified by direct calculation of the four commutators ($[N, E_{jk}] = 0$), using the anticommutation relations (1.23). The four
excitation operators $E_{jk}$ satisfy the commutation relations (1.6) of the general
linear Lie algebra gl(2). This can be verified by direct calculation using the
identities proved in Exercise 1.3. For example, the commutator of $E_{12}$ and $E_{21}$
can be evaluated as follows:

$$
\begin{aligned}
[E_{12}, E_{21}] \\
&= [a_p^+ a_n, a_n^+ a_p] \\
&= a_p^+ [a_n, a_n^+ a_p] + [a_p^+, a_n^+ a_p] a_n \\
&= a_p^+ (\{a_n, a_n^+\} a_p - a_n^+ \{a_n, a_p\}) + (\{a_p^+, a_n^+\} a_p - a_n^+ \{a_p^+, a_p\}) a_n \\
&= a_p^+ a_p - a_n^+ a_n = E_{11} - E_{22}.
\end{aligned}
$$

The full set of commutation relations is given by

$$
\begin{aligned}
[E_{12}, E_{21}] &= E_{11} - E_{22}, \\
[E_{12}, E_{11}] &= -E_{12}, \quad [E_{12}, E_{22}] = E_{12}, \\
[E_{21}, E_{11}] &= E_{21}, \quad [E_{21}, E_{22}] = -E_{21}, \\
[E_{11}, E_{22}] &= 0.
\end{aligned}
\tag{1.26}
$$

Now if we define operators $K_+$, $K_-$ and $K_3$ by

$$K_+ = E_{12} = a_p^+ a_n,$$
$$K_- = E_{21} = a_n^+ a_p,$$
$$K_3 = \tfrac{1}{2}(E_{11} - E_{22}) = \tfrac{1}{2}(a_p^+ a_p - a_n^+ a_n), \tag{1.27}$$

then these operators satisfy the commutation relations (see (1.19) to (1.21))

$$[K_+, K_-] = 2K_3, \quad [K_3, K_+] = K_+, \quad [K_3, K_-] = -K_-,$$

of the Lie algebra su(2). Thus we can "play the same game" [LI65] with them as with the angular momentum operators. The analysis given here parallels that for spin in a two-state system (spin up and spin down) and by analogy is called isospin in the nuclear case.

Boson systems, having annihilation and creation operators which satisfy commutation relations rather than anticommutation relations, can also be used to construct realizations (see Exercise 1.14). The reason is that in both the fermion and boson case only bilinear products of the annihilation and creation operators are used and these satisfy commutation relations (see Exercise 1.3).

## 1.5   Irreducible Representations

The problem of classifying all matrix representations of a given Lie algebra is an important but difficult one which reduces to the determination of all irreducible representations (irreps). There is a close relationship between the representation theory of Lie groups and Lie algebras in the sense that the purely algebraic problem of computing all irreducible representations of the Lie algebra of a group also gives, via exponentiation, the irreducible representations of the group. The basic terminology and definitions are the same as in the representation theory of finite groups.

The concept of an irreducible representation of a Lie algebra can be defined in terms of invariant subspaces of the representation space. If $T : \mathcal{L} \to \mathrm{gl}(\mathcal{V})$ is a representation of a Lie algebra $\mathcal{L}$ with representation space $\mathcal{V}$ and if $\mathcal{W}$ is a subspace of $\mathcal{V}$ then $\mathcal{W}$ is invariant under $T$ if $T(A)w \in \mathcal{W}$ for every $A \in \mathcal{L}$ and $w \in \mathcal{W}$. In other words $\mathcal{W}$ is mapped to itself by all linear operators in $\mathrm{gl}(\mathcal{V})$. A representation is said to be reducible if there is a proper subspace of $\mathcal{V}$ which is invariant under $T$. If this is not the case we say that the representation is irreducible.

Two representations $T' : \mathcal{L} \to \mathrm{gl}(\mathcal{V})$ and $T'' : \mathcal{L} \to \mathrm{gl}(\mathcal{V})$ having the same representation space $\mathcal{V}$ are said to be equivalent if there exists a non-singular linear transformation $S : \mathcal{V} \to \mathcal{V}$, called a similarity transformation, such that $T''(A) = ST'(A)S^{-1}$ for all $A \in \mathcal{L}$. In terms of matrix representations this simply means that a suitable change in the basis of $\mathcal{V}$ can be found which transforms the matrices $T'(A)$ into the matrices $T''(A)$.

The concept of an irreducible representation is more easily understood in terms of matrix representations. Let $T : \mathcal{L} \to \mathrm{gl}(n, \mathcal{F})$ be a matrix representation such that the $n \times n$ matrices $T(A)$, $A \in \mathcal{L}$, all have the identical block structure (or are equivalent to such a representation in the sense defined above)

$$T(A) = \begin{pmatrix} T'(A) & X(A) \\ Z & T''(A) \end{pmatrix}, \tag{1.28}$$

where $T'(A)$ is a $k \times k$ matrix, $T''(A)$ is an $(n-k) \times (n-k)$ matrix, $X(A)$ is an arbitrary $k \times (n-k)$ matrix and $Z$ is an $(n-k) \times k$ zero matrix. It follows that $T' : \mathcal{L} \to \mathrm{gl}(k, \mathcal{F})$ and $T'' : \mathcal{L} \to \mathrm{gl}(n-k, \mathcal{F})$ are smaller dimensional matrix representations of $\mathcal{L}$ and $T$ is reducible. Furthermore if a similarity transformation can be found such that $X(A)$ is the zero matrix for all $A \in \mathcal{L}$ then the matrices of the representation have the same block diagonal structure and we say that $T$ is completely reducible to the direct sum $T = T' \oplus T''$. We can continue this process and eventually in the completely reducible case $T$ can be expressed as the direct sum of irreducible representations $T = T_1 \oplus T_2 \oplus \cdots \oplus T_k$. For infinite dimensional representations this may be an infinite direct sum and some of the irreps may also be infinite dimensional.

For finite groups it can be shown that every finite dimensional unitary representation is completely reducible to a direct sum of unitary irreducible representations (unirreps). Also the irreps are all finite dimensional and every representation is equivalent to a unitary one. The irreps are the building blocks for more general representations. These important results also carry over to the compact Lie groups and their corresponding Lie algebras.

On the other hand non-compact Lie groups can have infinite dimensional irreps. In particular we shall consider the Lie algebra so(2,1) of the non-compact group SO(2,1) and show that all unirreps are infinite dimensional. This will be important in our algebraic study of the hydrogen atom where the scaled radial eigenfunctions $\{|n\ell\rangle : n = \ell + 1, \ldots, \infty\}$ form a basis for an infinite dimensional unirrep of so(2,1).

### 1.5.1 Example: a reducible representation

Consider the three $8 \times 8$ matrices

$$K_j = L_j \oplus L'_j = \begin{pmatrix} L_j & 0 \\ 0 & L'_j \end{pmatrix}, \quad j = 1, 2, 3, \tag{1.29}$$

where the matrices $L_j$ are given in (1.16) and the $L'_j$ are the $5 \times 5$ matrices

$$L'_1 = \frac{1}{2} \begin{pmatrix} 0 & 2 & 0 & 0 & 0 \\ 2 & 0 & \sqrt{6} & 0 & 0 \\ 0 & \sqrt{6} & 0 & \sqrt{6} & 0 \\ 0 & 0 & \sqrt{6} & 0 & 2 \\ 0 & 0 & 0 & 2 & 0 \end{pmatrix}, \quad L'_2 = -\frac{i}{2} \begin{pmatrix} 0 & 2 & 0 & 0 & 0 \\ -2 & 0 & \sqrt{6} & 0 & 0 \\ 0 & -\sqrt{6} & 0 & \sqrt{6} & 0 \\ 0 & 0 & -\sqrt{6} & 0 & 2 \\ 0 & 0 & 0 & -2 & 0 \end{pmatrix}$$

$$L_3' = \begin{pmatrix} 2 & 0 & 0 & 0 & 0 \\ 0 & 1 & 0 & 0 & 0 \\ 0 & 0 & 0 & 0 & 0 \\ 0 & 0 & 0 & -1 & 0 \\ 0 & 0 & 0 & 0 & -2 \end{pmatrix}. \tag{1.30}$$

The matrices $L_j'$ also satisfy the defining commutation relations (1.15) so they provide a matrix representation of so(3). Therefore the matrices $K_j$ form a reducible representation of so(3). In Chapter 3 we shall see how these matrices arise from the representation theory of so(3). We shall see that the sets $\{L_j\}$ and $\{L_j'\}$ are bases for irreducible representations corresponding to the values $\ell = 1$ and $\ell = 2$, respectively, of the orbital angular momentum quantum number. Thus the matrices $L_j'$ cannot be reduced further to a block diagonal form with lower dimensional matrices.

## 1.6    Exercises

◈ **Exercise 1.1** It follows from rule (4) that $[A, A] = 0$. Conversely, show that $[A, A] = 0$ for all $A$ implies rule (4). Therefore rule (4) is often replaced by the rule $[A, A] = 0$.

◈ **Exercise 1.2** Show that the commutator definition

$$[A, B] = AB - BA$$

of the Lie multiplication satisfies the Jacobi identity, rule (5).

◈ **Exercise 1.3** Show that the commutator satisfies the identities

$$[AB, C] = A[B, C] + [A, C]B,$$
$$[A, BC] = [A, B]C + B[A, C],$$
$$[AB, C] = A\{B, C\} - \{A, C\}B,$$
$$[A, BC] = \{A, B\}C - B\{A, C\},$$

where $\{A, B\} \equiv AB + BC$ is the anticommutator of $A$ and $B$. These identities are useful for simplifying commutation relations.

◈ **Exercise 1.4** Consider the three-dimensional vector space $\mathbb{R}^3$ with orthonormal basis vectors $e_1$, $e_2$ and $e_3$ so that any vector $x$ can be expressed as $x = (x_1, x_2, x_3) = \sum_j x_j e_j$. Define a Lie multiplication in terms of the vector cross product by $[e_j, e_k] = e_j \times e_k$. Show that the defining rules (1) to (5) of a Lie algebra are satisfied and that the defining commutation relations are $[e_1, e_2] = e_3$, $[e_2, e_3] = e_1$ and $[e_3, e_1] = e_2$.

◈ **Exercise 1.5** Derive the defining commutation relations for the general linear Lie algebra gl($n$, $\mathcal{F}$).

◈ **Exercise 1.6** Show that the Pauli spin matrices satisfy the following identities

$$(a) \quad \sigma_j\sigma_k = i\sum_\ell \epsilon_{jk\ell}\sigma_\ell + \delta_{jk},$$

$$(b) \quad [\sigma_j,\sigma_k] = 2i\sum_\ell \epsilon_{jk\ell}\sigma_\ell,$$

$$(c) \quad \{\sigma_j,\sigma_k\} = 2\delta_{jk},$$

$$(d) \quad (\boldsymbol{u}\cdot\boldsymbol{\sigma})(\boldsymbol{v}\cdot\boldsymbol{\sigma}) = \boldsymbol{u}\cdot\boldsymbol{v} + i(\boldsymbol{u}\times\boldsymbol{v})\cdot\boldsymbol{\sigma},$$

$$(e) \quad [\boldsymbol{u}\cdot\boldsymbol{\sigma},\boldsymbol{v}\cdot\boldsymbol{\sigma}] = 2i(\boldsymbol{u}\times\boldsymbol{v})\cdot\boldsymbol{\sigma},$$

$$(f) \quad \{\boldsymbol{u}\cdot\boldsymbol{\sigma},\boldsymbol{v}\cdot\boldsymbol{\sigma}\} = 2\boldsymbol{u}\cdot\boldsymbol{v}.$$

Here $\boldsymbol{u}$ and $\boldsymbol{v}$ are vectors or vector operators whose components commute among themselves and also with the components, $\sigma_j$, of $\boldsymbol{\sigma}$. Also $\{A,B\} = AB + BA$ denotes the anticommutator of $A$ and $B$ (see Appendix A for the definition of $\epsilon_{jk\ell}$.)

◈ **Exercise 1.7** Show that components of the total one-electron angular momentum $\boldsymbol{J} = \boldsymbol{L} + \boldsymbol{S}$ satisfy the commutation relations (1.15).

◈ **Exercise 1.8** Show that the square of the orbital angular momentum $L^2 = L_1^2 + L_2^2 + L_3^2$ commutes with the so(3) basis vectors $L_1$, $L_2$ and $L_3$. An operator which commutes with all elements of a Lie algebra is called a Casimir operator.

◈ **Exercise 1.9** Derive the commutation relations (1.26)

◈ **Exercise 1.10** Show that the set of all $n \times n$ skew hermitian matrices is a real Lie algebra. (A matrix $A$ is skew hermitian if $A^\dagger = -A$ where $A^\dagger$ denotes the hermitian conjugate of $A$.)

◈ **Exercise 1.11** Show that the set of all $n \times n$ traceless skew hermitian matrices forms a real Lie algebra of dimension $n^2 - 1$ and show that a suitable basis set, defined in terms of the elementary matrices $E_{jk}$ (see Section 1.2.1) is given by the matrices

$$\begin{aligned}
G_{jk} &= i(E_{jk} + E_{kj}), & j < k, \\
H_{jk} &= E_{jk} - E_{kj}, & j < k, \\
D_j &= i(E_{jj} - E_{j+1,j+1}), & j = 1,\ldots,n-1.
\end{aligned}$$

The matrices $iG_{jk}$, $iH_{jk}$ and $iD_j$ are hermitian and are infinitesimal generators of the special unitary group $SU(n)$ of all unitary matrices with unit determinant so the Lie algebra is called su($n$). The special case $n = 2$ is important and is considered in Appendix B.

◈ **Exercise 1.12** For the case $n = 3$ in the preceding exercise there are 8 generators which can be denoted by

$$E_1 = G_{12} = \begin{pmatrix} 0 & i & 0 \\ i & 0 & 0 \\ 0 & 0 & 0 \end{pmatrix}, \quad E_2 = H_{12} = \begin{pmatrix} 0 & 1 & 0 \\ -1 & 0 & 0 \\ 0 & 0 & 0 \end{pmatrix},$$

$$E_3 = D_1 = \begin{pmatrix} i & 0 & 0 \\ 0 & -i & 0 \\ 0 & 0 & 0 \end{pmatrix}, \quad E_4 = G_{13} = \begin{pmatrix} 0 & 0 & i \\ 0 & 0 & 0 \\ i & 0 & 0 \end{pmatrix},$$

$$E_5 = H_{13} = \begin{pmatrix} 0 & 0 & 1 \\ 0 & 0 & 0 \\ -1 & 0 & 0 \end{pmatrix}, \quad E_6 = G_{23} = \begin{pmatrix} 0 & 0 & 0 \\ 0 & 0 & i \\ 0 & i & 0 \end{pmatrix},$$

$$E_7 = H_{23} = \begin{pmatrix} 0 & 0 & 0 \\ 0 & 0 & 1 \\ 0 & -1 & 0 \end{pmatrix}, \quad E_8 = \frac{1}{\sqrt{3}}\begin{pmatrix} i & 0 & 0 \\ 0 & i & 0 \\ 0 & 0 & -2i \end{pmatrix},$$

where $E_8 = \frac{1}{\sqrt{3}}(D_1 + 2D_2)$. Use the commutation relations (1.6) in Section 1.2.1 to obtain the structure constants $e_{jk\ell}$ defined by $[E_j, E_k] = \sum_\ell e_{jk\ell} E_\ell$. The choice given for $E_8$ (rather than $E_8 = D_2$) makes the structure constants antisymmetric in all pairs of indices (like the Levi-Civita symbol $\epsilon_{jk\ell}$ which gives the structure constants of su(2)).

Also show that the generators $E_1$, $E_2$ and $E_3$ form an su(2) subalgebra with defining commutation relations isomorphic to (1.9) and that the quadratic operator $C = \sum_{j=1}^{8} E_j^2$ is a Casimir operator for su(3) ($[C, E_j] = 0$, $j = 1, \ldots, 8$). The Lie algebra su(3) is important in the classification of elementary particles [GR89], [SC68].

◈ **Exercise 1.13** Show that if $T$ is a representation of a Lie algebra whose matrices have the block structure (1.28) then $T'$ and $T''$ are also representations.

◈ **Exercise 1.14** In Section 1.4.1 fermion annihilation and creation operators, defined using anticommutators, were used to construct a realization of su(2). Bosons can also be used. Consider the boson annihilation and creation operators $a_k^+$, $a_k$, $k = 1, 2$ for a two-state system which satisfy the commutation relations $[a_j^+, a_k^+] = [a_j, a_k] = 0$, $[a_j, a_k^+] = \delta_{jk}$. Show that the three operators

$$J_1 = \frac{1}{2}(a_1^+ a_2 + a_1 a_2^+),$$

$$J_2 = -\frac{i}{2}(a_1^+ a_2 - a_1 a_2^+),$$

$$J_3 = \frac{1}{2}(a_1^+ a_1 - a_2^+ a_2),$$

satisfy the su(2) commutation relations (1.14). The ladder operators $J_+ = a_1^\dagger a_2$ and $J_- = a_2^\dagger a_1$ represent excitations of a boson from state 2 to state 1 and from state 1 to state 2 respectively. This boson realization of su(2) can be expressed compactly in terms of the Pauli spin matrices (1.22) as [SC65], [WY74]

$$J_j = \frac{1}{2} a^\dagger \sigma_j a,$$

where

$$a = \begin{pmatrix} a_1 \\ a_2 \end{pmatrix}, \quad a^\dagger = \begin{pmatrix} a_1^\dagger & a_2^\dagger \end{pmatrix}.$$

# Chapter 2

# Commutator Gymnastics

## 2.1  Introduction

In later chapters we will require various commutators involving position, $r = (x_1, x_2, x_3)$, momentum $p = -i\nabla$ and angular momentum $L = r \times p$ so we collect these results together in this chapter. Derivations of selected identities are also given to provide the reader with experience in "commutator gymnastics" (the calculus of commutators). Derivations of the remaining identities are given in the solutions to the end of chapter exercises.

The derivation of many of the important commutation relations is tedious but not difficult. Many of the commutations relations can be expressed concisely in terms of the Levi-Civita symbol and the various identities associated with it (see Appendix A) are very useful in the derivations.

## 2.2  General Commutator Identities

The following two identities are useful for moving operators in products outside the commutator brackets:

$$[A, BC] = [A, B]C + B[A, C], \qquad (2.1)$$
$$[AB, C] = A[B, C] + [A, C]B. \qquad (2.2)$$

These rules are of fundamental importance and are used in virtually all calculations involving the simplification of commutators. They have the same structure and importance as the product rule for differentiation does in the calculus: in fact, defining $D_A(B) = [A, B]$, the first rule becomes $D_A(BC) = BD_A(C) + D_A(B)C$.

The following two rules are often useful for moving functions of an operator inside or outside a commutator:

$$f(A)[g(A), B] = [g(A), f(A)B], \qquad (2.3)$$
$$[A, g(B)]f(B) = [Af(B), g(B)]. \qquad (2.4)$$

Here we assume that a function of an operator is a polynomial or a formal power series in the operator without regard to the convergence of the power series. Thus if $f(x) = \sum_j a_j x^j$ is a formal power series in the real variable $x$ and $A$ denotes an operator then by $f(A)$ we mean the series obtained by formally replacing $x$ by $A$.

Finally, two identities involving powers and exponentials of an operator are

$$[A, B^n] = \sum_{k=0}^{n-1} B^k [A, B] B^{n-k-1}, \quad n \geq 1, \tag{2.5}$$

$$e^{-B} A e^B = A + [A, B] + \frac{1}{2!} [[A, B], B] + \cdots$$

$$= \sum_{n=0}^{\infty} \frac{1}{n!} [A, B]^{(n)}, \tag{2.6}$$

where $[A, B]^{(0)} = A$, $[A, B]^{(1)} = [A, B]$ and in general

$$[A, B]^{(n)} = [[\cdots [[A, B], B], \cdots], B]$$
$$= [[A, B]^{(n-1)}, B], \text{ for } n \geq 1.$$

Equation (2.5) can be proved by induction as follows. It is true for $n = 1$ so assume (2.5) is true. Using (2.1)

$$[A, B^{n+1}] = [A, BB^n] = B[A, B^n] + [A, B] B^n$$

$$= \sum_{k=0}^{n-1} B^{k+1} [A, B] B^{n-k-1} + [A, B] B^n$$

$$= \sum_{k=-1}^{n-1} B^{k+1} [A, B] B^{n-k-1}$$

$$= \sum_{k=0}^{n} B^k [A, B] B^{n-k},$$

which is just (2.5) with $n$ replaced by $n + 1$.

To obtain (2.6) define $f(\lambda) = e^{-\lambda B} A e^{\lambda B}$. The first two derivatives are given by

$$f'(\lambda) = -e^{-\lambda B} B A e^{\lambda B} + e^{-\lambda B} A B e^{\lambda B}$$
$$= e^{-\lambda B} [A, B] e^{\lambda B}$$
$$f''(\lambda) = -e^{-\lambda B} B [A, B] e^{\lambda B} + e^{-\lambda B} [A, B] B e^{\lambda B}$$
$$= e^{-\lambda B} [[A, B], B] e^{\lambda B},$$

and the result for the order $n$ derivative is $f^{(n)}(\lambda) = e^{-\lambda B} [A, B]^{(n)} e^{\lambda B}$, which can be proved by induction. Letting $\lambda = 0$ and substituting into the formal Taylor series

$$f(\lambda) = \sum_{n=0}^{\infty} \frac{1}{n!} f^{(n)}(0) \lambda^n$$

it follows that (2.6) is $f(1)$.

## 2.3   Commutators Involving r and p

In the coordinate representation in quantum mechanics the momentum is a differential operator $p = (p_1, p_2, p_3) = -i\nabla$ so commutators involving $p$ will contain differential operators. The simplest such nonvanishing commutator is $[x, d/dx]$ which can be evaluated by applying it to an arbitrary function $f(x)$:

$$\left[x, \frac{d}{dx}\right] f(x) = \left(x\frac{d}{dx} - \frac{d}{dx}x\right) f(x)$$

$$= xf'(x) - \frac{d}{dx}\big(xf(x)\big)$$

$$= xf'(x) - xf'(x) - f(x) = -f(x).$$

Therefore $[x, d/dx] = -1$. It is important to understand the difference between $\frac{d}{dx}x$ as an operator product which can act on functions of $x$ and $\frac{d}{dx}(x) = 1$ which is just the derivative of the function $f(x) = x$. Commutator identities involving derivatives are usually derived by applying them to an arbitrary function.

It follows that the basic commutation relations involving position and momentum are

$$[x_j, x_k] = 0, \tag{2.7}$$

$$[p_j, p_k] = 0, \tag{2.8}$$

$$[x_j, p_k] = i\delta_{jk}, \quad j, k = 1, 2, 3. \tag{2.9}$$

All other commutators and identities involving $r$ and $p$ can be obtained using them and general rules such as (2.1) and (2.2). In the following we shall sometimes use the implied summation convention over repeated indices in the same term unless otherwise stated. For example, setting $k = j$ and summing over $j$ in (2.9) gives the identity

$$x_j p_j - p_j x_j = r \cdot p - p \cdot r = 3i. \tag{2.10}$$

Defining the length of the position vector as $r = |r|$ we can obtain the following commutation relations involving the components of $p$:

$$[p_j, f(r)] = -ix_j r^{-1} f'(r) = -i\frac{\partial f}{\partial x_j}, \tag{2.11}$$

$$[p_j, r^{-n}] = inx_j r^{-n-2}, \tag{2.12}$$

$$[p_j, x_k r^{-n}] = ir^{-n}(nx_j x_k r^{-2} - \delta_{jk}), \tag{2.13}$$

$$[p_j, x_j r^{-n}] = i(n-3)r^{-n}, \text{ (summed over } j). \tag{2.14}$$

The first two of these equations can also be expressed in vector form:

$$[p, f(r)] = -i\nabla f, \tag{2.15}$$

$$[p, r^{-n}] = inr^{-n-2}r. \tag{2.16}$$

There are also several useful commutators involving $p^2 = \boldsymbol{p} \cdot \boldsymbol{p} = p_j p_j$ and $\boldsymbol{r} \cdot \boldsymbol{p} = x_j p_j$:

$$[\boldsymbol{r}, p^2] = 2i\boldsymbol{p}, \tag{2.17}$$

$$[f(r), p^2] = 2ir^{-1}f'(r)(\boldsymbol{r} \cdot \boldsymbol{p} - i) + f''(r), \tag{2.18}$$

$$[\boldsymbol{r} \cdot \boldsymbol{p}, p^2] = 2ip^2, \tag{2.19}$$

$$[\boldsymbol{r}, \boldsymbol{r} \cdot \boldsymbol{p}] = i\boldsymbol{r}, \tag{2.20}$$

$$[\boldsymbol{p}, \boldsymbol{r} \cdot \boldsymbol{p}] = -i\boldsymbol{p}, \tag{2.21}$$

$$[f(r), \boldsymbol{r} \cdot \boldsymbol{p}] = irf'(r). \tag{2.22}$$

Similar commutators involving $\boldsymbol{p} \cdot \boldsymbol{r}$ rather than $\boldsymbol{r} \cdot \boldsymbol{p}$ can be obtained using identity (2.10). We derive (2.18) here and the others in the solutions to the end of chapter exercises:

$$[f(r), p^2] = [f(r), p_j p_j]$$

$$= p_j [f(r), p_j] + [f(r), p_j]p_j, \quad \text{from (2.2)}$$

$$= p_j \left(ix_j r^{-1}f'(r)\right) + \left(ix_j r^{-1}f'(r)\right) p_j, \quad \text{from (2.11)}$$

$$= ip_j x_j r^{-1}f'(r) + ir^{-1}f'(r)x_j p_j$$

$$= ip_j r^{-1}f'(r)x_j + ir^{-1}f'(r)x_j p_j.$$

Now use (2.11) with $r^{-1}f'(r)$ in place of $f(r)$ to obtain

$$p_j r^{-1}f'(r) = r^{-1}f'(r)p_j + [p_j, r^{-1}f'(r)]$$

$$= r^{-1}f'(r)p_j - ix_j r^{-1}\frac{d}{dr}\left(r^{-1}f'(r)\right).$$

Substitute this result in the first term to obtain

$$[f(r), p^2]$$

$$= i\left(r^{-1}f'(r)p_j - ix_j r^{-1}\frac{d}{dr}\left(r^{-1}f'(r)\right)\right) x_j + ir^{-1}f'(r)x_j p_j$$

$$= ir^{-1}f'(r)p_j x_j + r\left(r^{-1}f''(r) - r^{-2}f'(r)\right) + ir^{-1}f'(r)\boldsymbol{r} \cdot \boldsymbol{p}.$$

Finally use (2.10) in the first term to obtain

$$[f(r), p^2]$$

$$= ir^{-1}f'(r)(\boldsymbol{r} \cdot \boldsymbol{p} - 3i) + f''(r) - r^{-1}f'(r) + ir^{-1}f'(r)\boldsymbol{r} \cdot \boldsymbol{p}$$

$$= 2ir^{-1}f'(r)\boldsymbol{r} \cdot \boldsymbol{p} + 2r^{-1}f'(r) + f''(r)$$

$$= 2ir^{-1}f'(r)(\boldsymbol{r} \cdot \boldsymbol{p} - i) + f''(r).$$

## 2.4  Commutators Involving L

The basic commutation relations involving the orbital angular momentum, obtained from $\boldsymbol{L} = \boldsymbol{r} \times \boldsymbol{p}$ and the properties of the Levi-Civita symbol (see Appendix A) are

$$[L_j, x_k] = i\epsilon_{jk\ell}x_\ell, \tag{2.23}$$

$$[L_j, p_k] = i\epsilon_{jk\ell}p_\ell, \tag{2.24}$$

$$[L_j, L_k] = i\epsilon_{jk\ell}L_\ell, \tag{2.25}$$

$$[L_j, f(r)] = 0. \tag{2.26}$$

The implied sums in the first three equations each contain only one term because of the antisymmetric properties of $\epsilon_{jk\ell}$. The components of $\boldsymbol{L}$ are given by $L_j = \epsilon_{jk\ell}x_k p_\ell$ or $x_j p_k - x_k p_j = \epsilon_{jk\ell}L_\ell$. Equations (2.23) to (2.25) can be expressed in vector form as

$$\boldsymbol{L} \times \boldsymbol{r} = i\boldsymbol{r}, \quad \boldsymbol{L} \times \boldsymbol{p} = i\boldsymbol{p}, \quad \boldsymbol{L} \times \boldsymbol{L} = i\boldsymbol{L}, \tag{2.27}$$

and (2.26) expresses the fact that functions having radial symmetry are invariant under rotation since the components of angular momentum are infinitesimal generators of the rotation group SO(3) (see Appendix B).

Equation (2.25) can be derived in two ways as follows. First derive $[L_1, L_2] = iL_3$ using specific subscripts. Then $[L_2, L_3] = iL_1$ and $[L_3, L_1] = iL_2$ follow by cyclic permutation of the indices:

$$
\begin{aligned}
[L_1, L_2] &= [x_2 p_3 - x_3 p_2, x_3 p_1 - x_1 p_3] \\
&= [x_2 p_3, x_3 p_1] - [x_2 p_3, x_1 p_3] - [x_3 p_2, x_3 p_1] + [x_3 p_2, x_1 p_3] \\
&= x_2 [p_3, x_3 p_1] + [x_2, x_3 p_1]p_3 - x_2 [p_3, x_1 p_3] - [x_2, x_1 p_3]x_2 \\
&\quad - x_3 [p_2, x_3 p_1] - [x_3, x_3 p_1]p_2 + x_3 [p_2, x_1 p_3] + [x_3, x_1 p_3]p_2 \\
&= x_2 [p_3, x_3]p_1 + x_1 [x_3, p_3]p_2 \\
&= i(x_1 p_2 - x_2 p_1) = iL_3.
\end{aligned}
$$

Here we have used the basic commutation relations (2.7) to (2.9) and the general rules (2.1) and (2.2). Alternatively we can derive (2.25) directly, using the properties of the Levi-Civita symbol (see Appendix A):

$$
\begin{aligned}
[L_j, L_k] \\
&= [\epsilon_{j\ell m}x_\ell p_m, \epsilon_{knq}x_n p_q] \\
&= \epsilon_{j\ell m}\epsilon_{knq}[x_\ell p_m, x_n p_q] \\
&= \epsilon_{j\ell m}\epsilon_{knq}(x_\ell [p_m, x_n p_q] + [x_\ell, x_n p_q]p_m) \\
&= \epsilon_{j\ell m}\epsilon_{knq}(x_\ell [p_m, x_n]p_q + x_n [x_\ell, p_q]p_m), \quad \text{from (2.1)} \\
&= \epsilon_{j\ell m}\epsilon_{knq}(-i\delta_{mn}x_\ell p_q + i\delta_{\ell q}x_n p_m), \quad \text{from (2.2),(2.8)} \\
&= -i\epsilon_{j\ell m}\epsilon_{kmq}x_\ell p_q + i\epsilon_{j\ell m}\epsilon_{kn\ell}x_n p_m.
\end{aligned}
$$

Now redefine the summation indices in the first term so that both terms contain $x_n p_m$ as a factor to obtain

$$
\begin{aligned}
[L_j, L_k] &= i(-\epsilon_{jn\ell}\epsilon_{k\ell m} + \epsilon_{j\ell m}\epsilon_{kn\ell})x_n p_m \\
&= i(-\epsilon_{mk\ell}\epsilon_{jn\ell} - \epsilon_{jm\ell}\epsilon_{kn\ell})x_n p_m \\
&= i\epsilon_{kj\ell}\epsilon_{mn\ell}x_n p_m, \quad \text{from (A.8)} \\
&= i\epsilon_{jk\ell}\epsilon_{nm\ell}x_n p_m \\
&= i\epsilon_{jk\ell}L_\ell.
\end{aligned}
$$

From these results commutators involving $L^2$ and $p^2$ can be obtained:

$$[L^2, r] = 2ir \times L + 2r, \tag{2.28}$$
$$[L^2, p] = 2ip \times L + 2p, \tag{2.29}$$
$$[L, p^2] = 0. \tag{2.30}$$

Equation (2.28) can be derived as follows

$$
\begin{aligned}
[L^2, x_j] &= [L_k L_k, x_j] \\
&= L_k[L_k, x_j] + [L_k, x_j]L_k, \quad \text{from (2.2)} \\
&= L_k(-i\epsilon_{jk\ell}x_\ell) + (-i\epsilon_{jk\ell}x_\ell)L_k, \quad \text{from (2.23)} \\
&= -i\epsilon_{jk\ell}(x_\ell L_k + L_k x_\ell) \\
&= -i\epsilon_{jk\ell}(x_\ell L_k + x_\ell L_k - i\epsilon_{\ell km}x_m), \quad \text{from (2.23)} \\
&= -2i\epsilon_{jk\ell}x_\ell L_k - \epsilon_{jk\ell}\epsilon_{\ell km}x_m \\
&= 2i\epsilon_{j\ell k}x_\ell L_k + \epsilon_{jk\ell}\epsilon_{mk\ell}x_m \\
&= 2i\epsilon_{j\ell k}x_\ell L_k + 2\delta_{jm}x_m, \quad \text{from (A.10)} \\
&= 2i\epsilon_{j\ell k}x_\ell L_k + 2x_j \\
&= 2i(r \times L)_j + 2x_j.
\end{aligned}
$$

Finally some useful vector identities involving $L$ are

$$r \cdot L = L \cdot r = 0, \tag{2.31}$$
$$p \cdot L = L \cdot p = 0, \tag{2.32}$$
$$
\begin{aligned}
p \times L &= rp^2 - p(r \cdot p - i) \\
&= p^2 r - (p \cdot r - i)p, 
\end{aligned}\tag{2.33}
$$
$$
\begin{aligned}
r \times L &= -pr^2 + (r \cdot p - i)r \\
&= -r^2 p + (r \cdot p + i)r \\
&= r(r \cdot p) - r^2 p, 
\end{aligned}\tag{2.34}
$$

$$r \times L + L \times r = 2ir, \tag{2.35}$$
$$p \times L + L \times p = 2ip. \tag{2.36}$$

The derivation of (2.31) illustrates another useful rule:

$$\boldsymbol{r} \cdot \boldsymbol{L} = x_j L_j = \epsilon_{jk\ell} x_j x_k p_\ell = (\boldsymbol{r} \times \boldsymbol{r}) \cdot \boldsymbol{p} = 0,$$

since $\boldsymbol{r} \times \boldsymbol{r} = 0$. This result also follows from the general rule:

If $A = \sum_{jk} a_{jk} s_{jk}$ is such that $a_{jk}$ is antisymmetric in $j$ and $k$ and $s_{jk}$ is symmetric in $j$ and $k$, then $A = 0$.

We can conclude that $\boldsymbol{r} \cdot \boldsymbol{L} = 0$ since $\epsilon_{jk\ell}$ is antisymmetric in $j$ and $k$ but $x_j x_k$ is symmetric in $j$ and $k$.

## 2.5  Exercises

◈ **Exercise 2.1** Prove that if $A = \sum_{jk} a_{jk} s_{jk}$ is such that $a_{jk}$ is antisymmetric in $j$ and $k$ and $s_{jk}$ is symmetric in $j$ and $k$, then $A = 0$. This result is often useful for evaluating commutation relations involving the Levi-Civita symbol $\epsilon_{jk\ell}$ and will be used several times in the following exercises and the exercises of later chapters.

◈ **Exercise 2.2** Derive commutation relations (2.11) to (2.14).

◈ **Exercise 2.3** Derive the commutation relations which involve $\boldsymbol{r} \cdot \boldsymbol{p}$ and $p^2$ that are given in (2.17) and (2.19) to (2.22).

◈ **Exercise 2.4** Derive the commutation relations (2.23), (2.24) and (2.26). The first two commutation relations show that $\boldsymbol{r}$ and $\boldsymbol{p}$ are $so(3)$ vector operators (see Chapter 3).

◈ **Exercise 2.5** Derive commutation relations (2.29) and (2.30).

◈ **Exercise 2.6** Derive identities (2.32) to (2.36).

# Chapter 3

# Angular Momentum Theory and so(3)

## 3.1  Introduction

The study of so(3) and its representations is the study of the familiar angular momentum theory but from the algebraic viewpoint.

Since angular momentum theory is covered in considerable detail in many specialized books and textbooks on quantum mechanics, using both the algebraic and differential equation approaches, we consider only the basic ideas which illustrate in a familiar context the algebraic approach used in subsequent chapters to study more complicated Lie algebras such as so(4) and so(4,2). Also the algebraic approach presented here is directly applicable to the derivation in the next chapter of the unirreps of so(2,1). which are much less familiar than the unirreps of so(3). In fact the algebraic approach for these two Lie algebras is almost identical yet the unirreps obtained are quite different: all unirreps of so(3) are finite dimensional but all unireps of so(2,1) are infinite dimensional.

The algebraic approach is also superior to methods based on the solution of differential equations (e.g., for so(3) the spherical harmonic functions obtained as solutions of a differential equation provide representations only for integral values of the angular momentum) since all representations are obtained and all relevant matrix elements of the operators of the Lie algebra in a representation can be evaluated without knowledge of the various special functions which arise as series solutions in the differential equation approach. In fact Lie algebraic methods can provide a unifying foundation for the study of many of the special functions of mathematical physics since they arise as particular realizations of the abstract basis vectors used in the algebraic approach.

The simplest example illustrating the two approaches is given by the one-dimensional harmonic oscillator considered in virtually all textbooks on quantum mechanics using annihilation and creation operators and as an application of the series solution of a second-order differential equation to obtain eigenfunctions in terms of Hermite polynomials.

The basic idea behind the algebraic approach is to start with the commutation relations for the generators $J_1$, $J_2$ and $J_3$ of the Lie algebra so(3) and an abstract vector space of states, whose basis vectors are chosen as simultaneous eigenfunctions of the maximal set $\{J^2, J_3\}$ of commuting operators, and construct the irreducible representations (irreps).

For angular momentum theory the most general irreps are not required. We need only the irreps corresponding to unitary representations of the Lie group SO(3) and these correspond to irreps of so(3) for which the generators $J_1$, $J_2$ and $J_3$ are hermitian with respect to some scalar product[1]. We find that all the unirreps are finite dimensional: a well known fact from angular momentum theory.

The important concept of a vector operator with respect to so(3) is introduced. In fact vector operators are rank-1 tensor operators and the results obtained here in a pedagogical and "classical" fashion for the matrix elements of a vector operator are a special case of the Wigner-Eckart theorem that is originally due to Dirac, Pauli and Guttinger (see Appendix 31 of [SL60]). The representation of a vector operator in an angular momentum basis is obtained as far as is possible since the complete representation requires additional properties of the vector operator. These results will be useful later for the representation theory of so(4).

In the final section an alternate derivation of the matrix representation of a vector operator is given using some of the more advanced aspects of angular momentum theory based on the Wigner-Eckart theorem. This section is not essential for the understanding of later chapters.

## 3.2  Ladder Operators for so(3)

We begin with the defining commutation relations

$$[J_1, J_2] = iJ_3, \quad [J_2, J_3] = iJ_1, \quad [J_3, J_1] = iJ_2, \tag{3.1}$$

and the Casimir operator $J^2 = J_1^2 + J_2^2 + J_3^2$ which commutes with the generators:

$$[J^2, J_k] = 0, \quad k = 1, 2, 3. \tag{3.2}$$

The Casimir operator does not belong to so(3) since it is not a linear combination of the generators[2].

It is more convenient to introduce the so-called ladder operators $J_+$ and $J_-$ instead of $J_1$ and $J_2$:

$$J_+ = J_1 + iJ_2, \quad J_- = J_1 - iJ_2. \tag{3.3}$$

---

[1]The word "unirrep" refers to representations of the Lie group by unitary matrices or operators but it is common to also use it to refer to representations of the Lie algebra.

[2]Casimir operators belong to the enveloping algebra of the Lie algebra.

These two definitions can be conveniently combined using the compact notation $J_\pm = J_1 \pm iJ_2$. Instead of using the generators $\{J_1, J_2, J_3\}$ we now use the equivalent set $\{J_+, J_-, J_3\}$ so we need the commutation relations

$$[J_+, J_-] = 2J_3, \tag{3.4}$$
$$[J_3, J_\pm] = \pm J_\pm, \tag{3.5}$$

which are easily derived from (3.1) and (3.3). The Casimir operator $J^2$ now satisfies

$$[J^2, J_\pm] = 0, \tag{3.6}$$

and

$$J^2 = J_+ J_- + J_3^2 - J_3 = J_- J_+ + J_3^2 + J_3, \tag{3.7}$$

which can be derived using (3.2) and (3.3) and the expressions $J_1 = (J_+ + J_-)/2$ and $J_2 = -i(J_+ - J_-)/2$.

The next step is to introduce an abstract vector space of states on which the operators $J_\pm$, $J_3$ and $J^2$ can act to produce matrix representations. From (3.2) it follows that we can choose $J^2$ and one of the $J_k$ as a complete set of commuting operators. It is conventional[3] to choose $J_3$. Since a set of commuting operators can be simultaneously diagonalized we can choose basis vectors $\psi_{\Lambda m}$ labeled by the eigenvalues $\Lambda$ and $m$ of $J^2$ and $J_3$, respectively:

$$J^2 \psi_{\Lambda m} = \Lambda \psi_{\Lambda m}, \tag{3.8}$$
$$J_3 \psi_{\Lambda m} = m \psi_{\Lambda m}. \tag{3.9}$$

The action of $J_+$ on an eigenvector can be obtained from (3.5) and (3.9)

$$(J_3 J_+ - J_+ J_3)\psi_{\Lambda m} = J_+ \psi_{\Lambda m},$$
$$J_3(J_+ \psi_{\Lambda m}) - m(J_+ \psi_{\Lambda m}) = J_+ \psi_{\Lambda m},$$
$$J_3(J_+ \psi_{\Lambda m}) = (m + 1)J_+ \psi_{\Lambda m}.$$

This important result shows that $J_+ \psi_{\Lambda m}$, if it is non-zero, is an eigenfunction of $J_3$ with eigenvalue $m + 1$ so the effect of $J_+$ is to raise the eigenvalue by one unit. Similarly $J_- \psi_{\Lambda m}$, if it is non-zero, is an eigenfunction of $J_3$ with eigenvalue $m - 1$. This follows again from (3.5) and (3.9):

$$(J_3 J_- - J_- J_3)\psi_{\Lambda m} = -J_- \psi_{\Lambda m},$$
$$J_3(J_- \psi_{\Lambda m}) - m(J_- \psi_{\Lambda m}) = -J_- \psi_{\Lambda m},$$
$$J_3(J_- \psi_{\Lambda m}) = (m - 1)J_- \psi_{\Lambda m}.$$

This is why $J_+$ and $J_-$ are often called raising and lowering operators.

---

[3]for so(2,1) the different choices give different classes of representations.

## 3.3    General Eigenvalue Spectrum of $J_3$

We can now define

$$J_+\psi_{\Lambda m} = A_{\Lambda m}\psi_{\Lambda,m+1}, \tag{3.10}$$

$$J_-\psi_{\Lambda m} = B_{\Lambda m}\psi_{\Lambda,m-1}. \tag{3.11}$$

The eigenvalue of $J^2$ is not changed by any of the operators in so(3). If we assume the existence of some eigenfunction $\psi_{\Lambda m_0}$ then the general eigenvalue spectrum of $J_3$ is a subset of

$$\mathcal{S}_{\Lambda m_0} = \{m : m = m_0 + k, \ k = 0, \pm 1, \pm 2, \ldots\}. \tag{3.12}$$

With no further restrictions on the so(3) generators there are four types of eigenvalue spectra:

1. **bounded below:** the set $\mathcal{S}_{\Lambda m_0}$ has a smallest element $m_1$ which means that $J_-\psi_{\Lambda m_1} = 0$,

2. **bounded above:** the set $\mathcal{S}_{\Lambda m_0}$ has a largest element $m_2$ which means that $J_+\psi_{\Lambda m_2} = 0$,

3. **unbounded:** $J_-\psi_{\Lambda m} \neq 0$ and $J_+\psi_{\Lambda m} \neq 0$ for all $m \in \mathcal{S}_{\Lambda m_0}$,

4. **bounded:** the set $\mathcal{S}_{\Lambda m_0}$ has both a smallest element $m_1$ and a largest element $m_2$ such that $J_-\psi_{\Lambda m_1} = 0$ and $J_+\psi_{\Lambda m_2} = 0$.

In each case the space of eigenfunctions of $J_3$ is invariant under the action of the so(3) generators and gives irreducible representations (irreps). In the first three cases these representations are infinite dimensional. A complete characterization of all irreps is not needed for the applications considered in this book (see [AD88b], [BA65]).

For our applications of angular momentum theory and so(3) to quantum mechanical problems only the unitary representations of the Lie group SO(3) are required and this imposes the condition that the generators $J_1$, $J_2$ and $J_3$ be hermitian with respect to some scalar product[4] and that the eigenfunctions $\psi_{\Lambda m}$ be normalizable with respect to this scalar product.

## 3.4    Unirreps of so(3)

If we introduce a scalar product $(\phi, \psi)$ and a norm $\|\psi\| = (\psi, \psi)^{1/2}$ then an operator $A$ is hermitian if $(A^\dagger\phi, \psi) = (\phi, A\psi)$, where $A^\dagger$ denotes the hermitian conjugate of $A$. It follows from $J_k^\dagger = J_k$, $k = 1, 2, 3$ that $J_+$ and $J_-$ are hermitian conjugates of each other

$$J_+^\dagger = J_-, \quad J_-^\dagger = J_+. \tag{3.13}$$

---

[4] A matrix $g = e^{iA}$ is unitary if and only if $A$ is hermitian.

We also assume that the abstract basis vectors are orthonormal with respect to the scalar product

$$(\psi_{\Lambda'm'}, \psi_{\Lambda m}) = \delta_{\Lambda'\Lambda}\delta_{m'm}. \tag{3.14}$$

Since the eigenvalues of an hermitian operator are real it follows that $\Lambda$ and $m$ are real.

The operators $J_+J_-$ and $J_-J_+$ are also positive definite hermitian operators since

$$(\psi_{\Lambda m}, J_-J_+\psi_{\Lambda m}) = (J_+\psi_{\Lambda m}, J_+\psi_{\Lambda m}) = \|J_+\psi_{\Lambda m}\|^2 \geq 0,$$
$$(\psi_{\Lambda m}, J_+J_-\psi_{\Lambda m}) = (J_-\psi_{\Lambda m}, J_-\psi_{\Lambda m}) = \|J_-\psi_{\Lambda m}\|^2 \geq 0.$$

Then from (3.10) and (3.11)

$$\|J_+\psi_{\Lambda m}\|^2 = |A_{\Lambda m}|^2 \geq 0, \text{ and } \|J_-\psi_{\Lambda m}\|^2 = |B_{\Lambda m}|^2 \geq 0.$$

On the other hand from (3.7) $J_+J_- = J^2 - J_3^2 + J_3$ and $J_-J_+ = J^2 - J_3^2 - J_3$ so

$$\|J_+\psi_{\Lambda m}\|^2 = (\psi_{\Lambda m}, (J^2 - J_3^2 - J_3)\psi_{\Lambda m}) = \Lambda - m(m+1),$$
$$\|J_-\psi_{\Lambda m}\|^2 = (\psi_{\Lambda m}, (J^2 - J_3^2 + J_3)\psi_{\Lambda m}) = \Lambda - m(m-1).$$

Therefore

$$|A_{\Lambda m}|^2 = \Lambda - m(m+1) \geq 0, \tag{3.15}$$
$$|B_{\Lambda m}|^2 = \Lambda - m(m-1) \geq 0. \tag{3.16}$$

Adding these results together gives $m^2 \leq \Lambda$ which means that the eigenvalue spectrum of $J_3$ is bounded above and below (case (4)) and the only unirreps are the finite dimensional ones.

Let $m_1$ be the smallest value of $m$ and let $m_2$ be the largest value of $m$. Then $J_+\psi_{\Lambda m_2} = 0$ and $J_-\psi_{\Lambda m_1} = 0$ which will be the case if $A_{\Lambda m_2} = 0$ and $B_{\Lambda m_1} = 0$. Therefore from (3.15) and (3.16) $m_1$ and $m_2$ satisfy $\Lambda - m_1(m_1 - 1) = 0$ and $\Lambda - m_2(m_2+1) = 0$. Eliminating $\Lambda$ gives $(m_1+m_2)(m_1-m_2-1) = 0$ so either $m_1 = -m_2$ or $m_1 = m_2 + 1$. The latter case is impossible since $m_1 \leq m_2$. If we let $j = m_2$ then $m_1 = -j$ and $\Lambda = j(j+1)$. Since $m_2 - m_1 = 2j$ must be a non-negative integer ($J_+$ and $J_-$ raise or lower $m$ in unit steps) it follows that the $J_3$ eigenvalue spectra are

$$\mathcal{S}_j = \{m = -j, -j+1, \ldots, j\}, \quad 2j = 0, 1, 2, \ldots \tag{3.17}$$

The standard choice of phase is obtained by taking the positive square root in (3.15) and (3.16) and we obtain the following matrix representations, using the Dirac notation $|jm\rangle = \psi_{\Lambda m}$:

$$J^2|jm\rangle = j(j+1)|jm\rangle, \tag{3.18}$$
$$J_3|jm\rangle = m|jm\rangle, \tag{3.19}$$
$$J_+|jm\rangle = \sqrt{(j-m)(j+m+1)}\,|j, m+1\rangle, \tag{3.20}$$
$$J_-|jm\rangle = \sqrt{(j+m)(j-m+1)}\,|j, m-1\rangle. \tag{3.21}$$

Strictly speaking the unirreps for integer values of $j$ (corresponding to the orbital angular momentum) are unirreps of so(3) and those for non-integer (half-integer) values are the so-called spinor representations of su(2).

## 3.5   so(3) Scalar Operators

An operator $S$ is an so(3) scalar operator if it commutes with the generators of so(3):

$$[S, J_k] = 0, \quad k = 1, 2, 3. \tag{3.22}$$

Since the $J_k$ generate infinitesimal rotations in SO(3) or SU(2) it follows that a scalar operator is invariant under rotation. The matrix elements of a scalar operator with respect to the angular momentum basis $\{|jm\rangle, m = -j, \ldots, j\}$ have a particularly simple diagonal form obtained by taking matrix elements of the commutation relations (3.22):

$$\begin{aligned} \langle j'm'|[S, J_3]|jm\rangle &= \langle j'm'|SJ_3 - J_3S|jm\rangle \\ &= m\langle j'm'|S|jm\rangle - m'\langle j'm'|S|jm\rangle \\ &= (m - m')\langle j'm'|S|jm\rangle = 0, \end{aligned}$$

and since $[S, J^2] = 0$,

$$\begin{aligned} \langle j'm'|[S, J^2]|jm\rangle &= \langle j'm'|SJ^2 - J^2S|jm\rangle \\ &= (j(j+1) - j'(j'+1))\langle j'm'|S|jm\rangle = 0. \end{aligned}$$

These results show that matrix elements of a scalar operator are represented by diagonal matrices in an angular momentum basis:

$$\langle j'm'|S|jm\rangle = \delta_{j'j}\delta_{m'm}\langle j\|S\|j\rangle. \tag{3.23}$$

The "double bar" matrix elements are called reduced matrix elements. They cannot be evaluated unless further algebraic or physical properties of $S$ are known. For example from (3.2) and (3.18) $J^2$ is a scalar operator with reduced matrix elements

$$\langle j\|J^2\|j\rangle = j(j+1). \tag{3.24}$$

## 3.6   so(3) Vector Operators

An operator $V$ is said to be a vector operator with respect to so(3) if its components satisfy the commutation relations

$$[J_j, V_k] = i\epsilon_{jk\ell}V_\ell, \quad j, k, \ell = 1, 2, 3, \tag{3.25}$$

where the summation over repeated indices is implied. If $J_k$ is substituted for $V_k$ we obtain the defining commutation relations for so(3) so $\boldsymbol{J}$ itself is a vector operator. The commutation relations (2.23) and (2.24) also show that $\boldsymbol{r}$ and $\boldsymbol{p}$ are vector operators.

We now obtain several identities relating $\boldsymbol{J}$ and $\boldsymbol{V}$ which will be used to determine matrix elements of components of $\boldsymbol{V}$. From (3.25)

$$\boldsymbol{J} \times \boldsymbol{V} + \boldsymbol{V} \times \boldsymbol{J} = 2i\boldsymbol{V}. \tag{3.26}$$

Since $J_1 V_1 = V_1 J_1$ and similarly for the other components,

$$\boldsymbol{J} \cdot \boldsymbol{V} = \boldsymbol{V} \cdot \boldsymbol{J}. \tag{3.27}$$

Also

$$[\boldsymbol{J} \cdot \boldsymbol{V}, J_j] = 0, \; j = 1, 2, 3, \tag{3.28}$$

which can be expressed in vector form as

$$[\boldsymbol{J} \cdot \boldsymbol{V}, \boldsymbol{J}] = 0. \tag{3.29}$$

Thus $\boldsymbol{J} \cdot \boldsymbol{V}$ is an so(3) scalar operator having matrix elements

$$\langle j'm'|\boldsymbol{J} \cdot \boldsymbol{V}|jm\rangle = \delta_{j'j}\delta_{m'm}\langle j\|\boldsymbol{J} \cdot \boldsymbol{V}\|j\rangle. \tag{3.30}$$

The following three identities are also important:

$$[J^2, \boldsymbol{V}] = i(\boldsymbol{V} \times \boldsymbol{J} - \boldsymbol{J} \times \boldsymbol{V}), \tag{3.31}$$

$$[J^2, \boldsymbol{J} \times \boldsymbol{V}] = 2i(J^2\boldsymbol{V} - (\boldsymbol{J} \cdot \boldsymbol{V})\boldsymbol{J}), \tag{3.32}$$

$$[J^2, [J^2, \boldsymbol{V}]] = 2(J^2\boldsymbol{V} - 2(\boldsymbol{J} \cdot \boldsymbol{V})\boldsymbol{J} + \boldsymbol{V}J^2). \tag{3.33}$$

They are derived in the end of chapter exercises. The important double commutator indentity (3.33) will be used in the next section to obtain the $j$-selection rules for a vector operator.

## 3.7  Selection Rules for a Vector Operator

We want to derive formulas analogous to (3.18) to (3.21) for the action of the components of $\boldsymbol{V}$ on the angular momentum basis vectors. Defining $V_\pm = V_1 \pm iV_2$, the 9 commutation relations (3.25) can be expressed in the equivalent form

$$\begin{aligned}
[J_+, V_-] &= 2V_3, \quad [J_-, V_+] = -2V_3, \\
[J_3, V_\pm] &= \pm V_\pm, \quad [J_\pm, V_3] = \mp V_\pm, \\
[J_+, V_+] &= [J_-, V_-] = [J_3, V_3] = 0.
\end{aligned} \tag{3.34}$$

Using these commutation relations and the identities derived in the preceding section we can evaluate matrix elements of the form $\langle j'm'|V_3|jm\rangle$ and

$\langle j'm'|V_\pm|jm\rangle$. The relationship between $m'$ and $m$ which corresponds to non-vanishing matrix elements is called the $m$-selection rule and similarly the relationship between $j'$ and $j$ is called the $j$-selection rule.

The $m$-selection rule is obtained by taking matrix elements of the appropriate commutation relations in (3.34):

$$
\begin{aligned}
\langle j'm'|[J_3, V_3]|jm\rangle &= \langle j'm'|J_3 V_3|jm\rangle - \langle j'm'|V_3 J_3|jm\rangle \\
&= m'\langle j'm'|V_3|jm\rangle - m\langle j'm'|V_3|jm\rangle \\
&= (m' - m)\langle j'm'|V_3|jm\rangle = 0,
\end{aligned}
$$

so matrix elements of $V_3$ are non-zero only if $m' = m$. Similarly

$$
\begin{aligned}
\langle j'm'|[J_3, V_\pm]|jm\rangle &= \langle j'm'|J_3 V_\pm|jm\rangle - \langle j'm'|V_\pm J_3|jm\rangle \\
&= m'\langle j'm'|V_\pm|jm\rangle - m\langle j'm'|V_\pm|jm\rangle \\
&= (m' - m)\langle j'm'|V_\pm|jm\rangle \\
&= \pm\langle j'm'|V_\pm|jm\rangle,
\end{aligned}
$$

so $(m' - m \mp 1)\langle j'm'|V_\pm|jm\rangle = 0$ and matrix elements of $V_\pm$ are non-zero only if $m' = m \pm 1$. Therefore the $m$-selection rules are

$$\langle j'm'|V_3|jm\rangle = 0, \text{ unless } m' = m, \tag{3.35}$$

$$\langle j'm'|V_\pm|jm\rangle = 0, \text{ unless } m' = m \pm 1, \tag{3.36}$$

and we see that $V_+$ and $V_-$ act like raising and lowering operators for $m$. However, unlike $J_+$ and $J_-$, they can also change the value of $j$.

The $j$-selection rules can be obtained by taking matrix elements of the double commutator identity (3.33). The left side of (3.33) is

$$
\begin{aligned}
\langle j'm'|[J^2, [J^2, V]]|jm\rangle &= \langle j'm'|J^2[J^2, V] - [J^2, V]J^2|jm\rangle \\
&= j'(j'+1)\langle j'm'|[J^2, V]|jm\rangle - j(j+1)\langle j'm'|[J^2, V]|jm\rangle \\
&= (j'(j'+1) - j(j+1))\,\langle j'm'|[J^2, V]|jm\rangle \\
&= (j'(j'+1) - j(j+1))^2\,\langle j'm'|V|jm\rangle.
\end{aligned}
$$

The right side of (3.33) is

$$
\begin{aligned}
2\langle j'm'|J^2 V &- 2(J \cdot V)J + V J^2|jm\rangle \\
&= 2j'(j'+1)\langle j'm'|V|jm\rangle - 4\langle j'm'|(J \cdot V)J|jm\rangle \\
&\quad + 2j(j+1)\langle j'm'|V|jm\rangle \\
&= 2\,(j'(j'+1) + j(j+1))\,\langle j'm'|V|jm\rangle - 4\langle j'm'|(J \cdot V)J|jm\rangle.
\end{aligned}
$$

The term involving $J \cdot V$ can be expressed as

$$
\begin{aligned}
\langle j'm'|(J \cdot V)J|jm\rangle &= \sum_{j''m''} \langle j'm'|J \cdot V|j''m''\rangle\langle j''m''|J|jm\rangle \\
&= \langle j'\|J \cdot V\|j'\rangle\langle j'm'|J|jm\rangle,
\end{aligned}
$$

where we have used (3.23) for the matrix elements of $\boldsymbol{J} \cdot \boldsymbol{V}$ since it is a scalar operator. Collecting these results together

$$A(j',j)\langle j'm'|\boldsymbol{V}|jm\rangle = -4\langle j'\|\boldsymbol{J} \cdot \boldsymbol{V}\|j'\rangle\langle j'm'|\boldsymbol{J}|jm\rangle, \tag{3.37}$$

where

$$
\begin{aligned}
A(j',j) &= (j'(j'+1) - j(j+1))^2 - 2\,(j'(j'+1) + j(j+1)) \\
&= \left((j'-j)^2 - 1\right)\left((j'+j+1)^2 - 1\right).
\end{aligned} \tag{3.38}
$$

This result shows that the selection rules for any vector operator $\boldsymbol{V}$ follow from the selection rules for $\boldsymbol{J}$. There are two cases to consider: $j' = j$ and $j' \neq j$. If $j' = j$ in (3.37) then

$$\langle jm'|\boldsymbol{V}|jm\rangle = \frac{\langle j\|\boldsymbol{J} \cdot \boldsymbol{V}\|j\rangle}{j(j+1)}\langle jm'|\boldsymbol{J}|jm\rangle, \tag{3.39}$$

so unless $\langle j\|\boldsymbol{J} \cdot \boldsymbol{V}\|j\rangle$ is zero because of special properties of $\boldsymbol{V}$ in a particular case then matrix elements of components of $\boldsymbol{V}$ are proportional to the corresponding matrix elements of $\boldsymbol{J}$ in case $j' = j$. Later we consider the Laplace-Runge-Lenz vector $\boldsymbol{A}$ which satisfies $\boldsymbol{A} \cdot \boldsymbol{L} = \boldsymbol{L} \cdot \boldsymbol{A} = 0$, where $\boldsymbol{L}$ is the orbital angular momentum.

If $j' \neq j$ then $\langle j'm'|\boldsymbol{J}|jm\rangle = 0$ and (3.37) reduces to

$$A(j',j)\langle j'm'|\boldsymbol{V}|jm\rangle = 0. \tag{3.40}$$

From (3.38) $A(j',j) = 0$ for $j' \neq j$ only if $j' = j \pm 1$ so the off-diagonal matrix elements of $\boldsymbol{V}$ can be non-zero only if $j' = j \pm 1$. Combining both cases the $j$-selection rule is

$$\langle j'm'|\boldsymbol{V}|jm\rangle = 0, \quad \text{unless } j' = j-1, j, j+1. \tag{3.41}$$

## 3.8  Matrix Elements of a Vector Operator

Using the $m$- and $j$-selection rules we can obtain the action of the components $V_{\pm}$, $V_3$ of a vector operator on the angular momentum basis vectors. For $V_+$ we obtain, using (3.36) and (3.41),

$$
\begin{aligned}
V_+|jm\rangle = &\ \langle j+1, m+1|V_+|jm\rangle|j+1, m+1\rangle \\
&+ \langle j, m+1|V_+|jm\rangle|j, m+1\rangle \\
&+ \langle j-1, m+1|V_+|jm\rangle|j-1, m+1\rangle. \tag{3.42}
\end{aligned}
$$

The three matrix elements can be partially evaluated by taking matrix elements of the commutator identity $J_+V_+ - V_+J_+ = 0$. Therefore

$$\langle j'm'|J_+V_+|jm\rangle = \langle j'm'|V_+J_+|jm\rangle,$$

or using (3.20) and (3.21)

$$\sqrt{(j' + m')(j' - m' + 1)}\langle j', m' - 1|V_+|jm\rangle$$
$$= \sqrt{(j - m)(j + m + 1)}\langle j'm'|V_+|j, m + 1\rangle.$$

Since $m' = m + 2$ for non-vanishing matrix elements

$$\frac{\langle j', m + 1|V_+|jm\rangle}{\sqrt{(j - m)(j + m + 1)}} = \frac{\langle j', m + 2|V_+|j, m + 1\rangle}{\sqrt{(j' + m + 2)(j' - m - 1)}}. \tag{3.43}$$

Substitute $j' = j + 1$ to obtain

$$\frac{\langle j + 1, m + 1|V_+|jm\rangle}{\sqrt{(j - m)(j + m + 1)}} = \frac{\langle j + 1, m + 2|V_+|j, m + 1\rangle}{\sqrt{(j + m + 3)(j - m)}}.$$

Cancel the common factor and divide both sides by $\sqrt{j + m + 2}$ to obtain

$$\frac{\langle j + 1, m + 1|V_+|jm\rangle}{\sqrt{(j + m + 1)(j + m + 2)}} = \frac{\langle j + 1, m + 2|V_+|j, m + 1\rangle}{\sqrt{(j + m + 2)(j + m + 3)}}.$$

This important result shows that these two ratios are independent of $m$ (replacing $m$ by $m + 1$ in the left side gives the right side). Denoting this common ratio by $-d_j$ we obtain (the minus sign is conventional)

$$\langle j + 1, m + 1|V_+|jm\rangle = -\sqrt{(j + m + 1)(j + m + 2)}\, d_j. \tag{3.44}$$

Similarly by substituting $j' = j$ and $j' = j - 1$ into (3.43) we obtain

$$\langle j, m + 1|V_+|jm\rangle = -\sqrt{(j + m + 1)(j - m)}\, a_j, \tag{3.45}$$

$$\langle j - 1, m + 1|V_+|jm\rangle = \sqrt{(j - m - 1)(j - m)}\, c_j, \tag{3.46}$$

for some $j$-dependent factors $-a_j$ and $c_j$. The importance of these results is that the $m$ dependence of the matrix elements is completely known. The unknown factors $a_j$, $c_j$ and $d_j$ cannot be obtained without specifying further properties of the vector operator $\boldsymbol{V}$.

The action of $V_3$ on the basis vectors is obtained using (3.35) and (3.41):

$$V_3|jm\rangle = \langle j + 1, m|V_3|jm\rangle|j + 1, m\rangle$$
$$+ \langle jm|V_3|jm\rangle|jm\rangle + \langle j - 1, m|V_3|jm\rangle|j - 1, m\rangle. \tag{3.47}$$

The matrix elements of $V_3$ can be obtained by taking matrix elements of the commutation relation $-2V_3 = [J_-, V_+]$:

$$-2\langle j'm'|V_3|jm\rangle = \langle j'm'|J_- V_+|jm\rangle - \langle j'm'|V_+ J_-|jm\rangle$$
$$= \sqrt{(j' - m')(j' + m' + 1)}\langle j', m' + 1|V_+|jm\rangle$$
$$- \sqrt{(j + m)(j - m + 1)}\langle j'm'|V_+|j, m - 1\rangle. \tag{3.48}$$

Substituting $m' = m$ and $j' = j + 1$ into (3.48) and using (3.44) gives

$$-2\langle j + 1, m|V_3|jm\rangle =$$
$$-d_j\sqrt{(j - m + 1)(j + m + 2)}\sqrt{(j + m + 1)(j + m + 2)}$$
$$+ d_j\sqrt{(j + m)(j - m + 1)}\sqrt{(j + m)(j + m + 1)}$$
$$= -2d_j\sqrt{(j - m + 1)(j + m + 1)}.$$

Therefore

$$\langle j + 1, m|V_3|jm\rangle = \sqrt{(j - m + 1)(j + m + 1)}\, d_j. \tag{3.49}$$

Similarly substituting $m' = m$ and $j' = j$ into (3.48) and using (3.45) gives

$$\langle jm|V_3|jm\rangle = -ma_j, \tag{3.50}$$

and substituting $m' = m$ and $j' = j - 1$ into (3.48) and using (3.46) gives

$$\langle j - 1, m|V_3|jm\rangle = \sqrt{(j - m)(j + m)}\, c_j. \tag{3.51}$$

Finally the action of $V_-$ on the basis vectors is obtained using (3.36) and (3.41):

$$V_-|jm\rangle = \langle j + 1, m - 1|V_-|jm\rangle|j + 1, m - 1\rangle$$
$$+ \langle j, m - 1|V_-|jm\rangle|j, m - 1\rangle$$
$$+ \langle j - 1, m - 1|V_-|jm\rangle|j - 1, m - 1\rangle. \tag{3.52}$$

The matrix elements of $V_-$ can be obtained by taking matrix elements of the commutation relation $V_- = [J_-, V_3]$:

$$\langle j'm'|V_-|jm\rangle = \langle j'm'|J_-V_3|jm\rangle - \langle j'm'|V_3J_-|jm\rangle$$
$$= \sqrt{(j' - m')(j' + m' + 1)}\langle j', m' + 1|V_3|jm\rangle$$
$$- \sqrt{(j + m)(j - m + 1)}\langle j'm'|V_3|j, m - 1\rangle. \tag{3.53}$$

Substituting $m' = m - 1$ and $j' = j + 1$ into (3.53) and using (3.49)

$$\langle j + 1, m - 1|V_-|jm\rangle =$$
$$d_j\sqrt{(j - m + 2)(j + m + 1)}\sqrt{(j - m + 1)(j + m + 1)}$$
$$- d_j\sqrt{(j + m)(j - m + 1)}\sqrt{(j - m + 2)(j + m)}.$$

Therefore

$$\langle j + 1, m - 1|V_-|jm\rangle = \sqrt{(j - m + 1)(j - m + 2)}\, d_j. \tag{3.54}$$

Similarly substituting $m' = m - 1$ and $j' = j$ into (3.53) and using (3.50) gives

$$\langle j, m - 1|V_-|jm\rangle = -\sqrt{(j - m + 1)(j + m)}\, a_j, \tag{3.55}$$

and substituting $m' = m$ and $j' = j - 1$ into (3.53) and using (3.51) gives

$$\langle j - 1, m - 1|V_-|jm\rangle = -\sqrt{(j + m - 1)(j + m)}\, c_j. \qquad (3.56)$$

Collecting these results together we obtain the following representation of a vector operator in an angular momentum basis:

$$\begin{aligned} V_3|jm\rangle &= \sqrt{(j - m)(j + m)}\, c_j|j - 1, m\rangle \\ &\quad - ma_j|jm\rangle \\ &\quad - \sqrt{(j - m + 1)(j + m + 1)}\, d_j|j + 1, m\rangle, \end{aligned} \qquad (3.57)$$

$$\begin{aligned} V_+|jm\rangle &= \sqrt{(j - m - 1)(j - m)}c_j|j - 1, m + 1\rangle \\ &\quad - \sqrt{(j + m + 1)(j - m)}\, a_j|j, m + 1\rangle \\ &\quad - \sqrt{(j + m + 1)(j + m + 2)}\, d_j|j + 1, m + 1\rangle, \end{aligned} \qquad (3.58)$$

$$\begin{aligned} V_-|jm\rangle &= -\sqrt{(j + m - 1)(j + m)}\, c_j|j - 1, m - 1\rangle \\ &\quad - \sqrt{(j - m + 1)(j + m)}\, a_j|j, m - 1\rangle \\ &\quad - \sqrt{(j - m + 1)(j - m + 2)}\, d_j|j + 1, m - 1\rangle. \end{aligned} \qquad (3.59)$$

If $V$ is assumed to be an hermitian vector operator ($V_+^\dagger = V_-$, $V_-^\dagger = V_+$ and $V_3^\dagger = V_3$) further simplification is obtained. Recall that the matrix elements of an operator $A$ satisfy $\langle a|A|b\rangle = \langle b|A^\dagger|a\rangle^*$. Therefore

$$\langle j + 1, m + 1|V_+|jm\rangle = \langle jm|V_-|j + 1, m + 1\rangle^*,$$

and using (3.44) and (3.56) (with $j$ replaced by $j + 1$ and $m$ replaced by $m + 1$) we obtain

$$d_j = c_{j+1}^*. \qquad (3.60)$$

Similarly, since $\langle jm|V_3|jm\rangle = \langle jm|V_3|jm\rangle^*$ it follows that

$$a_j^* = a_j. \qquad (3.61)$$

It is also possible to make a $j$-dependent phase change of basis (see Exercise 3.6):

$$|jm\rangle = e^{i\gamma_j}|jm\rangle',$$

such that $c_j$ and $d_j$ are real and

$$d_j = c_{j+1}. \qquad (3.62)$$

In general, without assuming hermiticity, it is possible to make a suitable change of basis vectors of the form $|jm\rangle = \omega_j|jm\rangle'$ so that (3.62) is satisfied but the new basis vectors are orthogonal although not necessarily normalized.

## 3.9  Wigner-Eckart Theorem

In this section we provide an alternate derivation of the results of the preceding section for the matrix elements of a vector operator in an angular momentum basis. It is a much simpler derivation but requires some of the more advanced aspects of angular momentum theory, namely spherical tensor operators, Clebsch-Gordan coefficients and the Wigner-Eckart theorem ([SC68], [BO86], [DE63], [ME70], [AL73], [ME70b], [ED57], [BI81a], [BR68]).

The concept of a vector operator can be generalized to tensor operators. Thus we can define an irreducible spherical tensor operator $T^{(k)}$ of rank $k$ to be an operator with $2k + 1$ components $T_q^{(k)}$, $q = -k, -k + 1, \ldots, k$ which satisfy the following commutation relations with the components $J_3, J_+, J_-$ of an angular momentum $J$:

$$[J_3, T_q^{(k)}] = qT_q^{(k)}, \tag{3.63}$$

$$[J_\pm, T_q^{(k)}] = [k(k + 1) - q(q \pm 1)]^{1/2} T_{q\pm 1}^{(k)}. \tag{3.64}$$

The components $T_q^{(k)}$ are called spherical components by analogy with the spherical harmonic functions.

In case $k = 1$ we obtain the spherical components of a vector operator $V = T^{(k)}$, which are simply related to the components $V_3, V_+, V_-$ introduced earlier in Section 3.7, by

$$V_{-1}^{(1)} = \frac{1}{\sqrt{2}} V_-, \quad V_0^{(1)} = V_3, \quad V_1^{(1)} = -\frac{1}{\sqrt{2}} V_+. \tag{3.65}$$

With these definitions the commutation relations (3.63) and (3.64) for $k = 1$ and $q = -1, 0, 1$ reduce to those in (3.34) for the components $V_3$, $V_+$ and $V_-$.

The Wigner-Eckart theorem states that the matrix elements of a spherical tensor operator $T^{(k)}$ can be expressed in the form

$$\langle \gamma' j' m' | T_q^{(k)} | \gamma j m \rangle$$
$$= (-1)^{2k}(2j' + 1)^{-1/2} \langle jm\, kq | j'm' \rangle \langle \gamma' j' \| T^{(k)} \| \gamma j \rangle. \tag{3.66}$$

Here the states $|\gamma j m\rangle$ are angular momentum states (eigenfunctions of $J^2$ and $J_3$) and $\gamma$ specifies any additional quantum numbers or labels needed to fully specify the state (e.g., for the hydrogen atom $\gamma$ would be the principal quantum number). The coefficient $\langle jm\, kq | j'm' \rangle$ is a Clebsch-Gordan (CG) coefficient and $\langle \gamma' j' \| T^{(k)} \| \gamma j \rangle$ is called the reduced matrix element of the tensor operator $T^{(k)}$. Unfortunately there are two common definitions of reduced matrix elements depending on whether the factor $(-1)^{2k}(2j' + 1)^{-1/2}$ is absorbed into the reduced matrix element or not (for example, [SC68], [ME70b], [ED57], [AL73], [DE63] use (3.66) whereas [AD88b], [BI81a], [BO86], [ME70], use the other definition). Our definition (3.66) is more suitable since it can be conveniently expressed in terms of the Wigner 3-j symbol

$$\langle \gamma' j' m' | T_q^{(k)} | \gamma j m \rangle = (-1)^{j'-m'} \begin{pmatrix} j' & k & q \\ -m' & q & m \end{pmatrix} \langle \gamma' j' \| T^{(k)} \| \gamma j \rangle. \tag{3.67}$$

In calculations the Wigner 3-j symbol is more convenient than the CG coefficients since it has higher symmetry properties: it is invariant under cyclic permutations of the three columns, multiplied by the phase factor $(-1)^{j'+k+j}$ under interchange of any two columns or under a change of sign of all entries in the second row. The connection between the two types of coefficients is simply given by

$$\langle j_1 m_1\, j_2 m_2 | jm \rangle = (-1)^{j_1-j_2+m}(2j+1)^{1/2} \begin{pmatrix} j_1 & j_2 & j \\ m_1 & m_2 & -m \end{pmatrix}, \qquad (3.68)$$

or conversely

$$\begin{pmatrix} j_1 & j_2 & j \\ m_1 & m_2 & m \end{pmatrix} = (-1)^{j_1-j_2-m}(2j+1)^{-1/2}\langle j_1 m_1\, j_2 m_2 | j, -m \rangle. \qquad (3.69)$$

The Wigner-Eckart theorem is an important result in the angular momentum theory since it states that the matrix element factors into a simple geometrical part, the CG coefficient or the Wigner 3-j symbol, which contains all the dependence of the components of the tensor operator on $m$, $m'$ and $q$ and is the same for any tensor operators of rank $k$, and a physical part or reduced matrix element which is independent of the $2k+1$ components of the tensor operator, depending only on the particular tensor operator.

The CG coefficients $\langle j_1 m_1\, j_2 m_2 | jm \rangle^5$ are more convenient than the 3-j symbols in discussing the coupling problem for two angular momenta, $\boldsymbol{J} = \boldsymbol{J}_1 \oplus \boldsymbol{J}_2$. They vanish unless $m = m_1 + m_2$ and $j_1$, $j_2$ and $j$ satisfy the triangle inequality $|j_1-j_2| \le j \le j_1+j_2$. This immediately gives the selection rules for the nonvanishing matrix elements of $T_q^{(k)}$ in (3.66): $m' = m+q$ and $|j-k| \le j' \le j+k$. In the vector operator case $(k=1)$, in which we are primarily interested, the selection rules (3.35), (3.36) and (3.41) follow.

We can now use the Wigner-Eckart theorem to obtain the matrix representation of a vector operator given in (3.57) to (3.59):

$$V_q^{(1)}|\gamma j m\rangle = \sum_{\gamma'j'm'} |\gamma'j'm'\rangle \langle \gamma'j'm'|V_q^{(1)}|\gamma j m\rangle \qquad (3.70)$$

$$= \sum_{\gamma'j'm'} (-1)^{j'-m'} \begin{pmatrix} j' & 1 & j \\ -m' & q & m \end{pmatrix} \langle \gamma'j' \| \boldsymbol{V} \| \gamma j \rangle |\gamma'j'm'\rangle$$

$$= \sum_{\gamma'j'} (-1)^{j'-m-q} \begin{pmatrix} j' & 1 & j \\ -m-q & q & m \end{pmatrix} \langle \gamma'j' \| \boldsymbol{V} \| \gamma j \rangle |\gamma'j', m+q\rangle.$$

Only terms with $j' = j-1$, $j$, or $j+1$ contribute. In the simplest case the reduced matrix elements are diagonal in the additional labels represented by $\gamma$ and there is no summation over $\gamma'$. We have in fact assumed this case in Section 3.7 and Section 3.8 so the three results $(q = -1, 0, 1)$ in (3.70) correspond to (3.57) to (3.59).

---

[5]The notation $\langle j_1 j_2\, m_1 m_2 | jm \rangle$ is also common.

Useful tabulations of the CG coefficients are available and they are usually expressed in terms of the 3-j symbol because of its higher symmetry properties ([ED57], [BR68]).  The results we need can all be obtained from the four standard values [BR68], [RO59]

$$\begin{pmatrix} j & j & 1 \\ m & -m-1 & 1 \end{pmatrix} = (-1)^{j-m} \left[ \frac{(j-m)(j+m+1)}{2j(j+1)(2j+1)} \right]^{1/2}, \tag{3.71}$$

$$\begin{pmatrix} j & j & 1 \\ m & -m & 0 \end{pmatrix} = (-1)^{j-m} \frac{m}{[j(j+1)(2j+1)]^{1/2}}, \tag{3.72}$$

$$\begin{pmatrix} j & j+1 & 1 \\ m & -m-1 & 1 \end{pmatrix} = (-1)^{j-m} \left[ \frac{(j+m+1)(j+m+2)}{(2j+1)(2j+2)(2j+3)} \right]^{1/2}, \tag{3.73}$$

$$\begin{pmatrix} j & j+1 & 1 \\ m & -m & 0 \end{pmatrix} = (-1)^{j-m-1} \left[ \frac{(j-m+1)(j+m+1)}{(j+1)(2j+1)(2j+3)} \right]^{1/2}, \tag{3.74}$$

and the symmetry properties of the 3-j symbols.  If we define

$$c_j = -\frac{\langle \gamma, j-1 \| V \| \gamma j \rangle}{\sqrt{j(2j-1)(2j+1)}}, \tag{3.75}$$

$$a_j = -\frac{\langle \gamma j \| V \| \gamma j \rangle}{\sqrt{j(j+1)(2j+1)}}, \tag{3.76}$$

$$d_j = -\frac{\langle \gamma, j+1 \| V \| \gamma j \rangle}{\sqrt{(j+1)(2j+1)(2j+3)}}, \tag{3.77}$$

and use $V_+ = -\sqrt{2}V_1^{(1)}$, $V_- = \sqrt{2}V_{-1}^{(1)}$, $V_3 = V_0^{(1)}$ then (3.67) reduces to the previous results (3.57) to (3.59).

## 3.10   Exercises

◈ **Exercise 3.1** Derive identity (3.26) by multiplying commutation relations (3.25) by $\epsilon_{mjk}$, summing over $j$ and $k$, and using the definition $(A \times B)_k = \epsilon_{k\ell m} A_\ell B_m$ for the components of the cross product of two vectors (or vector operators) $A$ and $B$.

◈ **Exercise 3.2** Derive identity (3.28) which shows that $J \cdot V$ is a scalar operator.

◈ **Exercise 3.3** Derive identity (3.31).

◈ **Exercise 3.4** Derive identity (3.32) using (3.25), (3.26), and (3.31).

◈ **Exercise 3.5** Using (3.26), (3.27), (3.29), (3.31), and (3.32) derive the important double commutator identity given in (3.33)

◈ **Exercise 3.6** Show that if $V$ is an hermitian vector operator then a $j$-dependent phase change of the basis vectors can be made such that the $c_j$ and $d_j$ are real in (3.57) to (3.59) and $d_j = c_{j+1}$.

◈ **Exercise 3.7** Consider a 1-state boson system with state vectors denoted by $|n\rangle$ where $n$ is the number of bosons in the state.  Let $a$ and $a^+$ be the annihilation and creation operators satisfying the commutation relation $[a, a^+] = 1$. Assuming an orthonormal basis such that $\langle m|n\rangle = \delta_{mn}$ show that

$$\text{(a)} \quad |n\rangle = \tfrac{1}{\sqrt{n!}}(a^+)^n|0\rangle,$$

$$\text{(b)} \quad a^+|n\rangle = \sqrt{n+1}\,|n+1\rangle,$$

$$\text{(c)} \quad a|n\rangle = \sqrt{n}\,|n-1\rangle.$$

◈ **Exercise 3.8** This exercise is a continuation of Exercise 1.15.  Consider a 2-state boson system with state vectors $|n_1, n_2\rangle$ where $n_1$ is the number of bosons in state 1 and $n_2$ is the number of bosons in state 2. The annihilation and creation operators are defined in Exercise 1.15. The operators $J_+$, $J_-$ and $J_3$ satisfy the su(2) commutation relations (1.19) to (1.21). Obtain the irreps of su(2) using the boson states $|n_1, n_2\rangle$. This shows that boson realizations of a Lie algebra can be used to find general representations ([SA86], [SC65]).

# Chapter 4

# Representations and Realizations of so(2,1)

## 4.1 Introduction

The Lie algebra so(2,1) plays a fundamental role in the algebraic reformulation and study of the radial Schrödinger equation for many of the important model problems in quantum mechanics such as the hydrogen atom, harmonic oscillator and one-electron diatomic ions. In this context so(2,1) appears as a spectrum-generating algebra in the sense that energy formulas such as the Bohr formula are all obtained in a unified and purely algebraic manner without resorting to the series solutions of second order radial differential equations.

In this chapter we first consider the so(2,1) representation theory following the same approach used in the previous chapter for so(3). These two algebras have a similar structure and the general representation theories for both are closely related. However we are only interested in the unitary irreducible representations (unirreps) which are obtained by requiring that the generators be hermitian with respect to a suitable scalar product. In this case the unirreps of so(2,1) are infinite dimensional in contrast to the unirreps of so(3) which are finite dimensional.

Next we consider the realizations of the three generators $T_1$, $T_2$ and $T_3$ of so(2,1) in the coordinate representation in terms of position and momentum. It is more difficult to motivate the realizations we shall obtain than it is for the more familiar orbital angular momentum realizations of the so(3) generators. The reason is that the well-known classical expressions for the angular momentum components can easily be extended to their quantum mechanical forms by the correspondence principle but there are no classical expressions for the so(2,1) generators which can be extended in a similar fashion.

Instead the motivation for the realization of the so(2,1) generators arises from the desire to express the radial eigenvalue problems for various quantum mechanical problems such as the hydrogen atom as eigenvalue problems expressed in terms of the so(2,1) generators. This means that we should look for realizations in terms of $r$, $p^2$ and the radial momentum $p_r$.

Next we use the matrix representation of so(2,1), the realization obtained for the so(2,1) generators and a simple scaling transformation to show that the radial Schrödinger equation for the 3-dimensional hydrogen atom and isotropic harmonic oscillator can be expressed as eigenvalue problems for the generator $T_3$. Since the classification of the eigenvalue spectra of this generator are known from the matrix representation theory the formulas for the energy levels are directly obtained. These results are then generalized to $D$ dimensions.

An important aspect of the transformation of the Schrödinger equation into a $T_3$ eigenvalue problem using a scaling transformation is that the transformation is not unitary with respect to the usual scalar product so we also show that a new scalar product is easily introduced such that the so(2,1) generators are hermitian with respect to it. Moreover the fact that we can now describe bound states using only a discrete basis of $T_3$ eigenfunctions (completeness property for bound states) leads to a particularly useful form of algebraic perturbation theory for hydrogenic systems in later chapters, which can be applied to many important perturbations such as charmonium, harmonium, the Zeeman and Stark perturbations, and the one-electron diatomic ion, since there will be no integrations over intermediate continuum states as is the case with perturbation theory based on the original hydrogenic hamiltonian.

The explicit basis functions with respect to our particular realization of the so(2,1) generators are not needed in the calculation of matrix elements of operators expressible in terms of these generators, since they can easily be obtained using the matrix representation. However, we consider them as an example of the use of raising and lowering operators to obtain the solutions of second order differential equations by purely algebraic methods.

We also consider other simple quantum systems such as the Klein-Gordon and the relativistic Dirac-Coulomb equations and show how they can be expressed as $T_3$ eigenvalue problems.

Finally we show how to separate the Schrödinger equation for a 3-dimensional hydrogenic atom in parabolic coordinates and we obtain a parabolic realization of the so(2,1) generators. This realization will be useful in the application of perturbation theory to the Stark effect.

## 4.2  Ladder Operators for so(2,1)

Denoting the three generators of so(2,1) by $T_j$, $j = 1, 2, 3$, the defining commutation relations are

$$[T_1, T_2] = -iT_3, \quad [T_2, T_3] = iT_1, \quad [T_3, T_1] = iT_2, \tag{4.1}$$

which differ from the so(3) commutation relations (3.1) only in the minus sign in the first commutation relation. It is not possible to find a real linear transformation of the so(2,1) generators which transforms these commutation relations into those of so(3). Thus so(2,1) and so(3) are not isomorphic real Lie algebras.

To obtain the Casimir operator for so(2,1) we use the following trick: define the complex linear transformation $J_1 = iT_1$, $J_2 = iT_2$ and $J_3 = T_3$ of the so(2,1) generators. It follows that the $J_k$ satisfy the so(3) commutation relations with Casimir operator $J^2 = J_1^2 + J_2^2 + J_3^2$. This suggests that the operator

$$T^2 = T_3^2 - T_1^2 - T_2^2 \tag{4.2}$$

is the analogous Casimir operator for so(2,1). It is easily verified that this is the case:

$$[T^2, T_k] = 0, \quad k = 1, 2, 3. \tag{4.3}$$

We use the notation $T^2$ even though we shall not think of $T^2$ as the square or dot product of some vector with itself.

Introducing the ladder operators

$$T_+ = T_1 + iT_2, \quad T_- = T_1 - iT_2, \tag{4.4}$$

we can try and "play the same game" with so(2,1) that we did with so(3) in the previous chapter. The set $\{T_+, T_-, T_3\}$ of operators satisfy the commutation relations

$$[T_+, T_-] = -2T_3, \tag{4.5}$$
$$[T_3, T_\pm] = \pm T_\pm, \tag{4.6}$$

and the Casimir operator can be expressed as

$$T^2 = -T_+T_- + T_3^2 - T_3 = -T_-T_+ + T_3^2 + T_3. \tag{4.7}$$

In order to obtain representations it is necessary to choose $T^2$ and one of the components $T_k$ as a set of commuting operators. Now however, because of the lack of cyclic symmetry in the commutation relations (4.1), the different choices give rise to different classes of irreps. If $\{T^2, T_1\}$ or $\{T^2, T_2\}$ is chosen we obtain unirreps which would be suitable for a description of continuum or scattering states whereas the choice of $\{T^2, T_3\}$ is suitable for a description of bound states.

Therefore we make the latter choice and introduce an abstract vector space of states such that $T^2$ and $T_3$ are diagonal:

$$T^2\psi_{Qq} = Q\psi_{Qq}, \tag{4.8}$$
$$T_3\psi_{Qq} = q\psi_{Qq}. \tag{4.9}$$

As in Chapter 3 for so(3) it follows from (4.7), (4.8) and (4.9) that $T_+$ and $T_-$ are raising and lowering operators for the $T_3$ eigenvalue $q$:

$$T_3(T_+\psi_{Qq}) = (q+1)T_+\psi_{Qq},$$
$$T_3(T_-\psi_{Qq}) = (q-1)T_-\psi_{Qq}.$$

Therefore as in the so(3) case we can define

$$T_+\psi_{Qq} = A_{Qq}\psi_{Q,q+1}, \tag{4.10}$$
$$T_-\psi_{Qq} = B_{Qq}\psi_{Q,q-1}, \tag{4.11}$$

and the general classification of the $J_3$ eigenvalue spectra given in Section 3.3 also applies to the $T_3$ eigenvalue spectra so the general representation theory is the same for both so(2,1) and so(3). The fundamental differences occur only when we restrict them to unitary representations.

## 4.3   Unirreps of so(2,1)

If we introduce a scalar product such that the $T_k$ are hermitian then

$$T_+^\dagger = T_-, \quad T_-^\dagger = T_+, \tag{4.12}$$

and we can go through the same steps as in Section 3.4 using $T_+T_-$ and $T_-T_+$ obtained from (4.7). It follows that

$$(\psi_{Qq}, T_-T_+\psi_{Qq}) = (T_+\psi_{Qq}, T_+\psi_{Qq}) = \|T_+\psi_{Qq}\|^2 \geq 0,$$
$$(\psi_{Qq}, T_+T_-\psi_{Qq}) = (T_-\psi_{Qq}, T_-\psi_{Qq}) = \|T_-\psi_{Qq}\|^2 \geq 0,$$

and

$$\|T_+\psi_{Qq}\|^2 = |A_{Qq}|^2 \geq 0, \quad \|T_-\psi_{Qq}\|^2 = |B_{Qq}|^2 \geq 0,$$

and

$$\|T_+\psi_{Qq}\|^2 = (\psi_{Qq}, (-T^2 + T_3^2 + T_3)\psi_{Qq}) = -Q + q(q+1),$$
$$\|T_-\psi_{Qq}\|^2 = (\psi_{Qq}, (-T^2 + T_3^2 - T_3)\psi_{Qq}) = -Q + q(q-1).$$

Instead of inequalities (3.15) and (3.16) we obtain

$$|A_{Qq}|^2 = -Q + q(q+1) \geq 0, \tag{4.13}$$
$$|B_{Qq}|^2 = -Q + q(q-1) \geq 0. \tag{4.14}$$

Adding these results gives $q^2 \geq Q$ which shows that all unirreps of so(2,1) are infinite dimensional.

To analyze the different cases it is convenient to define

$$Q = k(k+1), \tag{4.15}$$

so that the above inequalities can be expressed in the factored forms

$$-(k-q)(k+q+1) \geq 0, \tag{4.16}$$
$$-(k+q)(k-q+1) \geq 0. \tag{4.17}$$

Also $Q$ and $q$ must be real since they are eigenvalues of the hermitian operators $T^2$ and $T_3$.

The standard phase choice is again obtained by choosing $A_{Qq}$ and $B_{Qq}$ to be real in (4.13) and (4.14) and taking the positive square root. Therefore we have the following matrix representation

$$T^2|qk\rangle = k(k+1)|qk\rangle, \tag{4.18}$$

$$T_3|qk\rangle = q|qk\rangle, \tag{4.19}$$

$$T_+|qk\rangle = \sqrt{-(k-q)(k+q+1)}\,|q{+}1,k\rangle, \tag{4.20}$$

$$T_-|qk\rangle = \sqrt{-(k+q)(k-q+1)}\,|q{-}1,k\rangle. \tag{4.21}$$

In our applications of so(2,1) we only need the class of unirreps whose $T_3$ eigenvalue spectra are bounded below and given by

$$\mathcal{D}^+(k) = \{q = -k + \mu,\ \mu = 0,1,2,...\},\ k < 0. \tag{4.22}$$

Thus each negative value of $k$ defines a unirrep with lowest eigenvalue $q_0 = -k$. It is clear from (4.20) and (4.21) that $T_-|q_0 k\rangle = 0$ but $T_+|qk\rangle \neq 0$ for $q \in \mathcal{D}^+(k)$. We must have $k < 0$ to ensure that (4.16) and (4.17) are satisfied.

It also follows that $k_1 = -k - 1$ defines an equivalent unirrep since substituting $k_1 = -k - 1$ for $k$ does not change (4.15), (4.16) and (4.17). Therefore we can use the notation

$$\mathcal{D}^+(-k - 1) = \{q = k + 1 + \mu,\ \mu = 0,1,2,...\},\ k > -1 \tag{4.23}$$

to express the same class of $T_3$ eigenvalue spectra. There is a similar clase of unirreps whose $T_3$ eigenvalue spectra are bounded above and two classes of unirreps called the principal and supplementary series having unbounded $T_3$ eigenvalue spectra (see [AD88b] and [BA65] for more details).

A simple example of a realization of the so(2,1) generators in terms of boson annihilation and creation operators, which gives two classes of $T_3$ eigenvalue spectra of type (4.23), is given in Exercise 4.1.

## 4.4   Realizations of so(2,1)

It is not as straightforward to obtain realizations of so(2,1) as it is for so(3). Given the correspondence principle it is clear how to extend the classical orbital angular momentum to its quantum mechanical counterpart in order to obtain a realization in the coordinate representation for the generators of so(3). However there is no similar classical analogue in the coordinate representation for the generators of so(2,1). Since our goal is to use so(2,1) as a spectrum generating algebra for simple quantum systems such as the hydrogen atom and the harmonic oscillator, we look for realizations which can be used to express the hamiltonians for these systems in terms of the generators $T_k$. Then we can express the radial Schrödinger equations for these systems as eigenvalue problems for the $T_3$ operator. The known results (4.18) to (4.20) can then be used to obtain energy eigenvalues. Moreover we also want to do algebraic perturbation theory based on these simple systems so it is necessary to be able to express the perturbation in terms of the so(2,1) generators. This will be the case for central perturbations (functions of $r = |\boldsymbol{r}|$ only).

Therefore we look for realizations that can be expressed in terms of $r$ and $p^2$ or equivalently in terms of $r$ and $p_r^2$ where $p_r$ is the radial momentum operator [CI77a]. Later we shall perform a scaling transformation from the physical set of operators $\{r, p_r\}$ to another model set $\{R, P_R\}$ in terms of which the realizations of the generators are expressed so in this section we develop the realizations in terms of $R$ and $P_R$.

In classical mechanics the radial momentum $P_R$ is given by the directional derivative $P_R = \hat{\boldsymbol{R}} \cdot \boldsymbol{P}$, where $\hat{\boldsymbol{R}} = \boldsymbol{R}/R$ is the unit vector in the radial direction, but the corresponding quantum mechanical operator is not hermitian. Instead it is necessary to use the symmetrized hermitian definition

$$P_R = \frac{1}{2}(\hat{\boldsymbol{R}} \cdot \boldsymbol{P} + \boldsymbol{P} \cdot \hat{\boldsymbol{R}}) \tag{4.24}$$

$$= -\frac{i}{R}\frac{\partial}{\partial R}R = -i\left(\frac{\partial}{\partial R} + \frac{1}{R}\right) = \frac{1}{R}(\boldsymbol{R} \cdot \boldsymbol{P} - i). \tag{4.25}$$

The connection between the momentum $\boldsymbol{P}$ and the radial momentum $P_R$ in 3-dimensional Euclidean space $\mathbb{R}^3$ is given by

$$P^2 = P_R^2 + \frac{L^2}{R^2}, \tag{4.26}$$

and the basic commutation relation is

$$[R, P_R] = i. \tag{4.27}$$

The identities (4.25) and (4.26) are valid only in $\mathbb{R}^3$ but we shall see that the symmetrized definition (4.24) and the commutation relation (4.27) are also valid in $D$-dimensional Euclidean space $\mathbb{R}^D$. In our derivation of realizations of the so(2,1) generators only (4.27) will be used so the results will also apply to $\mathbb{R}^D$ with $D \neq 3$. Equations (4.25) and (4.26) and their generalizations will only be needed when we consider specific hamiltonians and their expressions in terms of the so(2,1) generators.

We now consider operators of the form $R^m P_R^n$ in order to obtain three operators that close under commutation. Since hamiltonians are second order differential operators in the coordinate representation we can assume that $n = 0, 1, 2$ in the term $R^m P_R^n$. Therefore we choose the set $\{R^a, R^b P_R, R^c P_R^2\}$ and try to find $a, b$ and $c$ such that this set of operators is closed under commutation.

Evaluating the commutators in pairs using the basic commutator identities (2.1) and (2.2), we obtain

$$[R^a, R^b P_R] = R^b[R^a, P_R], \tag{4.28}$$

$$[R^b P_R, R^c P_R^2] = R^b [P_R, R^c] P_R^2 + R^c [R^b, P_R^2] P_R, \tag{4.29}$$

$$[R^a, R^c P_R^2] = R^c [R^a, P_R^2]. \tag{4.30}$$

The commutators of the form $[R^n, P_R]$ and $[R^n, P_R^2]$ can be evaluated using identity (2.5) and the basic commutation relation (4.27):

$$[R^n, P_R] = \sum_{k=0}^{n-1} R^k [R, P_R] R^{n-k-1}$$

$$= i \sum_{k=0}^{n-1} R^{n-1} = in R^{n-1}, \tag{4.31}$$

$$[R^n, P_R^2] = P_R [R^n, P_R] + [R^n, P_R] P_R$$
$$= P_R(in R^{n-1}) + (in R^{n-1}) P_R$$
$$= in(R^{n-1} P_R - i(n-1)R^{n-2}) + in R^{n-1} P_R$$
$$= 2in R^{n-1} P_R + n(n-1) R^{n-2}. \tag{4.32}$$

Substitute these two results into (4.28), (4.29) and (4.30) to obtain

$$[R^a, R^b P_R] = ia R^{a+b-1}, \tag{4.33}$$

$$[R^b P_R, R^c P_R^2] = i(2b - c)R^{b+c-1} P_R^2 + b(b-1)R^{b+c-2} P_R, \tag{4.34}$$

$$[R^a, R^c P_R^2] = 2ia R^{a+c-1} P_R + a(a-1)R^{a+c-2}. \tag{4.35}$$

We must have $R^a$ on the right side of (4.33) so $b = 1$ and we must have $a + c - 1 = b$ so $c = 2 - a$. This gives the Lie algebra with generators $\{R^a, RP_R, R^{2-a} P_R^2\}$ and commutation relations

$$[R^a, RP_R] = ia R^a, \tag{4.36}$$

$$[RP_R, R^{2-a} P_R^2] = ia R^{2-a} P_R^2, \tag{4.37}$$

$$[R^a, R^{2-a} P_R^2] = 2ia \left[ RP_R - \frac{i}{2}(a - 1) \right]. \tag{4.38}$$

The constant term in (4.38) can be subtracted from the operator $RP_R$ in (4.36) and (4.37) without changing these commutation relations. Similarly the parameter $a$ can be removed from the right sides of the commutation relations if we divide the first by $a$, the second by $a^3$ and the third by $a^2$. Thus if we define

$$V_1 = R^a, \tag{4.39}$$

$$V_2 = \frac{1}{a} \left[ RP_R - \frac{i}{2}(a - 1) \right], \tag{4.40}$$

$$V_3 = \frac{1}{a^2} R^{2-a} P_R^2, \tag{4.41}$$

then the $V_k$ satisfy the simple commutation relations

$$[V_1, V_2] = iV_1, \tag{4.42}$$

$$[V_2, V_3] = iV_3, \tag{4.43}$$

$$[V_1, V_3] = 2iV_2. \tag{4.44}$$

We can generalize these results somewhat by noting that (4.31) also holds for negative powers of $R$ (see Exercise 4.3):

$$[R^{-n}, P_R] = -in R^{-n-1}. \tag{4.45}$$

Then (4.36) becomes

$$[R^{-a}, RP_R] = -ia R^{-a}, \tag{4.46}$$

and from the definitions (4.39) and (4.40)

$$[V_2, V_1^{-1}] = iV_1^{-1}. \tag{4.47}$$

We can now combine (4.43) and (4.47) to obtain

$$[V_2, V_3] + \tau[V_2, V_1^{-1}] = iV_3 + i\tau V_1^{-1},$$

or

$$[V_2, V_3 + \tau V_1^{-1}] = i(V_3 + \tau V_1^{-1}),$$

where $\tau$ is any operator or constant which commutes with the $V_k$. Thus we obtain the commutation relations

$$[V_2, V_1] = -iV_1, \tag{4.48}$$
$$[V_2, V_3 + \tau V_1^{-1}] = i(V_3 + \tau V_1^{-1}), \tag{4.49}$$
$$[V_1, V_3 + \tau V_1^{-1}] = 2iV_2, \tag{4.50}$$

where (4.50) is obtained from (4.44) since $V_1$ and $V_1^{-1}$ commute with each other.  Now add and subtract (4.48) and (4.49) to obtain the commutation relations

$$[V_2, V_3 + V_1 + \tau V_1^{-1}] = i(V_3 - V_1 + \tau V_1^{-1}), \tag{4.51}$$
$$[V_2, V_3 - V_1 + \tau V_1^{-1}] = i(V_3 + V_1 + \tau V_1^{-1}), \tag{4.52}$$
$$[V_1, V_3 + \tau V_1^{-1}] = 2iV_2. \tag{4.53}$$

Finally if we define

$$T_1 = \frac{1}{2}(V_3 - V_1 + \tau V_1^{-1}),$$
$$T_2 = V_2,$$
$$T_3 = \frac{1}{2}(V_3 + V_1 + \tau V_1^{-1}),$$

then these operators satisfy the so(2,1) commutation relations (4.1) and we obtain the realizations

$$T_1 = \frac{1}{2}\left[\frac{1}{a^2}R^{2-a}P_R^2 + \frac{\tau}{R^a} - R^a\right], \tag{4.54}$$
$$T_2 = \frac{1}{a}\left[RP_R - \frac{i}{2}(a-1)\right], \tag{4.55}$$
$$T_3 = \frac{1}{2}\left[\frac{1}{a^2}R^{2-a}P_R^2 + \frac{\tau}{R^a} + R^a\right]. \tag{4.56}$$

These realizations are sufficiently general for our purposes. In fact we shall use $a = 1$ for the hydrogen atom and $a = 2$ for the harmonic oscillator. As mentioned above they are also valid in the $D$-dimensional Euclidean space $\mathbb{R}^D$ even when $D \neq 3$ since only the basic commutation relation (4.27) was used.

In order to make the connection between this realization of the so(2,1) generators and the matrix representation (4.18) to (4.21), which is independent of the particular realization, it is necessary to relate the eigenvalues, $Q = k(k+1)$ of $T^2$ and $q$ of $T_3$, to the parameters $a$ and $\tau$ in (4.54) to (4.56). To do this use the realization to evaluate $T^2$ as follows:

$$
\begin{aligned}
T^2 &= T_3^2 - T_1^2 - T_2^2 \\
&= (T_3 - T_1)(T_3 + T_1) - [T_3, T_1] - T_2^2 \\
&= R^a \left( \frac{1}{a^2} R^{2-a} P_R^2 + \tau R^{-a} \right) - iT_2 - T_2^2 \\
&= \frac{1}{a^2} R^2 P_R^2 + \tau - \frac{i}{a} RP_R - \frac{a-1}{2a} - T_2^2 .
\end{aligned}
$$

$T_2^2$ can be simplified using the basic commutation relation (4.27):

$$
\begin{aligned}
T_2^2 &= \frac{1}{a^2} \left[ RP_R - \frac{i}{2}(a-1) \right] \left[ RP_R - \frac{i}{2}(a-1) \right] \\
&= \frac{1}{a^2} RP_R RP_R - i \left( \frac{a-1}{a^2} \right) RP_R - \left( \frac{a-1}{2a} \right)^2 \\
&= \frac{1}{a^2} R(RP_R - i)P_R - i \left( \frac{a-1}{a^2} \right) RP_R - \left( \frac{a-1}{2a} \right)^2 \\
&= \frac{1}{a^2} R^2 P_R^2 - \frac{i}{a} RP_R - \left( \frac{a-1}{2a} \right)^2 .
\end{aligned}
$$

Therefore

$$
T^2 = \tau + \frac{1 - a^2}{4a^2} . \tag{4.57}
$$

Since $\tau$ is a constant or an operator which commutes with the so(2,1) generators it is also diagonal in the abstract basis $|qk\rangle$ so denoting its eigenvalue also by $\tau$ we obtain from (4.18)

$$
\tau + \frac{1 - a^2}{4a^2} = k(k+1) . \tag{4.58}
$$

Solving for $k$ gives

$$
k = \frac{1}{2} \left[ -1 \pm \sqrt{4\tau + \frac{1}{a^2}} \right] . \tag{4.59}
$$

The $T_3$ eigenvalues (4.23) have the form

$$
q = q_0 + \mu, \quad \mu = 0, 1, 2, \ldots \tag{4.60}
$$

with lowest eigenvalue

$$
q_0 = k + 1, \quad k > -1 . \tag{4.61}
$$

Before considering the general case of so(2,1) as a spectrum generating algebra for simple quantum systems and the generalization to $D$-dimensional cases we show how so(2,1) can be used to obtain the energy levels for the two most important cases: the 3-dimensional hydrogen atom and the 3-dimensional isotropic harmonic oscillator.

These two problems are equivalent in the sense that the radial differential equation for the $D$-dimensional harmonic oscillator can be transformed into that of the $D$-dimensional hydrogen atom by a change of both the dependent and independent variable (see Exercise 4.8). For our purposes it is easier and more instructive to consider them as separate problems with $a = 1$ for the hydrogen atom and $a = 2$ for the harmonic oscillator.

## 4.5   3-dimensional Hydrogenic Atom

The hamiltonian for the 3-dimensional hydrogen atom in atomic units ($m = e = \hbar = 1$) is

$$H = \frac{1}{2}p^2 - \frac{\mathcal{Z}}{r}, \tag{4.62}$$

where $\mathcal{Z}$ is the nuclear charge. From (4.26)

$$H = \frac{1}{2}p_r^2 + \frac{L^2}{2r^2} - \frac{\mathcal{Z}}{r}, \tag{4.63}$$

and the Schrödinger equation for the energy $E$ is

$$(H - E)\psi(\boldsymbol{r}) = 0. \tag{4.64}$$

The hamiltonian is not directly expressible in terms of the so(2,1) generators (4.54) to (4.56) for any value of the parameter $a$. However if we multiply (4.64) on the left by $r$ and introduce the scaling transformation from the physical operators $\{r, p_r\}$ to the model space operators $\{R, P_R\}$ given by

$$r = \gamma R, \quad p_r = \frac{1}{\gamma}P_R, \tag{4.65}$$

then (4.64) can be expressed as

$$\left(\frac{1}{2}rp_r^2 + \frac{L^2}{2r} - \mathcal{Z} - Er\right)\psi(\boldsymbol{r}) = 0,$$

$$\left(\frac{1}{2}\gamma rp_r^2 + \frac{\gamma L^2}{2r} - \gamma\mathcal{Z} - \gamma Er\right)\psi(\boldsymbol{r}) = 0,$$

$$\left[\frac{1}{2}\left(RP_R^2 + \frac{L^2}{R} - 2\gamma^2 ER\right) - \gamma\mathcal{Z}\right]\psi(\gamma\boldsymbol{R}) = 0.$$

If we use model space operators and choose $a = 1$ in the realization (4.54) to (4.56) to obtain

$$T_1 = \frac{1}{2}\left[RP_R^2 + \frac{\tau}{R} - R\right], \tag{4.66}$$

$$T_2 = RP_R, \tag{4.67}$$

$$T_3 = \frac{1}{2}\left[RP_R^2 + \frac{\tau}{R} + R\right], \tag{4.68}$$

and if we choose $\gamma$ and $\tau$ such that

$$2\gamma^2 E = -1, \tag{4.69}$$

$$\tau = L^2, \tag{4.70}$$

then a $T_3$ eigenvalue equation is obtained:

$$(T_3 - \gamma\mathcal{Z})\psi(\gamma\mathbf{R}) = 0. \tag{4.71}$$

Since the hydrogen atom wave functions $\psi(\mathbf{r})$ are separable in spherical coordinates

$$\psi(r, \vartheta, \varphi) = \phi(r)Y_{\ell m}(\vartheta, \varphi) \tag{4.72}$$

as products of a radial function and a spherical harmonic it follows that (4.64) or (4.71) can be reduced to eigenvalue equations for the radial functions if we replace $L^2$ by its eigenvalue $\ell(\ell+1)$. Therefore

$$(H - E)\phi(r) = 0, \tag{4.73}$$

$$(T_3 - \gamma\mathcal{Z})\Phi(R) = 0, \tag{4.74}$$

where we have defined

$$\Phi(R) = \phi(\gamma R). \tag{4.75}$$

It follows immediately from (4.59) that

$$k = \tfrac{1}{2}[-1 \pm (2\ell + 1)].$$

Since $\ell \geq 0$ and $k > -1$ we must choose the upper sign to obtain

$$k = \ell, \tag{4.76}$$

$$q = \ell + 1 + \mu, \quad \mu = 0, 1, 2, \ldots, \tag{4.77}$$

where $q$ is an eigenvalue of $T_3$, and from (4.69) and (4.71) the scaling factor and energy are

$$\gamma = \frac{q}{\mathcal{Z}}, \tag{4.78}$$

$$E = E_q = -\frac{\mathcal{Z}^2}{2q^2}. \tag{4.79}$$

This is the famous Bohr formula for the energy levels of the hydrogen atom if we identify $q$ with the principal quantum number $n$.

There are several important ideas related to the above derivations. The choice $2\gamma^2 E = 1$ instead of (4.69) would lead to positive energies corresponding to scattering states and a $T_1$ eigenvalue equation. We are only interested in bound states so (4.74) justifies our earlier choice of $\{T^2, T_3\}$ as the set of commuting operators to diagonalize. It is also important to realize that (4.66) to (4.68), with the specific choice (4.70) of $\tau$, do form a realization of so(2,1). This follows since the components $L_j$ of the orbital angular momentum (and hence $L^2$) commute with the so(2,1) generators:

$$[L_j, T_k] = 0, \quad j, k = 1, 2, 3. \tag{4.80}$$

This also shows the the $T_3$ eigenfunctions can be labeled as

$$\Psi_{n\ell m}(R, \vartheta, \varphi) = \Phi_{n\ell}(R) Y_{\ell m}(\vartheta, \varphi) \tag{4.81}$$

by identifying the abstract basis function $|qk\rangle$ with $\Phi_{n\ell}(R)$ and that they form a basis for the direct sum algebra so(2,1) $\oplus$ so(3), corresponding to the direct product SO(2,1) $\otimes$ SO(3) of the associated Lie groups, in the sense that the set of 6 operators $L_j, T_j$, $j = 1, 2, 3$ generate the Lie algebra so(2,1) $\oplus$ so(3). This idea of merging so(2,1) with other Lie algebras to form a larger Lie algebra will be pursued in Chapter 5 and Chapter 6.

Finally the most important aspect of the derivation of (4.74) from (4.73) is that the scaling transformation and pre-multiplication of the Schrödinger equation by $r$ is a non-unitary transformation. This is clear since the discrete state spectrum of $H$ is not complete for bound state wavefunctions (the continuum states also contribute) whereas the spectrum of $T_3$ is purely discrete and complete. Thus the eigenfunctions of (4.71) and (4.74) are not the usual hydrogenic ones so we call them the scaled hydrogenic eigenfunctions.

Moreover the so(2,1) generators given by the realization (4.66) to (4.68) are not hermitian with respect to the usual scalar product and the $T_3$ eigenfunctions do not form an orthonormal set with respect to it. This is not a disadvantage since we shall show that a new scalar product can be chosen with respect to which the $T_k$ are hermitian. The fact that we can now describe bound states using only a discrete basis has an enormous advantage when we consider perturbation theory using $T_3$ as the unperturbed "hamiltonian" since there will be no integrations to perform over intermediate continuum states as is the case with the conventional perturbation theory based on the original hydrogenic hamiltonian.

The scaling transformation used here to convert the original eigenvalue problem (4.64) into the $T_3$ eigenvalue problem (4.74) can also be performed from an active viewpoint using one coordinate system and the unitary transformation of operators and states given by the operator $e^{i\alpha T_2}$. This so-called "tilting transformation" is commonly used in the literature ([AD88b], [BA71a], [WY74]) but is more complicated than the simple scaling transformation (4.65). The details are given in Appendix C.

## 4.6  3-dimensional Harmonic Oscillator

The hamiltonian for the 3-dimensional isotropic harmonic oscillator in atomic units is

$$H = \frac{1}{2}p^2 + \frac{1}{2}\omega^2 r^2$$
$$= \frac{1}{2}p_r^2 + \frac{L^2}{2r^2} + \frac{1}{2}\omega^2 r^2, \tag{4.82}$$

using (4.26), and the Schrödinger equation for the energy $E$ is

$$(H - E)\psi(\mathbf{r}) = 0. \tag{4.83}$$

In this case there is no need to pre-multiply (4.83) by $r$ so we multiply by $\gamma^2/4$ and perform the scaling transformation (4.65) to obtain

$$\left(\frac{1}{8}\gamma^2 p_r^2 + \frac{\gamma^2 L^2}{8r^2} + \frac{1}{8}\gamma^2\omega^2 r^2 - \frac{1}{4}\gamma^2 E\right)\psi(\mathbf{r}) = 0,$$
$$\left(\frac{1}{8}P_R^2 + \frac{L^2}{8R^2} + \frac{1}{8}\gamma^4\omega^2 R^2 - \frac{1}{4}\gamma^2 E\right)\psi(\gamma\mathbf{R}) = 0,$$
$$\left[\frac{1}{2}\left(\frac{1}{4}P_R^2 + \frac{L^2}{4R^2} + \frac{1}{4}\gamma^4\omega^2 R^2\right) - \frac{1}{4}\gamma^2 E\right]\psi(\gamma\mathbf{R}) = 0.$$

If we use model space operators and choose $a = 2$ in the realization (4.54) to (4.56) to obtain

$$T_1 = \frac{1}{2}\left[\frac{1}{4}P_R^2 + \frac{\tau}{R^2} - R^2\right], \tag{4.84}$$

$$T_2 = \frac{1}{2}\left[RP_R - \frac{i}{2}\right], \tag{4.85}$$

$$T_3 = \frac{1}{2}\left[\frac{1}{4}P_R^2 + \frac{\tau}{R^2} + R^2\right], \tag{4.86}$$

and if we choose $\gamma$ and $\tau$ such that

$$\tfrac{1}{4}\gamma^4\omega^2 = 1, \tag{4.87}$$
$$\tau = \tfrac{1}{4}L^2, \tag{4.88}$$

then we obtain the $T_3$ eigenvalue equation

$$\left(T_3 - \tfrac{1}{4}\gamma^2 E\right)\Phi(R) = 0, \tag{4.89}$$

analogous to (4.74) for the 3-dimensional hydrogen atom where $\Phi(R)$ is again defined by (4.75).

From (4.59) and (4.88), again replacing $L^2$ by its eigenvalue $\ell(\ell + 1)$, it follows that

$$k = \tfrac{1}{2}[-1 \pm \tfrac{1}{2}(2\ell + 1)]$$

and the upper sign must be chosen since $\ell \geq 0$ and we require that $k > -1$. Therefore

$$k = \frac{1}{2}\ell - \frac{1}{4}, \tag{4.90}$$

$$q = k + 1 + \mu = \frac{1}{2}\ell + \frac{3}{4} + \mu, \quad \mu = 0, 1, 2, \ldots, \tag{4.91}$$

where $q$ is the eigenvalue of $T_3$. From (4.87) and (4.89) the scaling parameter and energy are

$$\gamma^2 = \frac{2}{\omega}, \tag{4.92}$$

$$E = E_q = \frac{4q}{\gamma^2} = 2\omega q = \left(\ell + 2\mu + \frac{3}{2}\right)\omega. \tag{4.93}$$

Defining the principal quantum number

$$n = \ell + 2\mu, \tag{4.94}$$

we obtain the familiar formula

$$E_n = (n + \tfrac{3}{2})\omega \tag{4.95}$$

for the energy levels of the 3-dimensional isotropic harmonic oscillator.

## 4.7 Generalization to D Dimensions

In $D$-dimensional Euclidean space $\mathbf{R}^D$ we can define

$$\boldsymbol{r} = (x_1, x_2, \ldots, x_D), \quad \boldsymbol{p} = (p_1, p_2, \ldots, p_D) \tag{4.96}$$

as the coordinate and conjugate momentum vectors and

$$r = \sqrt{x_1^2 + x_2^2 + \cdots + x_D^2} \tag{4.97}$$

as the radial distance.

There are now $\frac{1}{2}D(D-1)$ linearly independent orbital angular momentum operators given by [JO67]

$$L_{jk} = x_j p_k - x_k p_j, \quad j < k = 1, 2, \ldots, D. \tag{4.98}$$

These operators are antisymmetric,

$$L_{kj} = -L_{jk}, \tag{4.99}$$

and satisfy the commutation relations

$$[L_{jk}, L_{\ell m}] = i(\delta_{j\ell} L_{km} + \delta_{km} L_{j\ell} - \delta_{jm} L_{k\ell} - \delta_{k\ell} L_{jm}) \tag{4.100}$$

of the Lie algebra so($D$) which are derived in Exercise 4.4.

The angular momentum can only be expressed as $\boldsymbol{r} \times \boldsymbol{p}$ in $\mathbb{R}^3$ and (4.98), (4.99) and (4.100) provide the appropriate generalization of the vector cross product to higher dimensional spaces. In case $D = 3$ we have $L_1 = L_{23}$, $L_2 = L_{31}$ and $L_3 = L_{21}$. The Casimir operator is given by

$$L^2 = \sum_{j<k} L_{jk} L_{jk} = \frac{1}{2} \sum_{j,k} L_{jk} L_{jk}, \tag{4.101}$$

and it is shown in Exercise 4.5 using (4.100) that

$$[L^2, L_{jk}] = 0, \quad j < k = 1, 2, \ldots, D. \tag{4.102}$$

From (4.98) and Exercise 4.6 it follows that

$$L^2 = r^2 p^2 - (\boldsymbol{r} \cdot \boldsymbol{p})^2 + i(D-2)\boldsymbol{r} \cdot \boldsymbol{p}. \tag{4.103}$$

Now using

$$\boldsymbol{r} \cdot \boldsymbol{p} = -i \sum_j x_j \frac{\partial}{\partial x_j} = -ir \frac{\partial}{\partial r} \tag{4.104}$$

we obtain

$$L^2 = r^2 p^2 + r^2 \left( \frac{\partial^2}{\partial r^2} + \frac{D-1}{r} \frac{\partial}{\partial r} \right). \tag{4.105}$$

The symmetrized form of the radial momentum (4.25) remains valid in $D$ dimensions and using the generalization $\boldsymbol{r} \cdot \boldsymbol{p} - \boldsymbol{p} \cdot \boldsymbol{r} = Di$ of (2.10) it follows that

$$p_r = -\frac{i}{r^M} \frac{\partial}{\partial r} r^M = -i \left( \frac{\partial}{\partial r} + \frac{M}{r} \right) = \frac{1}{r}(\boldsymbol{r} \cdot \boldsymbol{p} - iM), \tag{4.106}$$

where

$$M = \frac{1}{2}(D-1). \tag{4.107}$$

Substituting into (4.105) and solving for $p^2$ gives the generalization of (4.26)

$$p^2 = p_r^2 + \frac{(D-1)(D-3)}{4r^2} + \frac{L^2}{r^2}. \tag{4.108}$$

The most important aspect of the generalization to $D$ dimensions is that the basic commutation relation (4.27) is still valid for the definition (4.106) (see Exercise 4.2). Therefore the realization of the so(2,1) generators given by (4.54) to (4.56) remains valid since only this commutation relation was used to derive it. Similarly the expression (4.57) for $T^2$ is still valid. However the eigenvalues of $L^2$ are no longer given by $\ell(\ell+1)$. Instead if $Y$ is an eigenfunction of $L^2$ then it is shown in Exercise 4.7 that

$$L^2 Y = \ell(\ell + D - 2)Y, \tag{4.109}$$

so the eigenvalues of $L^2$ are $\ell(\ell + D - 2)$. We can now obtain the energy levels for the $D$-dimensional hydrogen atom and isotropic harmonic oscillator.

## 4.8  D-dimensional Hydrogenic Atom

From (4.108) the $D$-dimensional hamiltonian is

$$H = \frac{1}{2}p_r^2 + \frac{(D-1)(D-3)}{8r^2} + \frac{L^2}{2r^2} - \frac{Z}{r}. \qquad (4.110)$$

The scaling transformation (4.65) now gives the scaled Schrödinger equation

$$\left[ \frac{1}{2}\left( RP_R^2 + \frac{(D-1)(D-3)}{4R} + \frac{L^2}{R} - 2\gamma^2 ER \right) - \gamma Z \right] \psi(\gamma \boldsymbol{R}) = 0. \qquad (4.111)$$

Therefore in the realization (4.66) to (4.68) we can choose

$$\tau = \frac{1}{4}(D-1)(D-3) + L^2, \qquad (4.112)$$

which is the generalization of (4.70)

Finally replacing $L^2$ by its eigenvalue $\ell(\ell + D - 2)$ and substituting into (4.59) gives

$$\begin{aligned}
k &= \frac{1}{2}\left[ -1 \pm \sqrt{(D-1)(D-3) + 4\ell(\ell+D-2)+1} \right] \\
&= \frac{1}{2}\left[ -1 \pm \sqrt{4\ell(\ell+D-2) + (D-2)^2} \right] \\
&= \frac{1}{2}\left[ -1 \pm (2\ell + D - 2) \right].
\end{aligned}$$

Again since $\ell \geq 0$ and $k > -1$ we must choose the upper sign to obtain the generalizations of (4.76), (4.77) and (4.79) ($\gamma$ is still defined by (4.78)):

$$k = \ell + \frac{D-3}{2}, \qquad (4.113)$$

$$q = k + 1 + \mu = \ell + \frac{D-1}{2} + \mu, \quad \mu = 0, 1, 2, \ldots, \qquad (4.114)$$

$$E = E_q = -\frac{Z^2}{2\left(\ell + \frac{D-1}{2} + \mu\right)^2}. \qquad (4.115)$$

We can identify $q$ as the generalization of the 3-dimensional principal quantum number $n$.

## 4.9  D-dimensional Harmonic Oscillator

From (4.108) the $D$-dimensional isotropic harmonic oscillator hamiltonian is

$$H = \frac{1}{2}p_r^2 + \frac{(D-1)(D-3)}{8r^2} + \frac{L^2}{2r^2} + \frac{1}{2}\omega^2 r^2. \qquad (4.116)$$

Again the scaling transformation (4.65) gives the scaled Schrödinger equation

$$\left[\frac{1}{2}\left(\frac{1}{4}P_R^2 + \frac{(D-1)(D-3)}{16R^2} + \frac{L^2}{4R^2} + \frac{\gamma^2\omega^2}{4}R^2\right) - \frac{\gamma^2 E}{4}\right]\psi(\gamma\boldsymbol{R}) = 0 \quad (4.117)$$

and now we can choose

$$\tau = \frac{1}{16}(D-1)(D-3) + \frac{1}{4}L^2 \tag{4.118}$$

in the realization (4.84) to (4.86) to obtain the generalization of (4.88).

Finally replacing $L^2$ by its eigenvalue $\ell(\ell + D - 2)$ and substituting into (4.59) gives

$$\begin{aligned}
k &= \frac{1}{2}\left[-1 \pm \sqrt{\tfrac{1}{4}(D-1)(D-3) + \ell(\ell + D - 2) + \tfrac{1}{4}}\right] \\
&= \frac{1}{2}\left[-1 \pm \frac{1}{2}\sqrt{4\ell(\ell + D - 2) + (D-2)^2}\right] \\
&= \frac{1}{2}\left[-1 \pm \frac{1}{2}(2\ell + D - 2)\right].
\end{aligned}$$

The upper sign must be chosen and we obtain the generalizations of (4.90) to (4.93) ($\gamma$ is still defined by (4.91)):

$$k = \frac{1}{2}\ell - \frac{1}{4} + \frac{D-3}{4}, \tag{4.119}$$

$$q = k + 1 + \mu = \frac{1}{2}\ell + \frac{D}{4} + \mu, \quad \mu = 0,1,2,\ldots, \tag{4.120}$$

$$E = E_q = \left(\ell + 2\mu + \frac{D}{2}\right)\omega. \tag{4.121}$$

As in the 3-dimensional case we can identify $n = \ell + 2\mu$ as the principal quantum number to obtain $E_n = (n + \frac{D}{2})\omega$.

The one-dimensional harmonic oscillator can be treated as a special case. There is no orbital angular momentum term in the hamiltonian (4.116) and since $D = 1$ the first $r^{-2}$ term also vanishes. Using $x$ and $p$ as the coordinate and conjugate momentum satisfying $[x,p] = 1$, (4.116) reduces to

$$H = \frac{1}{2}p^2 + \frac{1}{2}\omega^2 x^2, \tag{4.122}$$

which is the hamiltonian for the 1-dimensional harmonic oscillator. The scaling transformation can be performed using $x = \gamma X$ and $p = \gamma^{-1}P$ and we obtain

$$\left[\frac{1}{2}\left(\frac{1}{4}P^2 + \frac{1}{4}\gamma^4\omega^2 X^2\right) - \frac{1}{4}\gamma^2 E\right]\psi(\gamma X) = 0. \tag{4.123}$$

Therefore we can choose $\tau = 0$ in the realization (4.84) to (4.86) and from (4.59) there are now two values, $k = -1/4$ and $k = -3/4$. The first gives the $T_3$ eigenvalues and energies

$$\begin{aligned}
q &= k + 1 + \mu = \tfrac{3}{4} + \mu, \\
E &= 2\omega q = (\tfrac{1}{2} + 2\mu + 1)\omega, \quad \mu = 0,1,2,\ldots,
\end{aligned}$$

and the second gives

$$q = k + 1 + \mu = \tfrac{1}{4} + \mu,$$
$$E = 2\omega q = (\tfrac{1}{2} + 2\mu)\omega, \quad \mu = 0, 1, 2, \ldots,$$

and these results can be combined to obtain the familiar result

$$E = (n + \tfrac{1}{2})\omega, \quad n = 0, 1, 2, \ldots \tag{4.124}$$

for the energy levels of the 1-dimensional harmonic oscillator. It is interesting that the use of so(2,1) as a spectrum generating algebra for the 1-dimensional harmonic oscillator produces the energy levels in two sequences whereas the usual algebraic approach using annihilation and creation operators gives the combined sequence (4.124) directly [ME70a].

## 4.10   Hermiticity of so(2,1) Generators

In order to use the matrix representation (4.18) to (4.21) for the so(2,1) generators given by the particular realization (4.54) to (4.56) it is necessary that the operators $T_k$ be hermitian with respect to some scalar product. In $D$-dimensional Euclidean space $\mathbb{R}^D$ the conventional scalar product of two wavefunctions $\psi_1(\mathbf{r})$ and $\psi_2(\mathbf{r})$ is given by

$$(\psi_1, \psi_2) = \int \psi_1(\mathbf{r})^* \psi_2(\mathbf{r}) dv, \tag{4.125}$$

where $dv = r^{D-1} dr\, d\Omega$ is the $D$-dimensional volume element separated into its radial and angular parts. The matrix elements of an operator $A$ are then defined by

$$(\psi_1, A\psi_2) = \int \psi(\mathbf{r})^* [A\psi_2(\mathbf{r})] dv. \tag{4.126}$$

To show that the operator $A$ is hermitian it is sufficient to show that its expectation values are real,

$$(\psi, A\psi) - (\psi, A\psi)^* = 0, \tag{4.127}$$

for any square integrable function $\psi$. The so(2,1) generators are not hermitian with respect to this scalar product. However if we use the model space operators and the volume element $R^{a-2} dV$, where $dV = R^{D-1} dR\, d\Omega$, to obtain a new scalar product such that the matrix elements of the operator $A$ are now defined by

$$\langle \Psi_1 | A | \Psi_2 \rangle = \int \Psi_1(\mathbf{R})^* [A\Psi_2(\mathbf{R})] R^{a-2} dV, \tag{4.128}$$

then the so(2,1) generators are hermitian with respect to this scalar product. In the important case $a = 1$ corresponding to the hydrogen atom this is sometimes called the "$1/R$ scalar product" and in case $a = 2$ corresponding to the harmonic oscillator the two scalar products are identical.

We now show that the generator $T_2$ defined by (4.55) is hermitian with respect to the new scalar product. Since the differential operators $T_k$ depend only on the radial distance $R$ we can assume that $\Psi(\boldsymbol{R}) = F(R)Y(\hat{\boldsymbol{R}})$ where $Y(\hat{\boldsymbol{R}})$ denotes a general angular function of $\hat{\boldsymbol{R}} = \boldsymbol{R}/R$ and we can assume that the angular functions are normalized to unity. In the 3-dimensional case this factorization is expressed by (4.72). After integration over the angular part the expectation value of $T_2$ is given by

$$
\begin{aligned}
\langle \Psi | T_2 | \Psi \rangle &= \int_0^\infty F(R)^* [T_2 F(R)] R^{D+a-3} dR \\
&= \frac{1}{a} \int_0^\infty F(R)^* \left[ -\frac{i}{R^{M-1}} \frac{d}{dR} R^M F(R) - \frac{i}{2}(a-1)F(R) \right] R^{D+a-3} dR \\
&= -\frac{i}{a} \int_0^\infty F(R)^* \left[ M F(R) + R F'(R) + \frac{1}{2}(a-1)F(R) \right] R^{D+a-3} dR \\
&= -\frac{i}{a} \int_0^\infty F(R)^* \left[ R F'(R) + \frac{1}{2}(D + a - 2)F(R) \right] R^{D+a-3} dR,
\end{aligned}
$$

where we have used (4.106) and (4.107). Therefore

$$
\begin{aligned}
\langle \Psi | T_2 | \Psi \rangle &- \langle \Psi | T_2 | \Psi \rangle^* \\
&= -\frac{i}{a} \int_0^\infty \Big[ (D + a - 2)R^{D+a-3} F(R)^* F(R) \\
&\qquad\qquad + R^{D+a-2} \left( F(R)^* F'(R) + F'(R)^* F(R) \right) \Big] dR \\
&= -\frac{i}{a} \int_0^\infty \frac{d}{dR} \left[ R^{D+a-2} F(R)^* F(R) \right] dR \\
&= -\frac{i}{a} \int_0^\infty \frac{d}{dR} \left| R^{\frac{1}{2}(D+a-2)} F(R) \right|^2 dR. \qquad (4.129)
\end{aligned}
$$

This integral will be zero for the class of functions $F(R)$ such that

$$
\lim_{R \to 0} R^{\frac{1}{2}(D+a-2)} F(R) = 0, \qquad (4.130)
$$

$$
\lim_{R \to \infty} R^{\frac{1}{2}(D+a-2)} F(R) = 0. \qquad (4.131)
$$

Since we are assuming that $\Psi$ is square-integrable,

$$
\int_0^\infty \left| R^{\frac{1}{2}(D+a-2)} F(R) \right|^2 dR < \infty, \qquad (4.132)
$$

then (4.131) is satisfied. Therefore $T_2$ is hermitian with respect to the class of square integrable functions $\Psi(\boldsymbol{R})$ whose radial part satisfies (4.130).

Even though $T_2$ is hermitian with respect to this class of functions, its eigenfunctions do not belong to this class so $T_2$ is not an observable. This is analogous to the well known result that the radial momentum $P_R$ is hermitian with respect to the conventional scalar product (4.126) but its eigenfunctions are not square-integrable [ME70a]. This is not a problem since the basis functions used in the matrix representation (4.18) to (4.21) are eigenfunctions of $T_3$ not $T_2$.

In a similar manner it can be shown that $T_1$ and $T_3$ are also hermitian with respect to the new scalar product. In the next section we shall obtain the explicit form of the eigenfunctions of $T_3$.

## 4.11   Explicit $T_3$ Eigenfunctions

The realization of the operators $T_k$ can be expressed directly as differential operators in terms of the radial distance $R$ using the expression (4.106) for the radial momentum. Since $[\frac{\partial}{\partial R}, R^M] = MR^{M-1}$ we obtain

$$
P_R^2 = -\frac{1}{R^M}\frac{\partial^2}{\partial R^2}R^M
$$

$$
= -\frac{1}{R^M}\frac{\partial}{\partial R}\left(R^M\frac{\partial}{\partial R} + MR^{M-1}\right)
$$

$$
= -\frac{1}{R^M}\left(R^M\frac{\partial}{\partial R} + MR^{M-1}\right)\frac{\partial}{\partial R}
$$

$$
\qquad - \frac{M}{R^M}\left(R^{M-1}\frac{\partial}{\partial R} + (M-1)R^{M-2}\right),
$$

and using $M = (D-1)/2$

$$
P_R^2 = -\frac{\partial^2}{\partial R^2} - \frac{D-1}{R}\frac{\partial}{\partial R} - \frac{(D-1)(D-3)}{4R^2}. \tag{4.133}
$$

Therefore the so(2,1) generators can be expressed as the differential operators

$$
T_1 = -\frac{1}{2a^2}R^{2-a}\left(\frac{\partial^2}{\partial R^2} + \frac{D-1}{R}\frac{\partial}{\partial R}\right)
$$

$$
+ \frac{1}{2}\left(\tau - \frac{(D-1)(D-3)}{4a^2}\right)\frac{1}{R^a} - \frac{1}{2}R^a, \tag{4.134}
$$

$$
T_2 = -\frac{i}{a}\left(R\frac{\partial}{\partial R} + \frac{D+a-2}{2}\right), \tag{4.135}
$$

$$
T_3 = -\frac{1}{2a^2}R^{2-a}\left(\frac{\partial^2}{\partial R^2} + \frac{D-1}{R}\frac{\partial}{\partial R}\right)
$$

$$
+ \frac{1}{2}\left(\tau - \frac{(D-1)(D-3)}{4a^2}\right)\frac{1}{R^a} + \frac{1}{2}R^a. \tag{4.136}
$$

Here $\tau$ can be any constant or any operator commuting with functions of $R$, not just the particular value defined in (4.112) or (4.118).

Denoting the $T_3$ eigenfunctions in the coordinate representation by $F_{qk}(R)$ $= \langle R|qk\rangle$ this gives the eigenvalue problem

$$
T_3 F_{qk}(R) = qF_{qk}(R) \tag{4.137}
$$

analogous to (4.19) but expressed as a second order differential equation. We could now apply the series solution method to find the explicit eigenfunctions and eigenvalues of $T_3$. Of course this classical approach is not necessary since we have already obtained all the results in previous sections using the Lie algebraic method and the matrix representation of so(2,1) and it is not necessary to know the explicit form of the $T_3$ eigenfunctions in order to calculate matrix elements of operators expressible in terms of the so(2,1) generators.

Nevertheless the explicit eigenfunctions of $T_3$ can easily be obtained by the successive application of the raising operator $T_+$ to the lowest eigenfunction. The only differential equation which must be solved is a simple first order one for the lowest eigenfunction $F_{k+1,k}(R)$ of $T_3$. This approach using raising and lowering operators to factorize a second order differential operator into a product of two first order differential operators is not new and has been known since the beginning of quantum mechanics. In our case (4.7) expresses this factorization as $T_+T_- = T_3^2 - T_3 - T^2$ where from (4.19) and (4.57) the right side reduces to a constant.

Using the realization (4.54) to (4.57) for $T_1$, $T_2$ and $T_3$ and noting that $T_3 - T_1 = R^a$ it follows that

$$
\begin{aligned}
T_\pm &= T_1 \pm iT_2 \\
&= T_3 - R^a \pm iT_2 \\
&= T_3 - R^a \pm \frac{i}{a}RP_R \pm \frac{a-1}{2a} \\
&= T_3 - R^a \pm \frac{i}{a}R\left(-\frac{i}{R^M}\frac{\partial}{\partial R}R^M\right) \pm \frac{a-1}{2a} \\
&= T_3 - R^a \pm \frac{1}{aR^{M-1}}\left(R^M\frac{\partial}{\partial R} + MR^{M-1}\right) \pm \frac{a-1}{2a} \\
&= T_3 \pm \left(\frac{D+a-2}{2a}\right) - R^a \pm \frac{R}{a}\frac{\partial}{\partial R}.
\end{aligned}
$$

Therefore $T_\pm$ are given by the first order differential operators

$$
T_\pm = T_3 \pm G - R^a \pm \frac{R}{a}\frac{\partial}{\partial R}, \tag{4.138}
$$

where we have defined

$$
G = \frac{D+a-2}{2a}. \tag{4.139}
$$

Writing (4.20) in the form

$$
T_+F_{qk}(R) = A_{qk}F_{q+1,k}(R), \tag{4.140}
$$

where

$$
A_{qk} = \sqrt{(q-k)(q+k+1)}, \tag{4.141}
$$

it follows by successive application of $T_+$ that $F_{k+2,k}(R)$ is obtained from $F_{k+1,k}(R)$, then $F_{k+3,k}(R)$ is obtained from $F_{k+2,k}(R)$ and so on. Therefore

$$F_{qk}(R) = C_{qk}(T_+)^{q-k-1}F_{k+1,k}(R), \tag{4.142}$$

where

$$C_{qk} = [A_{k+1,k}A_{k+2,k}\cdots A_{q-1,k}]^{-1}$$
$$= \frac{\sqrt{(2k+1)!}}{\sqrt{(q-k-1)!(q+k)!}}. \tag{4.143}$$

Since $F_{k+1,k}(R)$ is the lowest eigenvector then from (4.138) it satisfies the first order separable differential equation

$$T_-F_{k+1,k}(R) = \left(k+1-G-R^a-\frac{R}{a}\frac{d}{dR}\right)F_{k+1,k}(R) = 0 \tag{4.144}$$

whose general solution is

$$F_{k+1,k}(R) = c_0 R^{a(k+1-G)}e^{-R^a}. \tag{4.145}$$

It is clear that this solution satisfies (4.130)

$$\lim_{R\to\infty} R^{aG}F_{k+1,k}(R) = 0,$$

and it also satisfies (4.131)

$$\lim_{R\to 0} R^{aG}F_{k+1,k}(R) = \lim_{R\to 0} c_0 R^{a(k+1)} = 0,$$

so long as $k > -1$ which is just the condition already obtained in (4.23) in the study of the unirreps of so(2,1). The normalization constant $c_0$ is determined so that the integral in (4.132) is unity and the result is

$$c_0 = \frac{2^{k+1}\sqrt{a}}{\sqrt{(2k+1)!}}. \tag{4.146}$$

The final step is to simplify the expression (4.142) for $F_{qk}(R)$. In the expression (4.138) for $T_+$ we can replace $T_3$ by its eigenvalues $k+1$, $k+2$, ..., $k+(q-k-1)$ as we operate from right to left with $T_+$ in (4.147). Therefore

$$F_{qk}(R) = C_{qk}T_+^{(q-1+G)}T_+^{(q-2+G)}\cdots T_+^{(k+1+G)}F_{k+1,k}(R), \tag{4.147}$$

where we have defined $T_+^{(\alpha)} = \alpha - R^a + \frac{R}{a}\frac{d}{dR}$. The $R^a$ term in each of these non-commuting factors can be eliminated using the operator identity

$$e^{-R^a}T_+^{(\alpha)} = e^{-R^a}\left(\alpha - R^a + \frac{R}{a}\frac{d}{dR}\right)$$
$$= (\alpha - R^a)e^{-R^a} + \frac{R}{a}\left(\frac{d}{dR}e^{-R^a} + aR^{a-1}e^{-R^a}\right)$$
$$= \left(\alpha + \frac{R}{a}\frac{d}{dR}\right)e^{-R^a} \equiv U_+^{(\alpha)}e^{-R^a}.$$

Substituting this identity into (4.147) gives

$$\begin{aligned}
F_{qk}(R) &= C_{qk}e^{R^a}e^{-R^a}T_+^{(q-1+G)}T_+^{(q-2+G)}\cdots T_+^{(k+1+G)}F_{k+1,k}(R)\\
&= C_{qk}e^{R^a}U_+^{(q-1+G)}e^{-R^a}T_+^{(q-2+G)}\cdots T_+^{(k+1+G)}F_{k+1,k}(R)\\
&\ \ \vdots\\
&= C_{qk}e^{R^a}U_+^{(q-1+G)}U_+^{(q-2+G)}\cdots U_+^{(k+1+G)}e^{-R^a}F_{k+1,k}(R).
\end{aligned}$$

The operator factors now commute so we can reverse their order to obtain

$$F_{qk}(R) = C_{qk}e^{R^a}U_+^{(k+1+G)}U_+^{(k+2+G)}\cdots U_+^{(q-1+G)}e^{-R^a}F_{k+1,k}(R).$$

Next substitute $\rho = e^{R^a}$ into this result and use the operator identity

$$\frac{R}{a}\frac{d}{dR} = \frac{R}{a}\frac{d\rho}{dR}\frac{d}{d\rho} = \rho\frac{d}{d\rho}$$

to obtain the operators $U_+^{(\alpha)} = \alpha + \rho\frac{d}{d\rho}$.

Now we can use $[\frac{d}{d\rho},\rho^\alpha] = \alpha\rho^{\alpha-1}$ to obtain $\rho^{\alpha-1}U_+^{(\alpha)} = \frac{d}{d\rho}\rho^\alpha$ and

$$\begin{aligned}
\rho^{k+G}&U_+^{(k+1+G)}U_+^{(k+2+G)}\cdots U_+^{(q-1+G)}\\
&= \frac{d}{d\rho}\rho^{k+1+G}U_+^{(k+2+G)}\cdots U_+^{(q-1+G)}\\
&\ \ \vdots\\
&= \frac{d}{d\rho}\frac{d}{d\rho}\cdots\frac{d}{d\rho}\rho^{q-2+G}U_+^{(q-1+G)}\\
&= \left(\frac{d}{d\rho}\right)^{q-k-1}\rho^{q-1+G}.
\end{aligned}$$

Therefore (4.147) simplifies to an expression of Rodrigues' type

$$F_{qk}(R) = c_0 C_{qk}e^{\rho}\rho^{-k-G}\left(\frac{d}{d\rho}\right)^{q-k-1}\rho^{k+G}e^{-2\rho}. \tag{4.148}$$

Finally this result can be expressed in terms of the associated Laguerre polynomials whose Rodrigues' formula is given by

$$L_n^{(\alpha)} = \frac{1}{n!}e^{x}x^{-\alpha}\left(\frac{d}{dx}\right)^{n}x^{n+\alpha}e^{-x}. \tag{4.149}$$

Identifying $n$ with $q - k - 1$, $\alpha$ with $2k + 1$ and $x$ with $2\rho$ we obtain the normalized $T_3$ eigenfunctions

$$F_{qk}(R) = D_{qk}e^{-R^a}(2R^a)^{k-G+1}L_{q-k-1}^{(2k+1)}(2R^a) \tag{4.150}$$

with the normalization factor

$$D_{qk} = 2^G \sqrt{\frac{a(q-k-1)!}{(q+k)!}}. \tag{4.151}$$

There exist at least two different definitions of the associated Laguerre polynomials. Our choice is defined by [PO61], [AB65], [GR80][1]

$$L_n^{(\alpha)} = \binom{n+\alpha}{n} {}_1F_1(-n, \alpha+1, x) \tag{4.152}$$

where ${}_1F_1(a, b, x)$, also denoted by $M$, is the confluent hypergeometric function defined by

$$_1F_1(a, b, x) = 1 + \frac{ax}{b} + \frac{(a)_2 x^2}{(b)_2 2!} + \cdots + \frac{(a)_n x^n}{(b)_n n!} + \cdots, \tag{4.153}$$

where

$$(a)_n = a(a+1)(a+2)\cdots(a+n-1), \quad (a)_0 = 1. \tag{4.154}$$

We now determine the relationship between the normalized $T_3$ eigenfunctions (4.150) and the conventional radial eigenfunctions $f_{qk}(r)$. Because of the simple scaling transformation (4.65) the scaled radial eigenfunctions are related to the conventional ones by

$$F_{qk}(R) = c_a f_{qk}(\gamma R), \tag{4.155}$$

where the constant $c_a$ is to be chosen so that the functions $F_{qk}(R)$ are normalized with respect to to scalar product (4.128) and the functions $f_{qk}(r)$ are normalized with respect to the conventional scalar product (4.125). In case $a = 1$ corresponding to the hydrogen atom

$$\begin{aligned}
1 &= \int_0^\infty f_{qk}(r) f_{qk}(r) r^{D-1} dr \\
&= \frac{\gamma^D}{c_1^2} \int_0^\infty F_{qk}(R) F_{qk}(R) R^{D-1} dR \\
&= \frac{\gamma^D}{c_1^2} \int_0^\infty F_{qk}(R) R F_{qk}(R) R^{D-2} dR \\
&= \frac{\gamma^D}{c_1^2} \langle F_{qk} | R | F_{qk} \rangle.
\end{aligned}$$

From the realizations (4.156) and (4.56) of $T_1$ and $T_3$ in case $a = 1$ this result can be expressed as

$$c_1^2 = \gamma^D \langle F_{qk} | T_3 - T_1 | F_{qk} \rangle.$$

---

[1] The other common choice is defined by $\mathcal{L}_{n+\alpha}^\alpha(x) = (-1)^\alpha (n+\alpha)! L_n^{(\alpha)}(x)$

However from the matrix representation (4.20) and (4.21) and $T_1 = (T_+ + T_-)/2$ it follows that the diagonal matrix elements of $T_1$ are zero so

$$c_1^2 = \gamma^D \langle F_{qk}|T_3|F_{qk}\rangle = \gamma^D q.$$

In case $a = 2$ the integral

$$\int_0^\infty F_{qk}(R)F_{qk}(R)R^{D-1}dR$$

is unity so we combine both cases to obtain

$$c_a = \begin{cases} \gamma^{D/2}q^{1/2} & \text{if } a = 1 \\ \gamma^{D/2} & \text{if } a = 2. \end{cases} \tag{4.156}$$

Therefore the conventional eigenfunctions are given from (4.155) by

$$f_{qk}(r) = \frac{1}{c_a} F_{qk}\left(\frac{r}{\gamma}\right). \tag{4.157}$$

In case $a = 1$ and $D = 3$ corresponding to the 3-dimensional hydrogen atom we have $G = 1$, $\gamma = q/Z$, $q = n$, $k = \ell$ and

$$f_{n\ell}(r) = \frac{Z^{3/2}}{n^2} F_{n\ell}\left(\frac{Zr}{n}\right) \tag{4.158}$$

$$= \frac{Z^{3/2}}{n^2} D_{n\ell} e^{-\frac{Zr}{n}} \left(\frac{2Zr}{n}\right)^\ell L_{n-\ell-1}^{(2\ell+1)}\left(\frac{2Zr}{n}\right), \tag{4.159}$$

where

$$D_{n\ell} = 2\sqrt{\frac{(n-\ell-1)!}{(n+\ell)!}}. \tag{4.160}$$

Similarly, if $a = 2$ and $D = 3$ corresponding to the 3-dimensional harmonic oscillator we have $\gamma^2 = 2/\omega$ and

$$f_{qk}(r) = \left(\frac{\omega}{2}\right)^{3/4} F_{qk}\left(\sqrt{\frac{\omega}{2}}\,r\right). \tag{4.161}$$

Since $G = \frac{3}{4}$, $2q = n + \frac{3}{2}$ and $k = \frac{1}{2}\ell - \frac{1}{4}$ we obtain

$$f_{n\ell}(r) = \left(\frac{\omega}{2}\right)^{3/4} D_{qk} e^{-\omega r^2/2} \left(\omega r^2\right)^{\ell/2} L_{q-k-1}^{(2k+1)}\left(\omega r^2\right). \tag{4.162}$$

The conventional hydrogenic eigenfunctions (4.157) in the general case, or (159) in the 3-dimensional case, do not form a complete set of radial eigenfunctions. This is primarily due to the dependence of the exponential factors on $r/n$ rather than $r$ in (4.159). On the other hand the $T_3$ eigenfunctions (4.150) have no exponential dependence on $n$ and form a complete set for bound state eigenfunctions.

## 4.12  D-dimensional Relativistic Systems

In this section we obtain the energy levels for the Klein-Gordon and Dirac equations in a Coulomb potential. The relativistic results are also compared with those for the non-relativistic hydrogen atom obtained in Section 4.8.

### 4.12.1  Klein-Gordon equation

The Klein-Gordon equation (Schrödinger's relativistic equation) describes a relativistic particle with spin 0. It is obtained from the relativistic energy-momentum relationship

$$(E - eV)^2 = p^2 c^2 + m^2 c^4, \tag{4.163}$$

where $V$ is the potential. For a Coulomb potential $V = -\mathcal{Z}/r$, using atomic units ($\hbar = m = e = 1$), we obtain [SC68]

$$\left[ p^2 + \frac{c^4 - E^2}{c^2} + \frac{1}{c^2}\left( -\frac{2\mathcal{Z}E}{r} - \frac{\mathcal{Z}^2}{r^2} \right) \right] \psi(\boldsymbol{r}) = 0. \tag{4.164}$$

For this system of units the fine structure constant is $\alpha = 1/c \approx \frac{1}{137}$. Expressing $p^2$ in terms of the radial momentum using (4.108) we obtain

$$\left[ p_r^2 + \left( \frac{(D-1)(D-3)}{4} + L^2 - \mathcal{Z}^2\alpha^2 \right) \frac{1}{r^2} \right.$$
$$\left. + c^2 - E^2\alpha^2 - \frac{2\mathcal{Z}E\alpha^2}{r} \right] \psi(\boldsymbol{r}) = 0. \tag{4.165}$$

We can now multiply by $r$, divide by 2, and use the scaling transformation (4.65) to obtain

$$\left[ \frac{1}{2}\left\{ RP_R^2 + \left( \frac{(D-1)(D-3)}{4} + L^2 - \mathcal{Z}^2\alpha^2 \right) \frac{1}{R} \right. \right.$$
$$\left. \left. + (c^2 - E^2\alpha^2)\gamma^2 R \right\} - \gamma\mathcal{Z}E\alpha^2 \right] \psi(\gamma\boldsymbol{R}) = 0, \tag{4.166}$$

which can be expressed as a $T_3$ eigenvalue problem of the form $[T_3 - q]\Phi(R) = 0$ for the radial part of the scaled wavefunction if we use the realization (4.54) to (4.56) with $a = 1$ and

$$\tau = \frac{(D-1)(D-3)}{4} + L^2 - \mathcal{Z}^2\alpha^2, \tag{4.167}$$

$$(c^2 - E^2\alpha^2)\gamma^2 = 1, \tag{4.168}$$

$$q = \gamma\mathcal{Z}E\alpha^2. \tag{4.169}$$

The last two equations can be solved for $\gamma$ and $E$ in terms of the $T_3$ eigenvalue $q$ to obtain the positive energy solutions

$$\gamma^2 = \frac{q^2}{\mathcal{Z}^2}\left(1 + \frac{\mathcal{Z}^2\alpha^2}{q^2}\right),\tag{4.170}$$

$$E = c^2\left(1 + \frac{\mathcal{Z}^2\alpha^2}{q^2}\right)^{-1/2}.\tag{4.171}$$

Finally, replacing $L^2$ in (4.167) by its eigenvalue $\ell(\ell + D - 2)$, the lowest eigenvalue $q_0$ of $q$ is obtained from (4.59) using the upper sign and (4.61):

$$\begin{aligned}
q_0 &= k + 1\\
&= \frac{1}{2}\left[1 + \sqrt{(D-1)(D-3) + 4\ell(\ell + D - 2) + 1 - 4\mathcal{Z}^2\alpha^2}\right]\\
&= \frac{1}{2}\left[1 + \sqrt{(D + 2\ell - 2)^2 - 4\mathcal{Z}^2\alpha^2}\right].
\end{aligned}\tag{4.172}$$

Therefore the energy levels are given by

$$E = c^2\left(1 + \frac{\mathcal{Z}^2\alpha^2}{(q_0 + \mu)^2}\right)^{-1/2}, \qquad \mu = 0, 1, 2, \ldots\tag{4.173}$$

It is interesting to compare this result with that for the non-relativistic $D$-dimensional hydrogen atom given by (4.114) and (4.115). Using the superscript $NR$ to denote the non-relativistic values of the principal quantum numbers $q$ and $q_0$ we obtain

$$q = q^{NR} + \Delta,\tag{4.174}$$

where $\Delta = q - q^{NR} = q_0 - q_0^{NR}$ is the relativistic correction given from (4.172) and (4.114) by

$$\begin{aligned}
\Delta &= \frac{1}{2}\left[1 + \sqrt{(D + 2\ell - 2)^2 - 4\mathcal{Z}^2\alpha^2}\right] - \frac{D + 2\ell - 1}{2}\\
&= \frac{D + 2\ell - 2}{2}\left[\sqrt{1 - \frac{4\mathcal{Z}^2\alpha^2}{(D + 2\ell - 2)^2}} - 1\right].
\end{aligned}\tag{4.175}$$

Substituting into (4.171) we obtain the energy formula

$$\begin{aligned}
E &= c^2\left[1 + \frac{\mathcal{Z}^2\alpha^2}{(q^{NR})^2}\left(1 + \frac{\Delta}{q^{NR}}\right)^{-2}\right]^{-1/2}\\
&= c^2\left[1 - \frac{\mathcal{Z}^2\alpha^2}{2(q^{NR})^2} - \frac{(\mathcal{Z}^2\alpha^2)^2}{2(q^{NR})^4}\left(\frac{2q^{NR}}{D + 2\ell - 2} - \frac{3}{4}\right) + \cdots\right].
\end{aligned}\tag{4.176}$$

The first term on the right side is the rest mass, the second is the non-relativistic energy and the third is the first correction term which depends on the orbital angular momentum quantum number $\ell$.

### 4.12.2  Dirac-Coulomb equation

The $D$-dimensional Dirac hamiltonian for a Coulomb potential describes a
particle with spin 1/2 and can be expressed in the radial form [CO67], [SC68]

$$H = c\alpha_r p_r + i\frac{c}{r}\alpha_r \beta K + \beta c^2 - \frac{Z}{r}, \tag{4.177}$$

where $\alpha_r$ and $\beta$ anticommute:

$$\{\alpha_r, \beta\} = \alpha_r \beta + \beta \alpha_r = 0, \tag{4.178}$$

and the operator $K$ is diagonal with eigenvalues given by

$$k_D = \pm\left[\ell_D + \frac{D-1}{2}\right], \quad \ell_D = 0, 1, 2, \ldots \tag{4.179}$$

The eigenvalue problem for the radial functions can be expressed in the
two-component spinor form

$$Hf = Ef, \tag{4.180}$$

where

$$\alpha_r = \begin{pmatrix} 0 & -i \\ i & 0 \end{pmatrix}, \quad \beta = \begin{pmatrix} 1 & 0 \\ 0 & -1 \end{pmatrix}, \quad f = \begin{pmatrix} f_+ \\ f_- \end{pmatrix}. \tag{4.181}$$

Substituting (4.177) and (4.181) into (4.180) gives

$$\begin{pmatrix} c^2 - \dfrac{Z}{r} & -icp_r - \dfrac{cK}{r} \\[2ex] icp_r - \dfrac{cK}{r} & -c^2 - \dfrac{Z}{r} \end{pmatrix} \begin{pmatrix} f_+ \\ f_- \end{pmatrix} = E \begin{pmatrix} f_+ \\ f_- \end{pmatrix}, \tag{4.182}$$

which is a system of two first-order coupled differential equations for the spinor
components $f_\pm$.

In order to use our realization (4.54) to (4.56) of the so(2,1) generators,
which is expressed in terms of $p_r^2$, it is necessary to convert (4.182) to a second-order differential equation for one of the functions $f_\pm$. To do this we first write
(4.182) in the compact form

$$\left(\mp icp_r - \frac{cK}{r}\right)f_\mp + \left(\pm c^2 - \frac{Z}{r} - E\right)f_\pm = 0. \tag{4.183}$$

Now premultiply by $-r$ to obtain

$$W^\mp f_\pm + c(K \pm irp_r)f_\mp = 0, \tag{4.184}$$

where we have defined

$$W^\mp = (E \mp c^2)r + Z. \tag{4.185}$$

Now premultiply (4.184) by $c(K \mp irp_r)$ to obtain

$$Uf_\pm + c^2(K^2 + rp_r rp_r)f_\mp = 0, \tag{4.186}$$

where we have defined the operator $U$ by

$$U = c(K \mp irp_r)W^\mp. \tag{4.187}$$

We can move $W^\mp$ to the left using the commutator result

$$[rp_r, W^\mp] = (E \mp c^2)r\,[p_r, r] = -i(E \mp c^2)r$$

to obtain

$$\begin{aligned}
U &= W^\mp cK \mp ic(W^\mp rp_r - i(E \mp c^2)r) \\
&= W^\mp c(K \mp irp_r) \mp c(E \mp c^2)r \\
&= W^\mp c(K \mp irp_r) \mp c[(E \mp c^2)r + \mathcal{Z}] \pm c\mathcal{Z} \\
&= W^\mp c(K \mp irp_r) \mp cW^\mp \pm c\mathcal{Z}. 
\end{aligned} \tag{4.188}$$

Substituting this result into (4.186) gives

$$W^\mp c(K \mp irp_r)f_\pm \mp cW^\mp f_\pm \pm c\mathcal{Z}f_\pm + c^2(K^2 + rp_r rp_r)f_\mp = 0.$$

Now substitute (4.184) with $-$ and $+$ reversed to obtain

$$-W^\mp W^\pm f_\mp \mp cW^\mp f_\pm \pm c\mathcal{Z}f_\pm + c^2(K^2 + rp_r rp_r)f_\mp = 0,$$

and substitute (4.184) again to obtain

$$-W^\mp W^\pm f_\mp \pm c^2(K \pm irp_r)f_\mp + c^2(K^2 + rp_r rp_r)f_\mp = \mp c\mathcal{Z}f_\pm,$$

which represents two second-order equations of the form

$$[M + c^2 K]f_- = -c\mathcal{Z}f_+, \tag{4.189}$$
$$[M - c^2 K]f_+ = c\mathcal{Z}f_-, \tag{4.190}$$

where

$$M = -W^\mp W^\pm + ic^2 rp_r + c^2(K^2 + rp_r rp_r). \tag{4.191}$$

We can now obtain an equation involving only $f_+$ using (4.190) in (4.189):

$$[M + c^2 K][M - c^2 K]f_+ = -(c\mathcal{Z})^2 f_+,$$

which can be expressed as

$$\frac{M^2}{c^4}f_+ = s^2 f_+, \tag{4.192}$$

where

$$s^2 = K^2 - \left(\frac{\mathcal{Z}}{c}\right)^2 = K^2 - \mathcal{Z}^2\alpha^2. \tag{4.193}$$

Finally taking square roots we obtain the second-order differential equation

$$\frac{M}{c^2}f_+ = \mp s f_+, \tag{4.194}$$

which can be transformed into a $T_3$ eigenvalue problem as follows. From (4.185) and (4.191)

$$\begin{aligned}
M &= -(E^2 - c^4)r^2 - 2\mathcal{Z}Er - \mathcal{Z}^2 + ic^2 r p_r + c^2 K^2 + c^2 r p_r r p_r \\
&= -(E^2 - c^4)r^2 - 2\mathcal{Z}Er - \mathcal{Z}^2 + ic^2 r p_r + c^2 K^2 + c^2 r(r p_r - i)p_r.
\end{aligned}$$

Substituting this result into (4.192) and premultiplying by $r^{-1}$ gives the eigenvalue problem for the radial function $f_+$

$$\left[ p_r^2 + \frac{s(s \pm 1)}{r^2} + (c^2 - E^2\alpha^2) - \frac{2\mathcal{Z}E\alpha^2}{r} \right] f_+(r) = 0, \tag{4.195}$$

which has the same structure as the radial Klein-Gordon equation (4.165), the only difference being the coefficient of $r^{-2}$. Therefore premultiplying (4.195) by $r$ and introducing the scaling transformation (4.63) gives

$$\left[ \frac{1}{2}\left( RP_R^2 + \frac{s(s \pm 1)}{R} + (c^2 - E^2\alpha^2)\gamma^2 R \right) - \gamma\mathcal{Z}E\alpha^2 \right] f_+(\gamma R) = 0. \tag{4.196}$$

The analysis is now identical to that of the scaled Klein-Gordon equation (4.166) except that in place of (4.167) we have

$$\tau = s(s \pm 1), \tag{4.197}$$

and the lowest eigenvalue of $q$ is now given by

$$\begin{aligned}
q_0 &= \frac{1}{2}\left[ 1 + \sqrt{4s(s \pm 1) + 1} \right] \\
&= \frac{1}{2}[1 + 2s \pm 1] \\
&= s \quad \text{or} \quad s + 1, \tag{4.198}
\end{aligned}$$

where from (4.193) and (4.179)

$$\begin{aligned}
s &= \sqrt{\left(\ell + \frac{D-1}{2}\right)^2 - \mathcal{Z}^2\alpha^2}, \quad \ell = 0, 1, 2, \ldots \\
&= \frac{1}{2}\sqrt{(D + 2j - 2)^2 - 4\mathcal{Z}^2\alpha^2}, \quad j = \tfrac{1}{2}, \tfrac{3}{2}, \ldots, \tag{4.199}
\end{aligned}$$

where we have defined

$$j = \ell + \tfrac{1}{2}. \tag{4.200}$$

Since $q = q_0 + \mu$, where $\mu$ is a non-negative integer, all eigenvalues can be obtained from the first case $q = s_0$ in (4.198).

Therefore we get the same formula (4.173) for the energy levels as in the Klein-Gordon case but with $q_0$ given by (4.199) instead of (4.172). The relativistic correction $\Delta$ defined in (4.174) is now given by

$$\Delta = \frac{1}{2}\sqrt{(D + 2j - 2)^2 - 4\mathcal{Z}^2\alpha^2} - \frac{D + 2\ell - 1}{2},$$

and substituting $\ell = j - 1/2$ in the second term

$$\Delta = \frac{D + 2j - 2}{2}\left[\sqrt{1 - \frac{4\mathcal{Z}^2\alpha^2}{(D + 2j - 2)^2}} - 1\right], \tag{4.201}$$

which is identical to the Klein-Gordon correction (4.175) with $\ell$ replaced by $j$.

## 4.13  General Radial Equations

We can summarize the results of the preceding sections by observing that so(2,1) with the realization of the generators given by (4.54) to (4.56) is a spectrum generating algebra for radial eigenvalue equations of the form

$$\left[\frac{1}{2a^2}p_r^2 + \frac{\tau}{2r^2} - \sigma r^{a-2} - \varepsilon r^{2a-2}\right] f(r) = 0. \tag{4.202}$$

Premultiply by $r^{2-a}$ and substitute the scaling transformation (4.65) to obtain

$$\left[\frac{1}{2}\left(\frac{1}{a^2}R^{2-a}P_R^2 + \frac{\tau}{R^a} - 2\varepsilon\gamma^{2a}R^a\right) - \sigma\gamma^a\right] F(R) = 0, \tag{4.203}$$

where $F(R) = f(\gamma R)$ is the scaled radial eigenfunction. This is a $T_3$ eigenvalue equation of the form $(T_3 - q)F(R) = 0$ if we choose

$$q = \sigma\gamma^a, \quad 2\varepsilon\gamma^{2a} = -1. \tag{4.204}$$

The values of $\tau$, $\sigma$ and $\varepsilon$ in (4.202) corresponding to the $D$-dimensional systems considered here are given in Table 4.1.

The table entries are obtained by comparing (4.202) with (4.110) for the $D$-dimensional hydrogen atom, (4.116) for the harmonic oscillator, (4.164) for the Klein-Gordon equation and (4.195) for the Dirac-Coulomb equation. Also we can replace $L^2$ by its $D$-dimensional eigenvalue $l(l + D - 2)$.

**Table 4.1**: Parameter values in (4.202) for the D-dimensional hydrogen atom, harmonic oscillator, Klein-Gordon equation and the Dirac-Coulomb equation.

| System | $a$ | $\tau$ | $\sigma$ | $\varepsilon$ |
|---|---|---|---|---|
| H-atom | 1 | $\frac{(D-1)(D-3)}{4} + L^2$ | $\mathcal{Z}$ | $E$ |
| Oscillator | 2 | $\frac{(D-1)(D-3)}{16} + \frac{L^2}{4}$ | $\frac{E}{4}$ | $-\frac{\omega^2}{8}$ |
| Klein-Gordon | 1 | $\frac{(D-1)(D-3)}{4} + L^2 - \mathcal{Z}^2\alpha^2$ | $\mathcal{Z}E\alpha^2$ | $E^2\alpha^2 - c^2$ |
| Dirac-Coulomb | 1 | $s(s \pm 1)$ | $\mathcal{Z}E\alpha^2$ | $E^2\alpha^2 - c^2$ |

# 4.14   Parabolic Hydrogenic Realization of so(2,1)

The Schrödinger equation for the 3-dimensional hydrogenic atom can also be separated in parabolic coordinates [BE57], [SC68], [LA77]. The advantage of this separation is that it is also valid for a hydrogenic atom in an electric field parallel to the $z$ direction (Stark effect) and provides a convenient framework for perturbation theory since the Stark perturbation matrix is diagonal in parabolic coordinates.

Since separation in parabolic coordinates is not as familiar as the usual separation in spherical coordinates we present here the details of the solution but our primary goal is to show how to obtain the equivalent results algebraically using a parabolic realization of so(2,1).

### 4.14.1   Parabolic coordinate system

We combine the transformation from parabolic to rectangular coordinates with the scaling transformation by using a scaled parabolic coordinate system defined by

$$\xi = \alpha(r + z), \quad 0 \leq \xi < \infty, \tag{4.205}$$

$$\eta = \alpha(r - z), \quad 0 \leq \eta < \infty, \tag{4.206}$$

$$\varphi = \tan^{-1}(y/x), \quad 0 \leq \varphi \leq 2\pi, \tag{4.207}$$

where $r = \sqrt{x^2 + y^2 + z^2}$ and $\alpha$ is the scaling parameter. The inverse transformation is

$$x = \alpha^{-1}\sqrt{\xi\eta}\,\cos\varphi, \tag{4.208}$$

$$y = \alpha^{-1}\sqrt{\xi\eta}\,\sin\varphi, \tag{4.209}$$

$$z = (2\alpha)^{-1}(\xi - \eta). \tag{4.210}$$

The volume element and Laplacian for a curvilinear coordinate system are (see [GR80], Sections 10.51 to 10.611)

$$dV = h_1 h_2 h_3 \, d\xi \, d\eta \, d\varphi, \tag{4.211}$$

$$\nabla^2 = \frac{1}{h_1 h_2 h_3} \left[ \frac{\partial}{\partial \xi} \left( \frac{h_2 h_3}{h_1} \frac{\partial}{\partial \xi} \right) + \frac{\partial}{\partial \eta} \left( \frac{h_3 h_1}{h_2} \frac{\partial}{\partial \eta} \right) + \frac{\partial}{\partial \varphi} \left( \frac{h_1 h_2}{h_3} \frac{\partial}{\partial \varphi} \right) \right], \tag{4.212}$$

where

$$h_1 = \sqrt{g_{11}}, \quad h_2 = \sqrt{g_{22}}, \quad h_3 = \sqrt{g_{33}}, \tag{4.213}$$

$$g_{11} = \left( \frac{\partial x}{\partial \xi} \right)^2 + \left( \frac{\partial y}{\partial \xi} \right)^2 + \left( \frac{\partial z}{\partial \xi} \right)^2, \tag{4.214}$$

$$g_{22} = \left( \frac{\partial x}{\partial \eta} \right)^2 + \left( \frac{\partial y}{\partial \eta} \right)^2 + \left( \frac{\partial z}{\partial \eta} \right)^2, \tag{4.215}$$

$$g_{33} = \left( \frac{\partial x}{\partial \varphi} \right)^2 + \left( \frac{\partial y}{\partial \varphi} \right)^2 + \left( \frac{\partial z}{\partial \varphi} \right)^2. \tag{4.216}$$

For the parabolic coordinate system

$$h_1 = \frac{1}{2\alpha} \sqrt{\frac{\xi + \eta}{\xi}}, \quad h_2 = \frac{1}{2\alpha} \sqrt{\frac{\xi + \eta}{\eta}}, \quad h_3 = \frac{1}{\alpha} \sqrt{\xi \eta}, \tag{4.217}$$

and we obtain ($h_1 h_2 / h_3$ is independent of $\varphi$)

$$dV = \frac{1}{4\alpha^3} (\xi + \eta) \, d\xi \, d\eta \, d\varphi, \tag{4.218}$$

$$\nabla^2 = \frac{4\alpha^2}{\xi + \eta} \left[ \frac{\partial}{\partial \xi} \left( \xi \frac{\partial}{\partial \xi} \right) + \frac{\partial}{\partial \eta} \left( \eta \frac{\partial}{\partial \eta} \right) \right] + \frac{\alpha^2}{\xi \eta} \frac{\partial^2}{\partial \varphi^2}. \tag{4.219}$$

### 4.14.2   Separation in parabolic coordinates

Now substitute into the Schrödinger equation (4.62) to obtain

$$\left[ \frac{\partial}{\partial \xi} \left( \xi \frac{\partial}{\partial \xi} \right) + \frac{\partial}{\partial \eta} \left( \eta \frac{\partial}{\partial \eta} \right) + \frac{\xi + \eta}{4 \xi \eta} \frac{\partial^2}{\partial \varphi^2} + \frac{\mathcal{Z}}{\alpha} + \frac{E(\xi + \eta)}{2\alpha^2} \right] \psi = 0. \tag{4.220}$$

Separation can be achieved by substituting a solution of the form

$$\psi(\xi, \eta, \varphi) = f_1(\xi) f_2(\eta) e^{im\varphi}, \quad m = 0, \pm 1, \pm 2, \ldots \tag{4.221}$$

into (4.220), cancelling $e^{im\varphi}$, and dividing by $f_1 f_2$ to obtain

$$\left[ \frac{1}{f_1} \frac{d}{d\xi} \left( \xi \frac{df_1}{d\xi} \right) - \frac{m^2}{4\xi} + \frac{E\xi}{2\alpha^2} \right]$$
$$+ \left[ \frac{1}{f_2} \frac{d}{d\eta} \left( \eta \frac{df_2}{d\eta} \right) - \frac{m^2}{4\eta} + \frac{E\eta}{2\alpha^2} \right] + \frac{\mathcal{Z}}{\alpha} = 0.$$

Each expression in square brackets must be a constant ($-\lambda_1$ and $-\lambda_2$ for example) and we obtain the two differential equations

$$\left[ -\frac{d}{d\xi}\left( \xi\frac{d}{d\xi} \right) + \frac{m^2}{4\xi} - \frac{E}{2\alpha^2}\xi - \lambda_1 \right] f_1(\xi) = 0, \tag{4.222}$$

$$\left[ -\frac{d}{d\eta}\left( \eta\frac{d}{d\eta} \right) + \frac{m^2}{4\eta} - \frac{E}{2\alpha^2}\eta - \lambda_2 \right] f_2(\eta) = 0, \tag{4.223}$$

and the separation constants must satisfy

$$\lambda_1 + \lambda_2 = \frac{Z}{\alpha}. \tag{4.224}$$

The two differential equations are identical so it is sufficient to solve one of them, say the first. It is interesting to note that (4.222) is not an eigenvalue problem for the energy $E$ but for the separation constant $\lambda_1$. To obtain the energy we choose the scale factor such that

$$E = -\frac{1}{2}\alpha^2. \tag{4.225}$$

After determining the separation constants, $\alpha$ can be found from (4.224) and $E$ can be found from (4.225).

The standard method of series solution can be used to obtain the bound state solutions of (4.222). The result is that the solution of (4.222) has the form

$$f(\xi) = e^{-\xi/2}\xi^{|m|/2}u(\xi), \tag{4.226}$$

where $u(\xi)$ must be a polynomial solution of

$$\xi\frac{d^2u}{d\xi^2} + (|m| + 1 - \xi)\frac{du}{d\xi} + \left( \lambda_1 - \frac{|m| + 1}{2} \right) u = 0. \tag{4.227}$$

This is a confluent hypergeometric equation which has a polynomial solution of the form (see (4.153))

$$u_{n_1,|m|}(\xi) = {}_1F_1(-n_1, |m| + 1, \xi) \tag{4.228}$$

only if $\lambda_1$ is given by

$$\lambda_1 = n_1 + \frac{|m| + 1}{2}, \quad n_1 = 0, 1, 2, \ldots \tag{4.229}$$

Similarly, we can solve (4.223) for $f(\eta)$ only if

$$\lambda_2 = n_2 + \frac{|m| + 1}{2}, \quad n_2 = 0, 1, 2, \ldots \tag{4.230}$$

Therefore we can define

$$n = \lambda_1 + \lambda_2 = n_1 + n_2 + |m| + 1 \tag{4.231}$$

as the principal quantum number and obtain the Bohr formula from (4.224)

$$E_n = -\frac{1}{2}\alpha^2 = -\frac{1}{2}\mathcal{Z}^2(\lambda_1 + \lambda_2)^{-2} = -\frac{\mathcal{Z}^2}{2n^2}. \tag{4.232}$$

The parabolic wavefunctions are labeled by the parabolic quantum numbers $n_1$ and $n_2$ and the usual magnetic quantum number $m$ and have the form

$$\psi_{n_1,n_2,m}(\xi,\eta,\varphi) = N_{n_1,n_2,m} f_{n_1,|m|}(\xi) f_{n_2,|m|}(\eta) e^{im\varphi}, \tag{4.233}$$

where $f_{n_1,|m|}(\xi)$ can be expressed in terms of the Laguerre polynomials (4.152) as

$$f_{n_1,|m|}(\xi) = N_{n_1,|m|} e^{-\xi/2} \xi^{|m|/2} L_{n_1}^{(|m|)}(\xi), \tag{4.234}$$

or in terms of the confluent hypergeometric function (4.153) by

$$f_{n_1,|m|}(\xi) = (|m|! N_{n_1,|m|})^{-1} e^{-\xi/2} \xi^{|m|/2} {}_1F_1(-n_1, |m|+1, \xi), \tag{4.235}$$

where the normalization factors

$$N_{n_1,|m|} = \sqrt{\frac{n_1!}{(n_1 + |m|)!}}, \tag{4.236}$$

$$N_{n_1,n_2,m} = \frac{Z^{3/2}}{n^2\sqrt{\pi}}, \tag{4.237}$$

are chosen so that (see Exercise 4.9)

$$\frac{1}{4\alpha^3} \int_0^\infty d\xi \int_0^\infty d\eta \int_0^{2\pi} d\varphi (\xi + \eta) |\psi_{n_1,n_2,m}|^2 = 1, \tag{4.238}$$

$$\int_0^\infty f_{n_1,|m|}(\xi)^2 d\xi = 1. \tag{4.239}$$

If we fix the principal quantum number $n$ then $|m|$ can have the values

$$|m| = 0, 1, 2, \ldots, n-1. \tag{4.240}$$

If we fix both $n$ and $|m|$ then $n_1$ can have the $n - |m|$ values $0, 1, \ldots, n - |m| - 1$. Since $m$ can have the two values $|m|$ and $-|m|$ if $m \neq 0$ then the total number of states for a given $n$ is

$$n - |0| + 2 \sum_{m=1}^{n-1} (n - m) = n^2,$$

which is in agreement with the standard result for the spherical wavefunctions $\psi_{n,\ell,m}(r,\vartheta,\varphi)$.

### 4.14.3   Algebraic solution: method 1

Using the scaling factor (4.225) the differential equation (4.222) can be expressed as

$$\left[ -\frac{d}{d\xi}\left( \xi \frac{d}{d\xi} \right) + \frac{m^2}{4\xi} + \frac{\xi}{4} - \lambda_1 \right] f(\xi) = 0. \tag{4.241}$$

Make the simple change of variable

$$\xi = 2\rho \tag{4.242}$$

to obtain

$$\left[ -\frac{1}{2}\frac{d}{d\rho}\left( \rho \frac{d}{d\rho} \right) + \frac{m^2}{8\rho} + \frac{\rho}{2} - \lambda_1 \right] F(\rho) = 0, \tag{4.243}$$

or equivalently

$$\left[ -\frac{1}{2}\rho\frac{d^2}{d\rho^2} - \frac{1}{2}\frac{d}{d\rho} + \frac{m^2}{8\rho} + \frac{\rho}{2} - \lambda_1 \right] F(\rho) = 0, \tag{4.244}$$

where

$$F(\rho) = \sqrt{2}\, f(2\rho). \tag{4.245}$$

The $\sqrt{2}$ factor is chosen so that if $f(\xi)$ is normalized to unity according to (4.239) then

$$\int_0^\infty F(\rho)^2 \, d\rho = 1. \tag{4.246}$$

If we now compare (4.244) with the $D$-dimensional hydrogenic realization (4.134) to (4.136) it is clear that (4.244) is a $T_3$ eigenvalue problem

$$(T_3 - q_1)F(\rho) = 0 \tag{4.247}$$

if we choose $a = 1$, $R = \rho$ and

$$D = 2, \quad \tau = \frac{1}{4}(m^2 - 1), \quad q_1 = \lambda_1, \tag{4.248}$$

to obtain the the parabolic hydrogenic realization

$$T_1 = -\frac{1}{2}\left( \rho\frac{d^2}{d\rho^2} + \frac{d}{d\rho} \right) + \frac{m^2}{8\rho} - \frac{\rho}{2}, \tag{4.249}$$

$$T_2 = -i\left( \rho\frac{d}{d\rho} + \frac{1}{2} \right), \tag{4.250}$$

$$T_3 = -\frac{1}{2}\left( \rho\frac{d^2}{d\rho^2} + \frac{d}{d\rho} \right) + \frac{m^2}{8\rho} + \frac{\rho}{2}. \tag{4.251}$$

The lowest $T_3$ eigenvalue is $k+1$ where from (4.59)

$$k = \frac{1}{2}\left[-1 \pm \sqrt{4\tau+1}\right] = \frac{|m|-1}{2}, \qquad (4.252)$$

and the upper sign has been chosen since we require that $k > -1$. Therefore the $T_3$ eigenvalue spectrum is

$$q_1 = \lambda_1 = k+1+n_1 = n_1 + \frac{|m|+1}{2}, \qquad (4.253)$$

in agreement with (4.229). We can use a similar realization for (4.223) in terms of $\eta$ to obtain

$$q_2 = \lambda_2 = k+1+n_2 = n_2 + \frac{|m|+1}{2}, \qquad (4.254)$$

so again the Bohr formula for the energy is easily obtained.

It follows from the results of Section 4.10 that the parabolic so(2,1) generators are hermitian with respect to the scalar product $(D + a - 3 = 0)$

$$\langle F_1|A|F_2\rangle = \int_0^\infty F_1(\rho)^* A F_2(\rho)\, d\rho, \qquad (4.255)$$

in agreement with (4.246) and (4.239).

Finally we can also obtain the explicit form of the eigenfunctions by substituting the values of $k$ and $q_1$ from (4.252) and (4.253) and $G = 1/2$ into (4.15) and (4.151) to obtain

$$F_{n_1,|m|}(\rho) = 2^{1/2} N_{n_1,|m|} e^{-\rho} (2\rho)^{|m|/2} L_{n_1}^{(|m|)}(2\rho), \qquad (4.256)$$

which agrees with (4.234) if we substitute $\rho = \xi/2$ and use (4.245) to obtain $f_{n_1,|m|}(\xi)$.

### 4.14.4  Algebraic solution: method 2

An alternate approach to the algebraic solution in parabolic coordinates, which more closely parallels that used in Section 4.5 for the 3-dimensional hydrogenic atom, is to derive the results of Section 4.14.1 and Section 4.14.2 without the scaling factor (put $\alpha = 1$). Then (4.222) would be

$$\left[-\frac{d}{d\xi}\left(\xi\frac{d}{d\xi}\right) + \frac{m^2}{4\xi} - \frac{E}{2}\xi - \lambda_1\right] f_1(\xi) = 0, \qquad (4.257)$$

and similarly for (4.223) where the separation constants now satisfy $\lambda_1 + \lambda_2 = \mathcal{Z}$ instead of (4.224).

Now define the scaling transformation (compare with (4.65))

$$\xi = 2\gamma X. \qquad (4.258)$$

Multiply (4.257) by $\gamma$ to obtain

$$\left[ -\frac{1}{2}\frac{d}{dX}\left( X\frac{d}{dX} \right) + \frac{m^2}{8X} - \gamma^2 EX - \gamma\lambda_1 \right] F(X) = 0. \qquad (4.259)$$

Now if we choose the scaling parameter such that

$$\gamma^2 E = -\frac{1}{2} \qquad (4.260)$$

and use the so(2,1) realization (4.249) to (4.251) with $\rho$ replaced by $X$ then (4.259) is a $T_3$ eigenvalue problem with eigenvalues

$$q_1 = \gamma\lambda_1 = n_1 + \frac{|m|+1}{2}, \qquad (4.261)$$

and similarly for the $\eta$ equation

$$q_2 = \gamma\lambda_2 = n_2 + \frac{|m|+1}{2}. \qquad (4.262)$$

Since $\lambda_1 + \lambda_2 = \mathcal{Z}$ then $q_1 + q_2 = n = \gamma\mathcal{Z}$ and $E = -\gamma^{-2}/2 = -\mathcal{Z}^2/(2n^2)$.

## 4.15   Exercises

⬨ **Exercise 4.1** Consider a simple many-boson system having only one single boson state with creation operator $a^+$ and annihilation operator $a$ satisfying the commutation relation $[a, a^+] = 1$. Determine the constants $\alpha_j$ and $\beta_j$ such that the linear combinations of bilinear products

$$T_1 = \alpha_1 a^+ a^+ + \beta_1 aa, \; T_2 = \alpha_2 a^+ a^+ + \beta_2 aa, \; T_3 = \alpha_3 a^+ a + \beta_3 aa^+$$

satisfy the so(2,1) commutation relations (4.1) and hence are a realization of the so(2,1) generators. Also determine the eigenvalues of $T^2$ and $T_3$ and the representations (4.18) to (4.22) [LI65].

⬨ **Exercise 4.2** Show that the basic commutation relation (4.16) is valid in $D$-dimensional space.

⬨ **Exercise 4.3** Derive the commutation relation (4.45) using (a) the realization of $P_R$ as a differential operator given in (4.25) and its generalization (4.106) to $D$-dimensions and (b) using only the basic commutation relation (4.27).

⬨ **Exercise 4.4** Derive the commutation relations (4.100) for the $D$-dimensional orbital angular momentum operators defined by (4.98).

◈ **Exercise 4.5** Show that the operator $L^2$ defined in (4.101) commutes with each angular momentum operator (4.98) and is therefore the Casimir operator for so$(n)$.

◈ **Exercise 4.6** Derive the identity (4.103) for $L^2$ in $D$-dimensions.

◈ **Exercise 4.7** Show that the eigenvalues of $L^2$ in $D$-dimensions are given by $\ell(\ell + D - 2)$ using the fact that the angular momentum eigenfunctions for integer values of $\ell$ are homogeneous polynomial solutions of the $D$-dimensional Laplace equation $\nabla^2 u = 0$.

◈ **Exercise 4.8** Show that the radial differential equation for the $D$-dimensional harmonic oscillator (4.116) can be transformed into the radial differential equation for the $D$-dimensional hydrogen atom (4.110) using a change of variable $\rho^a = cr$ and radial function $R(r) = \rho^b P(\rho)$ for some constants $a$, $b$ and $c$ to be determined. Also obtain the energy levels (4.115) of the hydrogen atom from the energy levels (4.121) of the harmonic oscillator [CI77a], [JO80].

◈ **Exercise 4.9** Obtain the two normalization factors (4.236) and (4.237).

# Chapter 5

# Representations and Realizations of so(4)

## 5.1 Introduction

In this chapter we first develop two approaches to the general representation theory of the Lie algebra so(4). In the first approach the results of Chapter 3 for the matrix representation of a vector operator are extended by specifying the commutators $[V_j, V_k]$ which close out the commutation relations among the six components of $J$ and $V$ to give the defining commutation relations of three important Lie algebras: so(4), the Lie algebra of the 4-dimensional rotation group, so(3,1), the Lie algebra of the Lorentz group with three space and one time dimension, and sl(3), the Lie algebra of the 3-dimensional Euclidean group. The advantage of the first approach (Section 5.2) is that the representation theory of all three Lie algebras can be obtained in a unified manner. We are interested only in so(4) so our final results will be presented only for this case. A general discussion of so($n$) and so($p,q$) is given in Appendix B.

The additional commutators $[V_j, V_k]$ provide enough information to obtain explicit values for the undetermined coefficients $a_j$ and $c_j$ appearing in the matrix representation (3.57) to (3.59) of the vector operator $V$. This approach can be called the difference equation approach since the coefficients $a_j$ and $c_j$ are obtained by solving a pair of difference equations.

We then require that $J$ and $V$ be hermitian and obtain the unitary irreducible representations (unirreps) of so(4). They are all finite dimensional and can be labeled by a pair of two real quantum numbers $(j_0, \eta)$ related to the eigenvalues of the two independent Casimir operators. The final results show that under the action of the so(3) generators each unirrep of so(4) reduces to a direct sum of unirreps of so(3)

The second approach to the representation theory of so(4) is presented in Section 5.3. It is based on the observation that a simple linear transformation of the so(4) generators can be made to obtain defining commutation relations which show that so(4) is the direct sum of two so(3) Lie algebras. In other words we have a familiar coupling or addition problem for two angular momenta. For example, the abstract basis vectors for the matrix representation of so(4) can be obtained as direct products of angular momentum states defined in two different spaces corresponding to angular momenta $j_1$ and $j_2$. The coupled states are denoted by $|[j_1, j_2]jm\rangle$ and are obtained from the uncoupled states $|j_1 m_1 j_2 m_2\rangle$ by the well-known real unitary transformation given by the Clebsch-Gordan coefficients.

The coupled states can now be used to construct the matrix unirreps of so(4). Each unirrep is now labeled by the pair of angular momentum quantum numbers $[j_1, j_2]$ which are simply related to the pair $(j_0, \eta)$ arising in the difference equation approach.

Thus angular momentum theory provides a simple approach to the representation theory of so(4) but only at the expense of requiring some of the more advanced tools from angular momentum theory such as the Wigner-Eckart theorem, the Racah calculus, 6-j symbols and 9-j symbols. Since this approach is equivalent to the difference equation approach the derivations in Section 5.3 can be omitted if desired. Only the results will be referenced in later sections.

In Section 5.4 the Laplace-Runge-Lenz vector (LRL vector) is introduced. This vector has a long history and was first introduced in classical mechanics by Laplace in his "*Traite de Méchanique Celeste*" in 1799. He showed that the three components of the LRL vector were constants of motion for the classical Kepler problem (particle subject to a central inverse square force), in addition to the components of orbital angular momentum.

Next we obtain the quantum mechanical analogue of the LRL vector which was first used by Pauli in his important 1926 paper to obtain the energy levels of the hydrogenic atom in a purely algebraic manner with what we would nowadays refer to as "so(4) Lie algebraic methods" although Pauli did not use this terminology.

In Section 5.5 we show that a modified LRL vector and the orbital angular momentum vector can be used to obtain a realization of the so(4) generators which we refer to as Pauli's hydrogenic realization. This realization is energy dependent in the sense that the realization of the modified LRL vector depends explicitly on the energy level $E_n$. This is a disadvantage since we eventually want to merge so(2,1) and so(4) into a bigger Lie algebra so(4,2) in which all states of the hydrogenic atom are contained in one unirrep. However if we apply the scaling transformation introduced in Chapter 4 to the modified LRL vector we do obtain an energy independent realization which we refer to as the scaled hydrogenic realization and which we can successfully merge with so(2,1) in the next chapter.

Finally we obtain the scaled hydrogenic matrix unirreps of so(4) from the general ones derived in Section 5.2 and it follows that the scaled hydrogenic basis functions $|nlm\rangle$ introduced in Chapter 4 provide a matrix representation of both so(2,1) and our scaled hydrogenic realization of so(4).

The derivations of many of the identities and commutation relations used in this chapter require extensive commutator gymnastics and are worked out in the Exercises.

## 5.2  Representations of so(4)

In Chapter 3 we derived the matrix representation of a vector operator and the results in (3.57) to (3.59) involved three undetermined sets of coefficients $a_j$, $c_j$ and $d_j$. They were reduced to just two sets $a_j$ and $c_j$ by requiring that the vector operator be hermitian since in this case it was shown that a $j$-dependent phase change of the basis vectors could be made such that $d_j = c_{j+1}$ (see Exercise 3.6). Such a change of basis can also be made in the general case although the new basis vectors are not necessarily normalized to unity as they are in the hermitian case.

To determine the coefficients $a_j$ and $c_j$ it is necessary to supplement the commutation relations (3.1) and (3.25) with expressions for the commutators of the form $[V_j, V_k]$ in such a way that the six operators $J_k$, $V_k$, $k = 1, 2, 3$ close under commutation to form a Lie algebra. There are three important cases:

$$[V_j, V_k] = \sigma i \epsilon_{jk\ell} J_\ell, \tag{5.1}$$

where $\sigma = 1$ corresponds to the Lie algebra so(4) of the 4-dimensional rotation group, $\sigma = -1$ corresponds to the Lie algebra of the homogeneous Lorentz group [NA64] and $\sigma = 0$ corresponds to the 3-dimensional Euclidean group (with $V$ corresponding to the linear momentum). The representation theory for all three Lie algebras can be obtained in a uniform manner [BO86], [AD88b] although in our applications we are only interested in so(4).

Introducing the operators $J_\pm = J_1 \pm i J_2$ and $V_\pm = V_1 \pm i V_2$, as in Chapter 3, the complete set of commutation relations

$$[J_j, J_k] = i \epsilon_{jk\ell} J_\ell, \tag{5.2}$$
$$[J_j, V_k] = i \epsilon_{jk\ell} V_\ell, \tag{5.3}$$
$$[V_j, V_k] = \sigma i \epsilon_{jk\ell} J_\ell, \quad j, k, \ell = 1, 2, 3, \tag{5.4}$$

can be expressed in the equivalent form (see (3.34))

$$[J_+, J_-] = 2J_3, \quad [J_3, J_\pm] = \pm J_\pm, \tag{5.5}$$
$$[J_+, V_-] = 2V_3, \quad [J_-, V_+] = -2V_3, \tag{5.6}$$
$$[J_3, V_\pm] = \pm V_\pm, \quad [J_\pm, V_3] = \mp V_\pm, \tag{5.7}$$
$$[J_+, V_+] = [J_-, V_-] = [J_3, V_3] = 0, \tag{5.8}$$
$$[V_3, V_\pm] = \pm \sigma J_\pm, \quad [V_+, V_-] = 2\sigma J_3. \tag{5.9}$$

The additional commutation relations (5.9) can now be used to obtain the matrix representations of the three Lie algebras. As in the case of so(2,1) and so(3) we are only interested in the unitary irreducible representations (unirreps) for which the operators $\boldsymbol{J}$ and $\boldsymbol{V}$ are hermitian.

The basic idea is to apply the additional commutation relations (5.9) to the angular momentum state $|jm\rangle$ and use the matrix representation (3.57) to (3.59) of $\boldsymbol{V}$ and the condition (3.62) to obtain difference equations which can be solved for the coefficients $a_j$ and $c_j$. We only need to consider one of the three commutation relations in (5.9) since it follows from the Jacobi identity (rule (5) in Chapter 1) that

$$[V_3, V_\pm] = \tfrac{1}{2}[[V_+, V_-], J_\pm], \tag{5.10}$$

$$[V_+, V_-] = [[V_3, V_+], J_-] = [[V_3, V_-], J_+]. \tag{5.11}$$

Therefore any two of the three commutators in (5.9) can be expressed in terms of the remaining one. For example

$$[V_3, V_-] = \tfrac{1}{2}[[[V_3, V_+], J_-], J_-], \tag{5.12}$$

$$[V_+, V_-] = [[V_3, V_+], J_-], \tag{5.13}$$

and we need only consider the commutator $[V_3, V_+]$ (see Exercise 5.1).

First we write the matrix representation of $\boldsymbol{J}$ and $\boldsymbol{V}$ from (3.19) to (3.21) and (3.57) to (3.59) in the compact form

$$J_3|jm\rangle = m|jm\rangle, \tag{5.14}$$

$$J_+|jm\rangle = \omega_m^j|j, m+1\rangle, \tag{5.15}$$

$$J_-|jm\rangle = \omega_{-m}^j|j, m-1\rangle, \tag{5.16}$$

$$\begin{aligned} V_3|jm\rangle = {} & \alpha_m^j c_j|j-1, m\rangle - m a_j|jm\rangle \\ & + \alpha_m^{j+1} c_{j+1}|j+1, m\rangle, \end{aligned} \tag{5.17}$$

$$\begin{aligned} V_+|jm\rangle = {} & \beta_m^{j-1} c_j|j-1, m+1\rangle - \omega_m^j a_j|j, m+1\rangle \\ & - \gamma_m^{j+1} c_{j+1}|j+1, m+1\rangle, \end{aligned} \tag{5.18}$$

$$\begin{aligned} V_-|jm\rangle = {} & -\beta_{-m}^{j-1} c_j|j-1, m-1\rangle - \omega_{-m}^j a_j|j, m-1\rangle \\ & + \gamma_{-m}^{j+1} c_{j+1}|j+1, m-1\rangle, \end{aligned} \tag{5.19}$$

where we have defined

$$\alpha_m^j = \sqrt{(j-m)(j+m)}, \tag{5.20}$$

$$\beta_m^j = \sqrt{(j-m+1)(j-m)}, \tag{5.21}$$

$$\gamma_m^j = \sqrt{(j+m+1)(j+m)}, \tag{5.22}$$

$$\omega_m^j = \sqrt{(j-m)(j+m+1)}. \tag{5.23}$$

Now apply the commutation relation $[V_3, V_+] = \sigma J_+$ to the state $|jm\rangle$ to obtain after some simplification

$$([V_3, V_+] - \sigma J_+)|jm\rangle = \left(\beta_m^{j-1}\alpha_{m+1}^{j-1} - \alpha_m^j\beta_m^{j-2}\right)c_j c_{j-1}|j-2, m+1\rangle$$

$$+ \left[\left(\alpha_m^j \omega_m^{j-1} - (m+1)\beta_m^{j-1}\right)a_{j-1}\right.$$
$$+ \left(m\beta_m^{j-1} - \omega_m^j \alpha_{m+1}^j\right)a_j\Big]c_j|j-1,m+1\rangle$$
$$+ \left[\omega_m^j a_j^2 - \left(\gamma_m^{j+1}\alpha_{m+1}^{j+1} + \alpha_m^{j+1}\beta_m^j\right)c_{j+1}^2\right.$$
$$+ \left(\beta_m^{j-1}\alpha_{m+1}^j + \alpha_m^j \gamma_m^j\right)c_j^2 - \sigma\omega_m^j\Big]|j,m+1\rangle$$
$$+ \left[-\left(\omega_m^j \alpha_{m+1}^{j+1} + m\gamma_m^{j+1}\right)a_j\right.$$
$$+ \left(\alpha_m^{j+1}\omega_m^{j+1} + (m+1)\gamma_m^{j+1}\right)a_{j+1}\Big]c_{j+1}|j+1,m+1\rangle$$
$$+ \left(\alpha_m^{j+1}\gamma_m^{j+2} - \gamma_m^{j+1}\alpha_{m+1}^{j+2}\right)c_{j+1}c_{j+2}|j+2,m+1\rangle = 0.$$

Substitute (5.20) to (5.23) to obtain

$$([V_3, V_+] - \sigma J_+)|jm\rangle$$
$$= \beta_{m+1}^j\Big[(j-1)a_{j-1} - (j+1)a_j\Big]c_j|j-1,m+1\rangle$$
$$+ \omega_m^j\Big[a_j^2 - (2j+3)c_{j+1}^2 + (2j+1)c_j^2 - \sigma\Big]|j,m+1\rangle$$
$$- \gamma_{m+1}^j\Big[ja_j - (j+2)a_{j+1}\Big]c_{j+1}|j+1,m+1\rangle = 0.$$

The square root factors are non-zero since the states $|jm\rangle$ are defined for $m = -j, -j+1, \ldots, j$. Also the third term is the same as the first with $j$ replaced by $j+1$. Therefore the following pair of difference equations must be satisfied:

$$[ja_j - (j+2)a_{j+1}]c_{j+1} = 0, \tag{5.24}$$
$$a_j^2 - (2j+3)c_{j+1}^2 + (2j-1)c_j^2 = \sigma. \tag{5.25}$$

To solve (5.24) we first note, since $2j$ is a nonnegative integer, that there is a smallest $j$, say $j_0$, such that $j_0 \geq 0$. Therefore $c_{j_0} = 0$, otherwise from (5.19) $|j_0 - 1, m\rangle$ would belong to the representation space, a contradiction. The admissible values of $j$ are $j_0, j_0 + 1, j_0 + 2, \ldots$ and there are two cases to consider:

(I)   $c_{j_0} = 0$, $c_{j_0} \neq 0$ for $j = j_0 + 1, \ldots, j_0 + k - 1$, $c_{j_0+k} = 0$,
(II)   $c_{j_0} = 0$, $c_j \neq 0$ for all $j > j_0$.

Clearly case (I) gives finite dimensional representations with basis vectors having $j = j_0, j_0 + 1, \ldots, j_0 + k - 1$ and case (II) gives infinite dimensional representations. By construction the representations obtained in either case will be irreducible.

The difference equation (5.24) is now equivalent to

$$(j-1)a_{j-1} - (j+1)a_j = 0, \quad j > j_0, \tag{5.26}$$

and can be solved by multiplying it by $j$ and defining $\alpha_j = j(j+1)a_j$ to obtain $\alpha_j - \alpha_{j-1} = 0$. Therefore for all $j > j_0$ we have $\alpha_j = \alpha_{j_0}$, a constant, so the general solution of (5.26) is

$$a_j = \frac{j_0(j_0+1)a_{j_0}}{j(j+1)}, \quad j \geq j_0, \tag{5.27}$$

which can also be expressed in the form

$$a_j = -\frac{j_0\eta}{j(j+1)}, \quad j \geq j_0, \tag{5.28}$$

for some constant $-\eta$. This is also consistent in the special case $j_0 = 0$ since the substitution of $j = 1, 2, \ldots$ into (5.26) gives $a_j = 0$, $j \geq 0$ in agreement with (5.28).

The second difference equation (5.25) is somewhat more complicated to solve. First multiply it by $2j + 1$ and let $\tau_j = (2j-1)(2j+1)c_j^2$ to obtain

$$\begin{aligned}
\tau_{j+1} - \tau_j &= (2j+1)(a_j^2 - \sigma) \\
&= -[(j+1)^2 - j^2]\sigma + [(j+1)^2 - j^2]a_j^2 \\
&= -[(j+1)^2 - j^2]\sigma - j_0^2\eta^2 \left( \frac{1}{(j+1)^2} - \frac{1}{j^2} \right).
\end{aligned}$$

Now sum both sides from $j_0$ to $j-1$ to obtain

$$\sum_{k=j_0}^{j-1} (\tau_{k+1} - \tau_k) = -\sigma \sum_{k=j_0}^{j-1} [(k+1)^2 - k^2] - j_0^2\eta^2 \sum_{k=j_0}^{j-1} \left( \frac{1}{(k+1)^2} - \frac{1}{k^2} \right).$$

In each sum all terms cancel except for $\tau_j - \tau_{j_0}$ so

$$\begin{aligned}
\tau_j - \tau_{j_0} &= -\sigma(j^2 - j_0^2) - j_0^2\eta^2 \left( \frac{1}{j^2} - \frac{1}{j_0^2} \right) \\
&= (j^2 - j_0^2) \left( \frac{\eta^2}{j^2} - \sigma \right).
\end{aligned}$$

Since $c_{j_0} = 0$ then $\tau_{j_0} = 0$ and the solution is

$$c_j^2 = \frac{(j^2 - j_0^2)(\eta^2 - \sigma j^2)}{j^2(4j^2 - 1)}, \quad j \geq j_0. \tag{5.29}$$

We now see that the irreducible representations can be specified with the two labels $j_0$ and $\eta$. Since the basis vectors for the representations now depend on this pair of additional labels we can denote them by $|\gamma j m\rangle = |(j_0, \eta)j m\rangle$ and the representation spaces are given by the span (all linear combinations) of these basis vectors,

$$V_{(j_0,\eta)} = \mathrm{span}\{|(j_0,\eta)j m\rangle \colon j \geq j_0, \ m = -j, \ldots, j\}. \tag{5.30}$$

We can express the fact that our representation spaces are "symmetry adapted" with respect to the so(3) subalgebra by writing the reduction

$$V_{(j_0,\eta)} \xrightarrow[so(3)]{} V_{j_0} \oplus V_{j_0+1} \oplus V_{j_0+2} \oplus \cdots \qquad (5.31)$$

in terms of the so(3) irrep spaces

$$V_j = \mathrm{span}\{|jm\rangle\colon m = -j, -j+1, \ldots, j\}. \qquad (5.32)$$

This indicates that under the action of the generators of the so(3) subalgebra the irreps of so(4), so(3,1), or e(3), as the case may be, reduce to a direct sum of irreps of so(3). Each irrep of so(3) appears once.

To complete the general derivation of the matrix representations we need to obtain the Casimir operators. Recall that they commute with the generators of the Lie algebra. For so(3), $J^2$ is the only Casimir operator and for so(4), so(3,1) or e(3) there are two such independent operators defined by (see Appendix B)

$$\mathcal{C}_1 = \sigma J^2 + V^2, \qquad (5.33)$$

$$\mathcal{C}_2 = \tfrac{1}{2}(\boldsymbol{J}\cdot\boldsymbol{V} + \boldsymbol{V}\cdot\boldsymbol{J}) = \boldsymbol{J}\cdot\boldsymbol{V} = \boldsymbol{V}\cdot\boldsymbol{J}, \qquad (5.34)$$

(see (3.27)) or in terms of $J_\pm$ and $V_\pm$ by

$$\mathcal{C}_1 = \sigma J_+J_- + V_+V_- + V_3^2 + \sigma J_3(J_3 - 2), \qquad (5.35)$$

$$\mathcal{C}_2 = \tfrac{1}{2}(V_+J_- + V_-J_+) + V_3J_3. \qquad (5.36)$$

The Casimir operators must be diagonal in the representation defined by (5.14) to (5.19), (5.28) and (5.29). We can let $\mathcal{C}_1$ and $\mathcal{C}_2$ act on the basis vector $|jm\rangle$ to obtain after some lengthy calculations

$$\begin{aligned}
\mathcal{C}_1|jm\rangle &= \alpha_m^j\big[(j-1)a_{j-1} - (j+1)a_j\big]c_j|j{-}1,m\rangle \\
&\quad + \big[(j+1)^2a_j^2 + (4j^2-1)c_j^2 + \sigma(j^2-1)\big]|jm\rangle \\
&\quad - (j-m+1)\big[a_j^2 + (2j-1)c_j^2 - (2j+3)c_{j+1}^2 - \sigma\big]|jm\rangle \\
&\quad + \alpha_m^{j+1}\big[ja_j - (j+2)a_{j+1}\big]c_{j+1}|j{+}1,m\rangle.
\end{aligned}$$

Since the difference equations (5.24) and (5.25) are satisfied

$$\mathcal{C}_1|jm\rangle = \big[(j+1)^2a_j^2 + (4j^2-1)c_j^2 + \sigma(j^2-1)\big]|jm\rangle, \qquad (5.37)$$

$$\mathcal{C}_2|jm\rangle = -j(j+1)a_j|jm\rangle. \qquad (5.38)$$

Finally, substituting for $a_j$ and $c_j$

$$\mathcal{C}_1|jm\rangle = (\sigma j_0^2 + \eta^2 - \sigma)|jm\rangle, \qquad (5.39)$$

$$\mathcal{C}_2|jm\rangle = j_0\eta|jm\rangle. \qquad (5.40)$$

We can now consider each Lie algebra separately. For so(4) we have $\sigma = 1$ and from (5.29) case (I) for finite dimensional irreps occurs only if $\eta^2 - j^2 = 0$ for some $j > j_0$, so we must have $\eta = \pm(j_0 + k)$ for some positive integer $k$. For so(3,1) we have $\sigma = -1$ and finite dimensional irreps occur only if $j^2 + \eta^2 = 0$ for some $j > j_0$, so we must have $\eta = \pm i(j_0 + k)$ for some positive integer $k$. Finally for $e(3)$ we have $\sigma = 0$ and only the trivial identity irrep, corresponding to $\eta = 0$, is finite dimensional.

As in the so(2,1) case we are only interested in the unirreps so the operators $\boldsymbol{J}$ and $\boldsymbol{V}$ must be hermitian and it follows from (3.60) to (3.62) that both $a_j$ and $c_j$ must be real for unirreps. Therefore $\eta$ must be real in (5.28) and $c_j^2 > 0$ in (5.29).

In the so(3,1) case finite dimensional unirreps are not possible since they would require that $\eta = \pm i(j_0 + k)$ but $\eta$ must be real (the case $j_0 = k = 0$ gives the trivial 1-dimensional identity irrep). Therefore $j$ is unbounded and only the infinite dimensional irreps of case (II) give rise to unirreps. Similarly for $e(3)$ only infinite dimensional nontrivial unirreps are possible since $c_j^2 \neq 0$ unless $\eta = 0$ or $j = j_0$, corresponding to the trivial identity representation.

In the so(4) case all unirreps are finite dimensional since $\eta = \pm(j_0 + k)$ is real and since $c_j^2 > 0$ requires that $j^2 < \eta^2$ so that $j$ is bounded (case (I)). The situation here is analogous to the so(3) and so(2,1) cases for which all unirreps of so(3) are finite dimensional but all unirreps of so(2,1) are infinite dimensional. In our applications we only require matrix representations of so(4) so we consider only this case from now on (see [BO86] and [NA64] for the representation theory of so(3,1) and e(3)).

It is more convenient in the so(4) case to let $j_0$ take on both positive and negative values and assume that $\eta$ is nonnegative. Thus the unirreps of so(4) are labeled by two real numbers $j_0$, where $2|j_0|$ is a nonnegative integer, and $\eta = |j_0| + k$ for some integer $k > 0$. The values of $j$ associated with this unirrep are $j = |j_0|, |j_0| + 1, \ldots, \eta - 1$ and the reduction under the action of the so(3) generators is given by

$$\mathcal{V}_{(j_0,\eta)} \xrightarrow[so(3)]{} \mathcal{V}_{|j_0|} \oplus \mathcal{V}_{|j_0|+1} \oplus \cdots \oplus \mathcal{V}_{\eta-1}. \tag{5.41}$$

The values $\pm|j_0|$ of $j_0$ for $j_0 \neq 0$ correspond to inequivalent unirreps since $a_j$ depends on the sign of $j_0$. If $j_0 = 0$ they are equivalent since $a_j = 0$ for either value.

We can now indicate the explicit dependence on $j_0$ and $\eta$ by denoting the basis vectors by $|\gamma j m\rangle = |(j_0, \eta) j m\rangle$ and writing $a_j^\gamma = a_j$ and $c_j^\gamma = c_j$. Thus we obtain the following final form of the matrix representation of so(4)

$$J^2|\gamma j m\rangle = j(j+1)|\gamma j m\rangle, \tag{5.42}$$

$$C_1|\gamma j m\rangle = \left(j_0^2 + \eta^2 - 1\right)|\gamma j m\rangle, \tag{5.43}$$

$$C_2|\gamma j m\rangle = j_0\eta|\gamma j m\rangle, \tag{5.44}$$

$$J_3|\gamma j m\rangle = m|\gamma j m\rangle, \tag{5.45}$$

$$J_+|\gamma jm\rangle = \omega_m^j|\gamma,j,m{+}1\rangle, \tag{5.46}$$

$$J_-|\gamma jm\rangle = \omega_{-m}^j|\gamma,j,m{-}1\rangle, \tag{5.47}$$

$$V_3|\gamma jm\rangle = \alpha_m^j c_j^\gamma|\gamma,j{-}1,m\rangle - m a_j^\gamma|\gamma jm\rangle$$
$$+ \alpha_m^{j+1} c_{j+1}^\gamma|\gamma,j{+}1,m\rangle, \tag{5.48}$$

$$V_+|\gamma jm\rangle = \beta_m^{j-1} c_j^\gamma|\gamma,j{-}1,m{+}1\rangle - \omega_m^j a_j^\gamma|\gamma,j,m{+}1\rangle$$
$$- \gamma_m^{j+1} c_{j+1}^\gamma|\gamma,j{+}1,m{+}1\rangle, \tag{5.49}$$

$$V_-|\gamma jm\rangle = -\beta_{-m}^{j-1} c_j^\gamma|\gamma,j{-}1,m{-}1\rangle - \omega_{-m}^j a_j^\gamma|\gamma,j,m{-}1\rangle$$
$$+ \gamma_{-m}^{j+1} c_{j+1}^\gamma|\gamma,j{+}1,m{-}1\rangle, \tag{5.50}$$

where

$$a_j^\gamma = a_j^{(j_0,\eta)} = -\frac{j_0\eta}{j(j+1)}, \tag{5.51}$$

$$c_j^\gamma = c_j^{(j_0,\eta)} = \sqrt{\frac{(j^2 - j_0^2)(\eta^2 - j^2)}{j^2(4j^2 - 1)}}, \tag{5.52}$$

$$j = |j_0|, |j_0| + 1, \ldots, \eta - 1, \quad \eta = |j_0| + k, \ k = 1, 2, \ldots. \tag{5.53}$$

The set $\{J^2, J_3, \mathcal{C}_1, \mathcal{C}_2\}$ is a complete set of commuting operators for so(4). This is why the basis vectors are labeled by the four "quantum numbers" $j_0$, $\eta$, $j$ and $m$.

## 5.3  so(4) as a Coupling Problem

We now show how the matrix representations of so(4) can be obtained by viewing so(4) in terms of an addition problem for two angular momenta (sometimes called a coupling problem) and using some of the more advanced aspects of the Racah calculus for reduced matrix elements of coupled spherical tensor operators (see [BI81b], [BR68], and Appendix C of [ME70b]). Since this is an alternate approach to the results of the preceding section an understanding of the present section is not necessary for the remainder of the book (only the results are referenced in later sections).

The so(4) commutation relations (5.2) to (5.4) with $\sigma = 1$ can be transformed to a simpler canonical form. If we define the two vector operators

$$\boldsymbol{M} = \tfrac{1}{2}(\boldsymbol{J} + \boldsymbol{V}), \tag{5.54}$$

$$\boldsymbol{N} = \tfrac{1}{2}(\boldsymbol{J} - \boldsymbol{V}), \tag{5.55}$$

then the components of $\boldsymbol{M}$ and $\boldsymbol{N}$ satisfy the commutation relations

$$[M_j, M_k] = i\epsilon_{jk\ell}M_\ell, \tag{5.56}$$

$$[N_j, N_k] = i\epsilon_{jk\ell}N_\ell, \tag{5.57}$$

$$[M_j, N_k] = 0, \tag{5.58}$$

which are equivalent to the original ones since (5.54) and (5.55) represent a real linear transformation of the original generators. The new commutation relations are simpler since $M$ and $N$ are two commuting angular momentum vectors. Therefore so(4) is the direct sum $so(3) \oplus so(3)$ of two so(3) subalgebras. The first is generated by $M$ and the second is generated by $N$.

This new form of the so(4) commutation relations also shows that there are two Casimir operators for so(4) and that they are just the Casimir operators $M^2$ and $N^2$ of the two so(3) subalgebras. Our previous expressions (5.33) and (5.34) in terms of $J^2$ and $V^2$ are now seen to be just the linear combinations

$$\mathcal{C}_1 = 2(M^2 + N^2), \quad \mathcal{C}_2 = M^2 - N^2. \tag{5.59}$$

If we let $|j_1 m_1\rangle$ and $|j_2 m_2\rangle$ denote the basis vectors for the two so(3) representation spaces then the direct product basis vectors

$$|j_1 m_1 j_2 m_2\rangle = |j_1 m_1\rangle \otimes |j_2 m_2\rangle \tag{5.60}$$

determine a representation space for so(4). Since $M$ acts only in the first space then

$$M^2|j_1 m_1 j_2 m_2\rangle = j_1(j_1 + 1)|j_1 m_1 j_2 m_2\rangle, \tag{5.61}$$

$$M_3|j_1 m_1 j_2 m_2\rangle = m_1|j_1 m_1 j_2 m_2\rangle, \tag{5.62}$$

$$M_\pm|j_1 m_1 j_2 m_2\rangle = \omega_{\pm m_1}^{j_1}|j_1, m_1 \pm 1, j_2 m_2\rangle, \tag{5.63}$$

and similarly since $N$ acts only in the second space

$$N^2|j_1 m_1 j_2 m_2\rangle = j_2(j_2 + 1)|j_1 m_1 j_2 m_2\rangle, \tag{5.64}$$

$$N_3|j_1 m_1 j_2 m_2\rangle = m_2|j_1 m_1 j_2 m_2\rangle, \tag{5.65}$$

$$N_\pm|j_1 m_1 j_2 m_2\rangle = \omega_{\pm m_2}^{j_2}|j_1 m_1 j_2, m_2 \pm 1\rangle. \tag{5.66}$$

Thus the direct product basis vectors are simultaneous eigenfunctions of the commuting set $\{M^2, M_3, N^2, N_3\}$. To be more precise we should write $M$ and $N$ as the direct products $M \otimes I$ and $I \otimes N$ to indicate that $M$ acts as the identity operator on the second so(3) space and $N$ acts as the identity operator on the first so(3) space.

The $(2j_1 + 1)(2j_2 + 1)$ dimensional direct product space

$$\mathcal{V}_{j_1} \otimes \mathcal{V}_{j_2} = \{|j_1 m_1 j_2 m_2\rangle : m_k = -j_k, \ldots, j_k, k = 1, 2\} \tag{5.67}$$

is not irreducible. In fact its decomposition into a direct sum of irreducible so(3) representation spaces is just the familiar addition problem for two angular momenta. It is common to refer to the basis vectors $|j_1 m_1 j_2 m_2\rangle$ as the uncoupled basis vectors. The coupled basis vectors are denoted by $|\gamma j m\rangle = |[j_1, j_2]jm\rangle$ and they are eigenfunctions of $J^2$ and $J_3$ where, from (5.54) and (5.55), $J = M \oplus N$. Therefore

$$J^2|\gamma j m\rangle = j(j + 1)|\gamma j m\rangle, \tag{5.68}$$

$$J_3|\gamma j m\rangle = m|\gamma j m\rangle, \tag{5.69}$$

$$J_\pm|\gamma j m\rangle = \omega_{\pm m}^{j}|\gamma j, m \pm 1\rangle. \tag{5.70}$$

The coupled and uncoupled basis vectors are related by a real unitary transformation defined by the Clebsch-Gordan (CG) coefficients introduced in Chapter 3:

$$|\gamma jm\rangle = \sum_{\substack{m_1,m_2 \\ m_1+m_2=m}} \langle j_1 m_1 j_2 m_2 | jm \rangle \, |j_1 m_1 j_2 m_2\rangle. \tag{5.71}$$

The CG coefficients are nonzero only if $m = m_1 + m_2$ and if $j_1$, $j_2$ and $j$ satisfy the triangle inequality $|j_1 - j_2| \leq j \leq j_1 + j_2$. Conversely

$$|j_1 m_1 j_2 m_2\rangle = \sum_{j=|j_1-j_2|}^{j_1+j_2} \langle j_1 m_1 j_2 m_2 | jm \rangle \, |\gamma jm\rangle, \tag{5.72}$$

which gives the reduction (see (5.4))

$$\mathcal{V}_{[j_1,j_1]} = \mathcal{V}_{j_1} \otimes \mathcal{V}_{j_2} \xrightarrow[so(3)]{} \mathcal{V}_{|j_1-j_2|} \oplus \mathcal{V}_{|j_1-j_2|+1} \oplus \cdots \oplus \mathcal{V}_{j_1+j_2} \tag{5.73}$$

into a direct sum of irreducible so(3) spaces.

Our goal is to calculate matrix elements of $V_+$, $V_-$ and $V_3$ in the coupled basis using

$$V = 2M - J, \tag{5.74}$$

which follows from (5.54). Since the matrix elements of $J_+$, $J_-$ and $J_3$ are given by (5.69) and (5.70) we only need to calculate matrix elements of $M_+$, $M_-$ and $M_3$ in the coupled basis. It is more convenient to use the spherical components $V_q, q = -1, 0, +1$ where $V_\pm = \mp V_\pm/\sqrt{2}$, $V_0 = V_3$, with analogous formulas for the spherical components of $M$ or $J$.

To calculate matrix elements of $V_q$ we first use the Wigner-Eckart theorem (3.67) applied to a vector operator ($k = 1$):

$$\langle \gamma' j' m' | V_q | \gamma jm \rangle = (-1)^{j'-m'} \begin{pmatrix} j' & 1 & j \\ -m' & q & m \end{pmatrix} \langle \gamma' j' \| V \| \gamma j \rangle, \tag{5.75}$$

where the reduced matrix elements are

$$\langle \gamma' j' \| V \| \gamma j \rangle = 2\langle \gamma' j' \| M \| \gamma j \rangle - \langle \gamma' j' \| J \| \gamma j \rangle. \tag{5.76}$$

The reduced matrix elements of $J$ are easily obtained from the Wigner-Eckart theorem and the result is (see Exercise 5.2)

$$\langle \gamma' j' \| J \| \gamma j \rangle = \delta_{\gamma\gamma'} \delta_{j'j} [j(j+1)(2j+1)]^{1/2}. \tag{5.77}$$

To evaluate the reduced matrix elements of $M$ we need a result on the reduced matrix elements of coupled tensor operators.

If $\boldsymbol{T}^{(k_1)}$ and $\boldsymbol{U}^{(k_2)}$ are spherical tensor operators of ranks $k_1$ and $k_2$ respectively then we can define a coupled tensor product operator $[\boldsymbol{T}^{(k_1)} \otimes \boldsymbol{U}^{(k_2)}]^{(k)}$ of rank $k$ with components

$$[\boldsymbol{T}^{(k_1)} \otimes \boldsymbol{U}^{(k_2)}]_q^{(k)} = \sum_{q_1,q_2} \langle k_1 q_1\, k_2 q_2 | kq \rangle T_{q_1}^{(k_1)} U_{q_2}^{(k_2)} \tag{5.78}$$

in the same way that the coupled states are defined by (5.71) (see [ME70b], Appendix C).

From the Wigner-Eckart theorem

$$\langle \gamma' j' m' | [\boldsymbol{T}^{(k_1)} \otimes \boldsymbol{U}^{(k_2)}]_q^{(k)} | \gamma j m \rangle$$
$$= (-1)^{j'-m'} \begin{pmatrix} j' & k & j \\ -m' & q & m \end{pmatrix} \langle \gamma' j' \| [\boldsymbol{T}^{(k_1)} \otimes \boldsymbol{U}^{(k_2)}]^{(k)} \| \gamma j \rangle, \tag{5.79}$$

and the reduced matrix elements are given in terms of the reduced matrix elements of the uncoupled operators and the 9-j symbol by

$$\langle \gamma' j' \| [\boldsymbol{T}^{(k_1)} \otimes \boldsymbol{U}^{(k_2)}]^{(k)} \| \gamma j \rangle = [(2j+1)(2j'+1)(2k+1)]^{1/2}$$
$$\times \begin{Bmatrix} j_1' & j_1 & k_1 \\ j_2' & j_2 & k_2 \\ j' & j & k \end{Bmatrix} \langle j_1' \| \boldsymbol{T}^{(k_1)} \| j_1 \rangle \langle j_2' \| \boldsymbol{U}^{(k_2)} \| j_2 \rangle. \tag{5.80}$$

Since $\boldsymbol{M} = \boldsymbol{M} \otimes \boldsymbol{I}$ we can consider the reduced matrix elements of $\boldsymbol{M}$ in (5.76) as the special case $k_1 = 1$ (vector operator in the first space) and $k_2 = 0$ (identity operator in the second space is a tensor operator of rank 0). The coupled tensor operator must also have rank 1 since $|k_1 - k_2| \leq k \leq k_1 + k_2$ implies that $k = 1$. Therefore the reduced matrix elements of $\boldsymbol{M}$ are given by

$$\langle \gamma' j' \| \boldsymbol{M} \| \gamma j \rangle = [3(2j+1)(2j'+1)]^{1/2}$$
$$\times \begin{Bmatrix} j_1' & j_1 & 1 \\ j_2' & j_2 & 0 \\ j' & j & 1 \end{Bmatrix} \langle j_1' \| \boldsymbol{M} \| j_1 \rangle \langle j_2' \| \boldsymbol{I} \| j_2 \rangle. \tag{5.81}$$

A 9-j symbol with a zero entry reduces to a 6-j symbol:

$$\begin{Bmatrix} j_1' & j_1 & 1 \\ j_2' & j_2 & 0 \\ j' & j & 1 \end{Bmatrix} = (-1)^{j_1+j_2+j+1} [3(2j_2+1)]^{1/2} \begin{Bmatrix} j' & j & 1 \\ j_1 & j_1 & j_2 \end{Bmatrix}. \tag{5.82}$$

The uncoupled reduced matrix elements in (5.81) can easily be calculated from the Wigner-Eckart theorem and the results are (see Exercise 5.2)

$$\langle j_1' \| \boldsymbol{M} \| j_1 \rangle = \delta_{j_1' j_1} [(j_1(j_1+1)(2j_1+1)]^{1/2}, \tag{5.83}$$
$$\langle j_2' \| \boldsymbol{I} \| j_2 \rangle = \delta_{j_2' j_2} (2j_2+1)^{1/2}. \tag{5.84}$$

Collecting these results together the reduced matrix elements of $V$ are given by

$$\langle\gamma'j'\|V\|\gamma j\rangle$$
$$= \delta_{\gamma'\gamma}\Bigg[(-1)^{j_1+j_2+j+1}2[(2j+1)(2j'+1)j_1(j_1+1)(2j_1+1)]^{1/2}$$
$$\times\left\{\begin{array}{ccc} j' & j & 1 \\ j_1 & j_1 & j_2 \end{array}\right\} - \delta_{j'j}[j(j+1)(2j+1)]^{1/2}\Bigg]. \tag{5.85}$$

We need only consider the values $j' = j$ and $j' = j-1$ since it follows from (5.85) and the symmetry properties of the 6-j symbol that

$$\langle\gamma,j+1\|V\|\gamma j\rangle = -\langle\gamma j\|V\|\gamma,j+1\rangle. \tag{5.86}$$

We need the following two formulas obtained from tables of 6-j symbols [BR68], [ED57], [RO59]

$$\left\{\begin{array}{ccc} a & a & 1 \\ b & b & c \end{array}\right\} = (-1)^{s+1}\frac{a(a+1)+b(b+1)-c(c+1)}{[4a(a+1)(2a+1)d]^{1/2}}, \tag{5.87}$$

$$\left\{\begin{array}{ccc} a-1 & a & 1 \\ b & b & c \end{array}\right\} = (-1)^{s}\left[\frac{(s+1)(s-2b)(s-2c)(s-2a+1)}{4a(2a+1)(2a-1)d}\right]^{1/2}, \tag{5.88}$$

where $s = a+b+c$ and $d = b(b+1)(2b+1)$. Substituting into (5.85) and (5.86)

$$\langle\gamma,j-1\|V\|\gamma j\rangle = -\left[\frac{[j^2-(j_1-j_2)^2][(j_1+j_2+1)^2-j^2]}{4j}\right]^{1/2}, \tag{5.89}$$

$$\langle\gamma,j\|V\|\gamma j\rangle = -(j_1+j_2+1)(j_2-j_1)\left[\frac{2j+1}{j(j+1)}\right]^{1/2}. \tag{5.90}$$

These results and (5.86) can be substituted into (5.75) to obtain the matrix elements of $V_q$. In case $q = 0$, $V_0 = V_3$ and

$$\langle\gamma,j-1,m|V_3|\gamma jm\rangle = \sqrt{(j-m)(j+m)}\,c_j^\gamma, \tag{5.91}$$
$$\langle\gamma jm|V_3|\gamma jm\rangle = -ma_j^\gamma, \tag{5.92}$$
$$\langle\gamma,j+1,m|V_3|\gamma jm\rangle = \sqrt{(j-m+1)(j+m+1)}\,c_{j+1}^\gamma, \tag{5.93}$$

where we have defined

$$a_j^\gamma = \frac{(j_1+j_2+1)(j_2-j_1)}{j(j+1)}, \tag{5.94}$$
$$c_j^\gamma = \frac{[j^2-(j_1-j_2)^2][(j_1+j_2+1)^2-j^2]}{j^2(4j^2-1)}. \tag{5.95}$$

These results agree with the ones obtained previously in Section 5.1 (compare with (5.49), (5.51) and (5.52)) if we define

$$j_0 = j_1 - j_2, \quad \eta = j_1 + j_2 + 1. \tag{5.96}$$

Since $j_1$, $j_2$ and $j$ satisfy the triangle inequality $|j_1 - j_2| \leq j \leq j_1 + j_2$ it follows that $|j_0| \leq j \leq \eta - 1$ in agreement with (5.53).

The matrix elements of $V_+ = V_{+1}/\sqrt{2}$ and $V_- = -V_{-1}/\sqrt{2}$ can be obtained in a similar fashion using (5.75) and (5.85). Thus the basis vectors $|\gamma j m\rangle$ of the unirreps of so(4) can be labeled using the pair of labels $\gamma = (j_0, \eta)$ or the pair of labels $\gamma = [j_1, j_2]$. The latter pair is more convenient because of the connection between so(4) and the familiar coupling problem for two angular momenta. The approach used in this section is based on that of Biedenharn [BI61] who uses the equivalent labels $p = j_1 + j_2$ and $q = j_1 - j_2$ and a different definition of the reduced matrix elements.

## 5.4   Laplace-Runge-Lenz Vector

We want to apply our results on the matrix representation of so(4) to the quantum mechanical Kepler problem (hydrogenic atom). The orthogonal group $SO(4)$ is the geometrical symmetry group for the hydrogenic atom in the sense that the constants of motion provide a particular realization of the generators of the Lie algebra so(4). The appropriate realization of the angular momentum vector $\boldsymbol{J}$ is given by the orbital angular momentum $\boldsymbol{L}$ and a realization of the vector operator $\boldsymbol{V}$ is provided by the the Laplace-Runge-Lenz (LRL) vector.

### 5.4.1   Classical case

In order to motivate the introduction of the LRL vector we first discuss the classical mechanical form originally due to Laplace (see [GO80] for interesting historical remarks on the LRL vector; also see [SC68], [SA71] and [GR89]). The three components of the orbital angular momentum $\boldsymbol{L} = \boldsymbol{r} \times \boldsymbol{p}$ are constants of motion for the classical nonrelativistic motion of a particle in a central field described by Newton's second law

$$\frac{d\boldsymbol{p}}{dt} = f(r)\frac{\boldsymbol{r}}{r}. \tag{5.97}$$

Laplace discovered in the special case $f(r) = -k/r^2$, corresponding to the Kepler problem, that the three components of the LRL vector

$$\boldsymbol{U} = \frac{1}{m}(\boldsymbol{p} \times \boldsymbol{L}) - k\frac{\boldsymbol{r}}{r} \tag{5.98}$$

are additional constants of motion (see Exercise 5.3).

The following identities relating $\boldsymbol{L}$, $\boldsymbol{U}$, the orbital parameters and the energy $E$ can also be derived (see Exercise 5.4).

$$\boldsymbol{U} \cdot \boldsymbol{L} = 0, \tag{5.99}$$

$$mU^2 = 2EL^2 + mk^2, \tag{5.100}$$

$$mUr \cos \theta = L^2 - mkr, \tag{5.101}$$

$$U = ke. \tag{5.102}$$

The first pair of identities show that the six constants of motion provided by the six components of $\boldsymbol{L}$ and $\boldsymbol{U}$ are not independent (only four of them are independent).  The second pair of identities give the equation of the conic section describing the orbit where $r$ and $\theta$ are polar coordinates relative to a focus of the conic section and $e$ is the eccentricity of the orbit.  Thus the magnitude of the LRL vector is proportional to the orbit eccentricity and the the direction of the LRL vector is in the plane of the orbit (perpendicular to $\boldsymbol{L}$) and directed from a focus to the perihelion point.  It also follows from these identities that (see Exercise 5.5)

$$L^2 = mka(1 - e^2), \quad E = -\frac{k}{2a}, \tag{5.103}$$

where $a$ is the semi-major axis length.  Thus we obtain the classical orbital equation and energy in a purely algebraic manner without solving any differential equations or performing any integrations.

### 5.4.2  Quantum mechanical case

Pauli discovered in 1926 that this classical algebraic approach also carries over to the quantum mechanical Kepler problem (hydrogenic atom) via the correspondence principle (see [PA26] and the English translation in [VA68]).  In this important paper Pauli uses the LRL vector and some "heavy algebra" to obtain the energy levels of the hydrogen atom in a purely algebraic manner with what we would nowadays refer to as "so(4) Lie algebraic methods".

In fact Pauli's algebraic treatment of the hydrogen atom using matrix mechanics actually preceded Schrödinger's treatment using wave mechanics and differential equations [SC26], which is the approach presented in all textbooks on quantum mechanics, by a few months (see [VA68] for historical commentary[1]).

The early textbooks ignored Pauli's approach and even today the algebraic approach is largely ignored (see however [SC68], [BO86] and [GR89] for example).  On the other hand the 1-dimensional harmonic oscillator, because of its simplicity in both the matrix and wave mechanical approaches, was always used to show the equivalence of the two methods.  Thus the reason that the

---

[1]to quote van der Waerden (page 58) "Pauli's paper convinced most physicists that Quantum Mechanics was correct".

algebraic approach for the hydrogen atom fell into abeyance is largely due to
the familiarity of physicists at the time with differential equations and special
functions and their unfamiliarity with matrix methods and in particular with
Lie algebras such as so(4) and their matrix representations. Of course at the
time there was no accessible literature available on the subject of Lie algebras
and their matrix representations.

In classical mechanics $\boldsymbol{p} \times \boldsymbol{L} = \frac{1}{2}(\boldsymbol{p} \times \boldsymbol{L} - \boldsymbol{L} \times \boldsymbol{p})$ since $\boldsymbol{p} \times \boldsymbol{L} = -\boldsymbol{L} \times \boldsymbol{p}$
but in quantum mechanics this is not true (see (2.36)) so we need to include
both $\boldsymbol{p} \times \boldsymbol{L}$ and $\boldsymbol{L} \times \boldsymbol{p}$ in the quantum mechanical generalization of the classical LRL vector (5.98). This suggests the definition

$$U = \frac{1}{2}(\boldsymbol{p} \times \boldsymbol{L} - \boldsymbol{L} \times \boldsymbol{p}) - \mathcal{Z}\frac{\boldsymbol{r}}{r}, \tag{5.104}$$

where $\mathcal{Z}$ is the nuclear charge and atomic units are used ($\hbar = m = e = 1$).
This symmetrization is also necessary in order to make $U$ hermitian since
$(\boldsymbol{p} \times \boldsymbol{L})^{\dagger} = -\boldsymbol{L}^{\dagger} \times \boldsymbol{p}^{\dagger} = -\boldsymbol{L} \times \boldsymbol{p}$ and similarly $(\boldsymbol{L} \times \boldsymbol{p})^{\dagger} = -\boldsymbol{p} \times \boldsymbol{L}$.

We need to show that $\boldsymbol{L}$ and $U$ are constants of motion for the quantum
mechanical Kepler problem whose hamiltonian is

$$H = \frac{1}{2}p^2 - \frac{\mathcal{Z}}{r}. \tag{5.105}$$

Since the total time derivative of an operator $A$ is given by

$$\frac{dA}{dt} = -i\,[A, H] + \frac{\partial A}{\partial t}, \tag{5.106}$$

it follows that a time independent operator is a constant of motion if it commutes with the hamiltonian: $[A, H] = 0$. Since $\boldsymbol{L}$ commutes with functions of
$r$ and $[\boldsymbol{L}, p^2] = 0$ (see (2.26) and (2.30)) the components of $\boldsymbol{L}$ are constants
of motion.

To show that $[U, H] = 0$ we first express $U$ in a more convenient form
using the identities (2.36) and (2.33):

$$U = \frac{1}{2}(2\boldsymbol{p} \times \boldsymbol{L} - 2i\boldsymbol{p}) - \mathcal{Z}\frac{\boldsymbol{r}}{r}$$

$$= \boldsymbol{p} \times \boldsymbol{L} - i\boldsymbol{p} - \mathcal{Z}\frac{\boldsymbol{r}}{r}$$

$$= \boldsymbol{r}p^2 - \boldsymbol{p}(\boldsymbol{r} \cdot \boldsymbol{p} - i) - i\boldsymbol{p} - \mathcal{Z}\frac{\boldsymbol{r}}{r}$$

$$= \frac{1}{2}\boldsymbol{r}p^2 - \boldsymbol{p}(\boldsymbol{r} \cdot \boldsymbol{p}) + \boldsymbol{r}\left(\frac{1}{2}p^2 - \frac{\mathcal{Z}}{r}\right).$$

Therefore

$$U = \frac{1}{2}\boldsymbol{r}p^2 - \boldsymbol{p}(\boldsymbol{r} \cdot \boldsymbol{p}) + \boldsymbol{r}H. \tag{5.107}$$

Now use this expression to show that $[U, H] = 0$:

$$[U, H] = [\tfrac{1}{2}rp^2, H] - [p(r \cdot p), H] + [rH, H],$$

and evaluate each of these commutators using the commutator identities derived in Chapter 2:

$$\begin{aligned}
[\tfrac{1}{2}rp^2, H] &= \tfrac{1}{4}[rp^2, p^2] - \tfrac{1}{2}\mathcal{Z}[rp^2, r^{-1}] \\
&= \tfrac{1}{4}[r, p^2]p^2 - \tfrac{1}{2}\mathcal{Z}r[p^2, r^{-1}] \\
&= \tfrac{1}{4}(2ip)p^2 - \tfrac{1}{2}\mathcal{Z}r\left(2ir^{-3}(r \cdot p - i) - 2r^{-3}\right) \\
&= \tfrac{1}{2}ipp^2 - \mathcal{Z}ir^{-3}r(r \cdot p),
\end{aligned}$$

where we have used (2.17) and (2.18);

$$\begin{aligned}
[p(r \cdot p), H] &= \tfrac{1}{2}[p(r \cdot p), p^2] - \mathcal{Z}[p(r \cdot p), r^{-1}] \\
&= \tfrac{1}{2}p[r \cdot p, p^2] - \mathcal{Z}p[r \cdot p, r^{-1}] - \mathcal{Z}[p, r^{-1}](r \cdot p) \\
&= \tfrac{1}{2}p(2ip^2) - \mathcal{Z}p(ir^{-1}) - \mathcal{Z}ir^{-3}r(r \cdot p) \\
&= ipp^2 - \mathcal{Z}ipr^{-1} - \mathcal{Z}ir^{-3}r(r \cdot p),
\end{aligned}$$

where we have used (2.19), (2.22) and (2.12);

$$[rH, H] = [r, H]H = \tfrac{1}{2}[r, p^2]H = ipH,$$

where we have used (2.17). Collecting these results

$$[U, H] = -ip\left(\frac{1}{2}p^2 - \frac{\mathcal{Z}}{r}\right) + ipH = 0.$$

Therefore as in the classical case $L$ and $U$ are constants of motion.

Finally, the quantum mechanical generalizations of (5.99) (5.100) are given by

$$U \cdot L = L \cdot U = 0, \tag{5.108}$$
$$U^2 = 2H(L^2 + 1) + \mathcal{Z}^2. \tag{5.109}$$

The derivation of these results requires some "heavy algebra", especially for (5.109)(see Exercise 5.6 to Exercise 5.8).

## 5.5  Hydrogenic Realizations of so(4)

We now use the LRL vector $U$ to obtain two hydrogenic realizations of so(4). One realization, first obtained by Pauli in his important 1926 paper and which we refer to as Pauli's hydrogenic realization, can be used to obtain the Bohr formula for the energy levels of the hydrogenic atom and to explain the so called "accidental degeneracy" of the energy levels. The other realization, obtained by applying the scaling transformation introduced in Chapter 4 (see (4.65) and Appendix C) to $U$, will be important in the merging of so(2,1) and so(4) to obtain the Lie algebra so(4,2).

### 5.5.1  Pauli's hydrogenic realization of so(4)

We need to evaluate commutators involving the components of $L$ and $U$. After a considerable amount of calculation it can be shown that (see Exercise 5.9 and Exercise 5.10)

$$[L_j, L_k] = i\epsilon_{jk\ell}L_\ell, \tag{5.110}$$

$$[L_j, U_k] = i\epsilon_{jk\ell}U_\ell, \tag{5.111}$$

$$[U_j, U_k] = (-2H)i\epsilon_{jk\ell}L_\ell. \tag{5.112}$$

The first two sets of commutation relations show that $U$ is an so(3) vector operator. The appearance of the hydrogenic hamiltonian (5.105) in (5.112) shows that $U$ and $L$ do not close under commutation to form a Lie algebra.

However, if we restrict ourselves to a particular bound state energy level with energy $E_n$ and replace $H$ by $E_n$ in (5.112) then we can define

$$V = \frac{1}{\sqrt{-2E_n}}U = \frac{1}{\sqrt{-2E_n}}\left(\tfrac{1}{2}rp^2 - p(r \cdot p) + rE_n\right), \tag{5.113}$$

so $L$ and $V$ satisfy the so(4) commutation relations ((5.2) to (5.4) with $\sigma = 1$) and provide a realization of the so(4) generators called Pauli's hydrogenic realization. Alternatively we could consider continuum states and replace $H$ by a positive energy and use (5.113) without the minus sign to obtain the so(3,1) commutation relations ((5.2) to (5.4) with $\sigma = -1$).

Substituting (5.113) into (5.108) and (5.109) we obtain the important identities

$$V \cdot L = L \cdot V = 0, \tag{5.114}$$

$$V^2 = -(L^2 + 1) - \frac{\mathscr{Z}^2}{2E_n}, \tag{5.115}$$

which can be used to obtain the Bohr formula for the energy levels as follows. From (5.54) and (5.55), using $L$ in place of $J$ and (5.114), we have

$$M^2 = N^2 = \tfrac{1}{4}(L^2 + V^2). \tag{5.116}$$

Since the eigenvalues of $M^2$ and $N^2$ are $j_1(j_1 + 1)$ and $j_2(j_2 + 1)$, respectively, it follows that $j_1 = j_2$. Therefore our realization of $V$ using the modified LRL vector does not give all irreducible representations of so(4); only the so called diagonal ones with $\gamma = [j_1, j_1]$. This is analogous to the use of $L$ as a realization of the so(3) generators. Only integral values of the angular momentum are realized. The values $j = 1/2, 3/2, \dots$ are not possible with the realization $L = r \times p$ in terms of coordinates and momenta.

From (5.115)

$$L^2 + V^2 = -1 - \frac{\mathscr{Z}^2}{2E_n},$$

or substituting (5.116)

$$4M^2 = -1 - \frac{\mathcal{Z}^2}{2E_n}.$$

Replacing $M^2$ by its eigenvalue $j_1(j_1 + 1)$ and solving for $E_n$ we obtain the Bohr formula

$$E_n = -\frac{\mathcal{Z}^2}{2(2j_1 + 1)^2} \tag{5.117}$$

for the energy levels of the hydrogenic atom. Comparing with the usual formula $E_n = -\mathcal{Z}^2/(2n^2)$ we see that the principal quantum number $n$ is given by

$$n = 2j_1 + 1. \tag{5.118}$$

We can also relate the principal quantum number to our original labeling scheme for the unirreps of so(4) given by $\gamma = (j_0, \eta)$ and the Casimir operators $C_1$ and $C_2$ (see Section 5.2). From (5.96)

$$j_0 = 0, \quad \eta = 2j_1 + 1 = n, \tag{5.119}$$

and the coefficients (5.51) and (5.52) defining the matrix representations are given by

$$a_\ell^\gamma = a_\ell^n = 0, \tag{5.120}$$

$$c_\ell^\gamma = c_\ell^n = \sqrt{\frac{n^2 - \ell^2}{4\ell^2 - 1}}, \quad c_0^n = 0, \tag{5.121}$$

$$\ell = 0, 1, \ldots, n - 1, \quad n = 1, 2, 3, \ldots, \tag{5.122}$$

using $\ell$ in place of $j$ to indicate that $\boldsymbol{L}$ is now being used as the realization of $\boldsymbol{J}$.

We also see that there is nothing "accidental" about the so called "accidental degeneracy" of the energy levels. This term arose because the degeneracy of the $2\ell + 1$ hydrogenic states $|n\ell m\rangle$ for fixed $n$ and $\ell$, corresponding to energy $E_n$, could be explained since $[\boldsymbol{L}, H] = 0$, but the full degeneracy of the energy level, $n^2$, could not be explained on the basis of so(3) alone. However, it is so(4) which is the geometrical symmetry group, not so(3). In fact it follows from (5.67) with $j_1 = j_2$ that the degeneracy of $E_n$ is $n^2$ since there are $n^2 = (2j_1 + 1)^2$ basis vectors in the direct product space corresponding to energy $E_n$.

The single quantum number $n$ now suffices to label the unirreps of the hydrogenic realization of so(4) and all states corresponding to a given energy level belong to a single unirrep. The states can be denoted either by

$$|n\ell m\rangle = |(0, n)\ell m\rangle, \quad \ell = 0, 1, \ldots, n - 1, \tag{5.123}$$

or by

$$|n\ell m\rangle = |[j_1, j_2]\ell m\rangle, \quad \ell = 0, 1, \ldots, 2j_1. \tag{5.124}$$

Finally, we should note that if we are only interested in using so(4) to obtain the Bohr formula and to explain the energy level degeneracy then it is not necessary to obtain the complete matrix form of the unirreps as we did in Section 5.2 and Section 5.3.

### 5.5.2   Scaled hydrogenic realization of so(4)

The main problem with Pauli's hydrogenic realization is that it is energy dependent and not suitable for merging with so(2,1) to form a larger Lie algebra. The explicit dependence of the modified LRL vector in (5.113) on the energy $E_n$ means that this realization is restricted to a single energy level at a time so it is not possible to merge it with the so(2,1) realization considered in Chapter 4 which contains raising and lowering operators $T_\pm$ (see (4.18) to (4.21) with $k$ replaced by $\ell$ and $q$ replaced by $n$ and Section 4.5) that change the principal quantum number $n$ and hence the energy level. Since so(4) provides raising and lowering operators for the $\ell$ and $m$ quantum numbers (using $L_\pm$ and $V_\pm$) this merging will produce a Lie algebra with raising and lowering operators for all three quantum numbers.

A suitable realization of so(4) can be obtained by applying the scaling transformation introduced in Chapter 4 to $\boldsymbol{L}$ and $\boldsymbol{V}$. Thus generalizing (4.65) to scale $\boldsymbol{r}$ and $\boldsymbol{p}$ we can again introduce the set of model space operators $\{\boldsymbol{R}, \boldsymbol{P}\}$ related to the set of physical operators $\{\boldsymbol{r}, \boldsymbol{p}\}$ by the scaling transformation

$$\boldsymbol{r} = \gamma \boldsymbol{R}, \quad \boldsymbol{p} = \frac{1}{\gamma}\boldsymbol{P}, \quad r = \gamma R. \tag{5.125}$$

For the hydrogenic atom the scaling parameter L and the energy $E_n$ are related by (4.69) so

$$\gamma = \frac{1}{\sqrt{-2E_n}}. \tag{5.126}$$

Applying (5.125) to the modified LRL vector (5.113) we obtain the scaled LRL vector

$$\boldsymbol{A} = \gamma \left( \tfrac{1}{2}\gamma^{-1}\boldsymbol{R}P^2 - \gamma^{-1}\boldsymbol{P}(\boldsymbol{R}\cdot\boldsymbol{P}) + \gamma \boldsymbol{R}(-\tfrac{1}{2}\gamma^{-2}) \right),$$

which no longer depends on the particular energy level.

Since $\boldsymbol{L}$ is invariant under scaling we obtain the following realization of the so(4) generators

$$\boldsymbol{L} = \boldsymbol{R} \times \boldsymbol{P}, \tag{5.127}$$

$$\boldsymbol{A} = \tfrac{1}{2}\boldsymbol{R}P^2 - \boldsymbol{P}(\boldsymbol{R}\cdot\boldsymbol{P}) - \tfrac{1}{2}\boldsymbol{R}, \tag{5.128}$$

which satisfy the so(4) commutation relations

$$[L_j, L_k] = i\epsilon_{jk\ell}L_\ell, \tag{5.129}$$
$$[L_j, A_k] = i\epsilon_{jk\ell}A_\ell, \tag{5.130}$$
$$[A_j, A_k] = i\epsilon_{jk\ell}L_\ell. \tag{5.131}$$

The same results are also obtained from the active viewpoint using the tilting transformation (see Appendix C).

To complete the scaled realization generated by $\boldsymbol{L}$ and $\boldsymbol{A}$ we need to obtain the identities corresponding to (5.114) and (5.115). The scaled version of (5.115) gives the analogous result

$$\boldsymbol{A} \cdot \boldsymbol{L} = \boldsymbol{L} \cdot \boldsymbol{A} = 0. \tag{5.132}$$

The scaled version of (5.115) is obtained by substituting the Bohr formula $E_n = -\mathcal{Z}^2/(2n^2)$ to obtain

$$A^2 + L^2 + 1 = n^2. \tag{5.133}$$

This result suggests that the correct energy independent relationship is given in terms of the $T_3$ generator of so(2,1) whose eigenvalues are just the principal quantum numbers $n$ (see (4.78), with $q$ replaced by $n$, and (4.74)). In fact with the hydrogenic so(2,1) realization (4.66) to (4.68) with $\tau = L^2$

$$T_1 = \frac{1}{2}\left(RP_R^2 + \frac{L^2}{R} - R\right) = \frac{1}{2}\left(RP^2 - R\right), \tag{5.134}$$

$$T_2 = RP_R = \boldsymbol{R} \cdot \boldsymbol{P} - i, \tag{5.135}$$

$$T_3 = \frac{1}{2}\left(RP_R^2 + \frac{L^2}{R} + R\right) = \frac{1}{2}\left(RP^2 + R\right), \tag{5.136}$$

it can be shown that (see Exercise 5.11 and Exercise 5.12)

$$L^2 = T_3^2 - T_1^2 - T_2^2 = T^2, \tag{5.137}$$
$$A^2 + L^2 + 1 = T_3^2, \tag{5.138}$$

so (5.138) is the appropriate scaled version of (5.115). Therefore we obtain a scaled hydrogenic realization of so(4) given by the generators (5.127) and (5.128) satisfying the special identities (5.132) and (5.138).

It is interesting to observe the close connection between the set $\{H, \boldsymbol{L}, \boldsymbol{V}\}$ and the set $\{T_3, \boldsymbol{L}, \boldsymbol{A}\}$ obtained from it via the scaling transformation. The so(4) Lie algebra generated by the original set $\{\boldsymbol{L}, \boldsymbol{V}\}$ is the geometrical symmetry algebra (also referred to as the dynamical invariance algebra) of the hydrogenic hamiltonian $H$ since $\boldsymbol{L}$ and $\boldsymbol{V}$ are constants of motion

$$[L_j, H] = [V_j, H] = 0. \tag{5.139}$$

On the other hand the scaled set $\{\boldsymbol{L}, \boldsymbol{A}\}$ is the geometrical symmetry algebra of the $T_3$ operator, which is just the scaled version of $rH$ (see Section 4.5). This suggests that

$$[L_j, T_3] = [A_j, T_3] = 0, \tag{5.140}$$

which can be verified directly from the definitions (see Exercise 5.13).

We can now obtain the matrix representations of so(4) for the scaled hydrogenic realization from the general results (5.42) to (5.53) of Section 5.2. The Casimir operators are given by

$$\mathcal{C}_1 = L^2 + A^2 = T_3^2 - 1, \tag{5.141}$$
$$\mathcal{C}_2 = \boldsymbol{A} \cdot \boldsymbol{L} = 0. \tag{5.142}$$

Using the notation (5.123) or (5.124) for the scaled hydrogenic states $|n\ell m\rangle$ and the values (5.120) to (5.122) for the coefficients $a_\ell^n$ and $c_\ell^n$ we obtain

$$L^2|n\ell m\rangle = \ell(\ell+1)|n\ell m\rangle, \tag{5.143}$$
$$\mathcal{C}_1|n\ell m\rangle = (n^2 - 1)|n\ell m\rangle, \tag{5.144}$$
$$\mathcal{C}_2|n\ell m\rangle = 0, \tag{5.145}$$
$$L_3|n\ell m\rangle = m|n\ell m\rangle, \tag{5.146}$$
$$L_+|n\ell m\rangle = \omega_m^\ell|n\ell, m{+}1\rangle, \tag{5.147}$$
$$L_-|n\ell m\rangle = \omega_{-m}^\ell|n\ell, m{-}1\rangle, \tag{5.148}$$
$$A_3|n\ell m\rangle = \alpha_m^\ell c_\ell^n|n, \ell{-}1, m\rangle + \alpha_m^{\ell+1} c_{\ell+1}^n|n, \ell{+}1, m\rangle, \tag{5.149}$$
$$A_+|n\ell m\rangle = \beta_m^{\ell-1} c_\ell^n|n, \ell{-}1, m{+}1\rangle - \gamma_m^{\ell+1} c_{\ell+1}^n|n, \ell{+}1, m{+}1\rangle, \tag{5.150}$$
$$A_-|n\ell m\rangle = -\beta_{-m}^{\ell-1} c_\ell^n|n, \ell{-}1, m{-}1\rangle + \gamma_{-m}^{\ell+1} c_{\ell+1}^n|n, \ell{+}1, m{-}1\rangle. \tag{5.151}$$

An important property of the scaled hydrogenic basis vectors $|n\ell m\rangle$ is that they also provide a matrix unirrep of so(2,1) since from (4.18) to (4.20) and the results of Section 4.5 (with $k = \ell$ and $q = n$)

$$T^2|n\ell m\rangle = \ell(\ell+1)|n\ell m\rangle, \tag{5.152}$$
$$T_3|n\ell m\rangle = n|n\ell m\rangle, \tag{5.153}$$
$$T_+|n\ell m\rangle = \omega_\ell^n|n{+}1, \ell m\rangle, \tag{5.154}$$
$$T_-|n\ell m\rangle = \omega_\ell^{-n}|n{-}1, \ell m\rangle. \tag{5.155}$$

Each unirrep is labeled by the values of $\ell$ and $m$. This strongly suggests that we will be able to merge so(2,1) and so(4) to obtain a larger Lie algebra.

## 5.6　Exercises

◈ **Exercise 5.1** Derive commutation relations (5.10) for a vector operator $\boldsymbol{V}$.

◈ **Exercise 5.2** Derive the formula (5.77) for the reduced matrix elements of $J$ and the formulas (5.83) and (5.84) for the reduced matrix elements of $M$ and $I$.

◈ **Exercise 5.3** Show that the classical angular momentum $L$ and LRL vector $U$ defined in (5.98) are constants of motion for the classical Kepler problem defined by (5.97) with $f(r) = -k/r^2$.

◈ **Exercise 5.4** Derive identities (5.99) to (5.102) relating $L$, $U$ and the orbital parameters.

◈ **Exercise 5.5** Derive identities (5.103) defining $L^2$ and the energy $E$ in terms of orbital parameters.

◈ **Exercise 5.6** Derive the quantum mechanical identity (5.108) using the commutation relations and identities of Chapter 2.

◈ **Exercise 5.7** Derive the following identities which are needed in the next exercise to derive (5.109)

$$
\begin{aligned}
&\text{(a)} \quad (\boldsymbol{p} \times \boldsymbol{L}) \cdot (\boldsymbol{p} \times \boldsymbol{L}) = p^2 L^2 \\
&\text{(b)} \quad (\boldsymbol{p} \times \boldsymbol{L}) \cdot \boldsymbol{p} = 2ip^2 \\
&\text{(c)} \quad \boldsymbol{p} \cdot (\boldsymbol{p} \times \boldsymbol{L}) = 0 \\
&\text{(d)} \quad (\boldsymbol{p} \times \boldsymbol{L}) \cdot \boldsymbol{r} = L^2 + 2i\boldsymbol{p} \cdot \boldsymbol{r} \\
&\text{(e)} \quad \boldsymbol{r} \cdot (\boldsymbol{p} \times \boldsymbol{L}) = L^2
\end{aligned}
$$

◈ **Exercise 5.8** Derive the identity (5.109) for the square of the LRL vector using the identities of the preceding exercise.

◈ **Exercise 5.9** Derive the commutation relations

$$
[L_j, U_k] = i\epsilon_{jk\ell} U_\ell
$$

which show that the LRL vector $U$ is an so(3) vector operator.

◈ **Exercise 5.10** Derive the commutation relations

$$
[U_j, U_k] = (-2H)i\epsilon_{jk\ell} J_\ell
$$

for the components of the LRL vector.

◈ **Exercise 5.11** Derive the identity (5.137):

$$
L^2 = T_3^2 - T_1^2 - T_2^2
$$

◈ **Exercise 5.12** Define the operators

$$C_\alpha = \tfrac{1}{2}RP^2 - P(R \cdot P) - \tfrac{1}{2}\alpha R$$

and show that

$$C_\alpha \cdot C_\beta = \tfrac{1}{4}(RP^2 - \alpha R)(RP^2 - \beta R) + \tfrac{1}{2}(\alpha + \beta)(T_2^2 - 1)$$
$$+ i(\alpha - \beta)T_2$$

Use this result to derive (5.138):

$$A^2 + L^2 + 1 = T_3^2$$

◈ **Exercise 5.13** Show that $[L_j, T_3] = 0$ and $[A_j, T_3] = 0$.

# Chapter 6

# Scaled Hydrogenic Realization of so(4,2)

## 6.1 Introduction

In this chapter we complete our study of Lie algebras by merging so(2,1) and so(4) to obtain the Lie algebras so(4,1) and so(4,2). The origin and basic properties of the Lie algebras so($p,q$) and the corresponding Lie groups SO($p,q$) are given in Appendix B.

First we merge so(4) with the so(2,1) generator $T_2$ which generates the scaling transformation (see Appendix B) and gives the scaled hydrogenic realization of the ten generators of so(4,1). We find that a new vector operator $\boldsymbol{B}$, similar to the scaled LRL vector $\boldsymbol{A}$, is required to close out the commutation relations. The importance of so(4,1) is that all scaled hydrogenic states form a basis for a single infinite dimensional unirrep of so(4,1).

Next we include the remaining so(2,1) generators $T_1$ and $T_3$. Again the so(4,1) generators and these two operators do not close under commutation and a new vector operator $\boldsymbol{\Gamma}$ is needed to close out the commutation relations. Thus we finally obtain the scaled hydrogenic realization of the fifteen generators of so(4,2) and the merging process is complete. The scaled hydrogenic states are also a basis for an infinite dimensional unirrep of so(4,2) [BE73].

The importance of so(4,2) for hydrogenic perturbation theory is that we have the important identities $\boldsymbol{r} = \boldsymbol{B} - \boldsymbol{A}$ and $r = T_3 - T_1$, giving simple expressions for the coordinates $x_j$ and radial distance in terms of the generators of so(4,2), so that any perturbation potential can be expressed in terms of the so(4,2) generators. Even more important is the fact that the scaled hydrogenic realization is a complete discrete basis (no continuum states) for hydrogenic perturbation theory based on the unperturbed hydrogenic hamiltonian (see Chapter 4).

The general representation theory of so(4,1) and so(4,2) is rather complex and is not considered here. Instead we find that our scaled hydrogenic realizations correspond to important special cases which have been discussed elsewhere so it is not difficult in our case to verify, by evaluating the Casimir operators, that we are really dealing with a single infinite dimensional unirrep of both so(4,1) and so(4,2).

Finally we work out and summarize the complete matrix representation of the fifteen generators of the scaled hydrogenic realization and show that any state $|n\ell m\rangle$ can be obtained from the ground state $|100\rangle$ by the appropriate application of the raising and lowering operators $A_+$ and $L_-$, derived from so(4), and from $T_+$.

## 6.2 Scaled Hydrogenic Realization of so(4,1)

In order to merge so(4) and so(2,1) to obtain a bigger Lie algebra we first merge so(4) with the so(2,1) generator $T_2$, which generates the active version of the scaling transformation (see Appendix C) from the physical space operators $\{r, p\}$ to the model space operators $\{R, P\}$.

Our realization of the so(2,1) generators acts on the space of scaled hydrogenic wavefunctions whose basis vectors are products of a radial function and a spherical harmonic function (see (4.80) and (4.81)). Since the so(2,1) generators act only on the radial part and the so(3) generators $L_j$ act only on the angular part, we expect in particular that $[L_j, T_2] = 0$. This result can in fact be verified directly (see Exercise 6.1).

Next we must evaluate commutators of the form $[T_2, A]$. It follows that the operator set $\{T_2, A, L\}$ does not close under commutation. In fact (see Exercise 6.2)

$$[T_2, A] = iB, \tag{6.1}$$

where $B$ is a new vector operator defined by

$$B = \tfrac{1}{2}RP^2 - P(R \cdot P) + \tfrac{1}{2}R. \tag{6.2}$$

The expression for $B$ is closely related to that of the scaled LRL vector $A$ defined in (5.128), differing only in the sign of the final term. Therefore we must include $B$ to obtain the operator set $\{T_2, L, A, B\}$.

To check for closure we now need to evaluate commutators of $B$ with itself and the other generators. We find that these ten operators close under commutation (see Exercise 6.3) to give the Lie algebra so(4,1) with defining commutation relations

$$[L_j, L_k] = i\epsilon_{jk\ell}L_\ell, \tag{6.3}$$

$$[L_j, A_k] = i\epsilon_{jk\ell}A_\ell, \tag{6.4}$$

$$[A_j, A_k] = i\epsilon_{jk\ell}L_\ell, \tag{6.5}$$

which are just the commutation relations of the so(4) subalgebra and

$$[L_j, B_k] = i\epsilon_{jk\ell}B_\ell, \tag{6.6}$$

$$[B_j, B_k] = -i\epsilon_{jk\ell}L_\ell, \tag{6.7}$$

$$[A_j, B_k] = i\delta_{jk}T_2, \tag{6.8}$$

$$[T_2, L_j] = 0, \tag{6.9}$$

$$[T_2, A_j] = iB_j, \tag{6.10}$$

$$[T_2, B_j] = iA_j. \tag{6.11}$$

Commutation relations (6.6) show that $\boldsymbol{B}$ is an so(3) vector operator and (6.3), (6.6), (6.7) also show that $\boldsymbol{L}$ and $\boldsymbol{B}$ generate an so(3,1) subalgebra (compare with (5.2) to (5.4) with $\sigma = -1$).

One of the most important aspects of this hydrogenic realization of so(4,1) is that the coordinate vector $\boldsymbol{R}$ is given by the simple expression

$$\boldsymbol{R} = \boldsymbol{B} - \boldsymbol{A}. \tag{6.12}$$

The simplicity of this expression will be of great advantage in our applications of algebraic perturbation theory in later chapters.

The general representation theory of the Lie algebra so(4,1) and the classification of the unirreps is more complicated than that of either so(2,1) or so(4) [TH41], [NE50], [DI61]. In fact there are 9 distinct classes of unirreps [KI65], [ST65]. However for our particular realization of the so(4) subalgebra generated by $\boldsymbol{L}$ and $\boldsymbol{A}$ there are only two possibilities [BO66] and we shall show that only one of them is actually obtained. Thus, as expected, we are dealing with a very special realization of the so(4,1) generators.

We need two important results from the representation theory of so(4,1). The first result is that the unirreps of so(4,1) are labeled by the eigenvalues of the two independent Casimir operators (Appendix B.8)

$$Q = T_2^2 + B^2 - A^2 - L^2, \tag{6.13}$$

$$W = D^2 - (\boldsymbol{L} \cdot \boldsymbol{A})^2 + (\boldsymbol{L} \cdot \boldsymbol{B})^2, \tag{6.14}$$

where

$$\boldsymbol{D} = T_2\boldsymbol{L} - \boldsymbol{A} \times \boldsymbol{B}. \tag{6.15}$$

It also follows from the commutation relations that (see Exercise 6.5)

$$\boldsymbol{L} \cdot \boldsymbol{A} = \boldsymbol{A} \cdot \boldsymbol{L}, \quad \boldsymbol{B} \cdot \boldsymbol{L} = \boldsymbol{L} \cdot \boldsymbol{B}. \tag{6.16}$$

It can be shown by direct calculation (see Exercise 6.4) that for our scaled hydrogenic realization of so(4,1)

$$W = 0. \tag{6.17}$$

This important result can also be obtained from the identities (see Exercise 6.5)

$$B(\boldsymbol{L} \cdot \boldsymbol{A}) - (\boldsymbol{L} \cdot \boldsymbol{A})B = -i\boldsymbol{D}, \tag{6.18}$$

$$A(\boldsymbol{L} \cdot \boldsymbol{B}) - (\boldsymbol{L} \cdot \boldsymbol{B})A = i\boldsymbol{D}, \tag{6.19}$$

$$\boldsymbol{A} \cdot \boldsymbol{D} - \boldsymbol{D} \cdot \boldsymbol{A} = -3i(\boldsymbol{L} \cdot \boldsymbol{B}), \tag{6.20}$$

$$\boldsymbol{B} \cdot \boldsymbol{D} - \boldsymbol{D} \cdot \boldsymbol{B} = -3i(\boldsymbol{L} \cdot \boldsymbol{A}), \tag{6.21}$$

which can be derived using only the so(4,1) commutation relations (6.3) to (6.11) and are therefore independent of our particular realization. These identities show that if any one of $\boldsymbol{L} \cdot \boldsymbol{A}$, $\boldsymbol{L} \cdot \boldsymbol{B}$ and $\boldsymbol{D}$ are zero then the other two are also zero. Since $\boldsymbol{L} \cdot \boldsymbol{A} = 0$ for our scaled hydrogenic realization then $\boldsymbol{L} \cdot \boldsymbol{B} = 0$, $\boldsymbol{D} = 0$ and from (6.14) $W = 0$.

The second result which we need from the representation theory of so(4,1) is that for the special case $W = 0$ there are only two classes of unirreps, both infinite dimensional, adapted to the subalgebra decomposition $so(4,1) \supset so(4) \supset so(3)$. They are labeled by the value of the second order Casimir operator $Q$ and are denoted by $\mathcal{D}^I$ and $\mathcal{D}^{II}$. The unirreps of class $I$ are defined by

$$Q \geq 0, \quad W = 0, \tag{6.22}$$
$$\mathcal{D}^I \xrightarrow[so(4)]{} \mathcal{V}_{[0,0]} \oplus \mathcal{V}_{[1/2,1/2]} \oplus \mathcal{V}_{[1,1]} \oplus \cdots,$$

where $\mathcal{V}_{[j_1,j_2]}$ is an irreducible representation space of so(4) (see (5.73) or equivalently (5.41)). The unirreps of class $II$ are defined by

$$Q = -(s-1)(s+2), \quad s = 1,2,3,\ldots, \quad W = 0, \tag{6.23}$$
$$\mathcal{D}^{II} \xrightarrow[so(4)]{} \mathcal{V}_{[s/2,s/2]} \oplus \mathcal{V}_{[(s+1)/2,(s+1)/2]} \oplus \mathcal{V}_{[(s+2)/2,(s+2)/2]} \oplus \cdots.$$

To determine which class we are dealing with for the scaled hydrogenic realization it is only necessary to evaluate the Casimir operator $Q$. The result is (see Exercise 6.6)

$$Q = 2 \tag{6.24}$$

and since $n = 2j_1 + 1$ is the principal quantum number (see (5.118)) we obtain the important conclusion that the entire infinite set of scaled hydrogenic states $|n\ell m\rangle = |[j_1,j_1]\ell m\rangle$ belongs to a single unirrep of so(4,1) and the reduction under the action of the so(4) subalgebra is given by (6.22) where $[0,0]$ corresponds to $n = 1$, $[1/2,1/2]$ corresponds to $n = 2$ and so on.

## 6.3    Scaled Hydrogenic Realization of so(4,2)

We need to include the remaining two so(2,1) generators $T_1$ and $T_3$ along with the existing ten generators of so(4,1) in order to have the simple expression $r = T_3 - T_1$ for $r$ in terms of the generators of our final Lie algebra. This expression will be most important for our later applications to algebraic perturbation theory based on the scaled hydrogenic hamiltonian. Another reason for including $T_1$ and $T_3$ is to obtain the raising and lowering operators $T_\pm = T_1 \pm iT_2$ for the principal quantum number $n$ (see (5.154) and (5.155)) so that we will have raising and lowering operators for all three quantum numbers $n$, $\ell$ and $m$. Then we can express matrix elements of any perturbation between any two scaled hydrogenic states in terms of the matrix elements of our Lie algebra generators.

Thus the final step in our merging process is to evaluate commutators among the ten so(4,1) generators $\{L, A, B, T_2\}$ and the two so(2,1) operators $T_1$ and $T_3$ to determine if the twelve operators close under commutation. It is shown in Exercise 5.13 and Exercise 6.1 that the so(2,1) generators commute with the so(3) generators:

$$[T_j, L_k] = 0. \tag{6.25}$$

The commutators of $T_1$ and $T_3$ with the vector operators do not close under commutation. Instead

$$[T_1, A_j] = i\Gamma_j, \tag{6.26}$$
$$[T_3, A_j] = 0, \tag{6.27}$$
$$[T_1, B_j] = 0, \tag{6.28}$$
$$[T_3, B_j] = -i\Gamma_j, \tag{6.29}$$

where

$$\boldsymbol{\Gamma} = R\boldsymbol{P} \tag{6.30}$$

is a new vector operator (see Exercise 6.7).

Therefore $\boldsymbol{\Gamma}$ must be included in our set of operators and it remains to determine if commutators of $\boldsymbol{\Gamma}$ with $\{L, A, B, T_1, T_2, T_3\}$ are closed. This is indeed the case and the results are (see Exercise 6.8)

$$[L_j, \Gamma_k] = i\epsilon_{jk\ell}\Gamma_\ell, \tag{6.31}$$

which shows that $\boldsymbol{\Gamma}$ is an so(3) vector operator and

$$[\Gamma_j, A_k] = -i\delta_{jk}T_1, \tag{6.32}$$
$$[\Gamma_j, B_k] = -i\delta_{jk}T_3, \tag{6.33}$$
$$[\Gamma_j, T_1] = -iA_j, \tag{6.34}$$
$$[\Gamma_j, T_2] = 0, \tag{6.35}$$
$$[\Gamma_j, T_3] = -iB_j, \tag{6.36}$$
$$[\Gamma_j, \Gamma_k] = -i\epsilon_{jk\ell}L_\ell. \tag{6.37}$$

The set of fifteen operators $\{L, A, B, \Gamma, T_1, T_2, T_3\}$ closes under commutation to form the Lie algebra so(4,2). This realization, which we summarize below, is called the scaled hydrogenic realization:

$$\boldsymbol{L} = \boldsymbol{R} \times \boldsymbol{P}, \tag{6.38}$$
$$\boldsymbol{A} = \tfrac{1}{2}\boldsymbol{R}P^2 - \boldsymbol{P}(\boldsymbol{R}\cdot\boldsymbol{P}) - \tfrac{1}{2}\boldsymbol{R}, \tag{6.39}$$
$$\boldsymbol{B} = \tfrac{1}{2}\boldsymbol{R}P^2 - \boldsymbol{P}(\boldsymbol{R}\cdot\boldsymbol{P}) + \tfrac{1}{2}\boldsymbol{R}, \tag{6.40}$$

$$\boldsymbol{\Gamma} = R\boldsymbol{P}, \tag{6.41}$$

$$T_1 = \frac{1}{2}\left(RP_R^2 + \frac{L^2}{R} - R\right) = \tfrac{1}{2}(RP^2 - R), \tag{6.42}$$

$$T_2 = RP_R = \boldsymbol{R} \cdot \boldsymbol{P} - i, \tag{6.43}$$

$$T_3 = \frac{1}{2}\left(RP_R^2 + \frac{L^2}{R} + R\right) = \tfrac{1}{2}(RP^2 + R). \tag{6.44}$$

The complete set of scaled hydrogenic states $|n\ell m\rangle$ are contained in a single unirrep of so(4,1) and we expect this to be true for the larger Lie algebra so(4,2). This is indeed the case since for the realization (6.38) to (6.44) it follows that the three so(4,2) Casimir operators $Q_2$, $Q_3$ and $Q_4$ (see Appendix B.8) all reduce to the constant values (see Exercise 6.9)

$$Q_2 = L^2 + A^2 - B^2 - \Gamma^2 + T_3^2 - T_1^2 - T_2^2 = -3, \tag{6.45}$$

$$Q_3 = -T_1(\boldsymbol{B} \cdot \boldsymbol{L}) + T_2(\boldsymbol{\Gamma} \cdot \boldsymbol{L}) + T_3(\boldsymbol{A} \cdot \boldsymbol{L}) - \boldsymbol{A} \cdot (\boldsymbol{B} \times \boldsymbol{\Gamma}) = 0, \tag{6.46}$$

$$Q_4 = 0, \tag{6.47}$$

and the eigenvalues of these operators label the different unirreps of so(4,2). Therefore we do not have to deal with the rather complex classification of unirreps of so(4,2) (see [BA70] for the derivation of some important classes of unirreps).

The complicated set of commutation relations (6.25) to (6.37) can be cast into a more compact and memorable form if we make the correspondence (see Appendix B.8)

$$L_{jk} \Longleftrightarrow
\begin{bmatrix}
0 & L_3 & -L_2 & A_1 & B_1 & \Gamma_1 \\
  & 0 & L_1 & A_2 & B_2 & \Gamma_2 \\
  &   & 0 & A_3 & B_3 & \Gamma_3 \\
  &   &   & 0 & T_2 & T_1 \\
  &   &   &   & 0 & T_3 \\
  &   &   &   &   & 0
\end{bmatrix} \tag{6.48}$$

extended to the lower left half by antisymmetry so that $L_{jk} = -L_{kj}$, $j, k = 1, 2, \ldots, 6$. Then the nonzero commutators can be expressed in the simple form

$$[L_{jk}, L_{jm}] = ig_{jj}L_{km}, \quad j \neq k \neq m, \text{ (no sum on } j\text{)}, \tag{6.49}$$

where $g_{jj}$ is a diagonal matrix element of the diagonal metric matrix $G = \mathrm{diag}(1, 1, 1, 1, -1, -1)$ (see Appendix B). 

The correspondence (6.48) clearly exhibits the subalgebra structure

$$so(4,2) \supset so(4,1) \supset so(4) \supset so(3). \tag{6.50}$$

## 6.4  Matrix Representation of so(4,2)

We can now complete the matrix representation of the generators of the scaled hydrogenic realization of so(4,2). The matrix representation of the so(2,1) subalgebra is given by (5.152) to (5.155) and that of the so(4) subalgebra is given by (5.143) to (5.151) so we only need to work out the action of the vector operators $B$ and $\Gamma$ on the basis vectors $|n\ell m\rangle$. It is more convenient to use the components $B_\pm = B_1 \pm iB_2$, $B_3$, $\Gamma_\pm = \Gamma_1 \pm i\Gamma_2$, and $\Gamma_3$.

Since $[T_2, A_j] = iB_j$ and $T_+ - T_- = 2iT_2$ it follows that

$$B_\pm = -\tfrac{1}{2}[T_+ - T_-, A_\pm], \tag{6.51}$$

$$B_3 = -\tfrac{1}{2}[T_+ - T_-, A_3], \tag{6.52}$$

so the action of $B_\pm$ and $B_3$ on the basis vectors can be obtained from (5.154), (5.155) and (5.149) to (5.151):

$$-2B_3|n\ell m\rangle = (T_+ - T_-)A_3|n\ell m\rangle - A_3(T_+ - T_-)|n\ell m\rangle$$

$$= (T_+ - T_-)(\alpha_m^\ell c_\ell^n|n, \ell-1, m\rangle + \alpha_m^{\ell+1}c_{\ell+1}^n|n, \ell+1, m\rangle)$$
$$- A_3(\omega_\ell^n|n+1, \ell m\rangle - \omega_\ell^{-n}|n-1, \ell m\rangle)$$

$$= \alpha_m^\ell c_\ell^n(\omega_{\ell-1}^n|n+1, \ell-1, m\rangle - \omega_{\ell-1}^{-n}|n-1, \ell-1, m\rangle)$$
$$+ \alpha_m^{\ell+1}c_{\ell+1}^n(\omega_{\ell+1}^n|n+1, \ell+1, m\rangle - \omega_{\ell+1}^{-n}|n-1, \ell+1, m\rangle)$$
$$- \omega_\ell^n(\alpha_m^\ell c_\ell^{n+1}|n+1, \ell-1, m\rangle + \alpha_m^{\ell+1}c_{\ell+1}^{n+1}|n+1, \ell+1, m\rangle)$$
$$+ \omega_{-\ell}^n(\alpha_m^\ell c_\ell^{n-1}|n-1, \ell-1, m\rangle + \alpha_m^{\ell+1}c_{\ell+1}^{n-1}|n-1, \ell+1, m\rangle)$$

$$= \alpha_m^\ell(\omega_\ell^{-n}c_\ell^{n-1} - \omega_{\ell-1}^{-n}c_\ell^n)|n-1, \ell-1, m\rangle$$
$$+ \alpha_m^{\ell+1}(\omega_\ell^{-n}c_{\ell+1}^{n-1} - \omega_{\ell+1}^{-n}c_{\ell+1}^n)|n-1, \ell+1, m\rangle$$
$$+ \alpha_m^\ell(\omega_{\ell-1}^n c_\ell^n - \omega_\ell^n c_\ell^{n+1})|n+1, \ell-1, m\rangle$$
$$+ \alpha_m^{\ell+1}(\omega_{\ell+1}^n c_{\ell+1}^n - \omega_\ell^n c_{\ell+1}^{n+1})|n+1, \ell+1, m\rangle. \tag{6.53}$$

Similarly

$$-2B_\pm|n\ell m\rangle = (T_+ - T_-)A_\pm|n\ell m\rangle - A_\pm(T_+ - T_-)|n\ell m\rangle$$

$$= \pm\beta_{\pm m}^{\ell-1}(\omega_\ell^{-n}c_\ell^{n-1} - \omega_{\ell-1}^{-n}c_\ell^n)|n-1, \ell-1, m \pm 1\rangle$$
$$\mp \gamma_{\pm m}^{\ell+1}(\omega_\ell^{-n}c_{\ell+1}^{n-1} - \omega_{\ell+1}^{-n}c_{\ell+1}^n)|n-1, \ell+1, m \pm 1\rangle$$
$$\pm \beta_{\pm m}^{\ell-1}(\omega_{\ell-1}^n c_\ell^n - \omega_\ell^n c_\ell^{n+1})|n+1, \ell-1, m \pm 1\rangle$$
$$\mp \gamma_{\pm m}^{\ell+1}(\omega_{\ell+1}^n c_{\ell+1}^n - \omega_\ell^n c_{\ell+1}^{n+1})|n+1, \ell+1, m \pm 1\rangle. \tag{6.54}$$

Since $\Gamma_j = i[T_3, B_j]$ the matrix representation of $\Gamma$ can be obtained from the matrix representation of $B$ and of $T_3$ using

$$\Gamma_3 = i[T_3, B_3], \tag{6.55}$$

$$\Gamma_\pm = i[T_3, B_\pm], \tag{6.56}$$

to obtain

$$\Gamma_3|n\ell m\rangle = iT_3 B_3|n\ell m\rangle - iB_3 T_3|n\ell m\rangle$$
$$= i(T_3 - n)B_3|n\ell m\rangle, \tag{6.57}$$
$$\Gamma_\pm|n\ell m\rangle = iT_3 B_\pm|n\ell m\rangle - iB_\pm T_3|n\ell m\rangle$$
$$= i(T_3 - n)B_\pm|n\ell m\rangle. \tag{6.58}$$

Now substitute into (6.53) and (6.54).

For convenient reference we give the complete matrix representation of the fifteen generators of so(4,2):

$$T_3|n\ell m\rangle = n|n\ell m\rangle, \tag{6.59}$$
$$T_+|n\ell m\rangle = \omega_\ell^n|n{+}1,\ell m\rangle, \tag{6.60}$$
$$T_-|n\ell m\rangle = \omega_\ell^{-n}|n{-}1,\ell m\rangle, \tag{6.61}$$
$$L_3|n\ell m\rangle = m|n\ell m\rangle, \tag{6.62}$$
$$L_+|n\ell m\rangle = \omega_m^\ell|n\ell,m{+}1\rangle, \tag{6.63}$$
$$L_-|n\ell m\rangle = \omega_{-m}^\ell|n\ell,m{-}1\rangle, \tag{6.64}$$
$$A_3|n\ell m\rangle = \alpha_m^\ell c_\ell^n|n,\ell{-}1,m\rangle + \alpha_m^{\ell+1}c_{\ell+1}^n|n,\ell{+}1,m\rangle, \tag{6.65}$$
$$A_+|n\ell m\rangle = \beta_m^{\ell-1}c_\ell^n|n,\ell{-}1,m{+}1\rangle - \gamma_m^{\ell+1}c_{\ell+1}^n|n,\ell{+}1,m{+}1\rangle, \tag{6.66}$$
$$A_-|n\ell m\rangle = -\beta_{-m}^{\ell-1}c_\ell^n|n,\ell{-}1,m{-}1\rangle + \gamma_{-m}^{\ell+1}c_{\ell+1}^n|n,\ell{+}1,m{-}1\rangle, \tag{6.67}$$
$$B_3|n\ell m\rangle = \alpha_m^\ell u_\ell^n|n{-}1,\ell{-}1,m\rangle + \alpha_m^\ell v_\ell^n|n{+}1,\ell{-}1,m\rangle$$
$$+ \alpha_m^{\ell+1}v_{\ell+1}^{n-1}|n{-}1,\ell{+}1,m\rangle + \alpha_m^{\ell+1}u_{\ell+1}^{n+1}|n{+}1,\ell{+}1,m\rangle, \tag{6.68}$$
$$B_+|n\ell m\rangle = \beta_m^{\ell-1}u_\ell^n|n{-}1,\ell{-}1,m{+}1\rangle + \beta_m^{\ell-1}v_\ell^n|n{+}1,\ell{-}1,m{+}1\rangle$$
$$- \gamma_m^{\ell+1}v_{\ell+1}^{n-1}|n{-}1,\ell{+}1,m{+}1\rangle - \gamma_m^{\ell+1}u_{\ell+1}^{n+1}|n{+}1,\ell{+}1,m{+}1\rangle, \tag{6.69}$$
$$B_-|n\ell m\rangle = -\beta_{-m}^{\ell-1}u_\ell^n|n{-}1,\ell{-}1,m{-}1\rangle - \beta_{-m}^{\ell-1}v_\ell^n|n{+}1,\ell{-}1,m{-}1\rangle$$
$$+ \gamma_{-m}^{\ell+1}v_{\ell+1}^{n-1}|n{-}1,\ell{+}1,m{-}1\rangle + \gamma_{-m}^{\ell+1}u_{\ell+1}^{n+1}|n{+}1,\ell{+}1,m{-}1\rangle, \tag{6.70}$$
$$\Gamma_3|n\ell m\rangle = -i\alpha_m^\ell u_\ell^n|n{-}1,\ell{-}1,m\rangle + i\alpha_m^\ell v_\ell^n|n{+}1,\ell{-}1,m\rangle$$
$$- i\alpha_m^{\ell+1}v_{\ell+1}^{n-1}|n{-}1,\ell{+}1,m\rangle + i\alpha_m^{\ell+1}u_{\ell+1}^{n+1}|n{+}1,\ell{+}1,m\rangle, \tag{6.71}$$
$$\Gamma_+|n\ell m\rangle = -i\beta_m^{\ell-1}u_\ell^n|n{-}1,\ell{-}1,m{+}1\rangle + i\beta_m^{\ell-1}v_\ell^n|n{+}1,\ell{-}1,m{+}1\rangle$$
$$+ i\gamma_m^{\ell+1}v_{\ell+1}^{n-1}|n{-}1,\ell{+}1,m{+}1\rangle - i\gamma_m^{\ell+1}u_{\ell+1}^{n+1}|n{+}1,\ell{+}1,m{+}1\rangle, \tag{6.72}$$
$$\Gamma_-|n\ell m\rangle = i\beta_{-m}^{\ell-1}u_\ell^n|n{-}1,\ell{-}1,m{-}1\rangle - i\beta_{-m}^{\ell-1}v_\ell^n|n{+}1,\ell{-}1,m{-}1\rangle$$
$$- i\gamma_{-m}^{\ell+1}v_{\ell+1}^{n-1}|n{-}1,\ell{+}1,m{-}1\rangle + i\gamma_{-m}^{\ell+1}u_{\ell+1}^{n+1}|n{+}1,\ell{+}1,m{-}1\rangle, \tag{6.73}$$

where the coefficients are defined by

$$\alpha_m^\ell = \sqrt{(\ell - m)(\ell + m)}, \tag{6.74}$$
$$\beta_m^\ell = \sqrt{(\ell - m + 1)(\ell - m)}, \tag{6.75}$$
$$\gamma_m^\ell = \sqrt{(\ell + m + 1)(\ell + m)}, \tag{6.76}$$
$$\omega_m^\ell = \sqrt{(\ell - m)(\ell + m + 1)}, \tag{6.77}$$

$$c_\ell^n = \sqrt{\frac{(n-\ell)(n+\ell)}{(2\ell-1)(2\ell+1)}}, \tag{6.78}$$

$$u_\ell^n = \frac{1}{2}\sqrt{\frac{(n+\ell-1)(n+\ell)}{(2\ell-1)(2\ell+1)}}, \tag{6.79}$$

$$v_\ell^n = \frac{1}{2}\sqrt{\frac{(n-\ell)(n-\ell+1)}{(2\ell-1)(2\ell+1)}}. \tag{6.80}$$

These final results can be used to calculate the scaled hydrogenic matrix elements of any operator expressible in terms of the so(4,2) generators.

For our applications to perturbation theory the most important operators are the radial distance $R = T_3 - T_1$ and the coordinates $X_j = B_j - A_j$ whose matrix elements are easily obtained from

$$R = T_3 - \tfrac{1}{2}(T_+ + T_-), \tag{6.81}$$
$$X = X_1 = \tfrac{1}{2}(B_+ + B_- - A_+ - A_-), \tag{6.82}$$
$$Y = X_2 = \tfrac{1}{2}(B_+ - B_- - A_+ + A_-), \tag{6.83}$$
$$Z = X_3 = B_3 - A_3. \tag{6.84}$$

For example,

$$R|n\ell m\rangle = R_{n\ell}^{n+1,\ell}|n{+}1,\ell m\rangle + R_{n\ell}^{n\ell}|n\ell m\rangle + R_{n\ell}^{n-1,\ell m}|n{-}1,\ell m\rangle, \tag{6.85}$$

where

$$R_{n\ell}^{n+\mu,\ell} = \langle n+\mu, \ell m | R | n\ell m\rangle, \tag{6.86}$$

so $R$ is represented by a symmetric tridiagonal matrix (for fixed $\ell$ and $m$) whose nonzero matrix elements are

$$R_{n\ell}^{n+1,\ell} = -\tfrac{1}{2}\sqrt{(n-\ell)(n+\ell+1)},$$
$$R_{n\ell}^{n\ell} = n,$$
$$R_{n\ell}^{n-1,\ell} = -\tfrac{1}{2}\sqrt{(n+\ell)(n-\ell-1)}. \tag{6.87}$$

Similarly for the $Z$ coordinate operator

$$Z|n\ell m\rangle = Z_{n\ell m}^{n+1,\ell}|n{+}1,\ell{+}1,m\rangle + Z_{n\ell,m}^{n+1,\ell-1}|n{+}1,\ell{-}1,m\rangle$$
$$+ Z_{n\ell,m}^{n,\ell+1}|n,\ell{+}1,m\rangle + Z_{n\ell,m}^{n,\ell-1}|n,\ell{-}1,m\rangle$$
$$+ Z_{n\ell,m}^{n-1,\ell+1}|n{-}1,\ell{+}1,m\rangle + Z_{n\ell,m}^{n-1,\ell-1}|n{-}1,\ell{-}1,m\rangle, \tag{6.88}$$

where

$$Z_{n\ell m}^{n+\mu,\ell+\nu} = \langle n{+}\mu, \ell{+}\nu, m | Z | n\ell m\rangle, \tag{6.89}$$

so $Z$ connects at most six different states and has nonzero matrix elements

$$Z_{n\ell m}^{n+1,\ell+1} = \frac{1}{2}\sqrt{\frac{(\ell-m+1)(\ell+m+1)(n+\ell+1)(n+\ell+2)}{(2\ell+1)(2\ell+3)}},$$

$$Z_{n\ell m}^{n+1,\ell-1} = \frac{1}{2}\sqrt{\frac{(\ell-m)(\ell+m)(n-\ell+1)(n\ell)}{(2\ell-1)(2\ell+1)}},$$

$$Z_{n\ell m}^{n,\ell+1} = -\sqrt{\frac{(\ell-m+1)(\ell+m+1)(n-\ell-1)(n+\ell+1)}{(2\ell-1)(2\ell+1)}},$$

$$Z_{n\ell m}^{n,\ell-1} = -\sqrt{\frac{(\ell-m)(\ell+m)(n-\ell)(n+\ell)}{(2\ell-1)(2\ell+1)}},$$

$$Z_{n\ell m}^{n-1,\ell+1} = \frac{1}{2}\sqrt{\frac{(\ell-m+1)(\ell+m+1)(n-\ell-1)(n-\ell-2)}{(2\ell-1)(2\ell+1)}},$$

$$Z_{n\ell m}^{n-1,\ell-1} = \frac{1}{2}\sqrt{\frac{(\ell-m)(\ell+m)(n+\ell-1)(n+\ell)}{(2\ell-1)(2\ell+1)}}. \tag{6.90}$$

In our applications to hydrogenic perturbation theory matrix elements of $R$, $R^2$, $R^3$, $Z$, $Z^2$, $RZ$ and $R(R^2 - Z^2)$ are needed. They are derived and tabulated in Appendix D for reference.

## 6.5　Hydrogenic Tower of States

To complete our discussion of the scaled hydrogenic so(4,2) realization we show how it is possible to use the raising and lowering operators $A_+$, $T_+$ and $L_-$ to obtain any scaled hydrogenic state $|n\ell m\rangle$ from the ground state $|100\rangle$.

For fixed $n$ there are $n^2$ states for $\ell = 0, 1, \ldots n-1$ and $m = -\ell, -\ell+1, \ldots, \ell$ which form a basis for a unirrep of so(4) (see Chapter 5). These states can be conveniently represented as a triangular array of points, as shown in Figure 6.1 for $n = 3$, which we call an so(4) subtower.

The Figure shows that all states in the subtower can be obtained by first applying $A_+$ to move from the top state $|n00\rangle$ along the right side of the triangle to the appropriate row and then applying $L_-$ to move from right to left in this row until the desired state is reached.

To obtain the results indicated in Figure 6.1 replace $\ell$ by $\ell-1$ in (6.66) and let $m = \ell - 1$, the maximum value. Then (6.66) reduces to one term given by

$$A_+|n, \ell-1, \ell-1\rangle = k_{n\ell}|n\ell\ell\rangle, \tag{6.91}$$

where

$$k_{n\ell} = -\sqrt{\frac{2\ell(n^2 - \ell^2)}{2\ell+1}}, \tag{6.92}$$

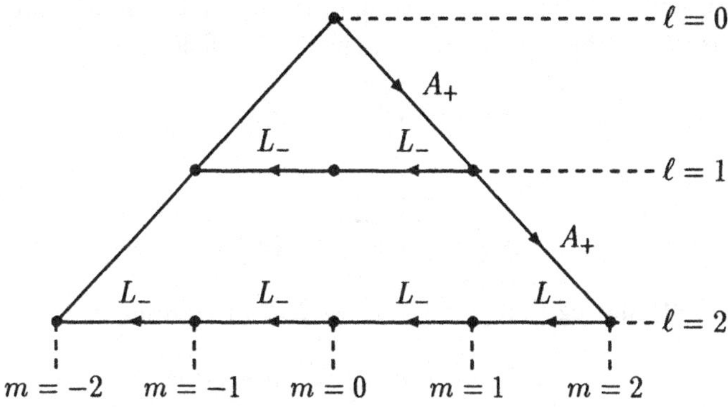


**Figure 6.1**: An so(4) subtower representing the $n^2$ scaled hydrogenic states, for $n = 3$, which form a basis for a unirrep of so(4). All states in the subtower can be obtained by applying the $A_+$ and $L_-$ operators to the top state, $|300\rangle$.

so $A_+$ is a raising operator connecting states with different $\ell$ values and maximal $m$ values. Iteration of (6.91) shows that

$$|n\ell\ell\rangle = K_{n\ell}(A_+)^\ell|n00\rangle, \tag{6.93}$$

where

$$K_{n\ell} = (k_{n1}k_{n2}\cdots k_{n\ell})^{-1}. \tag{6.94}$$

Similarly from (6.64)

$$L_-|n\ell, m+1\rangle = b_{\ell m}|n\ell m\rangle, \tag{6.95}$$

where

$$b_{\ell m} = \sqrt{(\ell + m + 1)(\ell - m)}, \tag{6.96}$$

so $L_-$ can be used to move to the left in any row of an so(4) subtower. Iteration of (6.95) shows that

$$|n\ell m\rangle = B_{\ell m}(L_-)^{\ell-m}|n\ell\ell\rangle, \tag{6.97}$$

where

$$B_{\ell m} = (b_{\ell m}b_{\ell,m+1}\cdots b_{\ell,\ell-1})^{-1}. \tag{6.98}$$

Combining these results any state in a subtower can be obtained from the top one using

$$|n\ell m\rangle = K_{n\ell}B_{\ell m}(L_-)^{\ell-m}(A_+)^\ell|n00\rangle. \tag{6.99}$$

We can now use the so(2,1) raising operator $T_+$ to connect the top of one so(4) subtower to the top of the next one. From (6.60)

$$T_+|n-1,\ell m\rangle = c_{n\ell}|n\ell m\rangle, \tag{6.100}$$

where

$$c_{n\ell} = \sqrt{(n-\ell-1)(n+\ell)}. \tag{6.101}$$

Iterate (6.100), noting that $n = \ell+1, \ell+2, \ldots$, to obtain

$$|n\ell m\rangle = c_{n\ell}(T_+)^{n-\ell-1}|\ell+1,\ell m\rangle, \tag{6.102}$$

where

$$C_{n\ell} = (c_{n\ell}c_{n-1,\ell}\cdots c_{\ell+2,\ell})^{-1}. \tag{6.103}$$

Combine (6.102), with $\ell = m = 0$, and (6.99) to obtain

$$|n00\rangle = C_{n0}(T_+)^{n-1}|100\rangle, \tag{6.104}$$

$$|n\ell m\rangle = N_{n\ell m}(L_-)^{\ell-m}(A_+)^{\ell}(T_+)^{n-1}|100\rangle, \tag{6.105}$$

$$N_{n\ell m} = K_{n\ell}B_{\ell m}C_{n0}$$

$$= \frac{(-1)^{\ell}}{2^{\ell}\ell!(n-1)!}\sqrt{\frac{(2\ell+1)(n-\ell-1)!(\ell+m)!}{(n+\ell)!(\ell-m)!}}. \tag{6.106}$$

This interesting result shows that all scaled hydrogenic states can be obtained using raising and lowering operators from the ground state. We can imagine the so(4) subtowers stacked on top of each other, for $n = 1, 2, 3, \ldots$, to form an so(4,2) tower of states and to obtain any state the $T_+$ operator is used to move to the top of the appropriate subtower and then the $A_+$ and $L_-$ operators are used to move to the state within the subtower.

## 6.6   Exercises

◈ **Exercise 6.1** Show that the so(2,1) generators $T_j$ defined by the hydrogenic realization (5.134) to (5.136) commute with the components of the orbital angular momentum.

◈ **Exercise 6.2** Derive the commutation relations

$$[T_2, A] = iB,$$

where

$$B = \tfrac{1}{2}RP^2 - P(R\cdot P) + \tfrac{1}{2}R$$

is a new vector operator needed in the merging of $T_2$ with the so(4) generators.

◈ **Exercise 6.3** Derive (6.6) to (6.8) and (6.11) for the commutators of the components of $B$ with the components of $A$ and $L$ and with $T_2$ which show that the ten operators $\{L, A, B, T_2\}$ close under commutation to form the Lie algebra so(4,1).

◈ **Exercise 6.4** Using the definition of $C_\alpha$ given in Exercise 5.12 show that

$$C_\alpha \times C_\beta = L[\tfrac{1}{2}(\alpha - \beta)(R \cdot P) + \beta i],$$

which gives the four identities

$$
\begin{aligned}
(a) \qquad & A \times A = C_1 \times C_1 = iL, \\
(b) \qquad & A \times B = C_1 \times C_{-1} = LT_2, \\
(c) \qquad & B \times A = C_{-1} \times C_1 = -LT_2, \\
(d) \qquad & B \times B = C_{-1} \times C_{-1} = -iL.
\end{aligned}
$$

Also show that $L \cdot B = 0$ follows from $L \cdot A = 0$ and conclude that $W = 0$ for the scaled hydrogenic realization of so(4,1).

◈ **Exercise 6.5** Derive the following identities using only the so(4,1) commutation relations (6.3) to (6.11)

$$
\begin{aligned}
(a) \qquad & L \cdot A = A \cdot L, \\
(b) \qquad & L \cdot B = B \cdot L, \\
(c) \qquad & [B, L \cdot A] = -iD, \\
(d) \qquad & [A, L \cdot B] = iD, \\
(e) \qquad & [A_j, D_k] = -i\delta_{jk} L \cdot B, \\
(f) \qquad & [B_j, D_k] = -i\delta_{jk} L \cdot A, \\
(g) \qquad & [T_2, L \cdot A] = iL \cdot B, \\
(h) \qquad & [T_2, L \cdot B] = iL \cdot A.
\end{aligned}
$$

Identities (c) and (d) correspond to (6.18) and (6.19) and setting $k = j$ in (e) and (f) and summing over $j$ gives (6.20) and (6.21).

◈ **Exercise 6.6** For the scaled hydrogenic realization of so(4,1) show that $Q = 2$ for the second order Casimir operator.

◈ **Exercise 6.7** Derive the commutation relations (6.26) to (6.29) among the so(2,1) generators $T_1$ and $T_3$ and the vector operators $A$ and $B$.

◈ **Exercise 6.8** Derive the commutation relations (6.31) to (6.37) between $\Gamma$ and the other twelve generators of so(4,2).

◈ **Exercise 6.9** Obtain expressions (6.45) and (6.46) for the so(4,2) Casimir operators $Q_2$, $Q_3$ given in Appendix B by (B.78) and (B.85).

# Chapter 7

# Lie Algebraic Perturbation Theory

## 7.1 Introduction

In this chapter we consider Rayleigh-Schrödinger perturbation theory (RSPT) in a general context for a nondegenerate state in preparation for several detailed examples of hydrogenic perturbation theory in the next chapters using the Lie algebraic methods we have developed for so(2,1) and so(4,2). The Stark effect is used as a simple example throughout the chapter and will be considered in more detail in the next chapter.

In Section 7.2 we develop the conventional RSPT formalism by expanding the wavefunction and energy as perturbation series and substituting them into the Schrödinger equation in order to obtain iterative formulas for the wavefunction and energy corrections to their unperturbed values.

Our interest is in hydrogenic perturbation theory for which the unperturbed Schrödinger equation is defined by the hydrogenic hamiltonian. In this important case it is well known that the perturbation theory for a bound state is incomplete in the sense that the unperturbed radial hydrogenic eigenfunctions for bound states do not form a complete set of states for the expansion of a bound state solution to the perturbed Schrödinger equation. The continuum states must also be taken into account in the perturbation expansions.

This problem with continuum states begins in second order and higher orders become unmanageable due to the multiple infinite summations over the discrete state contributions and the multiple integrations over the continuum state contributions. To illustrate these difficulties and show that the continuum state contributions cannot simply be ignored we calculate the discrete and continuum contributions to the second order energy correction to the ground state of a hydrogen atom in an electric field (Stark effect) in Section 7.2.4.

Various methods have been proposed to overcome the "problem with the continuum" in hydrogenic perturbation problems. In the method of Dalgarno and Lewis the inhomogeneous differential equations which characterize perturbation theory are directly solved.  In Section 7.2.5 and Exercise 7.3 to Exercise 7.5 we illustrate how this method can be used to calculate the energy corrections for the ground state Stark effect to fourth order.  This shows that while this method could in principle be extended to higher orders, it is not very systematic and not easily automated.

The most promising methods are based on the use of a scaling transformation to replace the unperturbed hamiltonian by a new unperturbed eigenvalue problem which has only discrete eigenvalues.  As we have already seen in Chapter 4, this is exactly what the so(2,1) and so(4,2) Lie algebraic methods do since the $T_3$ eigenvalue problem has a purely discrete spectrum ($q = 1, 2, 3, \ldots$).  Thus in Section 7.3 we extend the scaling transformation considered in Chapter 4 to include a perturbation.

The result is a scaled perturbation problem with $T_3$ as unperturbed hamiltonian that can be considered in two ways: (1) conventionally using the formalism of Section 7.2 but with an eigenvalue that is not the energy, so the energy expansion must be determined by solving implicit equations and (2) in a modified way with the energy directly given as the eigenvalue, which requires that a different perturbation formalism be developed.

We refer to the first method as conventional algebraic RSPT, since the usual RSPT formalism applies (for example, see [SC68]), and the second method as modified algebraic RSPT since a slightly different formalism must be derived (see Section 7.4).  Iterative equations are developed for the wavefunction and energy corrections.  They are not significantly more complicated than those of the conventional formalism presented in Section 7.2 and avoid the determination of the energy from an implicit equation.  A symmetric and more efficient formula for the energy is also developed which generalizes a well known result of conventional RSPT expressing energy corrections to order $2n + 1$ in terms of wavefunction corrections to order $n$.

## 7.2   Conventional RSPT

In order to understand and motivate the modified algebraic RSPT developed in Section 7.4 we first review the conventional Rayleigh-Schrödinger perturbation theory (RSPT).

Consider an unperturbed hamiltonian $H_0$ with a complete set of orthonormal eigenfunctions $\phi_a$ and eigenvalues $E_a$ satisfying

$$H_0\phi_a = E_a\phi_a. \tag{7.1}$$

We shall consider only the nondegenerate case here.  If the perturbed hamiltonian is denoted by $H = H_0 + \lambda V$ where $\lambda$ is a suitable perturbation parameter and $V$ is the perturbation then the problem is to solve the Schrödinger equation

$$(H_0 + \lambda V)\psi = E\psi \tag{7.2}$$

for the perturbed energy $E$ and wavefunction $\psi$ such that $\psi \to \phi_0$ and $E \to E_0$ as $\lambda \to 0$ where $\phi_0$ is any one of the nondegenerate eigenfunctions of $H_0$.

Approximate solutions to (7.2) are obtained from the wavefunction and energy expansions

$$\psi = \sum_{j=0}^{\infty} \psi^{(j)} \lambda^j, \quad \psi^{(0)} = \phi_0, \tag{7.3}$$

$$E = E_0 + \Delta E = \sum_{j=0}^{\infty} E^{(j)} \lambda^j, \quad E^{(0)} = E_0, \tag{7.4}$$

which imply that (7.2) has the form

$$(H_0 - E_0 + \lambda V - \Delta E)\psi = 0. \tag{7.5}$$

We also assume the intermediate normalization

$$\langle \phi_0 | \psi \rangle = 1, \tag{7.6}$$

which implies on substitution of (7.3) into (7.6) that

$$\langle \phi_0 | \psi^{(j)} \rangle = 0, \quad j > 0, \tag{7.7}$$

so all wavefunction corrections $\psi^{(j)}$ are orthogonal to the unperturbed wavefunction.

Substitution of the expansions (7.3) and (7.4) into (7.2) gives

$$\sum_{j=0}^{\infty} H_0 \psi^{(j)} \lambda^j + \sum_{j=1}^{\infty} V \psi^{(j-1)} \lambda^j = \sum_{j=0}^{\infty} \left( \sum_{k=0}^{j} E^{(j-k)} \psi^{(k)} \right) \lambda^j.$$

For $j = 0$ this is just (7.1) for $a = q$, and for $j > 0$

$$(E_0 - H_0)\psi^{(j)} = V \psi^{(j-1)} - \sum_{k=0}^{j-1} E^{(j-k)} \psi^{(k)}. \tag{7.8}$$

The first few equations are

$$(E_0 - H_0)\psi^{(1)} = V \phi_0 - E^{(1)} \phi_0, \tag{7.9}$$

$$(E_0 - H_0)\psi^{(2)} = V \psi^{(1)} - E^{(1)} \psi^{(1)} - E^{(2)} \phi_0, \tag{7.10}$$

$$(E_0 - H_0)\psi^{(3)} = V \psi^{(2)} - E^{(1)} \psi^{(2)} - E^{(2)} \psi^{(1)} - E^{(3)} \phi_0. \tag{7.11}$$

If we introduce the resolvent operator $G$ such that for any wavefunction $\psi$

$$G(E_0 - H_0)\psi = \psi - \langle \phi_0 | \psi \rangle \phi_0, \tag{7.12}$$

then $G$ is the inverse of $E_0 - H_0$ on the orthogonal complement of $\phi_0$. In terms of the unperturbed basis vectors $\phi_a$ we have the explicit expression

$$G = \left( \sum_{a \neq 0}^{\infty} + \int_{E_a > 0} \right) \frac{|\phi_a\rangle\langle\phi_a|}{E_0 - E_a} = \mathop{S}_{a \neq 0} \frac{|\phi_a\rangle\langle\phi_a|}{E_0 - E_a}, \tag{7.13}$$

where we use S to denote a generalized summation over discrete states, excluding the unperturbed state $\phi_0$, and the continuum states, if any. The energy differences in the denominator are nonzero in the nondegenerate case since $a \neq 0$ implies that $E_a \neq E_0$. It also follows from this definition that we can define

$$G\psi^{(0)} = G\phi_0 = 0. \tag{7.14}$$

### 7.2.1  Conventional iteration formula

Now apply $G$ to (7.8) and use (7.7), (7.12) and (7.14) to obtain

$$\psi^{(j)} = GV\psi^{(j-1)} - \sum_{k=1}^{j-1} E^{(j-k)}G\psi^{(k)}, \tag{7.15}$$

which is an iteration formula for wavefunction correction $\psi^{(j)}$ in terms of lower order ones. The energy corrections are obtained by projecting (7.8) onto $\phi_0$:

$$\langle\phi_0|E_0 - H_0|\psi^{(j)}\rangle = \langle\phi_0|V|\psi^{(j-1)}\rangle - \sum_{k=0}^{j-1} E^{(j-k)}\langle\phi_0|\psi^{(k)}\rangle.$$

From (7.1) the left side is zero and from (7.7) only the $k = 0$ term contributes to the sum so

$$E^{(j)} = \langle\phi_0|V|\psi^{(j-1)}\rangle. \tag{7.16}$$

Therefore (7.15) and (7.16) can be used to obtain the energy and wavefunction corrections to any desired order according to the calculational scheme $\phi_0 \to E^{(1)}, \psi^{(1)} \to E^{(2)}, \psi^{(2)} \to \cdots$.

### 7.2.2  Conventional low order formulas

If we define the shorthand notation $V_{ab} = \langle\phi_a|V|\phi_b\rangle$ for the perturbation matrix elements in the unperturbed basis then the energy corrections to fourth order are given by

$$E^{(1)} = V_{00}, \tag{7.17}$$

$$\psi^{(1)} = GV\phi_0, \tag{7.18}$$

$$E^{(2)} = \langle\phi_0|V|\psi^{(1)}\rangle = \langle\phi_0|VGV|\phi_0\rangle = \mathop{S}_{a\neq 0} \frac{V_{0a}V_{a0}}{E_0 - E_a}, \tag{7.19}$$

$$\psi^{(2)} = GV\psi^{(1)} - E^{(1)}G\psi^{(1)}, \tag{7.20}$$

$$E^{(3)} = \langle\phi_0|VGVGV|\phi_0\rangle - E^{(1)}\langle\phi_0|VG^2V|\phi_0\rangle$$

$$= \mathop{S}_{a\neq 0}\mathop{S}_{b\neq 0} \frac{V_{0a}V_{ab}V_{b0}}{(E_0 - E_a)(E_0 - E_b)} - E^{(1)}\mathop{S}_{a\neq 0}\frac{V_{0a}V_{a0}}{(E_0 - E_a)^2}, \tag{7.21}$$

$$\psi^{(3)} = GV\psi^{(2)} - E^{(1)}G\psi^{(2)} - E^{(2)}G\psi^{(1)}, \tag{7.22}$$

$$\begin{aligned}
E^{(4)} &= \langle\phi_0|VGVGVGV|\phi_0\rangle - E^{(1)}\langle\phi_0|VG^2VGV|\phi_0\rangle \\
&\quad - E^{(1)}\langle\phi_0|VGVG^2V|\phi_0\rangle + (E^{(1)})^2\langle\phi_0|VG^3V|\phi_0\rangle \\
&\quad - E^{(2)}\langle\phi_0|VG^2V|\phi_0\rangle \\
&= \sum_{a\neq0}\sum_{b\neq0}\sum_{c\neq0}\frac{V_{0a}V_{ab}V_{bc}V_{ca}}{(E_0-E_a)(E_0-E_b)(E_0-E_c)} \\
&\quad - E^{(1)}\sum_{a\neq0}\sum_{b\neq0}\frac{V_{0a}V_{ab}V_{b0}}{(E_0-E_a)^2(E_0-E_b)} \\
&\quad - E^{(1)}\sum_{a\neq0}\sum_{b\neq0}\frac{V_{0a}V_{ab}V_{b0}}{(E_0-E_a)(E_0-E_b)^2} \\
&\quad + (E^{(1)})^2\sum_{a\neq0}\frac{V_{0a}V_{a0}}{(E_0-E_a)^3} - E^{(2)}\sum_{a\neq0}\frac{V_{0a}V_{a0}}{(E_0-E_a)^2}. \tag{7.23}
\end{aligned}$$

### 7.2.3 Conventional high-order iteration scheme

To proceed to higher orders it is necessary to develop a practical iteration scheme rather than explicit formulas. This can easily be done by expanding the wavefunction corrections $\psi^{(j)}$ in terms of the unperturbed eigenfunctions $\phi_a$:

$$\psi^{(j)} = \sum_a c_a^{(j)}\phi_a. \tag{7.24}$$

We assume an expansion only over the discrete unperturbed states since it would be impractical to use the resulting formulas if there are continuum states (i.e., we assume that (7.24) is complete). Substitute the expansion into the iteration formula (7.15) and use (7.13) and (7.14) to obtain for $j > 0$

$$\begin{aligned}
\sum_{a\neq0} c_a^{(j)}\phi_a &= \sum_b c_b^{(j-1)}GV\phi_b - \sum_{k=1}^{j-1}E^{(j-k)}\sum_{a\neq0}c_a^{(k)}G\phi_a \\
&= \sum_b c_b^{(j-1)}\sum_{a\neq0}\frac{V_{ab}}{E_0-E_a}\phi_a - \sum_{k=1}^{j-1}E^{(j-k)}\sum_{a\neq0}\frac{c_a^{(k)}}{E_0-E_a}\phi_a.
\end{aligned}$$

Equating coefficients of $\phi_a$ and using the intermediate normalization condition (7.7) gives the following iteration scheme for the coefficients $c_a^{(j)}$ and the energy corrections $E^{(j)}$.

$$c_0^{(0)} = 1, \quad c_a^{(0)} = 0, \, a > 0, \tag{7.25}$$

$$c_0^{(j)} = 0, \, j > 0, \tag{7.26}$$

$$c_a^{(j)} = \frac{1}{E_0-E_a}\left[\sum_b V_{ab}c_b^{(j-1)} - \sum_{k=1}^{j-1}E^{(j-k)}c_a^{(k)}\right], \, a,j > 0, \tag{7.27}$$

$$E^{(j)} = \sum_a c_a^{(j-1)}V_{0a}. \tag{7.28}$$

### 7.2.4 Example: continuum difficulties for Stark effect

As an example of the difficulty in applying conventional perturbation theory to hydrogenic systems we briefly consider the quadratic Stark effect, using spherical coordinates, for the ground state of the hydrogen atom in an electric field in the $z$ direction. This problem will be considered in more detail in the next chapter using the conventional algebraic RSPT, applied to a scaled hydrogenic hamiltonian expressed in parabolic coordinates, and also using spherical coordinates and the modified algebraic RSPT to be developed in Section 7.4. For these methods it is possible to obtain exact results for the perturbation corrections in a systematic way, avoiding the continuum states. However the modified algebraic RSPT is more general since the applicability of conventional RSPT to the Stark effect relies on the separability of the Stark hamiltonian in parabolic coordinates [LA77], [AL74], [SI78], [PR80].

In atomic units the Schrödinger equation is given by

$$\left(\frac{1}{2}p^2 - \frac{1}{r} + \lambda z\right)\psi = E\psi, \tag{7.29}$$

where the perturbation parameter $\lambda$ is the electric field strength. The unperturbed eigenfunctions $\phi_{n\ell m} = R_{n\ell}(r)Y_{\ell m}(\vartheta, \varphi)$ have well defined parity $(-1)^\ell$ and the perturbation $z = r\cos\vartheta$ has odd parity so the first order energy correction (7.17) is zero (in fact all odd order energy corrections are zero for the ground state) and $E^{(2)}$ is given by (7.19) as the sum of contributions from the discrete states denoted by $E_d^{(2)}$ and the continuum states denoted by $E_c^{(2)}$. Thus

$$\Delta E_2^d = \sum_{n\ell m}{}' \frac{|\langle\phi_{n\ell m}|z|\phi_{100}\rangle|^2}{-\frac{1}{2} + \frac{1}{2n^2}} = -2\sum_{n\ell m}{}' \frac{|\langle\phi_{n\ell m}|z|\phi_{100}\rangle|^2 n^2}{n^2 - 1}, \tag{7.30}$$

where the summations do not include the $n = 1$ term and

$$\langle\phi_{n\ell m}|z|\phi_{100}\rangle = \int_0^\infty R_{n\ell}(r)r\,R_{10}(r)r^2\,dr \int_\Omega Y_{\ell m}^*(\vartheta,\varphi)\cos\vartheta\,Y_{00}(\vartheta,\varphi)\,d\Omega.$$

Since

$$Y_{00} = \frac{1}{\sqrt{4\pi}}, \qquad Y_{10} = \sqrt{\frac{3}{4\pi}}\cos\vartheta,$$

then from the orthogonality of the spherical harmonic functions we must have $\ell = 1$ and $m = 0$ so

$$\langle\phi_{n\ell m}|z|\phi_{100}\rangle = \frac{1}{\sqrt{3}}\delta_{\ell 1}\delta_{m0}\langle R_{n1}|r|R_{10}\rangle. \tag{7.31}$$

The radial integral can be evaluated [BE57],

$$\langle R_{n1}|r|R_{10}\rangle = \frac{2^4(n-1)^{n-5/2}n^{7/2}}{(n+1)^{n+5/2}}, \tag{7.32}$$

so for the discrete state contribution

$$E_d^{(2)} = -\frac{2^9}{3} \sum_{n=2}^{\infty} \frac{n^9 (n-1)^{2n-6}}{(n+1)^{2n+6}} \approx -1.8316289. \tag{7.33}$$

More than 4000 terms are needed to achieve the indicated accuracy.

For the continuum state contribution using positive energy eigenvalues $E_k = k^2/2$ we have

$$E_c^{(2)} = -\frac{2}{3} \int_0^{\infty} \frac{|\langle R_{k1} | r | R_{10} \rangle|^2}{k^2 + 1} \, dk. \tag{7.34}$$

The radial integral can be evaluated in closed form [BE57],

$$\langle R_{k1} | r | R_{10} \rangle = 16 \left( \frac{k}{1 - e^{-2\pi/k}} \right)^{1/2} \frac{e^{-\frac{2}{k} \tan^{-1} k}}{(k^2 + 1)^{5/2}}, \tag{7.35}$$

but the integral over $k$ must be evaluated numerically (using for example the QAGI subroutine in [KA89]):

$$E_c^{(2)} \approx -0.4183711. \tag{7.36}$$

Therefore

$$E^{(2)} = E_d^{(2)} + E_c^{(2)} \approx -2.25. \tag{7.37}$$

The exact result is $-9/4$ which we shall easily obtain algebraically in the next chapter without any infinite sums or integrations. The continuum state contribution to $E^{(2)}$ is more than 18%. For the Stark effect in the 2-dimensional hydrogen atom the contribution is even more significant, more than 48% [YA91].

It is clear from this example that conventional hydrogenic perturbation theory based on the expansions (7.3), (7.4) and (7.13) cannot be practically extended to higher orders due to the multiple infinite sums and the significant contributions from the multiple continuum state integrations.

### 7.2.5  Dalgarno and Lewis method for Stark effect

Since we are primarily interested in hydrogenic perturbation theory ($H_0$ is the hydrogenic hamiltonian), the continuum states can have important contributions, as we have seen in Section 7.2.4, so we can have difficulties with the continuum already in the second order of perturbation theory.

One way to circumvent these difficulties is to attempt to solve the perturbation equations (7.8) as inhomogeneous differential equations, rather than to deal directly with the formal perturbation series, thus avoiding an expansion in terms of unperturbed discrete and continuum eigenstates. For example, $\psi_1$ is the solution of the inhomogeneous differential equation and boundary condition

$$(E_0 - H_0)\psi^{(1)} = V\phi_0 - E^{(1)}\phi_0, \quad \langle \phi_0 | \psi^{(1)} \rangle = 0. \tag{7.38}$$

The formal generalization of this procedure to higher orders is the famous method of Dalgarno and Lewis first applied to long range forces in $H_2^+$ [DA55], [DA56a], [SC68] and in $HeH^{2+}$ [DA56b] and to other hydrogenic systems [SC59]. Unfortunately the Dalgarno and Lewis method is not systematic enough to be automated and extended to high orders (see Exercise 7.3, Exercise 7.4 and Exercise 7.5).

As a simple example consider the Stark effect and the Schrödinger equation (7.29). The inhomogeneous differential equation (7.38) is given by

$$\left( \nabla^2 + \frac{2}{r} - 1 \right) \psi_1 = \frac{2}{\sqrt{\pi}} r e^{-r} \cos \vartheta, \tag{7.39}$$

since $E_0 = -1/2$, $E^{(1)} = 0$ and $\phi_0 = \pi^{-1/2} e^{-r}$ and the solution is (see Exercise 7.3)

$$\psi^{(1)} = -\frac{1}{\sqrt{4\pi}} (2r + r^2) e^{-r} \cos \vartheta, \tag{7.40}$$

so the exact second order energy correction is

$$E^{(2)} = \langle \phi_0 | V | \psi^{(1)} \rangle = -\frac{2}{3} \int_0^\infty r^3 (2r + r^2) e^{-2r} \, dr = -\frac{9}{4}. \tag{7.41}$$

This procedure can be continued and the inhomogeneous differential equation for $\psi^{(2)}$ can be obtained by substituting (7.40) and (7.41) into (7.10). The solution is

$$\psi^{(2)} = \frac{1}{\sqrt{4\pi}} \left( -\tfrac{81}{8} + \tfrac{3}{2}r^2 + \tfrac{1}{2}r^3 + \tfrac{1}{12}r^4 \right) e^{-r}$$

$$+ \frac{1}{\sqrt{16\pi}} \left( \tfrac{5}{4}r^2 + \tfrac{5}{6}r^3 + \tfrac{1}{6}r^4 \right) (3 \cos^2 \vartheta - 1) e^{-r}. \tag{7.42}$$

The exact fourth order energy correction is obtained using the formula derived in Exercise 7.1 which does not require $\psi^{(3)}$

$$E^{(4)} = \langle \psi^{(1)} | V | \psi^{(2)} \rangle - E^{(2)} \langle \psi^{(1)} | \psi^{(1)} \rangle = -\frac{3555}{64}, \tag{7.43}$$

so the ground state energy correct to fourth order is

$$E = -\frac{1}{2} - \frac{9}{4}\lambda^2 - \frac{3555}{64}\lambda^4. \tag{7.44}$$

It is clear from this example that the generalization of the Dalgarno and Lewis method to higher orders is possible but not very systematic and not easily automated. The calculations involved in obtaining (7.40) to (7.44) are tedious and error prone. However, they can easily be performed using the Maple computer algebra system (see Exercise 7.5).

## 7.3  Scaled Hydrogenic Hamiltonian

In Chapter 4 we considered the $D$-dimensional harmonic oscillator and hydrogenic hamiltonians in a unified manner and converted them into eigenvalue problems for the so(2,1) generator $T_3$ using a scaling transformation. These results are easily extended to include a perturbation term and a perturbed $T_3$ eigenvalue problem is obtained. We shall consider only the hydrogenic case. The so(4,2) Lie algebraic methods are identical to the conventional perturbation theory in the harmonic oscillator case and offer no particular advantage since there are no continuum states.

Thus we consider the Schrödinger equation (7.2) where $H_0$ is the hydrogenic hamiltonian

$$H_0 = \frac{1}{2}p^2 - \frac{\mathcal{Z}}{r},\qquad(7.45)$$

and $\mathcal{Z}$ denotes the nuclear charge. Letting

$$E = E_0 + \Delta E,\qquad(7.46)$$

where $E_0$ is any eigenvalue of the unperturbed hamiltonian (7.39) we obtain the Schrödinger equation

$$(H_0 - E_0 + \lambda V(\boldsymbol{r}) - \Delta E)\,\psi(\boldsymbol{r}) = 0.\qquad(7.47)$$

Multiply on the left by $r$ to obtain

$$\left(\tfrac{1}{2}rp^2 - \mathcal{Z} - rE_0 + \lambda rV(\boldsymbol{r}) - r\Delta E\right)\psi(\boldsymbol{r}) = 0\qquad(7.48)$$

and apply the scaling transformation

$$\boldsymbol{r} = \gamma\boldsymbol{R},\quad \boldsymbol{p} = \gamma^{-1}\boldsymbol{P},\quad r = \gamma R\qquad(7.49)$$

to obtain

$$\left(\tfrac{1}{2}\gamma^{-1}RP^2 - \mathcal{Z} - \gamma RE_0 + \lambda\gamma RV(\gamma\boldsymbol{R}) - \gamma R\Delta E\right)\psi(\gamma\boldsymbol{R}) = 0.\qquad(7.50)$$

But from (4.78) and (4.79) in the 3-dimensional case or (4.115) in the $D$-dimensional case we have for the nuclear charge and energy

$$\mathcal{Z} = q\gamma^{-1},\quad E_0 = -\tfrac{1}{2}\gamma^{-2}.\qquad(7.51)$$

Substitute these values into (7.44) to obtain

$$\left(\tfrac{1}{2}RP^2 + \tfrac{1}{2}R - q + \lambda\gamma^2 RV(\gamma\boldsymbol{R}) - \gamma^2 R\Delta E\right)\Psi(\boldsymbol{R}) = 0,\qquad(7.52)$$

where $\Psi(\boldsymbol{R}) = \psi(\gamma\boldsymbol{R})$.

Using the realization (5.136) of the so(2,1) generator $T_3$ we finally have the scaled hydrogenic eigenvalue problem

$$(T_3 - q + \lambda W - \Delta E\, S)\Psi(\boldsymbol{R}) = 0, \tag{7.53}$$

where $q$ is the eigenvalue of $T_3$ corresponding to the unperturbed state,

$$W = \gamma^2 R V(\gamma \boldsymbol{R}) \tag{7.54}$$

is the scaled perturbation, and

$$S = \gamma^2 R. \tag{7.55}$$

In the important cases of a 3-dimensional hydrogenic atom in a magnetic field (Zeeman effect) and the one-electron diatomic ion, (7.53) is inherently a 2-dimensional problem (the angular dependence on $\varphi$ is trivial). It cannot be reduced to a 1-dimensional radial equation. The hydrogenic atom in an electric field (Stark effect) is special since it can be reduced to a 1-dimensional problem by separation of variables in parabolic coordinates.

All other perturbations which we shall consider are of the central field type ($V$ is a function of $R$ only) and (7.53) reduces to a radial equation valid for perturbation theory of the $D$-dimensional hydrogenic atom in any dimension $D$. In these cases the full power of the so(4,2) Lie algebra is not needed since a central field perturbation can always be expressed in term of only the so(2,1) generators using $R = T_3 - T_1$.

Except for the operator $S$, (7.53) has the same structure as the conventional equation (7.5). The presence of $S$ makes (7.53) look formally like perturbation theory with a nonorthogonal basis where $S$ would be interpreted as an overlap matrix. We have shown in Chapter 4 however that this is not the case since $T_3$ is hermitian with respect to the $1/R$ scalar product and its eigenfunctions are orthonormal.

More importantly the eigenvalues of $T_3$ are all discrete since $q = 1, 2, 3, \ldots$, and they form a complete set for the expansion of a bound state solution of (7.53) so perturbation theory based on (7.53) will involve no integrations over continuum states. Also for all of the perturbations we shall consider, the matrix representing the scaled perturbation $W$ in the unperturbed scaled hydrogenic basis is closely packed about the diagonal so the infinite sums over discrete states arising in the perturbation theory all reduce to small finite sums. Since the matrix elements of $S$ are essentially given by (6.85), $S$ is represented by a simple tridiagonal matrix in the scaled hydrogenic basis.

It is also possible to obtain a perturbed $T_3$ eigenvalue problem which is formally identical to the conventional one (7.2) or (7.5) as follows (see [SI81] for the example of the hydrogen atom in a magnetic field (Zeeman effect)). Define a different scaling factor $\gamma$ by

$$\gamma = (-2E)^{-1/2}, \tag{7.56}$$

where $E$ is now the exact energy eigenvalue of the perturbed problem (7.2) rather than the unperturbed energy eigenvalue as in (7.51). Then multiply (7.2) by $r$ as usual and use (7.49) to obtain

$$\left(\tfrac{1}{2}\gamma^{-1}RP^2 - \mathcal{Z} + \lambda\gamma RV(\gamma\boldsymbol{R}) - \gamma RE\right)\psi(\gamma\boldsymbol{R}) = 0,$$

$$\left(\tfrac{1}{2}\gamma^{-1}RP^2 - \mathcal{Z} + \lambda\gamma RV(\gamma\boldsymbol{R}) + \tfrac{1}{2}\gamma^{-1}R\right)\psi(\gamma\boldsymbol{R}) = 0,$$

$$\left(\tfrac{1}{2}RP^2 + \tfrac{1}{2}R + \lambda\gamma^2 RV(\gamma\boldsymbol{R}) - \mathcal{Z}\gamma\right)\psi(\gamma\boldsymbol{R}) = 0,$$

$$(T_3 + \lambda\gamma W(\boldsymbol{R}) - \mathcal{Z}\gamma)\,\Psi(\boldsymbol{R}) = 0, \tag{7.57}$$

which has the same form as the conventional equation (7.2) but with eigenvalue $\mathcal{Z}\gamma$ rather than $E$.

For example, for the Zeeman effect of a static magnetic field in the $z$ direction, $V = x^2 + y^2 = r^2 - z^2$ so that $W = \gamma^3 R(R^2 - Z^2)$, where $Z$ is the scaled $z$ coordinate, and $\Lambda = \lambda\gamma^3$ is the scaled perturbation parameter. The original perturbation parameter $\lambda$ is related to the magnetic field strength. Then we can obtain the eigenvalue expansion of $\mathcal{Z}\gamma$ using conventional RSPT,

$$\mathcal{Z}\gamma = n + a_1\Lambda + a_2\Lambda^2 + \cdots, \tag{7.58}$$

which is a perturbation expansion of the perturbed principal quantum number $n$. Then the energy expansion in terms of $\lambda$ can be obtained by assuming an expansion of the form

$$\gamma = \gamma_0 + \gamma_1\lambda + \gamma_2\lambda^2 + \cdots \tag{7.59}$$

and substituting it into both sides of (7.58), using $\Lambda = \lambda\gamma^3$, to obtain the coefficients $\gamma_k$. Finally the energy is obtained by substituting (7.59) into

$$E = -\frac{1}{2\gamma^2} = -\frac{\mathcal{Z}^2}{2n^2} + \lambda E^{(1)} + \lambda^2 E^{(2)} + \cdots. \tag{7.60}$$

The perturbation theory based on (7.57) with the added complication of unraveling the series expansions such as (7.58) and (7.59) to obtain (7.60) has the advantage that any conventional RSPT formalism can be applied without modification.

However we shall show in the next section that the modified algebraic RSPT based on (7.53) is not significantly more complicated than conventional RSPT in view of the fact that $S$ is a simple tridiagonal matrix.

## 7.4  Modified Algebraic RSPT

We refer to the perturbation theory based on the scaled hydrogenic hamiltonian (7.53) using the Lie algebras so(2,1) or so(4,2) as modified algebraic RSPT. Thus we consider a general perturbation problem of the form

$$(K_0 - \kappa_0 + \lambda W - \Delta E\, S)\Psi(\boldsymbol{R}) = 0, \tag{7.61}$$

where $K_0 = T_3$ and $\kappa_0 = q$ in the hydrogenic case. Letting $\Phi_a$ denote a scaled hydrogenic eigenfunction we have

$$K_0 \Phi_a = \kappa_a \Phi_a. \tag{7.62}$$

As in Section 7.2 we assume the perturbation expansions

$$\Psi = \sum_{j=0}^{\infty} \Psi^{(j)} \lambda^j, \quad \Psi^{(0)} = \Phi_0, \tag{7.63}$$

$$E = E_0 + \Delta E = \sum_{j=0}^{\infty} E^{(j)} \lambda^j, \quad E^{(0)} = E_0, \tag{7.64}$$

and the intermediate normalization which implies that

$$\langle \Phi_0 | \Psi \rangle = 1, \quad \langle \Phi_0 | \Psi^{(k)} \rangle = 0, \, k > 0. \tag{7.65}$$

Substituting these expansions into (7.61) gives

$$\sum_{j=0}^{\infty} (K_0 - \kappa_0) \Psi^{(j)} \lambda^j + \sum_{j=1}^{\infty} W \Psi^{(j-1)} \lambda^j = \left( S \sum_{j=1}^{\infty} E^{(j)} \lambda^j \right) \left( \sum_{k=0}^{\infty} \Psi^{(k)} \lambda^k \right)$$

$$= \sum_{j=1}^{\infty} \left( \sum_{k=0}^{j-1} E^{(j-k)} S \Psi^{(k)} \right) \lambda^j.$$

The $j = 0$ term is just (7.62) for $a = 0$, and for $j > 0$

$$(\kappa_0 - K_0) \Psi^{(j)} = W \Psi^{(j-1)} - \sum_{k=0}^{j-1} E^{(j-k)} S \Psi^{(k)}, \tag{7.66}$$

which should be compared with (7.8). The first few equations are

$$(\kappa_0 - K_0) \Psi^{(1)} = W \Phi_0 - E^{(1)} S \Phi_0, \tag{7.67}$$

$$(\kappa_0 - K_0) \Psi^{(2)} = W \Psi^{(1)} - E^{(2)} S \Phi_0 - E^{(1)} S \Psi^{(1)}, \tag{7.68}$$

$$(\kappa_0 - K_0) \Psi^{(3)} = W \Psi^{(2)} - E^{(3)} S \Phi_0 - E^{(2)} S \Psi^{(1)} - E^{(1)} S \Psi^{(2)}. \tag{7.69}$$

Again we introduce the resolvent operator $G$ such that

$$G(\kappa_0 - K_0) \Psi = \Psi - \langle \Phi_0 | \Psi \rangle \Psi, \tag{7.70}$$

but now $G$ can be expressed as a sum over the complete discrete set of scaled hydrogenic eigenfunctions $\Phi_a$, so assuming that $\kappa_0$ is nondegenerate we have

$$G = \sum_{a \neq 0} \frac{|\Phi_a\rangle \langle \Phi_a|}{\kappa_0 - \kappa_a}. \tag{7.71}$$

Since $|\Phi_a\rangle\langle\Phi_a|$ removes any component along $\Psi_0$ we can also define

$$G\Psi^{(0)} = G\Phi_0 = 0. \tag{7.72}$$

### 7.4.1  Modified iteration formula

Now apply $G$ to (7.66) to obtain the iterative formula

$$\Psi^{(j)} = GW\Psi^{(j-1)} - \sum_{k=0}^{j-1} E^{(j-k)}GS\Psi^{(k)}$$

$$= GW\Psi^{(j-1)} - \sum_{k=1}^{j} E^{(k)}GS\Psi^{(j-k)}, \qquad (7.73)$$

which is similar to (7.15) in the conventional case. The energy corrections are obtained by projecting (7.66) onto $\Phi_0$:

$$\langle\Phi_0|\kappa_0 - K_0|\Psi^{(j)}\rangle = \langle\Phi_0|W|\Psi^{(j-1)}\rangle - \sum_{k=0}^{j-1} E^{(j-k)}\langle\Phi_0|S|\Psi^{(k)}\rangle$$

$$= \langle\Phi_0|W|\Psi^{(j-1)}\rangle - E^{(j)}\langle\Phi_0|S|\Phi_0\rangle - \sum_{k=1}^{j-1} E^{(k)}\langle\Phi_0|S|\Psi^{(j-k)}\rangle.$$

The left side is zero so

$$\langle\Phi_0|S|\Phi_0\rangle E^{(j)} = \langle\Phi_0|W|\Psi^{(j-1)}\rangle - \sum_{k=1}^{j-1}\langle\Phi_0|S|\Psi^{(j-k)}\rangle E^{(k)}. \qquad (7.74)$$

Therefore (7.73) and (7.74) provide a practical scheme for algebraic perturbation theory analogous to (7.15) and (7.16) in the conventional case. In fact if $S$ is replaced by the identity operator these formulas are formally the same as the conventional ones.

### 7.4.2  Symmetric energy formula

As in the conventional case, (7.74) is an equation for the energy correction $E^{(j)}$ requiring the wavefunction corrections to order $j - 1$. However more efficient formulas exist. In fact it is possible to obtain energy corrections to order $2j+1$ using only wavefunction corrections to order $j$ with the symmetric formula

$$\langle\Phi_0|S|\Phi_0\rangle E^{(p+q+1)} = \langle\Psi^{(p)}|W|\Psi^{(q)}\rangle$$

$$- \sum_{r=1}^{p}\langle\Phi_0|S|\Psi^{(r)}\rangle E^{(p+q+1-r)} - \sum_{s=1}^{q}\langle\Phi_0|S|\Psi^{(s)}\rangle E^{(p+q+1-s)}$$

$$- \sum_{r=1}^{p}\sum_{s=1}^{q}\langle\Psi^{(r)}|S|\Psi^{(s)}\rangle E^{(p+q+1-r-s)} \qquad (7.75)$$

with $p = \lfloor j/2 \rfloor$ and $q = \lfloor (j-1)/2 \rfloor$ (see Exercise 7.6). This result is the generalization to modified perturbation theory of the well known conventional result (see for example [LO65], [DU60]) which can be obtained by replacing $S$ with the identity operator (the two single sums containing $\Phi_0$ disappear). In fact the results derived in Exercise 7.1 are also obtained from this symmetric formula in case $S$ is the identity operator.

As special cases substitute $j$ for $p$ and $j-1$ for $q$ to obtain the even order energy corrections

$$\langle\Phi_0|S|\Phi_0\rangle E^{(2j)} = \langle\Psi^{(j)}|W|\Psi^{(j-1)}\rangle - \langle\Phi_0|S|\Psi^{(j)}\rangle E^{(j)}$$
$$- 2\sum_{s=1}^{j-1}\langle\Phi_0|S|\Psi^{(s)}\rangle E^{(2j-s)} - \sum_{r=1}^{j}\sum_{s=1}^{j-1}\langle\Psi^{(r)}|S|\Psi^{(s)}\rangle E^{(2j-r-s)}, \qquad (7.76)$$

and substitute $j$ for $p$ and $j$ for $q$ to obtain the odd order corrections

$$\langle\Phi_0|S|\Phi_0\rangle E^{(2j+1)} = \langle\Psi^{(j)}|W|\Psi^{(j)}\rangle$$
$$-2\sum_{r=1}^{j}\langle\Phi_0|S|\Psi^{(r)}\rangle E^{(2j+1-r)} - \sum_{r=1}^{j}\sum_{s=1}^{j}\langle\Psi^{(r)}|S|\Psi^{(s)}\rangle E^{(2j+1-r-s)}. \qquad (7.77)$$

As mentioned previously the calculation of matrix elements of $S$ is easy since $S$ is a tridiagonal matrix with respect to the unperturbed eigenfunctions.

### 7.4.3　Modified low order formulas

The explicit formulas for the energy and wavefunction corrections are more complicated than the conventional ones. If we define $W_{ab} = \langle\Phi_a|W|\Phi_b\rangle$ for the scaled perturbation matrix elements in the unperturbed scaled hydrogenic basis and similarly define $S_{ab} = \langle\Phi_a|S|\Phi_b\rangle$ for matrix elements of the operator $S$ then the first few corrections are

$$E^{(1)} = S_{00}^{-1}W_{00}, \qquad (7.78)$$
$$\Psi^{(1)} = GW\Phi_0 - E^{(1)}GS\Phi_0, \qquad (7.79)$$
$$E^{(2)} = S_{00}^{-1}(\langle\Phi_0|W|\Psi^{(1)}\rangle - E^{(1)}\langle\Phi_0|S|\Psi^{(1)}\rangle)$$
$$= S_{00}^{-1}(\langle\Phi_0|WGW|\Phi_0\rangle - E^{(1)}\langle\Phi_0|WGS|\Phi_0\rangle$$
$$- E^{(1)}\langle\Phi_0|SGW|\Phi_0\rangle + (E^{(1)})^2\langle\Phi_0|SGS|\Phi_0\rangle)$$
$$= \frac{1}{S_{00}}\sum_{a\neq 0}\frac{1}{\kappa_0 - \kappa_a}\Big(W_{0a}W_{a0} - E^{(1)}W_{0a}S_{a0} - E^{(1)}S_{0a}W_{a0}$$
$$+ (E^{(1)})^2 S_{0a}S_{a0}\Big)$$
$$= \frac{1}{S_{00}}\sum_{a\neq 0}\frac{1}{\kappa_0 - \kappa_a}\Big(W_{a0}^2 - 2E^{(1)}W_{a0}S_{a0} + (E^{(1)})^2 S_{0a}^2\Big), \qquad (7.80)$$

where the final equality follows since $W$ and $S$ are hermitian and have real matrix elements. Higher order corrections are best obtained using a general iteration scheme analogous to the conventional one developed earlier.

### 7.4.4　Modified high order iteration scheme

Expand the wavefunction corrections $\Psi^{(j)}$ in terms of the complete and discrete orthonormal set of unperturbed basis functions:

$$\Psi^{(j)} = \sum_a C_a^{(j)}\Phi_a, \text{ where } C_a^{(j)} = \langle\Phi_a|\Psi^{(j)}\rangle. \qquad (7.81)$$

We can project (7.73) onto $\Phi_a$ to obtain for $j > 0$ and $a > 0$

$$\langle \Phi_a | \Psi^{(j)} \rangle = \langle \Phi_a | GW | \Psi^{(j-1)} \rangle - \sum_{k=1}^{j} E^{(k)} \langle \Phi_a | GS | \Psi^{(j-k)} \rangle,$$

$$\langle \Phi_a | GW | \Psi^{(j-1)} \rangle = \langle \Phi_a | G | \Phi_a \rangle \sum_b \langle \Phi_a | W | \Phi_b \rangle \langle \Phi_b | \Psi^{(j-1)} \rangle,$$

$$\langle \Phi_a | GS | \Psi^{(j-k)} \rangle = \langle \Phi_a | G | \Phi_a \rangle \sum_b \langle \Phi_a | S | \Phi_b \rangle \langle \Phi_b | \Psi^{(j-k)} \rangle,$$

where from (7.71) $\langle \Phi_a | G | \Phi_b \rangle = (\kappa_0 - \kappa_a)^{-1} \delta_{ab}$ if $a > 0$. For the energy use (7.74) to obtain

$$\langle \Phi_0 | S | \Phi_0 \rangle E^{(j)} = \sum_a \langle \Phi_0 | W | \Phi_a \rangle \langle \Phi_a | \Psi^{(j-1)} \rangle$$

$$- \sum_{k=1}^{j-1} E^{(k)} \sum_a \langle \Phi_0 | S | \Phi_a \rangle \langle \Phi_a | \Psi^{(j-k)} \rangle.$$

Therefore we obtain the following practical iteration scheme

$$C_0^{(0)} = 1, \quad C_a^{(0)} = 0,\, a > 0, \tag{7.82}$$

$$E^{(j)} = \frac{1}{S_{00}} \left[ \sum_a W_{0a} C_a^{(j-1)} - \sum_{k=1}^{j-1} E^{(k)} \sum_a S_{0a} C_a^{(j-k)} \right], \tag{7.83}$$

$$C_0^{(j)} = 0, \quad j > 0, \tag{7.84}$$

$$C_a^{(j)} = \frac{1}{\kappa_0 - \kappa_a} \left[ \sum_b W_{ab} C_b^{(j-1)} - \sum_{k=1}^{j} E^{(k)} \sum_b S_{ab} C_b^{(j-k)} \right], \tag{7.85}$$

where (7.82) follows from $\Psi^{(0)} = \Phi_0$ and (7.84) follows from the intermediate normalization $\langle \Phi_0 | \Psi^{(j)} \rangle = 0$ for $j > 0$. The summations involving $S$ have at most three terms since $S$ is a tridiagonal matrix (see (6.86) to (6.88)) so the modified iteration scheme is not significantly more complicated than the conventional one given by (7.27) and (7.28). Also for all our applications there are only a few nonzero perturbation matrix elements of the form $W_{ab}$ for each index $a$. For the Stark effect there are at most 10 nonzero matrix elements for each index $a$. Therefore there are at most 10 terms in (7.85) in the first sum over $b$ and at most 3 terms in the second sum over $b$.

Finally we can substitute (7.81) into the symmetric energy formula (7.75) to obtain

$$S_{00} E^{(p+q+1)} = \sum_a C_a^{(p)} \left( \sum_b W_{ab} C_b^{(q)} \right)$$

$$- \sum_{r=1}^{p} E^{(p+q+1-r)} \left( \sum_a S_{0a} C_a^{r} \right)$$

$$- \sum_{s=1}^{q} E^{(p+q+1-s)} \left( \sum_a S_{0a} C_a^{(s)} \right)$$

$$- \sum_{r=1}^{p} \sum_a C_a^{(r)} \left( \sum_{s=1}^{q} E^{(p+q+1-r-s)} \sum_b S_{ab} C_b^{(s)} \right). \tag{7.86}$$

If only the energy corrections are desired we can use (7.82) to (7.85) to some maximum order $j$ and then switch to formula (7.86) to calculate energy corrections for orders $j + 1$ to $2j + 1$.

### 7.4.5   Example: ground state Stark effect

As our first example of the application of so(4,2) Lie algebraic methods to perturbation theory we show how to obtain $E^{(2)}$ and $E^{(4)}$ for the Stark effect as it would be done "by hand".

For the ground state, $n = 1$, $\Phi_0 = |100\rangle$, $E_0 = -1/2$ and $\kappa_0 = 1$ so for nuclear charge $\mathcal{Z} = 1$, corresponding to the hydrogen atom, the scaling parameter defined in (7.51) is $\gamma = 1$. Therefore in (7.53) and (7.55) we have

$$S = R, \quad W = RZ. \tag{7.87}$$

The resolvent operator $G$ in (7.71) is

$$G = \sum_{\substack{n\ell m \\ n\neq 1}} \frac{|n\ell m\rangle\langle n\ell m|}{1 - n}. \tag{7.88}$$

Since $E^{(1)} = 0$, the first order correction to the wavefunction is given by (7.79)

$$\begin{aligned}
\Psi^{(1)} &= GW|100\rangle = G\left(W_{10}^{21}|210\rangle + W_{10}^{31}|310\rangle\right) \\
&= G\left(\sqrt{2}\,|210\rangle - \tfrac{\sqrt{2}}{2}|310\rangle\right) \\
&= -\sqrt{2}\,|210\rangle + \tfrac{\sqrt{2}}{4}|310\rangle,
\end{aligned} \tag{7.89}$$

where we have used the notation in Appendix D for the matrix elements of $W = RZ$ (see Table D.4 and the particular values in Appendix D.3.2). Therefore from (7.80), using $S_{00} = \langle 100|R|100\rangle = 1$,

$$\begin{aligned}
E^{(2)} &= \langle 100|W|\Psi^{(1)}\rangle \\
&= -\sqrt{2}\,\langle 100|W|210\rangle + \tfrac{\sqrt{2}}{4}\langle 100|W|310\rangle \\
&= -\sqrt{2}\,W_{21}^{10} + \tfrac{\sqrt{2}}{4}W_{31}^{10} = -\frac{9}{4}.
\end{aligned} \tag{7.90}$$

From (7.73) the wavefunction correction $\Psi^{(2)}$ is

$$\begin{aligned}
\Psi^{(2)} &= G\left(W\Psi^{(1)} - E^{(2)}S|100\rangle\right) \\
&= G\left(-\sqrt{2}\,W|210\rangle + \tfrac{\sqrt{2}}{4}W|310\rangle + \tfrac{9}{4}S|100\rangle\right),
\end{aligned} \tag{7.91}$$

where

$$S|100\rangle = R_{10}^{10}|100\rangle + R_{10}^{20}|200\rangle = |100\rangle - \tfrac{\sqrt{2}}{2}|200\rangle,$$

$$W|210\rangle = W_{21}^{10}\,|100\rangle + W_{21}^{20}\,|200\rangle + W_{21}^{30}\,|300\rangle$$
$$+ W_{21}^{32}\,|320\rangle + W_{21}^{40}\,|400\rangle + W_{21}^{42}\,|420\rangle$$

$$= \sqrt{2}\,|100\rangle - 3\,|200\rangle + \sqrt{6}\,|300\rangle + 2\sqrt{3}\,|320\rangle$$
$$- \tfrac{\sqrt{2}}{2}\,|400\rangle - \sqrt{2}\,|420\rangle,$$

$$W|310\rangle = W_{31}^{10}\,|100\rangle + W_{31}^{20}\,|200\rangle + W_{31}^{30}\,|300\rangle + W_{31}^{32}\,|320\rangle$$
$$+ W_{31}^{40}\,|400\rangle + W_{31}^{42}\,|420\rangle + W_{31}^{50}\,|500\rangle + W_{31}^{52}\,|520\rangle$$

$$= -\tfrac{\sqrt{2}}{2}\,|100\rangle + 4\,|200\rangle - 3\sqrt{6}\,|300\rangle - 3\sqrt{3}\,|320\rangle$$
$$+ 4\sqrt{2}\,|400\rangle + 5\sqrt{2}\,|420\rangle - \tfrac{\sqrt{10}}{2}\,|500\rangle - \sqrt{7}|520\rangle.$$

Therefore

$$\Psi^{(2)} = -\tfrac{23\sqrt{2}}{8}\,|200\rangle + \tfrac{7\sqrt{3}}{4}\,|300\rangle + \tfrac{11\sqrt{6}}{8}\,|320\rangle - |400\rangle$$
$$- \tfrac{3}{2}\,|420\rangle + \tfrac{\sqrt{5}}{16}\,|500\rangle + \tfrac{\sqrt{14}}{16}\,|520\rangle. \tag{7.92}$$

We can now use symmetric formula (7.76) with $E^{(1)} = E^{(3)} = 0$ to obtain

$$E^{(4)} = \langle\Psi^{(1)}|W|\Psi^{(2)}\rangle - E^{(2)}\langle\Psi^{(1)}|S|\Psi^{(1)}\rangle - E^{(2)}\langle100|S|\Psi^{(2)}\rangle,$$

where

$$\langle\Psi^{(1)}|W|\Psi^{(2)}\rangle = -\tfrac{23}{16}\,W_{20}^{31} + \tfrac{7\sqrt{6}}{16}\,W_{30}^{31} - \tfrac{\sqrt{2}}{4}\,W_{31}^{40} + \tfrac{\sqrt{10}}{64}\,W_{31}^{50}$$
$$+ \tfrac{11\sqrt{3}}{16}\,W_{31}^{32} - \tfrac{3\sqrt{2}}{8}\,W_{31}^{42} + \tfrac{\sqrt{7}}{32}\,W_{31}^{52} + \tfrac{23}{4}\,W_{20}^{21}$$
$$- \tfrac{7\sqrt{6}}{4}\,W_{21}^{30} + \sqrt{2}\,W_{21}^{40} - \tfrac{11\sqrt{3}}{4}\,W_{21}^{32} + \tfrac{3\sqrt{2}}{2}\,W_{21}^{42}$$
$$= -\tfrac{4743}{64},$$

$$\langle\Psi^{(1)}|S|\Psi^{(1)}\rangle = \tfrac{1}{8}\,R_{31}^{31} + 2\,R_{21}^{21} - R_{21}^{31} = \tfrac{43}{8},$$

$$\langle100|S|\Psi^{(2)}\rangle = -\tfrac{23\sqrt{2}}{8}\,R_{20}^{10} = \tfrac{23}{8}.$$

Therefore

$$E^{(4)} = -\frac{4743}{64} - \left(-\frac{9}{4}\right)\left(\frac{43}{8}\right) - \left(-\frac{9}{4}\right)\left(\frac{23}{8}\right) = -\frac{3555}{64}. \tag{7.93}$$

These results represent the practical limit of "hand calculations" although the author did manage to obtain $\Psi^{(3)}$ and $E^{(6)}$ by hand before having access to symbolic computation programs:

$$\Psi^{(3)} = -\tfrac{305\sqrt{2}}{16}\,|210\rangle + \tfrac{1529\sqrt{2}}{64}\,|310\rangle - \tfrac{175\sqrt{5}}{16}\,|410\rangle - \tfrac{21\sqrt{5}}{4}\,|430\rangle$$
$$+ \tfrac{13\sqrt{10}}{4}\,|510\rangle + \tfrac{9\sqrt{10}}{4}\,|530\rangle - \tfrac{23\sqrt{70}}{80}\,|610\rangle - \tfrac{9\sqrt{5}}{10}\,|630\rangle$$
$$+ \tfrac{3\sqrt{7}}{32}\,|710\rangle + \tfrac{3\sqrt{6}}{32}\,|730\rangle, \tag{7.94}$$

and from (7.76), using $E^{(1)} = E^{(3)} = E^{(5)} = 0$,

$$E^{(6)} = \langle \Psi^{(2)}|W|\Psi^{(3)}\rangle - \langle \Psi^{(1)}|S|\Psi^{(1)}\rangle E^{(4)} - \langle \Psi^{(1)}|S|\Psi^{(3)}\rangle E^{(2)}$$
$$\qquad - \langle \Psi^{(2)}|S|\Psi^{(2)}\rangle E^{(2)} - 2\langle 100|S|\Psi^{(2)}\rangle E^{(4)}$$

$$= -\frac{6455757}{1024} - \left(\frac{43}{8}\right)\left(-\frac{3555}{64}\right) - \left(\frac{23433}{128}\right)\left(-\frac{9}{4}\right)$$
$$\qquad - \left(\frac{41735}{256}\right)\left(-\frac{9}{4}\right) - 2\left(\frac{23}{8}\right)\left(-\frac{3555}{64}\right)$$

$$= -\frac{2512779}{512}. \tag{7.95}$$

Therefore the energy correct to 6th order is

$$E = -\frac{1}{2} - \frac{9}{4}\lambda^2 - \frac{3555}{64}\lambda^4 - \frac{2512779}{512}\lambda^6. \tag{7.96}$$

## 7.5   Exercises

◈ **Exercise 7.1** Using (7.8), and in particular (7.9) to (7.11), show that the calculation of $E^{(3)}$ does not require the second order correction $\psi^{(2)}$ to the wavefunction, as formula (7.16) suggests, but only the first order correction. Similarly show that the calculation of $E^{(4)}$ and $E^{(5)}$ requires only the corrections $\psi^{(1)}$ and $\psi^{(2)}$ and not $\psi^{(3)}$ or $\psi^{(4)}$.

◈ **Exercise 7.2** Show how the explicit formulas in Section 7.2.2 for the energy corrections can be obtained using the bracketing technique of Brueckner and Huby [BR55], [HU61].

◈ **Exercise 7.3** Apply the Dalgarno and Lewis method to the Stark effect to obtain the exact value $E^{(2)} = -9/4$ by solving the inhomogeneous differential equation (7.38) using the hamiltonian (7.29).

◈ **Exercise 7.4** Continue the preceding exercise and find $E^{(4)}$.

◈ **Exercise 7.5** Obtain the results of Exercise 7.3 and Exercise 7.4 using the Maple computer algebra system.

◈ **Exercise 7.6** Derive the symmetric formula (7.75) for the energy corrections.

# Chapter 8

# Symbolic Calculation of the Stark Effect

## 8.1  Introduction

Here we apply the perturbation theory of the preceding chapter to the Stark effect in a hydrogenic system. First the symbolic calculation of the modified perturbation theory to large order for the ground state is carried out using the Maple computer algebra system [CH91a,b,c]. The ground state results are tabulated in Appendix E to order 100 in both rational and floating point form.

Next the general case is obtained based on the separation of the Schrödinger equation in parabolic coordinates using the so(2,1) Lie algebraic results of Section 4.14, and the conventional perturbation theory for the separation constants. The Stark effect is a special case since the separation in parabolic coordinates produces an eigenvalue problem with only a discrete set of eigenvalues for the separation constants. Therefore conventional perturbation theory is applicable. Here we first consider the classical fourth order calculation of Alliluev and Malkin [AL74] for a general parabolic state, which was the first correct calculation to fourth order, and we solve the problem symbolically using a simple Maple program.

Extension to higher orders for a general state was first done numerically by Silverstone to order 17 [SI78] and he was able to infer exact symbolic results to order 10 from his floating point calculations. We show how to obtain these results to 12th order with a systematic high order symbolic calculation using two Maple programs based on Silverstone's approach and to even higher orders in rational form for particular parabolic states. Since the energy series for the Stark ground state is divergent and asymptotic we also show how Padé approximants can be used to sum the series.

Finally, the Stark effect for a 2-dimensional hydrogenic atom is also important (see [YA91] and references therein) so we show using separation in a 2-dimensional parabolic coordinate system that the energy corrections can easily be obtained directly from the 3-dimensional results [AD92].

## 8.2  Symbolic Calculations for the Ground State

### 8.2.1  Modified high order formalism

In atomic units the Schrödinger equation for a hydrogenic atom in an electric field in the $z$ direction is [SC68], [GA91]

$$\left(\frac{1}{2}p^2 - \frac{\mathcal{Z}}{r} + \lambda z\right)\psi = E\psi, \tag{8.1}$$

where the perturbation parameter $\lambda$ is the electric field strength and $\mathcal{Z}$ is the nuclear charge.

From the general results of Section 7.3 and Section 7.4 the eigenvalue problem for the scaled hamiltonian is

$$(T_3 - n + \lambda W - S\Delta E)\Psi(\boldsymbol{R}) = 0, \tag{8.2}$$

$$W = \gamma^3 RZ = \left(\frac{n}{\mathcal{Z}}\right)^3 RZ, \tag{8.3}$$

$$S = \gamma^2 R = \left(\frac{n}{\mathcal{Z}}\right)^2 R. \tag{8.4}$$

We can let $\mathcal{Z} = 1$ since the nuclear charge can easily be put back into the final results for the wavefunction and energy corrections (see Exercise 8.1).

Now apply the modified algebraic RSPT formalism derived in Section 7.4 to the perturbed ground state ($K_0 = T_3$, $\kappa_0 = n = 1$ in (7.61) and $W = RZ$, $S = R$ in (8.3) and (8.4)). The unperturbed scaled hydrogenic eigenfunctions can be denoted by

$$\Phi_a = \Phi_{n\ell m} = |n\ell m\rangle. \tag{8.5}$$

Since the parity of $|n\ell m\rangle$ is $(-1)^\ell$ the unperturbed ground state $|100\rangle$ has even parity, and since the scaled perturbation $W = RZ$ has odd parity then the first order energy correction from (7.74) is

$$E^{(1)} = \frac{\langle 100|RZ|100\rangle}{\langle 100|S|100\rangle} = 0. \tag{8.6}$$

Then from (7.73) the first order wavefunction correction $\Psi^{(1)} = GW\Phi_0$ has odd parity. In general it follows from (7.73) and (7.74) that for all odd order energy corrections

$$E^{(2j+1)} = 0, \quad j \geq 0 \tag{8.7}$$

and that the even order wavefunction corrections $\Psi^{(2j)}$ have even parity while the odd ones $\Psi^{(2j+1)}$ have odd parity.

To apply the high order iteration scheme developed in Section 7.4.4 it is first necessary to obtain the matrix elements of $R$ and $W = RZ$

$$W_{ab} = W^{n'\ell'm'}_{n\ell m} = \langle n'\ell'm'|RZ|n\ell m\rangle, \tag{8.8}$$

$$S_{ab} = R^{n'\ell'm'}_{n\ell m} = \langle n'\ell'm'|R|n\ell m\rangle. \tag{8.9}$$

The general results for these and other kinds of perturbation matrix elements are given in Appendix D, Tables D.1 and D.4. These matrix elements are diagonal in the magnetic quantum number $m$. Only states $|n\ell m\rangle$ with $m = 0$ contribute to the perturbation expansion of the ground state so we can substitute $m = 0$ into the general results in Tables D.1 and D.4 to obtain the required formulas. The selection rules for nonzero matrix elements are (dropping $m$ in the subscript and superscript)

$$R_{n\ell}^{n+\mu,\ell'} = 0 \quad \text{unless} \quad |\mu| \le 1, \ \ell' = \ell, \tag{8.10}$$

$$W_{n\ell}^{n+\mu,\ell+\nu} = 0 \quad \text{unless} \quad |\mu| \le 2, \ \nu = -1, 1, \tag{8.11}$$

so the perturbation can raise or lower the principal quantum number by at most 2 from one order to the next and change $\ell$ by $\pm 1$.

Therefore the wavefunction corrections can be expanded in terms of the states $|n\ell 0\rangle$ (see (7.81)) as

$$\Psi^{(j)} = \sum_n \sum_\ell C_{n\ell}^{(j)} |n\ell 0\rangle, \tag{8.12}$$

where $n = 2, 3, \ldots, 2j + 1$, $\ell = 0, 2, 4, \ldots, j$ if $j$ is even, $\ell = 1, 3, 5, \ldots, j$ if $j$ is odd, and $0 \le \ell < n$. Thus only even values of $\ell$ contribute to even order corrections and only odd values contribute to odd order corrections.

The modified iteration scheme (7.82) to (7.85) for the calculation of the expansion coefficients $C_{n\ell}^{(j)}$ and energy corrections is

$$C_{10}^{(0)} = 1, \quad C_{NL}^{(0)} = 0, \ N > 1, \tag{8.13}$$

$$R_{10}^{10} E^{(j)} = W_{10}^{21} C_{21}^{(j-1)} + W_{10}^{31} C_{31}^{(j-1)} - R_{10}^{20} \sum_{k=1}^{j-1} E^{(k)} C_{20}^{(j-k)}, \tag{8.14}$$

$$C_{10}^{(j)} = 0, \quad j > 0, \tag{8.15}$$

$$C_{NL}^{(j)} = \frac{1}{1-N} \sum_{N'=N-2}^{N+2} \left( \sum_{L'=L-1,L+1} W_{NL}^{N'L'} C_{N'L'}^{(j-1)} \right)$$

$$- \frac{1}{1-N} \sum_{N'=N-1}^{N+1} R_{NL}^{N'L} \sum_{k=1}^{j} E^{(k)} C_{N'L}^{(j-k)}, \tag{8.16}$$

where

$$N = 2, 3, \ldots, 2j + 1, \quad L = j, j-2, \ldots, \quad 0 \le L < N. \tag{8.17}$$

Similarly the symmetric energy formula (7.86) is

$$R_{10}^{10} E^{(2j)} = \sum_{NL} C_{NL}^j \sum_{N'=N-2}^{N+2} \left( \sum_{L'=L-1,L+1} W_{NL}^{N'L'} C_{N'L'}^{(j-1)} \right)$$

$$- 2 R_{10}^{20} \sum_{r=1}^{j-1} E^{(2j-r)} C_{20}^{(r)} - R_{10}^{20} E^{(j)} C_{20}^{(j)}$$

$$- \sum_{r=1}^{j} \sum_{NL} C_{NL}^{(r)} \left( \sum_{s=1}^{j-1} E^{(2j-r-s)} \sum_{N'=N-1}^{N+1} R_{NL}^{N'L} C_{N'L}^{(s)} \right).$$

$$(8.18)$$

It is clear from the structure of the matrix elements of $R$ and $W = RZ$ that the expansion coefficients $C_{NL}^{(j)}$ will involve square roots (see example in Section 7.4.5). However it is possible to "renormalize" (8.12) to (8.18) so that all quantities are rational numbers. To do this first define new expansion coefficients

$$D_{n\ell}^{(j)} = N_{n\ell} C_{n\ell}^{(j)},$$

$$(8.19)$$

where $N_{n\ell}$ is some "renormalization" factor to be determined in the next subsection and chosen so that for the ground state $N_{10} = 1$. This is equivalent to defining unnormalized basis vectors $\overline{\Phi}^{n\ell}$ such that

$$\Psi^{(j)} = \sum_{n\ell} C_{n\ell}^{(j)} \Phi_{n\ell} = \sum_{n\ell} D_{n\ell}^{(j)} \frac{1}{N_{n\ell}} \Phi_{n\ell} = \sum_{n\ell} D_{n\ell}^{(j)} \overline{\Phi}_{n\ell}.$$

$$(8.20)$$

Now multiply (8.16) by $N_{NL}$ and use (8.19) to obtain

$$D_{10}^{(0)} = 1, \quad D_{NL}^{(0)} = 0, \, N > 1,$$

$$(8.21)$$

$$D_{NL}^{(j)} = \frac{1}{1-N} \sum_{N'=N-2}^{N+2} \left( \sum_{L'=L-1,L+1} \overline{W}_{NL}^{N'L'} D_{N'L'}^{(j-1)} \right)$$

$$- \frac{1}{1-N} \sum_{N'=N-1}^{N+1} \overline{R}_{NL}^{N'L} \sum_{k=1}^{j} E^{(k)} D_{N'L}^{(j-k)},$$

$$(8.22)$$

where we have defined the renormalized matrix elements

$$\overline{R}_{n\ell}^{n+\mu,\ell} = \frac{N_{n\ell}}{N_{n+\mu,\ell}} R_{n\ell}^{n+\mu,\ell},$$

$$(8.23)$$

$$\overline{W}_{n\ell}^{n+\mu,\ell+\nu} = \frac{N_{n\ell}}{N_{n+\mu,\ell+\nu}} W_{n\ell}^{n+\mu,\ell+\nu}.$$

$$(8.24)$$

The renormalized matrix elements are not symmetric in the upper and lower index pairs as are the matrix elements of $R$ and $W$.

The energy formula (8.14) can also be expressed as

$$\overline{R}_{10}^{10} E^{(j)} = \overline{W}_{10}^{21} D_{21}^{(j-1)} + \overline{W}_{10}^{31} D_{31}^{(j-1)} - \overline{R}_{10}^{20} \sum_{k=1}^{j-1} E^{(k)} D_{20}^{(j-k)}$$

$$(8.25)$$

and the symmetric energy formula (8.18) can be expressed as

$$\overline{R}_{10}^{10} E^{(2j)} = \sum_{NL} \frac{D_{NL}^{(j)}}{N_{NL}^2} \sum_{N'=N-2}^{N+2} \left( \sum_{L'=L-1,L+1} \overline{W}_{NL}^{N'L'} D_{N'L'}^{(j-1)} \right)$$

$$- 2\overline{R}_{10}^{20} \sum_{r=1}^{j-1} E^{(2j-r)} D_{20}^{(r)} - \overline{R}_{10}^{20} E^{(j)} D_{20}^{(j)}$$

$$- \sum_{r=1}^{j} \sum_{NL} \frac{D_{NL}^{(r)}}{N_{NL}^2} \left( \sum_{s=1}^{j-1} E^{(2j-r-s)} \sum_{N'=N-1}^{N+1} \overline{R}_{NL}^{N'L} D_{N'L}^{(s)} \right).$$

$$(8.26)$$

The formulas (8.21) to (8.26) define a practical scheme for the high order symbolic calculation of the energy and wavefunction corrections. The complete details including the Maple programs and their results are given in the following subsections.

### 8.2.2  Renormalized matrix elements

The following short Maple program, *genstarkdata*, reads the file *rzdata.m* of general matrix elements created by the program *rzmatrix* in Appendix D and creates the file *starkdata* of matrix elements of $R$ and $W = RZ$ in case $m = 0$ (see (8.8) and (8.9)).

```
01   # genstarkdata
02   # Generate matrix element data for ground state
03   # Stark effect from general results in rzdata.m
04
05   read `rzdata.m`;
06   R := op(r1);
07   W := subs(m=0, op(rz));
08   save R, W, starkdata;
09   quit;
```

The output file *starkdata* contains

$$
\begin{aligned}
R := \mathbf{table}(&[\\
(0) \quad &= n,\\
(-1) \quad &= -1/2*(n-1-\ell)\hat{\ }(1/2)*(n+\ell)\hat{\ }(1/2),\\
(1) \quad &= -1/2*(n-\ell)\hat{\ }(1/2)*(n+\ell+1)\hat{\ }(1/2)\\
])&:
\end{aligned}
$$

$$
\begin{aligned}
W := \mathbf{table}(&[\\
(-2,\,1) \quad &= -1/4*(n-3-\ell)\hat{\ }(1/2)*(n+\ell)\hat{\ }(1/2)/(2*\ell+1)\hat{\ }(1/2)\\
&\quad /(2*\ell+3)\hat{\ }(1/2)*(\ell+1)*(n-1-\ell)\hat{\ }(1/2)*(n-2-\ell)\hat{\ }(1/2),\\
(1,\,1) \quad &= 1/2*(\ell+1)*(n+\ell+1)\hat{\ }(1/2)*(n+\ell+2)\hat{\ }(1/2)*(2*n-\ell)\\
&\quad /(2*\ell+1)\hat{\ }(1/2)/(2*\ell+3)\hat{\ }(1/2),\\
(2,-1) \quad &= -1/4*(n+2-\ell)\hat{\ }(1/2)*(n+\ell+1)\hat{\ }(1/2)\\
&\quad /(2*\ell-1)\hat{\ }(1/2)/(2*\ell+1)\hat{\ }(1/2)*\ell\\
&\quad *(n-\ell+1)\hat{\ }(1/2)*(n-\ell)\hat{\ }(1/2),\\
(-1,\,1) \quad &= 1/2*(\ell+1)*(n-1-\ell)\hat{\ }(1/2)*(n-2-\ell)\hat{\ }(1/2)*(2*n+\ell)\\
&\quad /(2*\ell+1)\hat{\ }(1/2)/(2*\ell+3)\hat{\ }(1/2),\\
(1,-1) \quad &= 1/2*\ell*(n-\ell+1)\hat{\ }(1/2)*(n-\ell)\hat{\ }(1/2)*(2*n+\ell+1)\\
&\quad /(2*\ell-1)\hat{\ }(1/2)/(2*\ell+1)\hat{\ }(1/2),\\
(0,-1) \quad &= -3/2*n/(2*\ell-1)\hat{\ }(1/2)/(2*\ell+1)\hat{\ }(1/2)*\ell*(n-\ell)\hat{\ }(1/2)\\
&\quad *(n+\ell)\hat{\ }(1/2),\\
(0,\,1) \quad &= -3/2*n/(2*\ell+1)\hat{\ }(1/2)/(2*\ell+3)\hat{\ }(1/2)*(\ell+1)\\
&\quad *(n-1-\ell)\hat{\ }(1/2)*(n+\ell+1)\hat{\ }(1/2),\\
(-1,-1) \quad &= 1/2*\ell*(n+\ell-1)\hat{\ }(1/2)*(n+\ell)\hat{\ }(1/2)*(2*n-1-\ell)
\end{aligned}
$$

$$/(2*\ell-1)\hat{\ }(1/2)/(2*\ell+1)\hat{\ }(1/2),$$
$$(2,\,1)\quad = -1/4*(n-\ell)\hat{\ }(1/2)*(n+3+\ell)\hat{\ }(1/2)/(2*\ell+1)\hat{\ }(1/2)$$
$$/(2*\ell+3)\hat{\ }(1/2)*(\ell+1)*(n+\ell+1)\hat{\ }(1/2)*(n+\ell+2)\hat{\ }(1/2),$$
$$(-2,-1)\quad = -1/4*(n-1-\ell)\hat{\ }(1/2)*(n-2+\ell)\hat{\ }(1/2)/(2*\ell-1)\hat{\ }(1/2)$$
$$/(2*\ell+1)\hat{\ }(1/2)*\ell*(n+\ell-1)\hat{\ }(1/2)*(n+\ell)\hat{\ }(1/2)$$
$$]):$$

where $R_{n\ell}^{n+\mu,\ell} = R[mu]$ and $W_{n\ell}^{n+\mu,\ell+\nu} = W[mu,nu]$ are expressed as functions of $n$ and $\ell$ which in typeset form using Maple's *latex* function is given by

$$R[-1] = -\frac{\sqrt{n-1-\ell}\sqrt{n+\ell}}{2},$$

$$R[0] = n,$$

$$R[1] = -\frac{\sqrt{n-\ell}\sqrt{n+\ell+1}}{2},$$

$$W[-2,-1] = -\frac{\sqrt{n-1-\ell}\sqrt{n-2+\ell}\,\ell\sqrt{n+\ell-1}\sqrt{n+\ell}}{4\sqrt{2\ell-1}\sqrt{2\ell+1}},$$

$$W[-2,1] = -\frac{\sqrt{n-3-\ell}\sqrt{n+\ell}(\ell+1)\sqrt{n-1-\ell}\sqrt{n-2-\ell}}{4\sqrt{2\ell+1}\sqrt{2\ell+3}},$$

$$W[-1,-1] = \frac{\ell\sqrt{n+\ell-1}\sqrt{n+\ell}(2n-1-\ell)}{2\sqrt{2\ell-1}\sqrt{2\ell+1}},$$

$$W[-1,1] = \frac{(\ell+1)\sqrt{n-1-\ell}\sqrt{n-2-\ell}(2n+\ell)}{2\sqrt{2\ell+1}\sqrt{2\ell+3}},$$

$$W[0,-1] = -\frac{3n\ell\sqrt{n-\ell}\sqrt{n+\ell}}{2\sqrt{2\ell-1}\sqrt{2\ell+1}},$$

$$W[0,1] = -\frac{3n(\ell+1)\sqrt{n-1-\ell}\sqrt{n+\ell+1}}{2\sqrt{2\ell+1}\sqrt{2\ell+3}},$$

$$W[1,-1] = \frac{\ell\sqrt{n-\ell+1}\sqrt{n-\ell}(2n+\ell+1)}{2\sqrt{2\ell-1}\sqrt{2\ell+1}},$$

$$W[1,1] = \frac{(\ell+1)\sqrt{n+\ell+1}\sqrt{n+\ell+2}(2n-\ell)}{2\sqrt{2\ell+1}\sqrt{2\ell+3}},$$

$$W[2,-1] = -\frac{\sqrt{n+2-\ell}\sqrt{n+\ell+1}\,\ell\sqrt{n-\ell+1}\sqrt{n-\ell}}{4\sqrt{2\ell-1}\sqrt{2\ell+1}},$$

$$W[2,1] = -\frac{\sqrt{n-\ell}\sqrt{n+3+\ell}(\ell+1)\sqrt{n+\ell+1}\sqrt{n+\ell+2}}{4\sqrt{2\ell+1}\sqrt{2\ell+3}}.$$

Since the matrix elements of $R$ and the scaled perturbation $W = RZ$ involve square root factors it is convenient and much more efficient for symbolic computation to choose a renormalization factor $N_{n\ell}$ such that the renormalized matrix elements defined by (8.23) and (8.24) are rational numbers. Then all quantities in (8.21) to (8.26) will be rational numbers.

This can be done with the renormalization factor

$$N_{n\ell} = \sqrt{\frac{(n+\ell)!}{(n-\ell-1)!(2\ell+1)}}.$$

The following Maple program, *renorm*, calculates these renormalized matrix elements and saves the results in the file *renormdata* and in the file *renormdata.m*.

```
01   # renorm
02   read starkdata;
03   Norm2 := (n+ℓ)!/(n−ℓ−1)!/(2*ℓ+1);
04
05   # Renormalize R matrix
06
07   for mu from −1 to 1 do
08       Na2 := Norm2;
09       Nb2 := subs(n=n+mu, Norm2);
10       f2 := simplify(Na2/Nb2);
11       R[mu] := simplify(f2^(1/2)*R[mu]);
12   od:
13
14   # Renormalize the perturbation matrix
15
16   for mu from −2 to 2 do
17       for nu in [−1,1] do
18           Na2 := Norm2;
19           Nb2 := subs(n=n+mu, ℓ=ℓ+nu, Norm2);
20           f2 := simplify(Na2/Nb2);
21           W[mu,nu] := simplify(f2^(1/2)*W[mu,nu]);
22       od;
23   od:
24
25   save Norm2, R, W, renormdata;
26   save Norm2, R, W, `renormdata.m`;
27   quit;
```

The output file *renormdata* contains

```
Norm2 := (n+ℓ)!/(n−1−ℓ)!/(2*ℓ+1):
R := table([
    (0)       = n,
    (−1)      = −1/2*n−1/2*ℓ,
    (1)       = −1/2*n+1/2*ℓ
]):

W := table([
    (−1, 1)   = 1/2/(2*ℓ+1)*(ℓ+1)*(2*n+ℓ),
```

$$
\begin{aligned}
(0, 1) \quad &= -3/2/(2*\ell+1)*n*(\ell+1), \\
(1, 1) \quad &= 1/2/(2*\ell+1)*(\ell+1)*(2*n-\ell), \\
(1, -1) \quad &= 1/2/(2*\ell+1)*(n-\ell)*(n-\ell+1)*\ell*(2*n+\ell+1), \\
(-2, -1) \quad &= -1/4/(2*\ell+1)*(n-2+\ell)*(n+\ell-1)*(n+\ell)*\ell, \\
(-1, -1) \quad &= 1/2/(2*\ell+1)*(n+\ell-1)*(n+\ell)*\ell*(2*n-1-\ell), \\
(2, -1) \quad &= -1/4/(2*\ell+1)*(n-\ell)*(n-\ell+1)*(n+2-\ell)*\ell, \\
(0, -1) \quad &= -3/2/(2*\ell+1)*(n+\ell)*(n-\ell)*n*\ell, \\
(2, 1) \quad &= -1/4/(2*\ell+1)*(n-\ell)*(\ell+1), \\
(-2, 1) \quad &= -1/4/(2*\ell+1)*(n+\ell)*(\ell+1)
\end{aligned}
$$
$$]):$$

which can also be typeset to give the rational matrix elements

$$
R[-1] = -\frac{n}{2} - \frac{\ell}{2},
$$
$$
R[0] = n,
$$
$$
R[1] = \frac{\ell}{2} - \frac{n}{2},
$$
$$
W[-2, -1] = -\frac{(n-2+\ell)\,(n+\ell-1)\,(n+\ell)\,\ell}{8\ell+4},
$$
$$
W[-2, 1] = -\frac{(n+\ell)\,(\ell+1)}{8\ell+4},
$$
$$
W[-1, -1] = \frac{(n+\ell-1)\,(n+\ell)\,\ell\,(2n-1-\ell)}{4\ell+2},
$$
$$
W[-1, 1] = \frac{(\ell+1)\,(2n+\ell)}{4\ell+2},
$$
$$
W[0, -1] = -\frac{(3n+3\ell)\,(n-\ell)\,n\ell}{4\ell+2},
$$
$$
W[0, 1] = -\frac{3n\,(\ell+1)}{4\ell+2},
$$
$$
W[1, -1] = \frac{(n-\ell)\,(n-\ell+1)\,\ell\,(2n+\ell+1)}{4\ell+2},
$$
$$
W[1, 1] = \frac{(\ell+1)\,(2n-\ell)}{4\ell+2},
$$
$$
W[2, -1] = -\frac{(n-\ell)\,(n-\ell+1)\,(n+2-\ell)\,\ell}{8\ell+4},
$$
$$
W[2, 1] = -\frac{(n-\ell)\,(\ell+1)}{8\ell+4}.
$$

where we have omitted the bar on $R$ and $W$ since we will no longer need the
original matrix elements involving the square roots.

### 8.2.3  Program *gsstark* for $E^{(j)}$ and $\Psi^{(j)}$

The following Maple program implements the modified iteration scheme defined by (8.21) to (8.25). The energy corrections $E^{(j)}$ are denoted in the pro-

gram by $DE[j]$ and the expansion coefficients $D_{NL}^{(j)}$ are denoted by $DD[j,N,L]$.

Both the energy and expansion coefficients are represented by Maple table structures which are dynamically allocated so the Maple *type* function is quite useful for determining whether one of the expansion coefficients is defined or not. For example the expression *type(x, name)* used in lines *23, 38, 42* and *52* is true only if $x$ has not been assigned a value.

The procedure *gsstark* creates a file, called *estark6.m* if *jmax* is *6*, containing the tables *DD* and *DE* to order *jmax* in Maple's internal format, as indicated by the *m* suffix.

```
01    # gsstark
02    # Ground state Stark effect using renormalized matrix elements
03    gsstark := proc(jmax)
04
05        local W2110, W3110, R2010, j, k, x, s1, s2, s3, N, L, NP, LP;
06        # global vars (input): R, W
07        # global vars (output): DD, DE
08
09        DD[0,1,0] := 1;
10        W2110 := subs(n=1, ℓ=0, W[1,1]);
11        W3110 := subs(n=1, ℓ=0, W[2,1]);
12        R2010 := subs(n=1, ℓ=0, R[1]);
13        for j from 1 to jmax do
14            lprint(j);
15
16            # Calculate jth order energy correction if j is even.
17
18            if type(j, even) then
19                s1 := W2110 * DD[j−1, 2, 1] + W3110 * DD[j−1, 3, 1];
20                s2 := 0;
21                for k from 2 by 2 to j−1 do
22                    x := DD[j−k, 2, 0];
23                    if not type(x, name) then
24                        s2 := s2 + x*DE[k];
25                    fi;
26                od;
27                DE[j] := s1 − R2010*s2;
28            fi;
29
30            # Now calculate the jth order wavefunction coefficients
31
32            for N from 2 to 2*j+1 do
33                for L from j by −2 to 0 do
34                    if L<N then
35                        s1 := 0;
36                        for NP from N−2 to N+2 do
37                            x := DD[j−1, NP, L−1];
```

```
38                        if not type(x, name) then
39                            s1 := s1 + subs(n=N, ℓ=L, W[NP−N,−1])*x;
40                        fi;
41                    y := DD[j−1, NP, L+1];
42                    if not type(y, name) then
43                            s1 := s1 + subs(n=N, ℓ=L, W[NP−N, 1])*y;
44                        fi;
45                od;
46
47                s2 := 0;
48                for NP from N−1 to N+1 do
49                    s3 := 0;
50                    for k from 2 by 2 to j do
51                        x := DD[j−k, NP, L];
52                        if not type(x, name) then
53                            s3 := s3 + DE[k]*x;
54                        fi;
55                    od;
56                    s2 := s2 + subs(n=N, ℓ=L, R[NP−N])*s3;
57                od;
58                DD[j, N, L] := (s1 − s2)/(1−N);
59            fi;
60        od; # L
61      od; # N
62    od: # j
63    DE[0] := −1/2;
64    save DD, DE, `estark`.jmax.`.m`;
65  end:
66
67  # Example: Calculate wavefunction and energy corrections
68  # to order jmax and display results
69
70  jmax := 6;
71  read `renormdata.m`;
72  gsstark(jmax);
73  Energy := sum('DE[2*j]*x^(2*j)', 'j'=0..jmax/2);
74  quit;
```

For the example in lines *67* to *74* the output file *estark6.m* contains the following data in the tables for *DD* and *DE*.

```
DD := table([
    (0, 1, 0) = 1,
    (1, 2, 1) = −2,              (1, 3, 1) = 1,

    (2, 2, 0) = −23/4,           (2, 3, 0) = 21/4,
    (2, 3, 2)=33/2,              (2, 4, 0) = −2,
    (2, 4, 2) = −18,             (2, 5, 0) = 5/16,
```

$(2, 5, 2) = 21/4,$

$(3, 2, 1) = -305/8,$  $(3, 3, 1) = 1529/16,$
$(3, 4, 1) = -875/8,$  $(3, 4, 3) = -315,$
$(3, 5, 1) = 65,$  $(3, 5, 3) = 540,$
$(3, 6, 1) = -161/8,$  $(3, 6, 3) = -324,$
$(3, 7, 1) = 21/8,$  $(3, 7, 3) = 135/2,$

$(4, 2, 0) = -17085/64,$  $(4, 3, 0) = 13755/32,$
$(4, 3, 2) = 33243/32,$  $(4, 4, 0) = -14013/32,$
$(4, 4, 2) = -49593/16,$  $(4, 5, 0) = 39775/128,$
$(4, 5, 2) = 138783/32,$  $(4, 5, 4) = 43605/4,$
$(4, 6, 0) = -9585/64,$  $(4, 6, 2) = -13803/4,$
$(4, 6, 4) = -25650,$  $(4, 7, 0) = 1505/32,$
$(4, 7, 2) = 25785/16,$  $(4, 7, 4) = 47025/2,$
$(4, 8, 0) = -35/4,$  $(4, 8, 2) = -1665/4,$
$(4, 8, 4) = -9900,$  $(4, 9, 0) = 189/256,$
$(4, 9, 2) = 1485/32,$  $(4, 9, 4) = 6435/4,$

$(5, 2, 1) = -92989/32,$  $(5, 3, 1) = 1260485/128,$
$(5, 4, 1) = -592129/32,$  $(5, 4, 3) = -178011/4,$
$(5, 5, 1) = 1454007/64,$  $(5, 5, 3) = 644967/4,$
$(5, 6, 1) = -310905/16,$  $(5, 6, 3) = -4365765/16,$
$(5, 6, 5) = -1195425/2,$  $(5, 7, 1) = 1500191/128,$
$(5, 7, 3) = 8700885/32,$  $(5, 7, 5) = 3586275/2,$
$(5, 8, 1) = -314187/64,$  $(5, 8, 3) = -2712105/16,$
$(5, 8, 5) = -2220075,$  $(5, 9, 1) = 21855/16,$
$(5, 9, 3) = 524205/8,$  $(5, 9, 5) = 1412775,$
$(5, 10, 1) = -29205/128,$  $(5, 10, 3) = -57915/4,$
$(5, 10, 5) = -921375/2,$  $(5, 11, 1) = 4455/256,$
$(5, 11, 3) = 45045/32,$  $(5, 11, 5) = 61425,$

$(6, 2, 0) = -15184995/512,$  $(6, 3, 0) = 30408177/512,$
$(6, 3, 2) = 17125467/128,$  $(6, 4, 0) = -43569685/512,$
$(6, 4, 2) = -144241833/256,$  $(6, 5, 0) = 388979825/4096,$
$(6, 5, 2) = 1273836705/1024,$  $(6, 5, 4) = 45966195/16,$
$(6, 6, 0) = -84408237/1024,$  $(6, 6, 2) = -114430425/64,$
$(6, 6, 4) = -197267895/16,$  $(6, 7, 0) = 56935599/1024,$
$(6, 7, 2) = 460997721/256,$  $(6, 7, 4) = 790117065/32,$
$(6, 7, 6) = 190724625/4,$  $(6, 8, 0) = -7399553/256,$
$(6, 8, 2) = -167157675/128,$  $(6, 8, 4) = -237344895/8,$
$(6, 8, 6) = -174139875,$  $(6, 9, 0) = 11645739/1024,$
$(6, 9, 2) = 87426405/128,$  $(6, 9, 4) = 1480792365/64,$
$(6, 9, 6) = 4353496875/16,$  $(6, 10, 0) = -3358775/1024,$
$(6, 10, 2) = -16164225/64,$  $(6, 10, 4) = -190478925/16,$
$(6, 10, 6) = -232186500,$  $(6, 11, 0) = 671055/1024,$
$(6, 11, 2) = 32207175/512,$  $(6, 11, 4) = 125696025/32,$

$$(6,\, 11,\, 6) = 227721375/2, \qquad (6,\, 12,\, 0) = \text{-}10395/128,$$
$$(6,\, 12,\, 2) = \text{-}1216215/128, \qquad (6,\, 12,\, 4) = \text{-}757575,$$
$$(6,\, 12,\, 6) = \text{-}30362850, \qquad (6,\, 13,\, 0) = 19305/4096,$$
$$(6,\, 13,\, 2) = 675675/1024, \qquad (6,\, 13,\, 4) = 1044225/16,$$
$$(6,\, 13,\, 6) = 13735575/4,$$

]);

$$DE := \mathbf{table}([$$
$$(0) = \text{-}1/2,$$
$$(2) = \text{-}9/4,$$
$$(4) = \text{-}3555/64,$$
$$(6) = \text{-}2512779/512$$

]);

This corresponds to the wavefunction expansion given in (8.20). Using (8.19) it is easy to return to the original expansion in terms of the coefficients $C_{n\ell}^{(j)}$. The results to order 6 are given by

$$\Psi^{(1)} = -\sqrt{2}\,|21\rangle + \tfrac{\sqrt{2}}{4}\,|31\rangle \tag{8.27}$$

$$\Psi^{(2)} = -\tfrac{23\sqrt{2}}{8}\,|20\rangle + \tfrac{7\sqrt{3}}{4}\,|30\rangle + \tfrac{11\sqrt{6}}{8}\,|32\rangle - |40\rangle$$
$$\qquad -\tfrac{3}{2}\,|42\rangle + \tfrac{\sqrt{5}}{16}\,|50\rangle + \tfrac{\sqrt{14}}{16}\,|52\rangle \tag{8.28}$$

$$\Psi^{(3)} = -\tfrac{305\sqrt{2}}{16}\,|21\rangle + \tfrac{1529\sqrt{2}}{64}\,|31\rangle - \tfrac{175\sqrt{5}}{16}\,|41\rangle - \tfrac{21\sqrt{5}}{4}\,|43\rangle$$
$$\qquad + \tfrac{13\sqrt{10}}{4}\,|51\rangle + \tfrac{9\sqrt{10}}{4}\,|53\rangle - \tfrac{23\sqrt{70}}{80}\,|61\rangle - \tfrac{9\sqrt{5}}{10}\,|63\rangle$$
$$\qquad + \tfrac{3\sqrt{7}}{32}\,|71\rangle + \tfrac{3\sqrt{6}}{32}\,|73\rangle \tag{8.29}$$

$$\Psi^{(4)} = -\tfrac{17085\sqrt{2}}{128}\,|20\rangle + \tfrac{4585\sqrt{3}}{32}\,|30\rangle + \tfrac{11081\sqrt{6}}{128}\,|32\rangle - \tfrac{14013}{64}\,|40\rangle$$
$$\qquad - \tfrac{16531}{64}\,|42\rangle + \tfrac{7955\sqrt{5}}{128}\,|50\rangle + \tfrac{46261\sqrt{14}}{896}\,|52\rangle + \tfrac{2907\sqrt{70}}{448}\,|54\rangle$$
$$\qquad - \tfrac{3195\sqrt{6}}{128}\,|60\rangle - \tfrac{4601\sqrt{21}}{224}\,|62\rangle - \tfrac{855\sqrt{7}}{56}\,|64\rangle + \tfrac{215\sqrt{7}}{32}\,|70\rangle$$
$$\qquad + \tfrac{2865\sqrt{21}}{448}\,|72\rangle + \tfrac{285\sqrt{154}}{224}\,|74\rangle - \tfrac{35\sqrt{2}}{16}\,|80\rangle - \tfrac{185\sqrt{42}}{224}\,|82\rangle$$
$$\qquad - \tfrac{15\sqrt{154}}{56}\,|84\rangle + \tfrac{63}{256}\,|90\rangle + \tfrac{45\sqrt{77}}{896}\,|92\rangle + \tfrac{3\sqrt{2002}}{448}\,|94\rangle \tag{8.30}$$

$$\Psi^{(5)} = -\tfrac{92989\sqrt{2}}{64}\,|21\rangle + \tfrac{1260485\sqrt{2}}{512}\,|31\rangle - \tfrac{592129\sqrt{5}}{320}\,|41\rangle$$
$$\qquad - \tfrac{59337\sqrt{5}}{80}\,|43\rangle + \tfrac{1454007\sqrt{10}}{1280}\,|51\rangle + \tfrac{214989\sqrt{10}}{320}\,|53\rangle$$
$$\qquad - \tfrac{8883\sqrt{70}}{32}\,|61\rangle - \tfrac{97017\sqrt{5}}{128}\,|63\rangle - \tfrac{3795\sqrt{7}}{32}\,|65\rangle$$
$$\qquad + \tfrac{214313\sqrt{7}}{512}\,|71\rangle + \tfrac{193353\sqrt{6}}{512}\,|73\rangle + \tfrac{3795\sqrt{21}}{64}\,|75\rangle$$
$$\qquad - \tfrac{104729\sqrt{42}}{1792}\,|81\rangle - \tfrac{5479\sqrt{66}}{128}\,|83\rangle - \tfrac{1265\sqrt{546}}{224}\,|85\rangle$$
$$\qquad + \tfrac{1457\sqrt{15}}{64}\,|91\rangle + \tfrac{1059\sqrt{110}}{128}\,|93\rangle + \tfrac{345\sqrt{13}}{32}\,|95\rangle$$

$$- \frac{177\sqrt{330}}{256} |10,1\rangle - \frac{9\sqrt{2145}}{32} |10,3\rangle - \frac{15\sqrt{195}}{32} |10,5\rangle$$
$$+ \frac{81\sqrt{110}}{1024} |11,1\rangle + \frac{7\sqrt{4290}}{512} |11,3\rangle + \frac{5\sqrt{39}}{64} |11,5\rangle \tag{8.31}$$

$$\Psi^{(6)} = -\frac{15184995\sqrt{2}}{1024} |20\rangle + \frac{10136059\sqrt{3}}{512} |30\rangle + \frac{5708489\sqrt{6}}{512} |32\rangle$$
$$- \frac{43569685}{1024} |40\rangle - \frac{48080611}{1024} |42\rangle + \frac{77795965\sqrt{5}}{4096} |50\rangle$$
$$+ \frac{424612235\sqrt{14}}{28672} |52\rangle + \frac{3064413\sqrt{70}}{1792} |54\rangle - \frac{28136079\sqrt{6}}{2048} |60\rangle$$
$$- \frac{38143475\sqrt{21}}{3584} |62\rangle - \frac{13151193\sqrt{7}}{1792} |64\rangle + \frac{8133657\sqrt{7}}{1024} |70\rangle$$
$$+ \frac{51221969\sqrt{21}}{7168} |72\rangle + \frac{52674471\sqrt{154}}{39424} |74\rangle + \frac{201825\sqrt{231}}{1408} |76\rangle$$
$$- \frac{7399553\sqrt{2}}{1024} |80\rangle - \frac{18573075\sqrt{42}}{7168} |82\rangle - \frac{15822993\sqrt{154}}{19712} |84\rangle$$
$$- \frac{184275\sqrt{66}}{704} |86\rangle + \frac{3881913}{1024} |90\rangle + \frac{2649285\sqrt{77}}{3584} |92\rangle$$
$$+ \frac{7593807\sqrt{2002}}{78848} |94\rangle + \frac{921375\sqrt{55}}{5632} |96\rangle - \frac{671755\sqrt{10}}{2048} |10,0\rangle$$
$$- \frac{163275\sqrt{33}}{512} |10,2\rangle - \frac{139545\sqrt{715}}{2816} |10,4\rangle - \frac{12285\sqrt{165}}{352} |10,6\rangle$$
$$+ \frac{61005\sqrt{11}}{1024} |11,0\rangle + \frac{25025\sqrt{858}}{2048} |11,2\rangle + \frac{92085\sqrt{286}}{5632} |11,4\rangle$$
$$+ \frac{2835\sqrt{2805}}{1408} |11,6\rangle - \frac{3465\sqrt{3}}{256} |12,0\rangle - \frac{405\sqrt{3003}}{512} |12,2\rangle$$
$$- \frac{555\sqrt{2002}}{704} |12,4\rangle - \frac{315\sqrt{1122}}{704} |12,6\rangle + \frac{1485\sqrt{13}}{4096} |13,0\rangle$$
$$+ \frac{225\sqrt{2002}}{4096} |13,2\rangle + \frac{45\sqrt{17017}}{2816} |13,4\rangle + \frac{45\sqrt{3553}}{2816} |13,6\rangle. \tag{8.32}$$

### 8.2.4 Program for symmetric energy formula

The following Maple program, *symstark*, implements the symmetric energy formula (8.18) which can be used in conjunction with program *gsstark* to calculate further energy corrections. If *gsstark(6)* has been used to calculate energy and wavefunction coefficients to order *6* then *symstark(8,12)* can be used to calculate the energy corrections *8, 10* and *12*. The energy corrections *0, 2, ..., 12* are saved in the file *DE12*.

```
01    # symstark
02    # Symmetric energy formula for the ground state
03    # Stark effect with renormalized matrix elements
04    symEnergy := proc(kmin, kmax)
05
06       local R2010, j, k, r, s, x, z, s1, s2, s3, s4, s5, N, L,
07            NP, LP, Na2;
08    # global vars (input): DD, DE, R, W, Norm2
09    # global vars (output): DE
10
11       R2010 := subs(n=1, ℓ=0, R[1]);
12    for k from kmin by 2 to kmax do
13        lprint(k);
14        j := k/2;
15
```

```
16              # Do terms involving the perturbation matrix
17
18              s1 := 0;
19              for N from 2 to 2*j+1 do
20                  for L from j by -2 to 0 do
21                      if L < N then
22                          Na2 := subs(n=N, l=L, Norm2);
23                          s2 := 0;
24                          z := DD[j, N, L];
25                          if not type(z, name) then
26                              for NP from N-2 to N+2 do
27                                  x := DD[j-1, NP, L-1];
28                                  if not type(x, name) then
29                                      s2 := s2 + subs(n=N, l=L, W[NP-N,-1])*x;
30                                  fi;
31                                  y := DD[j-1, NP, L+1];
32                                  if not type(y, name) then
33                                      s2 := s2 + subs(n=N, l=L, W[NP-N, 1])*y;
34                                  fi;
35                              od;
36                              s1 := s1 + z*s2/Na2;
37                          fi;
38                      fi; # L < N
39                  od; # L
40              od; # N
41
42              # do terms involving lower order energy corrections
43
44              s2 := 0;
45              for r from 2 by 2 to j-1 do
46                  s2 := s2 + DE[2*j-r]*DD[r, 2, 0];
47              od;
48              s1 := s1 - 2 * R2010 * s2;
49              if type(j, even) then
50                  s1 := s1 - DE[j] * R2010 * DD[j, 2, 0];
51              fi;
52
53              s2 := 0;
54              for r from 1 to j do
55                  s3 := 0;
56                  for N from 2 to 2*r+1 do
57                      for L from r by -2 to 0 do
58                          if L < N then
59                              Na2 := subs(n=N, l=L, Norm2);
60                              s4 := 0;
61                              for s from 1 to j-1 do
62                                  if type(2*j-r-s, even) then
```

```
63                               s5 := 0;
64                                for NP from N−1 to N+1 do
65                                 x := DD[s, NP, L];
66                                 if not type(x, name) then
67                                  s5 := s5 +
                                      subs(n=N, ℓ=L, R[NP−N])*x;
68                                 fi;
69                               od;
70                                s4 := s4 + DE[2*j−r−s]*s5;
71                       fi;
72                     od; # s
73                      s3 := s3 + DD[r, N, L]*s4/Na2;
74                fi; # L < N
75              od; # L
76           od; # N
77            s2 := s2 + s3;
78         od; # r
79        DE[k] := simplify(s1 − s2);
80     od: # k
81      save DE, `DE`.kmax;
82   end:
83
84   # Example: Given wavefunction and energy corrections to
85   # order jmax calculate the energy to order 2*jmax
86
87   read `renormdata.m`; # get R, W, Norm2
88   read `estark6.m`; # get DD, DE to order 6
89   symEnergy(8,12): # calculate DE[8] to DE[12]
90   Energy := sum('DE[2*j]*x^(2*j)', 'j'=0..6);
91   quit;
```

For the example shown in lines $84$ to $91$ the resulting energy table saved in file $DE12$ is

```
DE := table([
    (0)       = -1/2,
    (2)       = -9/4,
    (4)       = -3555/64,
    (6)       = -2512779/512,
    (8)       = -13012777803/16384,
   (10)       = -25497693122265/131072
   (12)       = -1389636659571727791/2097152,
]);
```

corresponding to the energy expansion

$$E = -\frac{1}{2} - \frac{9}{4}\lambda^2 - \frac{3555}{64}\lambda^4 - \frac{2512779}{512}\lambda^6 - \frac{13012777803}{16384}\lambda^8$$

$$- \frac{25497693122265}{131072}\lambda^{10} - \frac{1389636659571727791}{2097152}\lambda^{12} \dots$$

The energy corrections to order 100 in both rational and floating point form are given in Appendix E.

## 8.3   Summing the Ground State Series

The presence of the linear field term $\lambda z$ in (8.1) means that there are no true bound states for a hydrogenic atom in an electric field: the infinite Coulomb barrier is removed and all states are analogous to resonances with a finite lifetime characterized by both an energy level position and a level width. Thus it is not surprising that the ground state energy series

$$E(\lambda) = -\frac{1}{2} + \sum_{n=1}^{\infty} E^{(2n)} \lambda^{2n} = \sum_{n=0}^{\infty} a_n \lambda^n \qquad (8.33)$$

with coefficients to order 100 given in Appendix E, is divergent (see [SI82] and references therein).

### 8.3.1   Direct summation of the series

Direct summation of a divergent asymptotic series is only feasible for very small values of the parameter. A characteristic feature of such a series, if it is summed term by term, is that the error initially decreases and then rapidly becomes unbounded so only limited accuracy can be obtained and only for relatively small values of the parameter. The limited usefulness of direct summation of the ground state Stark series is summarized in Table 8.1 (see [SI78] for more detailed results).

**Table 8.1**: Direct summation of the ground state stark series (8.33). Order is the highest power of $\lambda$ used in the sum. The non-perturbative results are from [AL69] and [HE74].

| $\lambda$ | $E$ from (8.33) | Order | $E$ (non-perturbative) |
|---|---|---|---|
| 0.001 | $-0.50000\ 22500\ 5555$ | 16 | $-0.50000\ 22500\ 5556$ |
| 0.005 | $-0.50005\ 62847\ 9379$ | 16 | $-0.50005\ 62847\ 937$ |
| 0.01 | $-0.50022\ 55604\ 5796$ | 16 | $-0.50022\ 55604\ 59$ |
| 0.03 | $-0.50207\ 42636$ | 16 | $-0.50207\ 4273$ |
| 0.06 | $-0.50918$ | 8 | $-0.50920\ 4$ |
| 0.08 | $-0.5167$ | 4 | $-0.51756$ |
| 0.1 | $-0.5281$ | 4 | $-0.5275$ |
| 0.12 | $-0.532$ | 2 | $-0.5374$ |

To extend the parameter range and obtain better results for both the level position and level width it is necessary to find a summability method to extract information from (8.33) using the known coefficients $E^{(2n)}$ to a given order and possibly also an asymptotic formula for these coefficients which can be used for higher orders. A detailed rigorous analysis of these methods is beyond the scope of this book (see [SI82] and references therein). It has been shown that the series (8.33) is asymptotic [HE79] and Borel summable [GR78] for imaginary values of $\lambda$. In fact the leading asymptotic behaviour of the ground state energy coefficients is

$$E^{(2n)} \sim -\frac{6}{\pi}\left(\frac{3}{2}\right)^{2n}(2n)!\left[1+O\left(\frac{1}{n}\right)\right]. \tag{8.34}$$

A comparison of this formula with exact values, and more accurate versions of it valid for a general parabolic state, has been given in [SI79].

## 8.3.2  Summation using Padé approximants

The method of Padé approximants can often be used to sum a divergent series and extend the useful parameter range to larger values of $\lambda$. Given an infinite series whose first $N+1$ terms are known,

$$E(\lambda) = \sum_{n=0}^{N} a_n \lambda^n + O(\lambda^{N+1}), \tag{8.35}$$

it is possible to obtain the rational function approximations (Padé approximants) [BA75]

$$R^{[P,Q]}(\lambda) = \frac{\sum_{n=0}^{P} p_n \lambda^n}{\sum_{n=0}^{Q} q_n \lambda^n}, \quad q_0 = 1 \tag{8.36}$$

for degrees $P$ and $Q$ satisfying $P+Q \le N$. For example the diagonal sequence

$$R^{[P,P]}(\lambda), \quad P = 0, 1, \ldots, \lfloor N/2 \rfloor \tag{8.37}$$

for some given $\lambda$ may converge much faster than the original series or if the original series is divergent it may converge and provide a summability method.

Thus it is tempting to calculate diagonal sequences for the Stark series (8.33). In fact, the diagonal Padé sequences do not converge for real values of $\lambda$ because their poles are real [RE82]. However the diagonal Padé sequences do converge for imaginary values of $\lambda$ and this suggests that we make an analytic continuation of the series (8.33) about a new expansion point on the imaginary axis, say $\lambda = 0.01\,i$, and then evaluate this new expansion for the real values of $\lambda$ in which we are interested (see [RE82] for more details).

If we use the notation

$$E(\lambda, \lambda_0) = \sum_{k=0}^{\infty} a_k(\lambda_0)(\lambda - \lambda_0)^k \tag{8.38}$$

then the given series is $E(\lambda) = E(\lambda, 0)$ and we know the coefficients $a_n(0)$ to a certain order. Therefore we can calculate the new coefficients $a_n(\lambda_0)$ in terms of the $a_n(0)$ with the usual Taylor series formula

$$
\begin{aligned}
a_k(\lambda_0) &= \frac{1}{k!} \left. \frac{d^k E(\lambda, 0)}{d\lambda^k} \right|_{\lambda=\lambda_0} \\
&= \sum_{n=0}^{\infty} \frac{(n+1)(n+2)\cdots(n+k)}{k!} a_{n+k}(0)\lambda_0^n \\
&= \sum_{n=0}^{\infty} a_n^{(k)} \lambda_0^n.
\end{aligned}
\tag{8.39}
$$

Now if we know the first $N+1$ coefficients $a_k(0)$, $k = 0,\ldots,N$ then we can calculate the first $N - k + 1$ coefficients $a_n^{(k)}$, $n = 0,\ldots,N-k$ in the series (8.39). Therefore to sum each series of the form (8.39) (e.g. for $\lambda_0 = 0.01\,i$) we calculate its sequence of diagonal Padé approximants. Then using the converged values $a_k(\lambda_0)$ we can calculate the sequence of diagonal Padé approximants for (8.38) for some real value of $\lambda$.

**Maple program and results for $a_k(\lambda_0)$**

It is easy to develop a simple Maple program, *reinhardt*, which can be used to test the convergence properties of the infinite series for the $a_k(\lambda_0)$:

```
01    # reinhardt
02    test := proc(n)
03        et := evalf(taylor(coeff(Enew, x-x0, n), x0, pmax+1));
04        if type(n,even) then kmin := 2; else kmin := 3; fi;
05        for k from kmin by 2 to (pmax-n)/2 do
06            etpade := convert(et, ratpoly, k, k):
07            etvalue := evalf(evalc(subs(x0=0.01*I, etpade)));
08            print(n, k, etvalue);
09        od:
10    end:
11
12    read DE100: # get Stark energy coefficients
13    Digits := 32: # 32 digit precision
14    jmax := 50:
15    pmax := 2*jmax:
16    Energy := sum('DE[2*j]*x^(2*j)', 'j'=0..jmax):
17    Enew := taylor(Energy, x=x0, pmax+1):
18    pmax := 2*jmax:
19
20    for k from 0 to 30 do test(k); od;
```

Here we are assuming that the file *DE100* contains the ground state energy coefficients *DE[0]*, *DE[2]*, ..., *DE[100]* where *DE[2*j]* corresponds to $E^{(2j)}$. Also *x* and *x0* correspond to $\lambda$ and $\lambda_0$.

The series manipulations in (8.38) and (8.39) are easily performed by the *taylor* and *coeff* functions. Thus in lines *16* and *17* the energy coefficients are converted into the Taylor series (8.38) and in line *03* the Taylor series (8.39) are obtained and evaluated in floating point form using *evalf*. Since only even order energy corrections are non-zero then only even order terms appear in (8.39) if $k$ is even and only odd order terms appear if $k$ is odd. The Maple *convert* function with the *ratpoly* option is then used in line *06* to obtain the Padé approximants and in line *07* they are evaluated at the point $0.01\,i$ on the imaginary axis. It should be noted that Maple will work automatically with complex numbers ($i$ is denoted by $I$) and the *evalc* function is used to reduce a complex value to the standard form $a + bi$. Finally in line *20* the convergence of the series for $a_k(\lambda_0)$ for $0 \leq k \leq 30$ is tested. The series coefficients are real for $k$ even and purely imaginary for $k$ odd. It is to be expected as $k$ increases that the convergence is slower since we have fewer and fewer terms in (8.39). The results are summarized in Table 8.2 for some selected values of $k$.

Thus with coefficients $E^{(2n)}$ to order 100 we can calculate the coefficients in (8.38) but with decreasing accuracy. The converged results are given in Table 8.3.

## Maple program and results for $E(\lambda, 0.01\,i)$

Finally the following Maple program can be used to compute the diagonal Padé approximants for (8.38) for real values of $\lambda$. In the program we are assuming that the file *sumdata* contains the coefficients in Table 8.3, denoted by *a[0]* to *a[30]* in the program.

```
01    # sumit
02    makepade := proc(jmin,jmax)
03        local j, Ecomplex, et;
04        Ecomplex := sum('a[j]*z^j', 'j'=0..30):
05        et := evalf(taylor(Ecomplex,z,31)):
06        for j from jmin to jmax do
07            lprint(j);
08            epade[j] := convert(et,ratpoly,j,j);
09        od:
10        save epade, `epade`.jmax.`.m`;
11    end:
12
13    evalpade := proc(jmin, jmax, x)
14        local j, v;
15        for j from jmin to jmax do
16            subs(z=x-0.01*I, epade[j]);
17            v := evalc(");
18            print(j, evalf(v,16));
19        od:
20    end:
21
```

```
22    Digits := 32:
23    read sumdata:
24    makepade(10,15);
25    for x in [0.03,0.04,0.05,0.06,0.07,0.08,0.09,0.1] do
26       print(x);
27       evalpade(10,15,x);
28    od:
29    quit;
```

Some results for several values of $\lambda$ are given in Table 8.4.

**Table 8.2**: Padé summation of the coefficients $a_k(\lambda_0)$ in (8.39) for $\lambda_0 = 0.01\,i$ and selected values of $k$ illustrating the convergence behaviour.

| $k$ | $P$ | $a_k(\lambda_0) \sim R^{[P,P]}(\lambda_0)$ |
|---|---|---|
| 0 | 10 | $-0.49977\ 55506\ 38520\ 32110\ 21537\ 12680\ 93$ |
|  | 20 | $-0.49977\ 55506\ 38520\ 32110\ 07781\ 74864\ 62$ |
|  | 22 | $-0.49977\ 55506\ 38520\ 32110\ 07781\ 74864\ 04$ |
|  | 40 | $-0.49977\ 55506\ 38520\ 32110\ 07781\ 74864\ 08$ |
| 5 | 15 | $-0.25441\ 96872\ 56041\ 95773\ 09790\ 20723 \times 10^3\,i$ |
|  | 35 | $-0.25441\ 96872\ 56041\ 88548\ 26379\ 88902 \times 10^3\,i$ |
|  | 37 | $-0.25441\ 96872\ 56041\ 88548\ 26379\ 88808 \times 10^3\,i$ |
|  | 39 | $-0.25441\ 96872\ 56041\ 88548\ 26379\ 88801 \times 10^3\,i$ |
|  | 41 | $-0.25441\ 96872\ 56041\ 88548\ 26379\ 88800 \times 10^3\,i$ |
| 10 | 20 | $0.42157\ 36518\ 08538\ 48514\ 67956 \times 10^8$ |
|  | 30 | $0.42157\ 36518\ 08558\ 58935\ 56509 \times 10^8$ |
|  | 32 | $0.42157\ 36518\ 08558\ 58944\ 94260 \times 10^8$ |
|  | 34 | $0.42157\ 36518\ 08558\ 58944\ 07914 \times 10^8$ |
|  | 42 | $0.42157\ 36518\ 08558\ 58943\ 20766 \times 10^8$ |
|  | 44 | $0.42157\ 36518\ 08558\ 58943\ 20814 \times 10^8$ |
| 20 | 26 | $-0.20024\ 44260\ 90607 \times 10^{22}$ |
|  | 28 | $-0.20024\ 44260\ 57846 \times 10^{22}$ |
|  | 30 | $-0.20024\ 44265\ 35599 \times 10^{22}$ |
|  | 38 | $-0.20024\ 44265\ 46676 \times 10^{22}$ |
|  | 40 | $-0.20024\ 44265\ 46675 \times 10^{22}$ |
| 30 | 24 | $-0.63679 \times 10^{36}$ |
|  | 26 | $-0.61231 \times 10^{36}$ |
|  | 32 | $-0.61160 \times 10^{36}$ |
|  | 34 | $-0.61144 \times 10^{36}$ |

These results are not too sensitive to the expansion point. The choice $\lambda_0 = 0.005 + 0.005\,i$ gives results of similar accuracy [RE82] and there is agreement with more accurate results [BE80]. Comparison with Table 8.1 also shows that

**Table 8.3**: Converged series coefficients $a_k(\lambda_0)$ in (8.38) for $k = 0, \ldots, 30$ for the analytic continuation of (8.33) about the imaginary point $\lambda_0 = 0.01\, i$.

| $k$ | $a_k(\lambda_0)$ |
|---|---|
| 0 | $-0.49977\ 55506\ 38520\ 32110\ 07781\ 74864\ 1$ |
| 1 | $-0.44780\ 69549\ 36730\ 64457\ 16683\ 95590\ 6 \times 10^{-1}\, i$ |
| 2 | $-0.22173\ 86636\ 28131\ 55283\ 00144\ 73513\ 3 \times 10^1$ |
| 3 | $-0.21279\ 47435\ 33595\ 56142\ 71124\ 35763\ 0 \times 10^1\, i$ |
| 4 | $-0.48703\ 34056\ 92367\ 39290\ 82190\ 47099\ 5 \times 10^2$ |
| 5 | $-0.25441\ 96872\ 56041\ 88548\ 26379\ 88800 \times 10^3\, i$ |
| 6 | $-0.30389\ 93086\ 45119\ 50946\ 03590\ 50903 \times 10^4$ |
| 7 | $-0.44581\ 56137\ 87836\ 40592\ 93616\ 34 \times 10^5\, i$ |
| 8 | $-0.17491\ 00615\ 35298\ 60222\ 64981\ 9 \times 10^6$ |
| 9 | $-0.93506\ 85752\ 37433\ 88464\ 1495 \times 10^7\, i$ |
| 10 | $\phantom{-}0.42157\ 36518\ 08558\ 58943\ 208 \times 10^8$ |
| 11 | $-0.19259\ 72158\ 88386\ 34316\ 60 \times 10^{10}\, i$ |
| 12 | $\phantom{-}0.33566\ 64715\ 07272\ 66904 \times 10^{11}$ |
| 13 | $-0.20070\ 39805\ 27906\ 633 \times 10^{12}\, i$ |
| 14 | $\phantom{-}0.14929\ 32510\ 32930\ 862 \times 10^{14}$ |
| 15 | $\phantom{-}0.15099\ 24885\ 49190\ 67 \times 10^{15}\, i$ |
| 16 | $\phantom{-}0.39945\ 67702\ 26195\ 8 \times 10^{16}$ |
| 17 | $\phantom{-}0.14653\ 05032\ 79968 \times 10^{18}\, i$ |
| 18 | $-0.86163\ 90781\ 200 \times 10^{18}$ |
| 19 | $\phantom{-}0.63836\ 01134\ 50 \times 10^{20}\, i$ |
| 20 | $-0.20024\ 44262\ 4668 \times 10^{22}$ |
| 21 | $-0.82181\ 23587\ 2 \times 10^{22}\, i$ |
| 22 | $-0.11665\ 58899\ 6 \times 10^{25}$ |
| 23 | $-0.38482\ 0735 \times 10^{26}\, i$ |
| 24 | $\phantom{-}0.22316\ 4786 \times 10^{27}$ |
| 25 | $-0.24645\ 154 \times 10^{29}\, i$ |
| 26 | $\phantom{-}0.98836\ 67 \times 10^{30}$ |
| 27 | $\phantom{-}0.11066\ 3 \times 10^{32}\, i$ |
| 28 | $\phantom{-}0.54647 \times 10^{33}$ |
| 29 | $\phantom{-}0.3093 \times 10^{35}\, i$ |
| 30 | $-0.611 \times 10^{36}$ |

we obtain more accurate results than direct summation and have extended the useful parameter range from $\lambda \approx 0.12$ to $\lambda \approx 0.5$.

Our numerical analysis of the ground state perturbation series using Padé approximants is an excellent example of the use of a symbolic computation language such as Maple for simple exploratory analysis. Having determined the success of the method it would be more efficient for a more extensive analysis to use a numerical language such as FORTRAN and specialized routines for the efficient one-point evaluation of high order Padé approximants.

**Table 8.4**: Ground state Stark resonances for selected real values of $\lambda$ obtained by Padé summation of (8.38) for $\lambda_0 = 0.01\,i$ using the converged coefficients in Table 8.3. Here $|\epsilon| < 10^{-18}$. The real parts are the energy levels and the imaginary parts give the level widths (one-half the ionization rates).

| $\lambda$ | $E(\lambda, 0.01\,i)$ |
|---|---|
| 0.001 | $-0.50000\ 22500\ 55551\ 8 + \epsilon\,i$ |
| 0.01 | $-0.50022\ 55604\ 57959\ 9 + \epsilon\,i$ |
| 0.03 | $-0.50207\ 42726\ 076 - 0.00000\ 00111\ 874\,i$ |
| 0.04 | $-0.50377\ 15910\ 643 - 0.00000\ 19463\ 39873\,i$ |
| 0.05 | $-0.50610\ 54250\ 74 - 0.00003\ 85911\ 38\,i$ |
| 0.06 | $-0.50920\ 34508 - 0.00025\ 75462\ 7\,i$ |
| 0.07 | $-0.51307\ 67511 - 0.00092\ 36551\,i$ |
| 0.08 | $-0.51756\ 07157 - 0.00226\ 98365\,i$ |
| 0.09 | $-0.52241\ 2756 - 0.00439\ 22182\,i$ |
| 0.1 | $-0.52741\ 7696 - 0.00726\ 9028\,i$ |
| 0.2 | $-0.57011\ 101 - 0.06060\ 315\,i$ |
| 0.3 | $-0.59681\ 12 - 0.13068\ 53\,i$ |
| 0.4 | $-0.61310\ 46 - 0.20499\ 50\,i$ |
| 0.5 | $-0.62261\ 1 - 0.27995\ 48\,i$ |
| 1.0 | $-0.62572 - 0.64347\,i$ |

## 8.4 Stark Effect in Parabolic Coordinates

### 8.4.1 Separation and perturbation formalism

We have seen in Section 4.14 that the Schrödinger equation for the 3-dimensional hydrogen atom is separable in parabolic coordinates. This separability holds even when the Stark perturbation $\lambda z$ is included. Moreover the perturbation of any parabolic state, not just the ground state, can be treated using nondegenerate RSPT since the perturbation matrix is diagonal in the parabolic quantum numbers $n_1$, $n_2$ and $m$.

If we repeat the transformation to parabolic coordinates of Section 4.14.1 but with the Schrödinger equation (8.1) containing the perturbation term the result is the two separated equations

$$\left[-\frac{d}{d\xi}\left(\xi\frac{d}{d\xi}\right)+\frac{m^2}{4\xi}-\frac{E}{2\alpha^2}\xi+\frac{\lambda}{4\alpha^3}\xi^2-\lambda_1\right]f_1(\xi)=0, \tag{8.40}$$

$$\left[-\frac{d}{d\eta}\left(\eta\frac{d}{d\eta}\right)+\frac{m^2}{4\eta}-\frac{E}{2\alpha^2}\eta-\frac{\lambda}{4\alpha^3}\eta^2-\lambda_2\right]f_2(\eta)=0, \tag{8.41}$$

where

$$\lambda_1+\lambda_2=\frac{Z}{\alpha}. \tag{8.42}$$

These two equations are identical in form except for the sign of the perturbation so it will be sufficient to solve one of them, say the first.

If we let $\alpha=1$ and $Z=1$ and apply the scaling transformation $\xi=2\gamma X$ as in Section 4.14.4 then the scaled version of (8.40) is

$$\left[-\frac{1}{2}\frac{d}{dX}\left(X\frac{d}{dX}\right)+\frac{m^2}{8X}+\frac{X}{2}+\gamma^3\lambda X^2-\gamma\lambda_1\right]F_1(X)=0, \tag{8.43}$$

where the scaling parameter has been chosen so that

$$E=-\frac{1}{2\gamma^2}. \tag{8.44}$$

We can use the so(2,1) realization (4.249) to (4.251):

$$T_1=-\frac{1}{2}\left(X\frac{d^2}{dX^2}+\frac{d}{dX}\right)+\frac{m^2}{8X}-\frac{X}{2}, \tag{8.45}$$

$$T_2=-i\left(X\frac{d}{dX}+\frac{1}{2}\right), \tag{8.46}$$

$$T_3=-\frac{1}{2}\left(X\frac{d^2}{dX^2}+\frac{d}{dX}\right)+\frac{m^2}{8X}+\frac{X}{2}. \tag{8.47}$$

The Casimir operator is

$$T^2=\tau, \tag{8.48}$$

where (see Section 4.14.3)

$$\tau=\frac{m^2-1}{4}=k(k+1), \tag{8.49}$$

$$k=\frac{|m|-1}{2}, \tag{8.50}$$

and the eigenvalues of $T_3$ are defined in terms of $k$ (or equivalently, $|m|$) and the usual parabolic quantum number $n_1$ by

$$q_1=k+1+n_1, \quad n_1=0,1,2,\dots$$
$$=n_1+\frac{1+|m|}{2}. \tag{8.51}$$

Similarly for the $\eta$ equation (8.41), applying the scaling transformation $\eta = 2\gamma Y$

$$\left[ -\frac{1}{2}\frac{d}{dY}\left(Y\frac{d}{dY}\right) + \frac{m^2}{8Y} + \frac{Y}{2} - \gamma^3\lambda Y^2 - \gamma\lambda_2 \right] F_2(Y) = 0, \qquad (8.52)$$

and we can define a realization $S_1$, $S_2$ and $S_3$ of so(2,1) equivalent to the one in terms of $X$ but with $X$ replaced by $Y$ and with $S_3$ eigenvalue spectrum

$$q_2 = k + 1 + n_2, \quad n_2 = 0, 1, 2, \ldots$$
$$= n_2 + \frac{1 + |m|}{2}. \qquad (8.53)$$

Therefore we have two conventional perturbation problems

$$\left[ T_3 + \Lambda X^2 - Q_1 \right] F_1(X) = 0, \qquad (8.54)$$
$$\left[ S_3 - \Lambda Y^2 - Q_2 \right] F_2(Y) = 0, \qquad (8.55)$$

where

$$\Lambda = \gamma^3\lambda, \quad Q_1 = \gamma\lambda_1, \quad Q_2 = \gamma\lambda_2, \qquad (8.56)$$

and

$$Q_1 + Q_2 = \gamma. \qquad (8.57)$$

These are equations for the perturbed separation constants $Q_1$ and $Q_2$ rather than for the energy $E$ so we assume expansions of the form

$$Q_1 = Q_{10} + Q_{11}\Lambda + Q_{12}\Lambda^2 + \cdots,$$
$$Q_2 = Q_{20} + Q_{21}\Lambda + Q_{22}\Lambda^2 + \cdots,$$

where the coefficients depend on $q_1$, $q_2$ and $k$ (or equivalently, $n_1$, $n_2$ and $m$) corresponding to the unperturbed reference state. Therefore

$$Q_1 = f^{(0)}(q_1, k) + f^{(1)}(q_1, k)\Lambda + \cdots = \sum_{j=0}^{\infty} f^{(j)}(q_1, k)\Lambda^j, \qquad (8.58)$$

$$Q_2 = f^{(0)}(q_2, k) - f^{(1)}(q_2, k)\Lambda + \cdots = \sum_{j=0}^{\infty} (-1)^j f^{(j)}(q_2, k)\Lambda^j, \qquad (8.59)$$

which express the fact that (8.55) is the same as (8.54) except for the sign of $\Lambda$.

First we determine the functions $f^{(j)}(q, k)$ from conventional perturbation theory based on (8.54) and then we unravel the equations (8.57), (8.56) and (8.44) to determine the desired energy expansion

$$E = -\frac{1}{2n^2} + E^{(1)}\lambda + E^{(2)}\lambda^2 + \cdots \qquad (8.60)$$

in terms of the original perturbation parameter $\lambda$, the electric field strength.

We need the matrix elements of the perturbation $V = X^2$ with respect to the $T_3$ eigenfunctions $|qk\rangle$. They are given in Table D.7 (we are using $X$ rather than $R$ as the variable defining the realization) so the nonzero matrix elements are (dropping the subscript on $q$ so that $q = k + 1 + n_1$)

$$\langle q+2,k|X^2|qk\rangle = \frac{1}{4}\sqrt{(q-k)(q-k+1)(q+k+1)(q+k+2)},$$

$$\langle q+1,k|X^2|qk\rangle = -\frac{1}{2}(2q+1)\sqrt{(q-k)(q+k+1)},$$

$$\langle qk|X^2|qk\rangle = \frac{1}{2}\left[3q^2 - k(k+1)\right],$$

$$\langle q-1,k|X^2|qk\rangle = -\frac{1}{2}(2q-1)\sqrt{(q-k-1)(q+k)},$$

$$\langle q-2,k|X^2|qk\rangle = \frac{1}{4}\sqrt{(q-k-2)(q-k-1)(q+k-1)(q+k)},$$

$$(8.61)$$

and we can use the shorthand notation

$$V_{\mu\nu} = \langle q+\mu,k|X^2|q+\nu,k\rangle. \tag{8.62}$$

We can now apply the conventional perturbation theory developed in Section 7.2. This is easily done by hand only to second order and the results are

$$f^{(0)}(q,k) = q, \tag{8.63}$$

$$f^{(1)}(q,k) = V_{00} = \frac{1}{2}[3q^2 - k(k+1)]$$

$$= \frac{1}{4}(6q^2 - 2\tau), \tag{8.64}$$

$$f^{(2)}(q,k) = \sum_{\mu\neq 0} \frac{V_{0\mu}V_{\mu 0}}{q-(q+\mu)} = \sum_{\mu\neq 0} \frac{V_{\mu 0}^2}{(-\mu)}$$

$$= -\frac{1}{2}V_{2,0}^2 - V_{1,0}^2 + V_{-1,0}^2 + \frac{1}{2}V_{-2,0}^2$$

$$= \frac{1}{8}\left[-34\,q^3 - 5\,q + 18\,qk(k+1)\right]$$

$$= \frac{1}{8}(-34q^3 - 5q + 18q\tau). \tag{8.65}$$

The formulas (7.21) and (7.23) for $f^{(3)}$ and $f^{(4)}$ are

$$f^{(3)}(q,k) = \sum_{\mu\neq 0}\sum_{\nu\neq 0} \frac{V_{0\mu}V_{\mu\nu}V_{\nu 0}}{(-\mu)(-\nu)} - f^{(1)}\sum_{\mu\neq 0}\frac{V_{\mu 0}^2}{(-\mu)^2}, \tag{8.66}$$

$$f^{(4)}(q,k) = \sum_{\mu\neq 0}\sum_{\nu\neq 0}\sum_{\sigma\neq 0} \frac{V_{0\mu}V_{\mu\nu}V_{\nu\sigma}V_{\sigma 0}}{(-\mu)(-\nu)(-\sigma)}$$

$$-2f^{(1)}\sum_{\mu\neq0}\sum_{\nu\neq0}\frac{V_{0\mu}V_{\mu\nu}V_{\nu0}}{(-\mu)^2(-\nu)}+(f^{(1)})^2\sum_{\mu\neq0}\frac{V_{\mu0}^2}{(-\mu)^3}$$

$$-f^{(2)}\sum_{\mu\neq0}\frac{V_{\mu0}^2}{(-\mu)^2}. \tag{8.67}$$

They would be quite tedious to work out by hand.

### 8.4.2  Maple program for $f^{(0)}$ to $f^{(4)}$

The following Maple program easily produces the corrections $f^{(0)}$ to $f^{(4)}$ for the separation constant $Q_1$. The matrix elements of $V = X^2$ are defined as functions of $q$ and $k$ by a Maple table such that

$$V_{\mu0} = \langle q+\mu, k|X^2|qk\rangle \longleftrightarrow V[mu]$$
$$V_{\mu\nu} = \langle q+\mu, k|X^2|q+\nu, k\rangle \longleftrightarrow subs(q=q+nu, V[mu\text{-}nu])$$

and similarly for the other matrix elements.

```
01    # alliluev
02    # Explicit 4th order calculation
03
04    V := table([
05          (2) = sqrt((q−k)*(q−k+1)*(q+k+1)*(q+k+2))/4,
06          (1) = −(2*q+1)*sqrt((q−k)*(q+k+1))/2,
07          (0) = (3*q^2−k*(k+1))/2,
08          (−1) = −(2*q−1)*sqrt((q−k−1)*(q+k))/2,
09          (−2) = sqrt((q−k−2)*(q−k−1)*(q+k−1)*(q+k))/4
10       ]):
11
12    # 0th to 2nd order calculation
13
14    f[0] := q;
15    f[1] := V[0];
16
17    f[2] := 0:
18    for mu in [−2,−1,1,2] do
19        f[2] := f[2] + V[mu]^2/(−mu);
20    od:
21    f[2] := simplify(f[2]);
22
23    # 3rd order calculation
24
25    s1 := 0:
26    s2 := 0:
27    for mu in [−2,−1,1,2] do
28        for nu in [−2,−1,1,2] do
29            if abs(mu−nu) <= 2 then
```

```
30          s1 := s1 + V[mu]*subs(q=q+nu, V[mu-nu])
31                * V[nu]/(-mu)/(-nu);
32        fi;
33      od;
34      s2 := s2 + V[mu]^2/(-mu)^2;
35  od:
36  f[3] := simplify(s1 - f[1]*s2);
37
38  # 4th order calculation
39
40  s1 := 0:
41  for mu in [-2,-1,1,2] do
42      for nu from mu-2 to mu+2 do
43          if nu <> 0 and abs(mu-nu) <= 2 then
44              for sigma in [-2,-1,1,2] do
45                  if abs(nu-sigma) <= 2 then
46                      s1 := s1 + V[mu] * subs(q=q+nu,V[mu-nu])
47                          * subs(q=q+sigma, V[nu-sigma])
48                          * V[sigma]/((-mu)*(-nu)*(-sigma));
49                  fi;
50              od;
51          fi;
52      od;
53  od:
54
55  s2 := 0:
56  s3 := 0:
57  s4 := 0:
58  for mu in [-2,-1,1,2] do
59      for nu in [-2,-1,1,2] do
60          if abs(mu-nu) <= 2 then
61              s2:= s2 + V[mu]*subs(q=q+nu, V[mu-nu])
62                  * V[nu]/((-mu)^2*(-nu));
63          fi;
64      od;
65      s3 := s3 + V[mu]^2/(-mu)^3;
66      s4 := s4 + V[mu]^2/(-mu)^2;
67  od:
68  f[4] := simplify(s1 - 2*f[1]*s2 + f[1]^2*s3 - f[2]*s4);
69
70  # use q and m rather than q and k
71
72  for j from 0 to 4 do
73      f[j] := simplify(subs(k=-1/2+m/2, f[j]));
74  od;
75  save f, fqm;
```

The results for $f^{(1)}$ to $f^{(4)}$ have been expressed in terms of $q$ and $m$ rather than $q$ and $k$ using $k = (m - 1)/2$ to facilitate comparison with Alliluev and Malkin's results (the notation $k$ is used in [AL74] for our $q$). The results are

$$4\,f^{(1)} = 6\,q^2 + \tfrac{1}{2} - \tfrac{1}{2}\,m^2, \tag{8.68}$$

$$4^2\,f^{(2)} = -19q - 68\,q^3 + 9\,qm^2, \tag{8.69}$$

$$4^3\,f^{(3)} = \tfrac{131}{4} + 918\,q^2 - \tfrac{71}{2}\,m^2 + 1500\,q^4 + \tfrac{11}{4}\,m^4 \\ - 258\,q^2 m^2, \tag{8.70}$$

$$4^4\,f^{(4)} = -\tfrac{22709}{4}\,q - 46810\,q^3 + \tfrac{7389}{2}\,qm^2 + 8910\,q^3 m^2 \\ - \tfrac{909}{4}\,qm^4 - 42756\,q^5. \tag{8.71}$$

We can also compare with Silverstone's results by using $q$ and $\tau = (m^2 - 1)/4$ (the notation $k$ and $M$ is used in [SI78] for our $q$ and $\tau$) and the results are

$$4\,f^{(1)} = 6\,q^2 - 2\,\tau, \tag{8.72}$$

$$4^2\,f^{(2)} = -10\,q - 68\,q^3 + 36\,q\tau, \tag{8.73}$$

$$4^3\,f^{(3)} = 660\,q^2 - 120\,\tau + 1500\,q^4 + 44\,\tau^2 - 1032\,q^2\tau, \tag{8.74}$$

$$4^4\,f^{(4)} = -2210\,q - 37900\,q^3 + 12960\,q\tau + 35640\,q^3\tau \\ - 3636\,q\tau^2 - 42756\,q^5. \tag{8.75}$$

### 8.4.3   Unravelling the energy expansion

We can now use the $f^{(j)}$ to obtain the energy expansion (8.60) as follows. If we substitute the $f^{(j)}$ into (8.57) and (8.58) then

$$Q_1 + Q_2 = A_0 + A_1\Lambda + A_2\Lambda^2 + \cdots, \tag{8.76}$$

where

$$A_j = f^{(j)}(q_1, k) + (-1)^j f^{(j)}(q_2, k). \tag{8.77}$$

But $Q_1 + Q_2 = \gamma$ and $\Lambda = \gamma^3 \lambda$ so

$$A_0 + A_1\left(\gamma^3\lambda\right) + A_2\left(\gamma^3\lambda\right)^2 + \cdots = \gamma, \tag{8.78}$$

which is an implicit equation defining $\gamma$ as a function of $\lambda$. If we assume the expansion

$$\gamma = g_0 + g_1\lambda + g_2\lambda^2 + \cdots \tag{8.79}$$

we can substitute it into (8.78) and obtain for $g_j$ the coefficient of $\lambda^j$ in the left side of (8.78). Then the $g_j$ are expressions involving the $A_j$. Finally (8.79) can be substituted into

$$E = -\frac{1}{2\gamma^2} \tag{8.80}$$

to obtain the energy expansion (8.60). For example $A_0 = n$ and $A_1 = 3n(q_1 - q_2)/2$ so $g_0 = A_0 = n$ and $g_1 = A_1 g_0^3 = 3n^4(q_1 - q_2)/2$ and $E^{(1)} = g_1/n^3 = 3n(q_1 - q_2)/2$.

The following continuation of the preceding Maple program easily performs all the calculations to fourth order. First expand $Q_1 + Q_2$ in a Taylor series to order 4 in $\lambda$ (denoted by $x$ in the program ) and solve the implicit equation for $\gamma$ (denoted by $G$ in the program). The Maple *taylor* function does the expansion including an order term. These manipulations are first done symbolically in terms of the coefficients $A[j]$.

```
76   G := sum('g[j]*x^j', 'j' = 0..4);
77   Q1Q2 := sum('A[j]*(G^3*x)^j', 'j' = 0..4);
78   eq := taylor(Q1Q2, x, 5);
```

Now $g[j]$ is just the coefficient of $x^j$ in $eq$ so it can be obtained and factored using the Maple *coeff* and *factor* functions.

```
79   for j from 0 to 4 do
80       g[j] := factor(coeff(eq, x, j));
81   od;
```

Next *taylor* is used to find the series expansion of the energy from (8.80) and *coeff* and *factor* are used to extract the energy coefficients $E^{(j)}$ (denoted by $e[j]$ in the program).

```
82   Energy := taylor(-(1/2)*G^(-2), x, 5):
83   for j from 0 to 4 do
84       e[j] := factor(coeff(Energy, x, j));
85   od;
```

Now $A_0$ to $A_4$ are computed as functions of $n$, $Q$ and $m$ where $n = q_1 + q_2$ is the principal quantum number and

$$Q = q_1 - q_2 = n_1 - n_2. \tag{8.81}$$

```
86   for j from 0 to 4 do
87       A[j] := (subs(q=q1, f[j]) + (-1)^j * subs(q=q2, f[j]))/n;
88       A[j] := subs(q1=(n+Q)/2, q2=(n-Q)/2, A[j]);
89       A[j] := simplify(A[j]);
90   od:
```

The energy coefficients can now be obtained as functions of $n$, $Q$ and $m$.

```
91   for j from 0 to 4 do
92       e[j] := factor(e[j]);
93   od;
94   save e, enQm;
```

The results are

$$E^{(1)} = \frac{3}{2}\,nQ, \tag{8.82}$$

$$E^{(2)} = \frac{1}{16}\,n^4(3\,Q^2 - 19 - 17\,n^2 + 9\,m^2), \tag{8.83}$$

$$E^{(3)} = \frac{3}{32}\,n^7 Q(-Q^2 + 39 + 23\,n^2 + 11\,m^2), \tag{8.84}$$

$$\begin{aligned}
E^{(4)} = \frac{1}{1024}\,n^{10}(&549\,m^4 - 16211 + 8622\,m^2 - 35182\,n^2 \\
&- 5754\,Q^2 - 5487\,n^4 - 147\,Q^4 - 1806\,n^2 Q^2 \\
&+ 3402\,m^2 n^2 + 1134\,m^2 Q^2).
\end{aligned} \tag{8.85}$$

As a check the values $Q = 0$, $m = 0$ and $n = 1$ corresponding to the ground state can be substituted.

```
95   for j from 0 to 4 do
96       eg[j] := subs(Q=0, m=0, n=1, e[j]);
97   od;
```

The results agree with (7.96) to fourth order.

### 8.4.4　Higher orders: general case

The difficulty of the hand calculations involved in obtaining the energy to fourth order is evident since it wasn't until 1973 that the correct results were obtained by Alliluev and Malkin in [AL74]. The first systematic high order calculation which extends these fourth order results was given by Silverstone. He used the general iteration scheme (7.25) to (7.28) of conventional RSPT combined with extensive bookkeeping and a double precision floating point calculation rather than a symbolic one to obtain formulas for the $f^{(j)}$ to order 17 which are exact to order 10 [SI78].

Here we show that modern computer algebra systems can easily produce these results using two simple Maple programs, one called *silverstone* to obtain the $f^{(j)}$ and the other called *unravel* to obtain the energy corrections $E^{(j)}$, which generalize program *alliluev* of Section 8.4.2 and Section 8.4.3.

We can denote the unperturbed eigenstates of (8.54) using a single index,

$$\phi_a = |a\rangle = |q + a, k\rangle, \tag{8.86}$$

so that the unperturbed reference state has index 0,

$$\phi_0 = |0\rangle = |q, k\rangle, \tag{8.87}$$

and is defined by the fixed values of $q$ and $k$. Then the order $j$ correction to the reference state is (see (7.24))

$$\psi_j = \sum_{a=-n_1}^{\infty} c_a^{(j)} \phi_a, \tag{8.88}$$

where $n_1 = q - k - 1$ is fixed. The unperturbed eigenvalues corresponding to
(7.1) are $\epsilon_a = q + a$ so the iteration scheme (7.25) to (7.28) for the $f^{(j)}$ is

$$c_0^{(0)} = 1, \quad c_0^{(j)} = 0, \, j > 0, \tag{8.89}$$

$$c_A^{(j)} = -\frac{1}{A}\left[\sum_{A'} V_{A'A} c_{A'}^{(j-1)} - \sum_{i=1}^{j-1} f^{(j-i)} c_A^{(i)}\right], \tag{8.90}$$

$$f^{(j)} = \sum_A c_A^{(j-1)} V_{A0}, \, j > 0, \tag{8.91}$$

where the matrix elements $V_{ab} = \langle a|X^2|b\rangle$ are given by (8.61).

Since these matrix elements involve square roots it would be convenient
and more efficient to choose a renormalization factor $N_a$ which converts them
to polynomials. Following the procedure of Section 8.2 we define

$$d_a^{(j)} = N_a c_a^{(j)} \tag{8.92}$$

to obtain the iteration scheme

$$d_0^{(0)} = N_0, \quad d_0^{(j)} = 0, \, j > 0, \tag{8.93}$$

$$d_A^{(j)} = -\frac{1}{A}\left(\sum_{A'=A-2}^{A+2} \overline{V}_{A'A} d_{A'}^{(j-1)} - \sum_{i=1}^{j-1} f^{(j-i)} d_A^{(i)}\right), \tag{8.94}$$

$$N_0 f^{(j)} = \sum_{A=-2}^{2} \overline{V}_{A0} d_A^{(j-1)}, \, j > 0, \tag{8.95}$$

where the index range in (8.94) and (8.95) is

$$j > 0, \quad -2j \le A \le 2j, \quad A \ne 0, \tag{8.96}$$

since the perturbation $X^2$ can change the state label by at most $\pm 2$ from one
order to the next, and the renormalized matrix elements are

$$\overline{V}_{ba} = \frac{N_a}{N_b} V_{ab} = \frac{N_a}{N_b} V_{ba}. \tag{8.97}$$

If we choose [SI78]

$$N_a = \prod_{i=0}^{a-1}[(q-k+i)(q+k+i+1)]^{1/2}, \quad a > 0, \tag{8.98}$$

$$N_0 = 1, \tag{8.99}$$

$$N_a = \prod_{i=1}^{|a|}[(q-k-i)(q+k-i+1)]^{-1/2}, \quad a < 0, \tag{8.100}$$

then

$$\overline{V}_{2,0} = \tfrac{1}{4},$$

$$\overline{V}_{1,0} = -\tfrac{1}{2}(2q+1),$$

$$\overline{V}_{0,0} = \tfrac{1}{2}(3q^2 - k(k+1)), \tag{8.101}$$

$$\overline{V}_{-1,0} = -\tfrac{1}{2}(2q-1)(q-k-1)(q+k),$$

$$\overline{V}_{-2,0} = \tfrac{1}{4}(q-k-2)(q-k-1)(q+k-1)(q+k),$$

and the general matrix elements $\overline{V}_{a+\mu,a}$ are obtained from these ones by replacing $q$ with $q+a$.

The following Maple procedure implements this iteration scheme.

```
01   # silverstone
02   # Stark effect for any parabolic reference state
03   silverstone := proc(jmax, q, k, f, d)
04       local V, i, j, A, AP, x, s1, s2, a;
05
06       # Define V = X^2 with q, k fixed
07
08       V := table([
09           (2)  = 1/4,
10           (1)  = -(2*q+2*a+1)/2,
11           (0)  = (3*(q+a)^2-k*(k+1))/2,
12           (-1) = -(2*q+2*a-1)*(q+a-k-1)*(q+a+k)/2,
13           (-2) = (q+a-k-2)*(q+a-k-1)*(q+a+k-1) *(q+a+k)/4
14       ]);
15
16       d[0,0] := 1;
17       for j from 1 to jmax do
18           lprint(j);
19
20           # Calculate f[j]
21
22           s1 := 0;
23           for A from -2 to 2 do
24               x := d[j-1, A];
25                   if not type(x, name) then
26                       s1 := s1 + subs(a=0, V[A])*x;
27                   fi;
28           od;
29           f[j] := simplify(s1);
30
31           # calculate jth order expansion coefficients
32
33           for A from -2*j to 2*j do
```

```
34              if A <> 0 then
35                  s1 := 0;
36                  for AP from A−2 to A+2 do
37                      x := d[j−1, AP];
38                      if not type(x, name) then
39                          s1 := s1 + subs(a=A, V[AP−A])*x
40                      fi;
41                  od;
42                  s2 := 0;
43                  for i from 1 to j−1 do
44                      x := d[i, A];
45                      if not type(x, name) then
46                          s2 := s2 + f[j−i]*x;
47                      fi;
48                  od;
49                  d[j, A] := −simplify((s1 − s2)/A);
50              fi; # A <> 0
51          od; # A
52      od; # j
53      f[0] := q;
54  end:
```

The following program uses *silverstone* to obtain the $f^{(j)}$ to order 6 as functions of $q$ and $k$. The results are saved in the files *fqk6* and *d6.m*.

```
01  # stest
02  read silverstone;
03  jmax := 6;
04  silverstone(jmax, 'q', 'k', 'f', 'd'):
05  save f, `fqk`.jmax;
06  save d, `d`.jmax.`.m`;
```

To compare with the Alliluev and Malkin results to order 4 (they use $k$ for our $q$) we can display the $f^{(j)}$ as functions of $q$ and $m$ where $k = (m − 1)/2$

```
07  for j from 1 to jmax do
08      f[j] := simplify(subs(k = (m−1)/2, f[j]));
09      print(4^j*f[j]);
10  od:
```

or to compare with Silverstone's results express the $f^{(j)}$ as functions of $q$ and $\tau$ where $\tau = (m^2 − 1)/4$ (Silverstone uses $k$ and $M$ for our $q$ and $\tau$ and tabulates $\beta^{(j)} = 4^j f^{(j)}$)

```
11  for j from 1 to jmax do
12      f[j] := simplify(subs(m = sqrt(4*tau+1), f[j]));
13      print(4^j*f[j]);
14  od:
```

The results to order 6 as functions of $q$ and $\tau$ are given here and the complete results to order 12 are given in Appendix E.

$$f^{(0)} = q,$$

$$4^1 f^{(1)} = 6\,q^2 - 2\,\tau,$$

$$4^2 f^{(2)} = 36\,\tau\,q - 68\,q^3 - 10\,q,$$

$$4^3 f^{(3)} = 44\,\tau^2 - 1032\,\tau\,q^2 - 120\,\tau + 1500\,q^4 + 660\,q^2,$$

$$4^4 f^{(4)} = 35640\,\tau\,q^3 - 3636\,\tau^2 q + 12960\,\tau\,q - 42756\,q^5$$
$$- 37900\,q^3 - 2210\,q,$$

$$4^5 f^{(5)} = 229200\,\tau^2 q^2 - 3680\,\tau^3 + 33712\,\tau^2 - 1364160\,\tau\,q^4$$
$$- 1029600\,\tau\,q^2 - 54240\,\tau + 1400784\,q^6 + 2093280\,q^4$$
$$+ 381264\,q^2,$$

$$4^6 f^{(6)} = 552912\,\tau^3 q - 13182480\,\tau^2 q^3 - 5881416\,\tau^2 q$$
$$+ 55719216\,\tau\,q^5 + 72439200\,\tau\,q^3 + 12051720\,\tau\,q$$
$$- 50118384\,q^7 - 113822520\,q^5 - 43002592\,q^3$$
$$- 1656500\,q. \tag{8.102}$$

The following Maple procedure does the unravelling to obtain the energy expansion in terms of the original perturbation parameter $\lambda$. It generalizes the method used in program *alliluev* to do the symbolic series expansions more efficiently.

```
01   # unravel
02   # use the f[j] to get the energy
03   unravel := proc(f, q, k, jmax, e, n, Q, m)
04       local j, q1, q2, A, x, G, G3, Q1Q2, g, g3, Energy;
05
06       G3 := sum('g3[j]*x^j', 'j' = 0..jmax);
07       Q1Q2 := A[0] + sum('A[j]*x^j *taylor(G3^j, x,
08                   jmax+1-j)', 'j' = 1..jmax);
09       Q1Q2 := taylor(Q1Q2, x, jmax+1);
10
11       G := sum('g[j]*x^j', 'j' = 0..jmax);
12       G3 := taylor(G^3, x, jmax+1);
13       for j from 0 to jmax do
14          g3[j] := simplify(coeff(G3, x, j));
15       od;
16       Q1Q2 := simplify(Q1Q2);
17       for j from 0 to jmax do
18          g[j] := factor(eval(coeff(Q1Q2, x, j)));
19       od;
20       G := eval(G);
21
```

```
22        Energy := taylor(-(1/2)*G^(-2), x, jmax+1);
23        for j from 0 tojmax do
24           e[j] := factor(coeff(Energy, x, j));
25        od;
26
27        for j from 0 to jmax do
28           A[j] := (subs(q=q1, f[j]) + (-1)^j * subs(q=q2, f[j]));
29           A[j] := subs(q1=(n+Q)/2, q2=(n-Q)/2, k=(m-1)/2, A[j]);
30           A[j] := simplify(A[j]);
31        od;
32
33        for j from 0 to jmax do
34        e[j] := factor(e[j]);
35        od;
36        RETURN(NULL);
37   end:
```

The following program uses *unravel* to obtain the $E^{(j)}$ to order 6 as functions of $\tau$ and $q$ using the results in file *fqk6* obtained from the *silverstone* procedure in program *stest*. The energy corrections are saved in the file *enQm6*.

```
01   # utest
02   # unravelling the energy
03   jmax := 6;
04   read unravel;
05   read `fqk`.jmax;
06   unravel(f, q, k, jmax, 'e', 'n', 'Q', 'm');
07   save e, `enQm`.jmax;
08
09   # Display energy
10
11   for j from 0 to jmax do
12       e[j] := e[j];
13   od;
14
15   # Check the ground state
16
17   for j from 0 to jmax do
18       eg[j] := subs(Q=0, m=0, n=1, e[j]);
19   od;
20   quit;
```

The results to order 6 are given here and the complete results to order 12 are given in Appendix E.

$$E^{(0)} = -\frac{1}{2\,n^2},$$

$$E^{(1)} = \frac{3}{2}\,nQ,$$

$$E^{(2)} = -\frac{1}{16}n^4(17\,n^2 - 3\,Q^2 - 9\,m^2 + 19),$$

$$E^{(3)} = \frac{3}{32}n^7 Q(23\,n^2 - Q^2 + 11\,m^2 + 39),$$

$$\begin{aligned}
E^{(4)} = -\frac{1}{1024}n^{10}(&5487\,n^4 + 1806\,n^2 Q^2 - 3402\,n^2 m^2 + 35182\,n^2 \\
&+ 147\,Q^4 - 1134\,Q^2 m^2 + 5754\,Q^2 - 549\,m^4 \\
&- 8622\,m^2 + 16211),
\end{aligned}$$

$$\begin{aligned}
E^{(5)} = \frac{3}{1024}n^{13} Q(&10563\,n^4 + 98\,n^2 Q^2 + 772\,n^2 m^2 + 90708\,n^2 \\
&- 21\,Q^4 + 220\,Q^2 m^2 + 780\,Q^2 + 725\,m^4 + 830\,m^2 \\
&+ 59293),
\end{aligned}$$

$$\begin{aligned}
E^{(6)} = -\frac{1}{8192}n^{16}(&547262\,n^6 + 685152\,n^4 Q^2 - 429903\,n^4 m^2 \\
&+ 9630693\,n^4 + 390\,n^2 Q^4 - 25470\,n^2 Q^2 m^2 \\
&+ 7787370\,n^2 Q^2 - 16200\,n^2 m^4 - 4786200\,n^2 m^2 \\
&+ 22691096\,n^2 - 372\,Q^6 + 765\,Q^4 m^2 + 1185\,Q^4 \\
&- 36450\,Q^2 m^4 - 62100\,Q^2 m^2 + 7028718\,Q^2 - 6951\,m^6 \\
&- 16845\,m^4 - 4591617\,m^2 + 7335413).
\end{aligned} \tag{8.103}$$

### 8.4.5  Higher orders: special cases

To obtain the energy corrections $E^{(j)}$ for a particular state it is necessary to
substitute the appropriate values of $n$, $Q = n_1 - n_2$ and $m$ into the preceding formulas.  However it is possible to proceed to much higher orders if the
particular values are substituted from the outset.

For example to do the excited state with $n_1 = 1$, $n_2 = 0$ and $m = 0$
corresponding to $Q = 1$ and $n = 2$, or to $q_1 = 3/2$, $q_2 = 1/2$ and $k = -1/2$, we
can use the following short program to calculate the two sets of $f^{(j)}$ coefficients
to order 20 since procedure *silverstone* works with either symbolic values of $q$
and $k$ or particular numeric values.  The $f^{(j)}$ are saved in files *fa20* and *fb20*
as rational numbers.

```
01   # s1test
02   # Using silverstone procedure for a particular state
03   jmax := 20;
04   read silverstone;
05   silverstone(jmax, 3/2, -1/2, 'fa', 'da'):
06   save fa, `fa`.jmax;
07   silverstone(jmax, 1/2, -1/2, 'fb', 'db'):
08   save fb, `fb`.jmax;
09   quit;
```

Now we need the following modified version of procedure *unravel* to obtain the
energy.

```
01    # unravel1
02    # version of unravel for a particular state
03    unravel1 := proc(f1, f2, jmax, e, x)
04       local j, A, G, G3, Q1Q2, g, g3, Energy;
05
06       for j from 0 to jmax do
07          A[j] := f1[j] + (-1)^j*f2[j];
08       od;
09
10       G3 := sum('g3[j]*x^j', 'j' = 0..jmax);
11       Q1Q2 := A[0] + sum('A[j]*x^j*taylor(G3^j,x,jmax+1-j)'
12                  ,'j'=1..jmax);
13       Q1Q2 := taylor(Q1Q2, x, jmax+1);
14
15       G := sum('g[j]*x^j', 'j' = 0..jmax);
16       G3 := taylor(G^3, x, jmax+1);
17       for j from 0 to jmax do
18          g3[j] := simplify(coeff(G3, x, j));
19       od;
20
21       # we use eval to obtain full evaluation so that
22       # G will be series with rational coefficients
23
24       Q1Q2 := simplify(Q1Q2);
25       for j from 0 to jmax do
26          g[j] := eval(coeff(Q1Q2, x, j));
27       od;
28       G := eval(G);
29
30       Energy := taylor(-(1/2)*G^(-2), x, jmax+1);
31
32       for j from 0 to jmax do
33          e[j] := coeff(Energy, x, j);
34       od;
35       RETURN(Energy);
36    end:
```

Finally, the following program calculates the energy to order 20. The use
of *simplify* in lines *18* and *24* is important (without it this order 20 calculation
requires almost 10 megabytes of storage rather than 2 megabytes).

```
01    # u1test
02    # Calculating energy for a particular state
03    jmax := 20;
04    read unravel1;
05    read `fa`.jmax;
06    read `fb`.jmax;
07    Energy := unravel1(fa, fb, jmax, 'e', 'x');
```

*08* **quit;**

Using the notation $E_{nQm}$ for the energy levels the result to order 20 is

$$
\begin{aligned}
E_{2,1,0} = &-1/8 + 3\,\lambda - 84\,\lambda^2 + 1560\,\lambda^3 - 257856\,\lambda^4 + 14214816\,\lambda^5 \\
&- 2690869248\,\lambda^6 + 243073886976\,\lambda^7 \\
&- 48538616082432\,\lambda^8 + 5925429002813952\,\lambda^9 \\
&- 1247119322547093504\,\lambda^{10} \\
&+ 187912057677787975680\,\lambda^{11} \\
&- 41965106251721294217216\,\lambda^{12} \\
&+ 7428858655861394409406464\,\lambda^{13} \\
&- 1770053421993608758538797056\,\lambda^{14} \\
&+ 357540210738996369153435500544\,\lambda^{15} \\
&- 91188875752496789111151613968384\,\lambda^{16} \\
&+ 20630993847975259684560563105562624\,\lambda^{17} \\
&- 5639641009410340886695833574878216192\,\lambda^{18} \\
&+ 1411018960842074180136775323864617975808\,\lambda^{19} \\
&- 413301274094239026850581583427438244790272\,\lambda^{20}.
\end{aligned}
$$

$$(8.104)$$

## 8.5   2-Dim Stark Effect in Parabolic Coordinates

The Schrödinger equation for the Stark effect in a 2-dimensional hydrogenic atom is (with $Z = 1$)

$$
\left(\tfrac{1}{2}p^2 - \frac{1}{r} + \lambda x\right)\psi = E\psi,
\tag{8.105}
$$

where $x = r\cos\varphi$, $r = \sqrt{x^2 + y^2}$ and

$$
-p^2 = \nabla^2 = \frac{\partial^2}{\partial x^2} + \frac{\partial^2}{\partial y^2}.
\tag{8.106}
$$

We can introduce 2-dimensional parabolic coordinates $\xi$, $\eta$ defined by

$$
x = \frac{1}{2}(\xi - \eta), \quad y = \sqrt{\xi\eta},
\tag{8.107}
$$

or inversely

$$
\xi = r + x, \quad \eta = r - x.
\tag{8.108}
$$

Then the Laplacian is given by

$$
\nabla^2 = \frac{4}{\xi + \eta}\left[\sqrt{\xi}\frac{\partial}{\partial\xi}\left(\sqrt{\xi}\frac{\partial}{\partial\xi}\right) + \sqrt{\eta}\frac{\partial}{\partial\eta}\left(\sqrt{\eta}\frac{\partial}{\partial\eta}\right)\right],
\tag{8.109}
$$

and letting $\psi(\xi,\eta) = f_1(\xi)f_2(\eta)$ the Schrödinger equation separates into the two equations

$$\left[ -2\sqrt{\xi}\frac{d}{d\xi}\left(\sqrt{\xi}\frac{d}{d\xi}\right) + \frac{\lambda}{2}\xi^2 - E\xi - 2\lambda_1 \right] f_1(\xi) = 0, \qquad (8.110)$$

$$\left[ -2\sqrt{\eta}\frac{d}{d\eta}\left(\sqrt{\eta}\frac{d}{d\eta}\right) - \frac{\lambda}{2}\eta^2 - E\eta - 2\lambda_2 \right] f_2(\eta) = 0, \qquad (8.111)$$

where

$$\lambda_1 + \lambda_2 = 1. \qquad (8.112)$$

As in the 3-dimensional case in Section 8.4.1 we introduce the scaling transformations $\xi = 2\gamma X$ and $\eta = 2\gamma Y$ and choose

$$E = -\frac{1}{2\gamma^2} \qquad (8.113)$$

to obtain for the $\xi$ equation

$$\left[ -\frac{\sqrt{X}}{2}\frac{d}{dX}\left(\sqrt{X}\frac{d}{dX}\right) + \frac{X}{2} + \gamma^3\lambda X^2 - \gamma\lambda_1 \right] F_1(X) = 0, \qquad (8.114)$$

which corresponds to the $D$-dimensional hydrogenic so(2,1) realization (4.134) to (4.136) if we choose $a = 1$ and

$$D = \frac{3}{2}, \quad \tau = -\frac{3}{16} \qquad (8.115)$$

to obtain

$$T_1 = -\frac{1}{2}\sqrt{X}\frac{d}{dX}\left(\sqrt{X}\frac{d}{dX}\right) - \frac{X}{2}, \qquad (8.116)$$

$$T_2 = -i\left(X\frac{d}{dX} + \frac{1}{4}\right), \qquad (8.117)$$

$$T_3 = -\frac{1}{2}\sqrt{X}\frac{d}{dX}\left(\sqrt{X}\frac{d}{dX}\right) + \frac{X}{2}. \qquad (8.118)$$

The $T_3$ eigenvalues have the form

$$q_1 = k + 1 + n_1, \quad n_1 = 0, 1, 2, \ldots, \qquad (8.119)$$

where

$$k = \frac{1}{2}\left[-1 \pm \sqrt{4\tau + 1}\right] = -\frac{1}{4}, \ -\frac{3}{4}. \qquad (8.120)$$

Here both values of $k$ satisy $k > -1$ and are admissible.

Similarly for the scaled $\eta$ equation

$$q_2 = k + 1 + n_2, \quad n_2 = 0, 1, 2, \ldots \qquad (8.121)$$

and in the unperturbed case

$$q_1 + q_2 = \gamma(\lambda_1 + \lambda_2) = \gamma = 2k + 2 + n_1 + n_2, \qquad (8.122)$$

so from (8.112) the unperturbed energy levels are

$$E = -\frac{1}{2(2k + 2 + n_1 + n_2)^2}. \qquad (8.123)$$

Since $2k + 2 = 1/2$ or $3/2$ both cases can be obtained using $2k + 2 = 1/2$. The result agrees with the general result (4.115) for the $D$-dimensional hydrogenic atom in spherical coordinates (put $D = 2$ in (4.115) and identify $\ell + \mu$ with $n_1 + n_2$). For the ground state $E = -2$, corresponding to $k = -3/4$, $n_1 = 0$, $n_2 = 0$.

The perturbation theory for (8.114) is now the same as in the 3-dimensional case since (8.114), using (8.118) for $T_3$, is formally the same as (8.51) to (8.57) except that (8.51) and (8.53) are to be replaced by $q_1 = 1/4 + n_1$ and $q_2 = 1/4 + n_2$ which is formally equivalent to replacing $|m|$ by $-1/2$ in (8.51) and (8.53). Therefore the formulas (8.103) for the energy corrections $E^{(1)}$ to $E^{(6)}$, and the more extensive ones to order 12 in Appendix E, are also valid in the 2-dimensional case if we make the formal identifications

$$n = q_1 + q_2 = \tfrac{1}{2} + n_1 + n_2, \qquad (8.124)$$
$$Q = q_1 - q_2 = n_1 - n_2, \qquad (8.125)$$
$$m^2 = \tfrac{1}{4}. \qquad (8.126)$$

As an example the following Maple programs use *silverstone* and *unravel1* to calculate the 2-dimensional ground state energy to order 30.

```
01   # s1test2d
02   jmax := 30;
03   read silverstone;
04   silverstone(jmax, 1/4, -3/4, 'f', 'd');
05   save f, `f2d`.jmax;
06   quit;
```

```
01   # u1test2d
02   jmax := 30;
03   read `f2d`.jmax;
04   Energy := unravel1(f, f, jmax, 'e', 'x');
05   save e, `e2x`.jmax;
06   quit;
```

The result for the energy to order 20 is

$$E = -2 - \frac{21}{2^8}\,\lambda^2 - \frac{22947}{2^{20}}\,\lambda^4$$

$$- \frac{48653931}{2^{31}}\,\lambda^6 - \frac{800908686795}{2^{44}}\,\lambda^8$$

$$- \frac{5223462120917049}{2^{55}}\,\lambda^{10}$$

$$- \frac{98276453573919203439}{2^{67}}\,\lambda^{12}$$

$$- \frac{1260656338200202681485219}{2^{78}}\,\lambda^{14}$$

$$- \frac{16959456391919086065720296 1483}{2^{92}}\,\lambda^{16}$$

$$- \frac{3627017569463341601735954216022765}{2^{103}}\,\lambda^{18}$$

$$- \frac{192534420895031072124861785132418925125}{2^{115}}\,\lambda^{20} \qquad (8.127)$$

## 8.6  Exercises

❖ **Exercise 8.1** Show how to obtain the energy expansions for the Stark effect for nuclear charge $\mathcal{Z}$ from those with $\mathcal{Z} = 1$.

# Chapter 9

# Symbolic Calculation of the Zeeman Effect

## 9.1   Introduction

Unlike the Stark effect, which is reducible to a one-dimensional problem by separation in parabolic coordinates, the Zeeman effect is inherently a two-dimensional problem: in spherical coordinates the $\varphi$ dependence can be separated but the $r$ and $\vartheta$ dependence cannot. Like the Stark effect, the perturbation theory of the Zeeman effect has had a long history (see [RO84] and [JA90] and references therein) but the first few energy corrections for the ground state are more difficult to obtain. In fact the first 5 orders were first calculated only in 1976 by Galindo and Pascual [GA76] using the method of Dalgarno and Lewis [DA55], [DA56a], [SC68].

At about the same time the so(4,2) Lie algebraic method and the modified algebraic perturbation theory were systematically applied to the ground state and the first high order calculation was done to order 40 with floating point arithmetic [VR77], [CI77b] and later to order 100 [AV79]. Rational results to order 10 were done by Clay using the FORMAC language [AV79], [CL79] and it was concluded at the time that higher orders were too expensive to generate in rational form.

Asymptotic formulas for the energy corrections of the ground state were also obtained [AV77], [AV79] and also for the $3s-3d_0$ degenerate excited states [AD80], [AV81a], [AV82]. Recently the large-order perturbation theory for the ground state has also been obtained by Janke from the connection with a four-dimensional anisotropic oscillator (see [JA90] and references therein).

In this chapter we show how the modified algebraic RSPT of Chapter 7 can be used to obtain the energy and wavefunction corrections to the ground state. First we obtain the energy corrections to order 5 "by hand" and then we show how Maple can be used to obtain exact rational results to order 60 for the ground state (see Appendix F for tables of results). The procedure is very similar to that followed in the previous chapter for the Stark effect. A simple application of Padé approximants is also given to sum the divergent energy series.

We also develop the modified algebraic RSPT for degenerate states in the special case of a doubly degenerate state by adapting the general case of conventional degenerate RSPT [SI81] and apply it to the $3s-3d_0$ sublevel which is the first case where degenerate RSPT is needed for the Zeeman effect.

Finally we show how the method of moments, which does not require the wavefunction, can also be used to obtain the ground state energy corrections. This method was developed and first applied to the Zeeman effect by Fernández and Castro [FE84]. Having two quite different methods for calculating the energy corrections provides a useful check on the results.

## 9.2  Low Order Calculations for the Ground State

In atomic units the Schrödinger equation for a hydrogenic atom in a uniform magnetic field in the $z$ direction has the form [GA91]

$$\left[\frac{1}{2}p^2 - \frac{Z}{r} + \frac{B}{2}L_3 + \frac{B^2}{8}(r^2 - z^2)\right]\psi = E\psi, \tag{9.1}$$

where $B$ is the magnetic field strength. For the ground state we have $m = 0$ and the linear term in $B$ does not contribute. Therefore we consider the Schrödinger equation

$$\left[\frac{1}{2}p^2 - \frac{Z}{r} + \lambda(r^2 - z^2)\right]\psi = E\psi \tag{9.2}$$

with perturbation parameter $\lambda = B^2/8$.

From the general results of Section 7.3 and Section 7.4 we obtain the scaled eigenvalue problem

$$(T_3 - n + \lambda W - \Delta E\, S)\Psi(\boldsymbol{R}) = 0, \tag{9.3}$$

$$W = \gamma^4 R(R^2 - Z^2) = \left(\frac{n}{Z}\right)^4 R(R^2 - Z^2), \tag{9.4}$$

$$S = \gamma^2 R = \left(\frac{n}{Z}\right)^2 R. \tag{9.5}$$

As in the case of the Stark effect we can let $Z = 1$ since the nuclear charge can always be put back via scaling if desired (see Exercise 9.1).

We can now use the modified algebraic RSPT formalism of Section 7.4, as we did for the Stark effect in Section 7.4.5, to obtain "by hand" the wavefunction corrections $\Psi^{(1)}$ and $\Psi^{(2)}$ and the energy corrections to order 5 using the symmetric energy formula (7.75). For the ground state we have $\Psi^{(0)} = \Phi_0 = |100\rangle$, $E^{(0)} = E_0 = -1/2$, $K_0 = T_3 - 1$, $\kappa_0 = 1$, $S = R$ and $W = R(R^2 - Z^2)$ so the first order energy correction is

$$E^{(1)} = \frac{\langle 100|W|100\rangle}{\langle 100|S|100\rangle} = \frac{W_{10}^{10}}{R_{10}^{10}} = 2, \tag{9.6}$$

where we have used the notation of Appendix D for matrix elements, the formulas in Table D.1 for matrix elements of $R$, and the formulas in Tables D.5 and D.6 for matrix elements of $W$ (denoted by $U$ in these tables).

From (7.71) and (7.73) the first order correction to the wavefunction is

$$
\begin{aligned}
\Psi^{(1)} &= GW\Phi_0 - E^{(1)}GS\Phi_0 \\
&= G(W_{10}^{20}\,|200\rangle + W_{10}^{30}\,|300\rangle + W_{10}^{32}\,|320\rangle + W_{10}^{40}\,|400\rangle \\
&\quad + W_{10}^{42}\,|420\rangle) - 2G(R_{10}^{10}\,|100\rangle + R_{10}^{20}\,|200\rangle) \\
&= G(2\sqrt{2}\,|200\rangle - 2\sqrt{3}\,|300\rangle + \sqrt{6}\,|320\rangle + |400\rangle - |420\rangle) \\
&= 2\sqrt{2}\,|200\rangle - \sqrt{3}\,|300\rangle + \tfrac{\sqrt{6}}{2}\,|320\rangle + \tfrac{1}{3}\,|400\rangle - \tfrac{1}{3}\,|420\rangle.
\end{aligned}
\tag{9.7}
$$

From (7.74) the second order energy correction is

$$
E^{(2)} = \langle\Phi_0|W|\Psi^{(1)}\rangle - E^{(1)}\langle\Phi_0|S|\Psi^{(1)}\rangle,
$$

where

$$
\begin{aligned}
\langle\Phi_0|W|\Psi^{(1)}\rangle &= \langle 100|W|\Psi^{(1)}\rangle \\
&= 2\sqrt{2}\,W_{20}^{10} - \sqrt{3}\,W_{30}^{10} + \tfrac{\sqrt{6}}{2}\,W_{32}^{10} + \tfrac{1}{3}\,W_{40}^{10} - \tfrac{1}{3}\,W_{42}^{10} \\
&= -\frac{65}{3}, \\
\langle\Phi_0|S|\Psi^{(1)}\rangle &= \langle 100|S|\Psi^{(1)}\rangle \\
&= 2\sqrt{2}\,\langle 100|R|200\rangle = 2\sqrt{2}\,R_{20}^{10} = -2.
\end{aligned}
$$

Therefore

$$
E^{(2)} = -\frac{65}{3} - 2(-2) = -\frac{53}{3}.
\tag{9.8}
$$

The third order energy correction can also be obtained using the symmetric formula (7.77)

$$
\begin{aligned}
E^{(3)} &= \langle\Psi^{(1)}|W|\Psi^{(1)}\rangle - E^{(1)}\langle\Psi^{(1)}|S|\Psi^{(1)}\rangle - 2E^{(2)}\langle\Phi_0|S|\Psi^{(1)}\rangle \\
&= \frac{7034}{9} - 2\left(\frac{817}{18}\right) - 2\left(-\frac{53}{3}\right)(-2) = \frac{5581}{9}.
\end{aligned}
\tag{9.9}
$$

The second order correction to the wavefunction is

$$
\Psi^{(2)} = GW\Psi^{(1)} - E^{(1)}GS\Psi^{(1)} - E^{(2)}GS\Phi_0,
$$

and after a lengthy calculation

$$
\begin{aligned}
\Psi^{(2)} &= -\tfrac{134\sqrt{2}}{3}\,|200\rangle + \tfrac{139\sqrt{3}}{3}\,|300\rangle - \tfrac{95\sqrt{6}}{6}\,|320\rangle - \tfrac{554}{9}\,|400\rangle \\
&\quad + \tfrac{380}{9}\,|420\rangle + \tfrac{157\sqrt{5}}{12}\,|500\rangle - \tfrac{77\sqrt{14}}{12}\,|520\rangle + \sqrt{70}\,|540\rangle \\
&\quad - \tfrac{19\sqrt{6}}{6}\,|600\rangle + \tfrac{163\sqrt{21}}{105}\,|620\rangle - \tfrac{10\sqrt{7}}{7}\,|640\rangle + \tfrac{\sqrt{7}}{3}\,|700\rangle \\
&\quad - \tfrac{4\sqrt{21}}{21}\,|720\rangle + \tfrac{\sqrt{154}}{21}\,|740\rangle.
\end{aligned}
\tag{9.10}
$$

The symmetric energy formulas (7.76) and (7.77) can now be used to calculate $E^{(4)}$ and $E^{(5)}$:

$$
\begin{aligned}
E^{(4)} &= \langle \Psi^{(1)} | W | \Psi^{(2)} \rangle - E^{(1)} \langle \Psi^{(1)} | S | \Psi^{(2)} \rangle - E^{(2)} \langle \Psi^{(1)} | S | \Psi^{(1)} \rangle \\
&\quad - E^{(2)} \langle \Phi_0 | S | \Psi^{(2)} \rangle - 2 E^{(3)} \langle \Psi^{(0)} | S | \Psi^{(1)} \rangle \\
&= -\frac{21577397}{540},
\end{aligned}
\tag{9.11}
$$

$$
\begin{aligned}
E^{(5)} &= \langle \Psi^{(2)} | W | \Psi^{(2)} \rangle - E^{(1)} \langle \Psi^{(2)} | S | \Psi^{(2)} \rangle - 2 E^{(2)} \langle \Psi^{(1)} | S | \Psi^{(2)} \rangle \\
&\quad - 2 E^{(3)} \langle \Phi_0 | S | \Psi^{(2)} \rangle - E^{(3)} \langle \Psi^{(1)} | S | \Psi^{(1)} \rangle - 2 E^{(4)} \langle \Phi_0 | S | \Psi^{(1)} \rangle \\
&= \frac{31283298283}{8100}.
\end{aligned}
\tag{9.12}
$$

Therefore to 5th order the ground state energy series is

$$
\begin{aligned}
E = &-\frac{1}{2} + 2\lambda - \frac{53}{3}\lambda^2 + \frac{5581}{9}\lambda^3 - \frac{21577397}{540}\lambda^4 \\
&+ \frac{31283298283}{8100}\lambda^5 + \cdots
\end{aligned}
\tag{9.13}
$$

This result represents the limit of "hand calculations".


## 9.3   Symbolic Calculations for the Ground State

### 9.3.1   Modified high order formalism

The symbolic calculation of the energy and wavefunction corrections using the modified iteration scheme (7.82) to (7.85) is analogous to that for the Stark effect in Section 8.2.1 except now the selection rules for non-zero matrix elements of the perturbation $W = R(R^2 - Z^2)$ are given by (see Appendix D)

$$
W_{n\ell}^{n+\mu,\ell+\nu} = 0, \quad \text{unless } |\mu| \le 3, \ \nu = -2, 0, 2,
\tag{9.14}
$$

instead of (8.11). As we proceed from one wavefunction correction $\Psi^{(j)}$ to the next, the perturbation operator $W$ can raise the principal quantum number $n$ by at most 3 and the maximum orbital angular momentum quantum number increases by 2.

Therefore the wavefunction corrections can be expanded as in the Stark case (8.12) but now $n = 2, 3, \ldots, 3j+1$ and $\ell = 0, 2, 4, \ldots, 2j$ so the modified iteration scheme (7.82) to (7.85) (see (8.13) to (8.17) for the Stark effect) now has the form

$$
C_{10}^{(0)} = 1, \quad C_{NL}^{(0)} = 0, \ N > 1,
\tag{9.15}
$$

$$
R_{10}^{10} E^{(j)} = \sum_{N=1}^{4} \left( \sum_{L=0,2} W_{10}^{NL} C_{NL}^{(j-1)} \right) - R_{10}^{20} \sum_{k=1}^{j-1} E^{(k)} C_{20}^{(j-k)},
$$

$$
\tag{9.16}
$$

$$C_{10}^{(j)} = 0, \quad j > 0, \tag{9.17}$$

$$C_{NL}^{(j)} = \frac{1}{1-N} \sum_{N'=N-3}^{N+3} \left( \sum_{L'=L-2,L,L+2} W_{NL}^{N'L'} C_{N'L'}^{(j-1)} \right)$$
$$- \frac{1}{1-N} \sum_{N'=N-1}^{N+1} R_{NL}^{N'L} \sum_{k=1}^{j} E^{(k)} C_{N'L}^{(j-k)}, \tag{9.18}$$

where

$$N = 2,3,\ldots,3j+1, \quad L = 0,2,\ldots,2j, \quad 0 \le L < N. \tag{9.19}$$

Similarly the symmetric energy formula (7.86) (see (8.18) for Stark effect) now has the form

$$R_{10}^{10} E^{(p+q+1)} = \sum_{NL} C_{NL}^{(p)} \sum_{N'=N-3}^{N+3} \left( \sum_{L'=L-2,L,L+2} W_{NL}^{N'L'} C_{N'L'}^{(q)} \right)$$
$$- R_{10}^{20} \left( \sum_{r=1}^{p} E^{(p+q+1-r)} C_{20}^{(r)} + \sum_{s=1}^{q} E^{(p+q+1-s)} C_{20}^{(s)} \right)$$
$$- \sum_{r=1}^{p} \sum_{NL} C_{NL}^{(r)} \left( \sum_{s=1}^{q} E^{(p+q+1-r-s)} \sum_{N'=N-1}^{N+1} R_{NL}^{N'L} C_{N'L}^{(s)} \right), \tag{9.20}$$

and we can calculate $E^{(j)}$ by choosing $p = \lfloor j/2 \rfloor$ and $q = \lfloor (j-1)/2 \rfloor$.

The "renormalization" procedure which removes all square roots from the calculations is the same as in the Stark effect (see (8.19) and (8.20)) so we obtain the iteration scheme

$$D_{10}^{(0)} = 1, \tag{9.21}$$

$$N_{10} \overline{R}_{10}^{10} E^{(j)} = \sum_{N=1}^{4} \left( \sum_{L=0,2} \overline{W}_{10}^{NL} D_{NL}^{(j-1)} \right) - \overline{R}_{10}^{20} \sum_{k=1}^{j-1} E^{(k)} D_{20}^{(j-k)}, \tag{9.22}$$

$$D_{10}^{(j)} = 0, \quad j > 0, \tag{9.23}$$

$$D_{NL}^{(j)} = \frac{1}{1-N} \sum_{N'=N-3}^{N+3} \left( \sum_{L'=L-2,L,L+2} \overline{W}_{NL}^{N'L'} D_{N'L'}^{(j-1)} \right)$$
$$- \frac{1}{1-N} \sum_{N'=N-1}^{N+1} \overline{R}_{NL}^{N'L} \sum_{k=1}^{j} E^{(k)} D_{N'L}^{(j-k)}, \tag{9.24}$$

and the symmetric energy formula

$$R_{10}^{10} E^{(p+q+1)} = \sum_{NL} \frac{D_{NL}^{(p)}}{N_{NL}^2} \sum_{N'=N-3}^{N+3} \left( \sum_{L'=L-2,L,L+2} \overline{W}_{NL}^{N'L'} D_{N'L'}^{(q)} \right)$$
$$- \overline{R}_{10}^{20} \left( \sum_{r=1}^{p} E^{(p+q+1-r)} D_{20}^{(r)} + \sum_{s=1}^{q} E^{(p+q+1-s)} D_{2,0}^{(s)} \right)$$
$$- \sum_{r=1}^{p} \sum_{NL} \frac{D_{NL}^{(r)}}{N_{NL}^2} \left( \sum_{s=1}^{q} E^{(p+q+1-r-s)} \sum_{N'=N-1}^{N+1} \overline{R}_{NL}^{N'L} D_{N'L}^{(s)} \right), \tag{9.25}$$

where the renormalized matrix elements are defined by (8.23) and (8.24).

Thus the iteration scheme (9.21) to (9.24) can be used to obtain the energy and wavefunction corrections to a certain order, say $j$, and then the symmetric formula can be used to obtain further energy corrections to order $2j + 1$.

### 9.3.2  Renormalized matrix elements

The calculation of the renormalized matrix elements $\overline{R}_{n\ell}^{n'\ell}$ and $\overline{W}_{n\ell}^{n'\ell'}$ is analogous to the Stark case so we give here only the Maple programs *genzeemandata* and *zrenorm* corresponding to the Stark programs *genstarkdata* and *renorm* in Section 8.2.2. The input file *rzdata.m* was created by the program *rzmatrix* given in Appendix D.

```
01   # genzeemandata
02   # Generate matrix element data for ground state
03   # Zeeman effect from general results in rzdata.m
04
05   read `rzdata.m`;
06   R := op(r1);
07   W := subs(m=0, op(u));
08   save R, W, zeemandata;
09   quit;
```

```
01   # zrenorm
02   read zeemandata;
03   Norm2 := (n+l)!/(n-l-1)!/(2*l+1);
04
05   # Renormalize R matrix
06
07   for mu from -1 to 1 do
08       Na2 := Norm2;
09       Nb2 := subs(n=n+mu, Norm2);
10       f2 := simplify(Na2/Nb2);
11       R[mu] := simplify(f2^(1/2)*R[mu]);
12   od:
13
14   # Renormalize the perturbation matrix
15
16   for mu from -3 to 3 do
17       for nu from -2 by 2 to 2 do
18           Na2 := Norm2;
19           Nb2 := subs(n=n+mu, l=l+nu, Norm2);
20           f2 := simplify(Na2/Nb2);
21           W[mu,nu] := simplify(f2^(1/2)*W[mu,nu]);
22       od;
23   od:
```

```
24
25   save Norm2, R, W, `zrenormdata`;
26   save Norm2, R, W, `zrenormdata.m`;
27   quit;
```

The renormalized matrix elements of $W = R(R^2 - Z^2)$ are $\overline{W}_{n\ell}^{n+\mu,\ell+\nu}$, denoted in the program by $W[mu,nu]$ and are given as functions of $n$ and $\ell$ by

$$W[-3,-2] = \frac{(n-4+\ell)(n-3+\ell)(n-2+\ell)(n+\ell-1)(n+\ell)(\ell-1)\ell}{(16\ell+8)(2\ell-1)},$$

$$W[-3,0] = -\frac{(n-2+\ell)(n+\ell-1)(n+\ell)(\ell^2+\ell-1)}{(8\ell+12)(2\ell-1)},$$

$$W[-3,2] = \frac{(n+\ell)(\ell+2)(\ell+1)}{(16\ell+8)(2\ell+3)},$$

$$W[-2,-2] = -\frac{(n-3+\ell)(n-2+\ell)(n+\ell-1)(n+\ell)(\ell-1)\ell(3n-2-2\ell)}{(8\ell+4)(2\ell-1)},$$

$$W[-2,0] = \frac{(3n+3\ell-3)(n+\ell)(\ell^2+\ell-1)(n-1)}{(4\ell+6)(2\ell-1)},$$

$$W[-2,2] = -\frac{(\ell+2)(\ell+1)(3n+2\ell)}{(8\ell+4)(2\ell+3)},$$

$$W[-1,-2] = \frac{(5n-10+5\ell)(n+\ell-1)(n+\ell)(n-\ell)(\ell-1)\ell(3n-1-\ell)}{(16\ell+8)(2\ell-1)},$$

$$W[-1,0] = -\frac{(3n+3\ell)(\ell^2+\ell-1)(5n^2-5n-\ell+2-\ell^2)}{(8\ell+12)(2\ell-1)},$$

$$W[-1,2] = \frac{(5\ell+10)(\ell+1)(3n+\ell)}{(16\ell+8)(2\ell+3)},$$

$$W[0,-2] = -\frac{(5n+5\ell-5)(n+\ell)(n-\ell)(n-\ell+1)n(\ell-1)\ell}{(4\ell+2)(2\ell-1)},$$

$$W[0,0] = \frac{n(\ell^2+\ell-1)(5n^2+1-3\ell-3\ell^2)}{(2\ell+3)(2\ell-1)},$$

$$W[0,2] = -\frac{5n(\ell+2)(\ell+1)}{(4\ell+2)(2\ell+3)},$$

$$W[1,-2] = \frac{(5n+5\ell)(n-\ell)(n-\ell+1)(n+2-\ell)(\ell-1)\ell(3n+\ell+1)}{(16\ell+8)(2\ell-1)},$$

$$W[1,0] = -\frac{(3n-3\ell)(\ell^2+\ell-1)(5n^2-\ell-\ell^2+5n+2)}{(8\ell+12)(2\ell-1)},$$

$$W[1,2] = \frac{(5\ell+10)(\ell+1)(3n-\ell)}{(16\ell+8)(2\ell+3)},$$

$$W[2,-2] = -\frac{(n-\ell)(n-\ell+1)(n+2-\ell)(n+3-\ell)(\ell-1)\ell(3n+2\ell+2)}{(8\ell+4)(2\ell-1)},$$

$$W[2,0] = \frac{(3n-3\ell)(n-\ell+1)(\ell^2+\ell-1)(n+1)}{(4\ell+6)(2\ell-1)},$$

$$W[2,2] = -\frac{(\ell+2)(\ell+1)(3n-2\ell)}{(8\ell+4)(2\ell+3)},$$

$$W[3,-2] = \frac{(n-\ell)(n-\ell+1)(n+2-\ell)(n+3-\ell)(n+4-\ell)(\ell-1)\ell}{(16\ell+8)(2\ell-1)},$$

$$W[3,0] = -\frac{(n-\ell)(n-\ell+1)(n+2-\ell)(\ell^2+\ell-1)}{(8\ell+12)(2\ell-1)},$$

$$W[3,2] = \frac{(n-\ell)(\ell+2)(\ell+1)}{(16\ell+8)(2\ell+3)}.$$

### 9.3.3  Program *gszeeman* for $E^{(j)}$ and $\Psi^{(j)}$

The Maple program implementing the iteration scheme (9.21) to (9.24) for the ground state energy corrections $E^{(j)}$ and the expansion coefficients $D^{(j)}_{NL}$ is very similar to program *gsstark* given in Section 8.2.3 for the Stark effect:

```
01   # gszeeman
02   # Ground state Zeeman effect using renormalized matrix elements
03   gszeeman := proc(jmax)
04
05   local R2010, j, k, x, s1, s2, s3, N, L, NP, LP;
06   # global vars (input): R, W
07   # global vars (output): DD, DE
08
09   DD[0,1,0] := 1:
10   R2010 := subs(n=1,l=0,R[1]);
11
12   for j from 1 to jmax do
13       lprint(j);
14
15       # Calculate jth order energy correction.
16
17       s1 := 0;
18       for N from 1 to 4 do
19           for L in [0,2] do
20               x := DD[j-1,N,L];
21               if not type(x, name) then
22                   s1 := s1 + subs(n=1,l=0,W[N-1,L])*x;
23               fi;
24           od;
25       od;
26       s2 := 0;
27       for k from 1 to j-1 do
28           x := DD[j-k,2,0];
29           if not type(x, name) then
30               s2 := s2 + x*DE[k];
31           fi;
32       od;
33       DE[j] := s1 - R2010*s2;
34
```

```
35        # Now calculate the jth order wavefunction coefficients
36
37        for N from 2 to 3*j+1 do
38           for L from 0 by 2 to 2*j do
39              if L<N then
40
41                  s1 := 0;
42                  for NP from N-3 to N+3 do
43                     for LP from L-2 by 2 to L+2 do
44                        x := DD[j-1,NP,LP];
45                        if not type(x,name) then
46                           s1 := s1 +
                                 subs(n=N,l=L, W[NP-N,LP-L])*x;
47                        fi;
48                     od;
49                  od;
50
51                  s2 := 0;
52                  for NP from N-1 to N+1 do
53                     s3 := 0;
54                     for k from 1 to j do
55                        x := DD[j-k,NP,L];
56                        if not type(x,name) then
57                           s3 := s3 + DE[k]*x;
58                        fi;
59                     od;
60                     s2 := s2 + subs(n=N,l=L,R[NP-N])*s3;
61                  od;
62                  DD[j,N,L] := (s1 - s2)/(1-N);
63               fi;
64            od; # L
65         od; # N
66      od: # j
67      DE[0] := -1/2;
68      save DD, DE, `ezeeman`.jmax.`.m`;
69   end:
70
71   # Example: Calculate wavefunction and energy corrections
72
73   jmax := 4;
74   read `zrenormdata.m`; # get R, W
75   gszeeman(jmax);
76   energy := sum('DE[j]*x^j', 'j'=0..jmax);
77   quit;
```

The procedure *gszeeman* creates a file called *ezeeman4.m* , if *jmax* is 4,
containing the tables of energy corrections *DE[j]* and wavefunction coefficients
*DD[j,N,L]* to order *jmax*.

The wavefunction corrections to order 4 in terms of the original expansion coefficients $C_{NL}^{(j)} = D_{NL}^{(j)}/N_{NL}$ are

$$\Psi^{(1)} = 2\sqrt{2}\,|20\rangle - \sqrt{3}\,|30\rangle + \tfrac{\sqrt{6}}{2}\,|32\rangle + \tfrac{1}{3}\,|40\rangle - \tfrac{1}{3}\,|42\rangle, \tag{9.26}$$

$$\begin{aligned}
\Psi^{(2)} = {}&-\tfrac{134\sqrt{2}}{3}\,|20\rangle + \tfrac{139\sqrt{3}}{3}\,|30\rangle - \tfrac{95\sqrt{6}}{6}\,|32\rangle - \tfrac{554}{9}\,|40\rangle \\
&+ \tfrac{380}{9}\,|42\rangle + \tfrac{157\sqrt{5}}{12}\,|50\rangle - \tfrac{77\sqrt{14}}{12}\,|52\rangle + \sqrt{70}\,|54\rangle \\
&- \tfrac{19\sqrt{6}}{6}\,|60\rangle + \tfrac{163\sqrt{21}}{105}\,|62\rangle - \tfrac{10\sqrt{7}}{7}\,|64\rangle + \tfrac{\sqrt{7}}{3}\,|70\rangle \\
&- \tfrac{4\sqrt{21}}{21}\,|72\rangle + \tfrac{\sqrt{154}}{21}\,|74\rangle,
\end{aligned} \tag{9.27}$$

$$\begin{aligned}
\Psi^{(3)} = {}&\tfrac{39017\sqrt{2}}{18}\,|20\rangle - \tfrac{536921\sqrt{3}}{180}\,|30\rangle + \tfrac{22079\sqrt{6}}{24}\,|32\rangle + \tfrac{3309529}{540}\,|40\rangle \\
&- \tfrac{407285}{108}\,|42\rangle - \tfrac{4775\sqrt{5}}{2}\,|50\rangle + \tfrac{73277\sqrt{14}}{70}\,|52\rangle - \tfrac{2627\sqrt{70}}{21}\,|54\rangle \\
&+ \tfrac{61724\sqrt{6}}{45}\,|60\rangle - \tfrac{135076\sqrt{21}}{225}\,|62\rangle + \tfrac{2143\sqrt{7}}{5}\,|64\rangle - \tfrac{10133\sqrt{7}}{18}\,|70\rangle \\
&+ \tfrac{361789\sqrt{21}}{1260}\,|72\rangle - \tfrac{77299\sqrt{154}}{1386}\,|74\rangle + \tfrac{950\sqrt{231}}{99}\,|76\rangle + \tfrac{18719\sqrt{2}}{60}\,|80\rangle \\
&- \tfrac{1781\sqrt{42}}{28}\,|82\rangle + \tfrac{7913\sqrt{154}}{385}\,|84\rangle - \tfrac{350\sqrt{66}}{33}\,|86\rangle - \tfrac{392}{5}\,|90\rangle \\
&+ \tfrac{61\sqrt{77}}{7}\,|92\rangle - \tfrac{458\sqrt{2002}}{385}\,|94\rangle + \tfrac{35\sqrt{55}}{11}\,|96\rangle + 2\sqrt{10}\,|10,0\rangle \\
&- \tfrac{10\sqrt{33}}{9}\,|10,2\rangle + \tfrac{2\sqrt{715}}{11}\,|10,4\rangle - \tfrac{20\sqrt{165}}{99}\,|10,6\rangle,
\end{aligned} \tag{9.28}$$

$$\begin{aligned}
\Psi^{(4)} = {}&-\tfrac{45622303\sqrt{2}}{270}\,|20\rangle + \tfrac{751019419\sqrt{3}}{2700}\,|30\rangle - \tfrac{29377693\sqrt{6}}{360}\,|32\rangle \\
&- \tfrac{6000306581}{8100}\,|40\rangle + \tfrac{701168101}{1620}\,|42\rangle + \tfrac{48134471\sqrt{5}}{120}\,|50\rangle \\
&- \tfrac{1635138553\sqrt{14}}{9800}\,|52\rangle + \tfrac{78031853\sqrt{70}}{4410}\,|54\rangle - \tfrac{936495169\sqrt{6}}{2700}\,|60\rangle \\
&+ \tfrac{971684503\sqrt{21}}{6750}\,|62\rangle - \tfrac{143628382\sqrt{7}}{1575}\,|64\rangle + \tfrac{31982983\sqrt{7}}{135}\,|70\rangle \\
&- \tfrac{15154310797\sqrt{21}}{132300}\,|72\rangle + \tfrac{87024199\sqrt{154}}{4410}\,|74\rangle - \tfrac{161395\sqrt{231}}{54}\,|76\rangle \\
&- \tfrac{794066911\sqrt{2}}{3150}\,|80\rangle + \tfrac{300703201\sqrt{42}}{6174}\,|82\rangle - \tfrac{119678371\sqrt{154}}{8575}\,|84\rangle \\
&+ \tfrac{12685\sqrt{66}}{2}\,|86\rangle + \tfrac{184037293}{1200}\,|90\rangle - \tfrac{208936331\sqrt{77}}{12936}\,|92\rangle \\
&+ \tfrac{316182061\sqrt{2002}}{161700}\,|94\rangle - \tfrac{50796\sqrt{55}}{11}\,|96\rangle + \tfrac{1400\sqrt{1430}}{11}\,|98\rangle \\
&- \tfrac{305373\sqrt{10}}{20}\,|10,0\rangle + \tfrac{22276501\sqrt{33}}{2772}\,|10,2\rangle - \tfrac{8082197\sqrt{715}}{6930}\,|10,4\rangle \\
&+ \tfrac{113512\sqrt{165}}{99}\,|10,6\rangle - \tfrac{1400\sqrt{715}}{11}\,|10,8\rangle + \tfrac{31657\sqrt{11}}{10}\,|11,0\rangle \\
&- \tfrac{1220866\sqrt{858}}{3465}\,|11,2\rangle + \tfrac{681927\sqrt{286}}{1540}\,|11,4\rangle - \tfrac{7598\sqrt{2805}}{99}\,|11,6\rangle \\
&+ \tfrac{84\sqrt{27170}}{11}\,|11,8\rangle - 814\sqrt{3}\,|12,0\rangle + \tfrac{854\sqrt{3003}}{33}\,|12,2\rangle \\
&- \tfrac{3486\sqrt{2002}}{143}\,|12,4\rangle + \tfrac{650\sqrt{1122}}{33}\,|12,6\rangle - \tfrac{280\sqrt{16302}}{143}\,|12,8\rangle \\
&+ \tfrac{220\sqrt{13}}{9}\,|13,0\rangle - \tfrac{200\sqrt{2002}}{99}\,|13,2\rangle + \tfrac{80\sqrt{17017}}{143}\,|13,4\rangle \\
&- \tfrac{80\sqrt{3553}}{99}\,|13,6\rangle + \tfrac{140\sqrt{38038}}{1287}\,|13,8\rangle.
\end{aligned} \tag{9.29}$$

### 9.3.4   Program for symmetric formula

Again the Maple program *symzeeman* for the symmetric energy formula (9.25) is very similar to the program *symstark* of Section 8.2.4. If *gszeeman* has been used to obtain the energy and wavefunction corrections to a certain order *jmax* then *symzeeman* can be used to calculate further energy corrections for orders *jmax+1* to *2*jmax+1*.

```
01    # symzeeman
02    # Symmetric energy formula for the ground state
03    # Zeeman effect with renormalized matrix elements
04    symEnergy := proc(kmin, kmax)
05
06        local R2010, k, p, q, r, s, s1, s2, s3, s4, s5, N, L,
07            NP, LP, Na2, x, z;
08        # global vars (input) DD, DE, R, W, Norm2
09        # global vars (output) DE
10
11        R2010 := subs(n=1,l=0,R[1]);
12        for k from kmin to kmax do
13            p := iquo(k,2);
14            q := iquo(k-1,2);
15            lprint(k,p,q);
16
17            # Do terms involving the perturbation matrix
18
19            s1 := 0;
20            for N from 2 to 3*p+1 do
21                for L from 0 by 2 to 2*p do
22                    if L < N then
23                        Na2 := subs(n=N,l=L,Norm2);
24                        s2 := 0;
25                        z := DD[p,N,L];
26                        if not type(z,name) then
27                            for NP from N-3 to N+3 do
28                                for LP from L-2 by 2 to L+2 do
29                                    x := DD[q,NP,LP];
30                                    if not type(x,name) then
31                                        s2 := s2 +
32                                            subs(n=N,l=L, W[NP-N,LP-L])*x;
33                                    fi;
34                                od; # LP
35                            od; # NP
36                            s1 := s1 + z*s2/Na2;
37                        fi;
38                    fi; # L < N
39                od; # L
```

```
40           od; # N
41
42           # do terms involving lower order energy corrections
43
44           s2 := 0;
45           for r from 1 to q do
46               s2 := s2 + DE[p+q+1-r]*DD[r,2,0];
47           od;
48           s1 := s1 - 2 * R2010 * s2;
49           s2 := 0;
50           for r from q+1 to p do
51               s2 := s2 + DE[p+q+1-r]*DD[r,2,0];
52           od;
53           s1 := s1 - R2010 * s2;
54
55           s2 := 0;
56           for r from 1 to p do
57               s3 := 0;
58               for N from 2 to 3*r+1 do
59                   for L from 0 by 2 to 2*r do
60                       if L < N then
61                           Na2 := subs(n=N,l=L,Norm2);
62                           s4 := 0;
63                           for s from 1 to q do
64                               s5 := 0;
65                               for NP from N-1 to N+1 do
66                                   x := DD[s,NP,L];
67                                   if not type(x,name) then
68                                       s5 := s5 +
69                                           subs(n=N,l=L,R[NP-N])*x;
70                                   fi;
71                               od;
72                               s4 := s4 + DE[p+q+1-r-s]*s5;
73                           od; # s
74                           s3 := s3 + DD[r,N,L]*s4/Na2;
75                       fi; # L<N
76                   od; # L
77               od; # N
78               s2 := s2 + s3;
79           od; # r
80           DE[k] := simplify(s1 - s2);
81       od: # k
82       save DE, `DE`.kmax;
83   end:
84
85   # Example: Given wavefunction and energy corrections to
86   # order jmax calculate the energy to order 2*jmax
```

```
87
88    read `zrenormdata.m`; # get R, W, Norm2
89    read `ezeeman4.m`; # get DD, DE to order 4
90    symEnergy(5,8): # calculate DE[5] to DE[8]
91    quit;
```

### 9.3.5  Summation of the ground state series

The ground state series has the form

$$E = -\frac{1}{2} + \sum_{n=1}^{\infty} E^{(n)} \lambda^n = -\frac{1}{2} + \sum_{n=1}^{\infty} E^{(n)} \left(\frac{B^2}{8}\right)^n, \tag{9.30}$$

where the energy corrections $E^{(n)}$ are given in Appendix F to order 60 in both rational and floating point form.

It has been shown that this series is divergent, asymptotic and Borel summable [AV81b], [SI82]. The conjectured large order behaviour of the energy corrections is given by

$$E^{(n)} = C^{(n)} \left(A_0 + \frac{A_1}{n} + \frac{A_2}{n^2} + \cdots\right), \tag{9.31}$$

where $A_0 = 1$ and

$$C^{(n)} = (-1)^{n+1} \left(\frac{4}{\pi}\right)^{5/2} \left(\frac{8}{\pi^2}\right)^n \Gamma\left(2n + \frac{3}{2}\right). \tag{9.32}$$

Unlike the Stark effect, there is no rigorous mathematical proof of (9.32) but it has been verified numerically [AV79].

As in the Stark effect (see Section 8.3.1) direct summation of the series is possible only for rather small values of $\lambda$. The diagonal Padé approximants can extend this range somewhat and the results are given in Table 9.1.

Some extrapolated results and exact variational results are also given in the last two columns. The variational results were obtained by Vrscay using the so(4,2) Lie algebraic methods and a complete basis [VR77] and the extrapolated results were obtained by Čížek and Vrscay using more sophisticated summation techniques involving continued fractions for the series (9.30) and the Thiele extrapolation method [CI82].

## 9.4  Degenerate RSPT for the Zeeman Effect

The modified algebraic RSPT of Section 7.4 for the so(4,2) Lie algebraic approach can also be developed for degenerate cases by adapting the general conventional formalism developed by Silverstone [SI81] to apply to (7.61).

For the Zeeman effect the need for degenerate RSPT first occurs for the doubly degenerate $n = 3$ subspace consisting of the states $|300\rangle$ and $|320\rangle$ (in spectroscopic notation these are the $3s$ and $3d_0$ excited states). For these states the degeneracy is removed in first order. Therefore we consider only this case.

**Table 9.1**: Summation of the ground state Zeeman series. $E_{\text{parsum}}$ represents the best value obtained by direct summation, $[4,4]$ and $[28,28]$ are the Padé approximants, $E_{\text{extrap}}$ and $E_{\text{exact}}$ are extrapolated and exact results [CI82]. Entries indicated by $*$ would have no correct figures.

| $\mathcal{B}$ | $E_{\text{parsum}}$ | $[4,4]$ | $[28,28]$ | $E_{\text{extrap}}$ | $E_{\text{exact}}$ |
|---|---|---|---|---|---|
| 0.1 | $-0.49753$ | $-0.49753$ | $-0.49753$ | $-0.49753$ | $-0.49753$ |
| 0.2 | $-0.49039$ | $-0.49038$ | $-0.49038$ | $-0.49038$ | $-0.49038$ |
| 0.3 | $-0.47886$ | $-0.47919$ | $-0.47919$ | $-0.47919$ | $-0.47919$ |
| 0.4 | $-0.46707$ | $-0.46462$ | $-0.46461$ | $-0.46461$ | $-0.46461$ |
| 0.5 | $-0.43750$ | $-0.44731$ | $-0.44721$ | $-0.44721$ | $-0.44721$ |
| 1.0 | $*$ | $-0.33924$ | $-0.33140$ | $-0.33117$ | $-0.33117$ |
| 2.0 | $*$ | $*$ | $*$ | $-0.02219$ | $-0.02223$ |
| 3.0 | $*$ | $*$ | $*$ | $+0.3357$ | $+0.3354$ |
| 4.0 | $*$ | $*$ | $*$ | $+0.7206$ | $+0.7192$ |
| 5.0 | $*$ | $*$ | $*$ | $+1.1235$ | $+1.1196$ |

### 9.4.1 Doubly-degenerate modified RSPT

As in the non-degenerate case of Section 7.4 we consider the general perturbation problem

$$(K_0 - \kappa_0 + \lambda W - \Delta E\, S)\Psi = 0, \tag{9.33}$$

and the unperturbed eigenvalue problem

$$K_0 \Phi_a = \kappa_a \Phi_a, \quad a = 0, 1, 2, \ldots, \tag{9.34}$$

but now we assume that $a = 0$ and $a = 1$ correspond to the doubly degenerate states so

$$\kappa_0 = \kappa_1, \quad \kappa_a \neq \kappa_0, \quad a \geq 2. \tag{9.35}$$

We shall also assume that the degeneracy is removed in the first order of perturbation theory and that the correct linear combinations

$$\Phi_{0'} = C_{00}\Phi_0 + C_{10}\Phi_1, \tag{9.36}$$

$$\Phi_{1'} = C_{01}\Phi_0 + C_{11}\Phi_1 \tag{9.37}$$

of the two degenerate states $\Phi_0$ and $\Phi_1$ are obtained by diagonalizing the $2 \times 2$ secular equation

$$W\Phi = E^{(1)} S\Phi, \tag{9.38}$$

where $E^{(1)}$ is the first order energy correction. Thus our assumption is that the two eigenvalues satisfy $E_0^{(1)} \neq E_1^{(1)}$. This is really an ordinary secular equation rather than a generalized one since $S$ is a multiple of the $2 \times 2$ identity matrix (see (9.68)).

To obtain perturbation equations we first define the resolvent operator on the space of solutions of (9.34) orthogonal to the degenerate subspace spanned by $\Phi_0$ and $\Phi_1$, or equivalently $\Phi_{0'}$ and $\Phi_{1'}$,

$$G = \sum_{a=2}^{\infty} \frac{|\Phi_a\rangle\langle\Phi_a|}{\kappa_0 - \kappa_a}, \tag{9.39}$$

and consider the perturbed problem in the form

$$(\kappa_0 - K_0)\Psi = (\lambda W - \Delta E\, S)\Psi. \tag{9.40}$$

Then the solution of this equation can be expressed as

$$\Psi = \Phi_{0'} + b\,\Phi_{1'} + G(\lambda W - \Delta E\, S)\Psi. \tag{9.41}$$

Here we are considering $\Phi_{0'}$ as the unperturbed reference state so this will give the perturbed wavefunction evolving from $\Phi_{0'}$.

We can obtain the other perturbed wavefunction from the formalism we shall develop simply by reversing the roles of $\Phi_{0'}$ and $\Phi_{1'}$. This is the approach taken by Silverstone in the conventional case [SI81] in order to make the degenerate formalism look formally similar to the non-degenerate case.

If we introduce the notation

$$\Psi = \Phi_{0'} + \Omega_1 + \Omega_0, \tag{9.42}$$

$$\Omega_1 = b\,\Phi_{1'}, \tag{9.43}$$

$$\Omega_0 = G(\lambda W - \Delta E\, S)\Psi, \tag{9.44}$$

then we have an iterative formula for $\Omega_0$, the component of $\Psi$ orthogonal to $\Phi_{0'}$ and $\Phi_{1'}$,

$$\Omega_0 = G(\lambda W - \Delta E\, S)(\Phi_{0'} + \Omega_1 + \Omega_0). \tag{9.45}$$

Now substitute the series expansions

$$\Omega_0 = \sum_{j=1}^{\infty} \Omega_0^{(j)} \lambda^j, \quad \Omega_0^{(j)} = \sum_{k=2}^{\infty} C_k^{(j)} \Phi_k, \tag{9.46}$$

$$\Omega_1 = \sum_{j=0}^{\infty} \Omega_1^{(j)} \lambda^j, \quad \Omega_1^{(j)} = \beta^{(j)} \Phi_{1'} \tag{9.47}$$

$$\Delta E = \sum_{j=1}^{\infty} E^{(j)} \lambda^j = \lambda \sum_{j=0}^{\infty} E^{(j+1)} \lambda^j, \tag{9.48}$$

to obtain for $j = 1$

$$\Omega_0^{(1)} = G(W - E^{(1)}S)(\Phi_{0'} + \beta_0 \Phi_{1'}), \tag{9.49}$$

and for $j \geq 2$

$$\Omega_0^{(j)} = GW(\beta^{(j-1)}\Phi_{1'} + \Omega_0^{(j-1)}) - GS \sum_{k=1}^{j-1} E^{(k)}\Omega_0^{(j-k)}$$

$$- E^{(j)}GS\Phi_{0'} - GS \sum_{k=1}^{j} E^{(k)}\beta^{(j-k)}\Phi_{1'}. \tag{9.50}$$

Thus $\Omega_0^{(j)}$ depends on lower order $\Omega_0^{(k)}$, $\Omega_1^{(k)}$ (through the coefficients $\beta^{(k)}$) and $E^{(k)}$ so we also need iteration formulas for the $\beta^{(k)}$ and the $E^{(k)}$. To obtain them we project the Schrödinger equation (9.40) onto the degenerate states $\Phi_{0'}$ and $\Phi_{1'}$. Therefore

$$\langle\Phi_{0'}|\lambda W - \Delta E\, S|\Phi_{0'} + b\,\Phi_{1'} + \Omega_0\rangle = 0, \tag{9.51}$$

$$\langle\Phi_{1'}|\lambda W - \Delta E\, S|\Phi_{0'} + b\,\Phi_{1'} + \Omega_0\rangle = 0. \tag{9.52}$$

Define the shorthand notation

$$S_{i'j'} = \langle\Phi_{i'}|S|\Phi_{j'}\rangle, \quad W_{i'j'} = \langle\Phi_{i'}|W|\Phi_{j'}\rangle, \quad i',j' = 0,1, \tag{9.53}$$

and substitute the series expansions (9.46) to (9.48) to obtain for $j = 0$

$$(W_{0'0'} - S_{0'0'}E^{(1)}) + (W_{0'1'} - S_{0'1'}E^{(1)})\beta^{(0)} = 0, \tag{9.54}$$

$$(W_{1'0'} - S_{1'0'}E^{(1)}) + (W_{1'1'} - S_{1'1'}E^{(1)})\beta^{(0)} = 0, \tag{9.55}$$

and for $j \geq 1$

$$(W_{0'1'} - S_{0'1'}E^{(1)})\beta^{(j)} - S_{0'1'}\sum_{k=1}^{j} E^{(k+1)}\beta^{(j-k)} - S_{0'0'}E^{(j+1)}$$

$$= -\langle\Phi_{0'}|W|\Omega_0^{(j)}\rangle + \sum_{k=0}^{j-1} E^{(k+1)}\langle\Phi_{0'}|S|\Omega_0^{(j-k)}\rangle, \tag{9.56}$$

$$(W_{1'1'} - S_{1'1'}E^{(1)})\beta^{(j)} - S_{1'1'}\sum_{k=1}^{j} E^{(k+1)}\beta^{(j-k)} - S_{1'0'}E^{(j+1)}$$

$$= -\langle\Phi_{1'}|W|\Omega_0^{(j)}\rangle + \sum_{k=0}^{j-1} E^{(k+1)}\langle\Phi_{1'}|S|\Omega_0^{(j-k)}\rangle. \tag{9.57}$$

Since $\Phi_{0'}$ and $\Phi_{1'}$ are eigenvectors of the $2 \times 2$ secular equation (9.38) we can assume that $W_{0'1'} = W_{1'0'} = S_{1'0'} = S_{0'1'}$ so the two first order energies are

$$E_0^{(1)} = W_{0'0'}/S_{0'0'}, \tag{9.58}$$

$$E_1^{(1)} = W_{1'1'}/S_{1'1'}. \tag{9.59}$$

Therefore (9.54) and (9.55) are satisfied if we choose $\beta^{(0)} = 0$.

The energy corrections $E_0^{(j+1)}$ can now be determined from (9.56) in terms of lower order energy corrections and the wavefunction corrections $\Omega_{0k}$ as

$$S_{0'0'}E_0^{(j+1)} = \langle\Phi_{0'}|W|\Omega_0^{(j)}\rangle - \sum_{k=0}^{j-1} E_0^{(k+1)}\langle\Phi_{0'}|S|\Omega_0^{(j-k)}\rangle, \tag{9.60}$$

which has the same form as the corresponding result for the non-degenerate case.

Finally the coefficients $\beta^{(j)}$, $j > 0$, which are needed for the calculation of $\Omega_0^{(j)}$ in (9.50), can be determined from (9.57) as

$$S_{1'1'}\beta^{(j)} = \frac{1}{E_0^{(1)} - E_1^{(1)}}\left[\langle\Phi_{1'}|W|\Omega_0^{(j)}\rangle - \sum_{k=0}^{j-1} E_0^{(k+1)}\langle\Phi_{1'}|S|\Omega_0^{(j-k)}\rangle \right.$$
$$\left. - S_{1'1'}\sum_{k=1}^{j-1} E_0^{(k+1)}\beta^{(j-k)}\right], \tag{9.61}$$

where we have used $W_{1'1'} - S_{1'1'}E_0^{(1)} = S_{1'1'}(E_1^{(1)} - E_0^{(1)})$ and we have $\beta^{(0)} = 0$.

Thus we obtain the following algorithm. First diagonalize $W$ in (9.38) to obtain the two first order energies $E_0^{(1)}$ and $E_1^{(1)}$. Call one of the eigenfunctions $\Phi_{0'}$ and the other $\Phi_{1'}$ as in (9.36) and (9.37). Then for $j = 1, 2, 3, \ldots$ calculate $\Omega_0^{(j)}$ from (9.50), $\beta^{(j)}$ from (9.61) using $\beta^{(0)} = 0$, and $E_0^{(j+1)}$ from (9.60). Repeat the calculations with the roles of $\Phi_{0'}$, $E_0^{(1)}$ and $\Phi_{1'}$, $E_1^{(1)}$ reversed to calculate the wavefunction and energy corrections for the other state.

For practical calculations it is necessary to substitute the expansion (9.46) for the wavefunction corrections $\Omega_0^{(j)}$ in terms of the unperturbed eigenfunctions $\Phi_k$ and expansion coefficients $C_k^{(j)}$. The results, analogous to (7.82) to (7.85) in the non-degenerate case, are

$$C_A^{(1)} = \frac{1}{\kappa_0 - \kappa_A}\left[W_{A0'} - S_{A0'}E_0^{(1)}\right], \, A \geq 2, \tag{9.62}$$

$$C_A^{(j)} = \frac{1}{\kappa_0 - \kappa_A}\left[W_{A1'}\beta^{(j-1)} + \sum_{B=2}^{\infty} W_{AB}C_B^{(j-1)} - S_{A0'}E_0^{(j)}\right.$$
$$\left. - S_{A1'}\left(\sum_{k=1}^{j-1} E_0^{(k)}\beta^{(j-k)}\right) - \sum_{B=2}^{\infty} S_{AB}\left(\sum_{k=1}^{j-1} E_0^{(k)}C_B^{(j-k)}\right)\right],$$
$$j > 1, \, A \geq 2, \tag{9.63}$$

$$S_{1'1'}\beta^{(j)} = \frac{1}{E_0^{(1)} - E_1^{(1)}}\left[\sum_{A=2}^{\infty} W_{1'A}C_A^{(j)} - \sum_{A=2}^{\infty} S_{1'A}\sum_{k=0}^{j-1} E_0^{(k+1)}C_A^{(j-k)}\right.$$
$$\left. - S_{1'1'}\sum_{k=1}^{j-1} E_0^{(k+1)}\beta^{(j-k)}\right], \tag{9.64}$$

$$E_0^{(j+1)} = \frac{1}{S_{0'0'}}\left[\sum_{A=2}^{\infty} W_{0'A}C_A^{(j)} - \sum_{A=2}^{\infty} S_{0'A}\sum_{k=0}^{j-1} E_0^{(k+1)}C_A^{(j-k)}\right]. \tag{9.65}$$

The matrix elements $S_{AB}$ and $W_{AB}$, $A, B \geq 2$, are defined in terms of the standard unperturbed eigenfunctions $\Phi_A$ and $\Phi_B$ as in the non-degenerate case and those with subscripts $0'$ or $1'$ are evaluated in terms of $\Phi_0$ and $\Phi_1$ using (9.36) and (9.47).

### 9.4.2   The $3s$–$3d_0$ degenerate level

For the Zeeman effect the $3s$–$3d_0$ sublevel is the first case where degenerate perturbation theory is required. For $n = 3$ the operators $S$ and $W$ are (see

(9.4),(9.5))

$$S = 3^2 R, \quad W = 3^4 R(R^2 - Z^2).$$ (9.66)

The $2 \times 2$ matrices of $S$ and $W$ in the subspace spanned by the degenerate unperturbed eigenfunctions

$$\Phi_0 = |300\rangle, \quad \Phi_1 = |320\rangle,$$ (9.67)

are

$$S = \begin{bmatrix} \langle 300|S|300\rangle & \langle 300|S|320\rangle \\ \langle 320|S|300\rangle & \langle 320|S|320\rangle \end{bmatrix} = \begin{bmatrix} S_{00} & S_{01} \\ S_{10} & S_{11} \end{bmatrix}$$
$$= \begin{bmatrix} 27 & 0 \\ 0 & 27 \end{bmatrix},$$ (9.68)

$$W = \begin{bmatrix} \langle 300|W|300\rangle & \langle 300|W|320\rangle \\ \langle 320|W|300\rangle & \langle 320|W|320\rangle \end{bmatrix} = \begin{bmatrix} W_{00} & W_{01} \\ W_{10} & W_{11} \end{bmatrix}$$
$$= \begin{bmatrix} 46 \cdot 3^4 & -10\sqrt{2} \cdot 3^4 \\ -10\sqrt{2} \cdot 3^4 & 20 \cdot 3^4 \end{bmatrix}.$$ (9.69)

Therefore the secular equation (9.38) is

$$\begin{bmatrix} 138 - E^{(1)} & -30\sqrt{2} \\ -30\sqrt{2} & 60 - E^{(1)} \end{bmatrix} \begin{bmatrix} C_0 \\ C_1 \end{bmatrix} = \begin{bmatrix} 0 \\ 0 \end{bmatrix}.$$ (9.70)

The eigenvalues are the first order energy corrections

$$E_0^{(1)} = 99 + 9\sqrt{41}, \quad E_1^{(1)} = 99 - 9\sqrt{41},$$ (9.71)

and the coefficients defining the linear combinations (9.36) and (9.37) are (see Exercise 9.2)

$$C_{00} = \alpha^{1/2}, \quad C_{10} = -\frac{5}{3}\sqrt{\frac{2}{41}}\alpha^{-1/2},$$ (9.72)

$$C_{01} = -C_{10}, \quad C_{11} = C_{00},$$ (9.73)

$$\alpha = \frac{1}{2} + \frac{13}{246}\sqrt{41}.$$ (9.74)

Using the renormalized coefficients

$$D_A^{(j)} = N_A C_a^{(j)},$$ (9.75)
$$D_{00} = N_0 C_{00}, \quad D_{10} = N_1 C_{10},$$ (9.76)
$$D_{01} = N_0 C_{01}, \quad D_{11} = N_1 C_{11},$$ (9.77)

and the renormalized matrix elements $\bar{S}_{AB}$ and $\overline{W}_{AB}$ defined as in the nondegenerate case, we obtain from (9.62) to (9.65) the iteration scheme

$$D_{NL}^{(1)} = \frac{1}{3 - N}\left[s_0 - s_2 E_0^{(1)}\right],$$ (9.78)

$$D_{NL}^{(j)} = \frac{1}{3-N} \left[ s_1 \beta^{(j-1)} - s_2 E_0^{(j)} - s_3 s_4 + s_5 - s_6 \right], \qquad (9.79)$$

$$\beta^{(j)} = \frac{1}{S_{1'1'}(E_0^{(1)} - E_1^{(1)})} \left[ D_{01}(s_7 - s_9) + D_{11}(s_8 - s_{10}) - s_{11} \right], \qquad (9.80)$$

$$E_0^{(j+1)} = \frac{1}{S_{0'0'}} \left[ D_{00}(s_7 - s_9) + D_{10}(s_8 - s_{10}) \right], \qquad (9.81)$$

where

$$s_0 = D_{00}\overline{W}_{NL}^{30} + D_{10}\overline{W}_{NL}^{32}, \qquad (9.82)$$

$$s_1 = D_{01}\overline{W}_{NL}^{30} + D_{11}\overline{W}_{NL}^{32}, \qquad (9.83)$$

$$s_2 = D_{00}\overline{S}_{NL}^{30} + D_{10}\overline{S}_{NL}^{32}, \qquad (9.84)$$

$$s_3 = D_{01}\overline{S}_{NL}^{30} + D_{11}\overline{S}_{NL}^{32}, \qquad (9.85)$$

$$s_4 = \sum_{k=1}^{j-1} E_0^{(k)} \beta^{(j-k)}, \qquad (9.86)$$

$$s_5 = \sum_{N'=N-3}^{N+3} \left( \sum_{L'=L-2,L,L+2} \overline{W}_{NL}^{N'L'} D_{N'L'}^{(j-1)} \right), \qquad (9.87)$$

$$s_6 = \sum_{N'=N-1}^{N+1} \overline{S}_{NL}^{N'L} \left( \sum_{k=1}^{j-1} E_0^{(k)} D_{N'L}^{(j-k)} \right), \qquad (9.88)$$

$$s_7 = \frac{1}{N_{30}^2} \sum_{N=1}^{6} \left( \sum_{L=0,2} \overline{W}_{30}^{NL} D_{NL}^{(j)} \right), \qquad (9.89)$$

$$s_8 = \frac{1}{N_{32}^2} \sum_{N=1}^{6} \left( \sum_{L=0,2,4} \overline{W}_{32}^{NL} D_{NL}^{(j)} \right), \qquad (9.90)$$

$$s_9 = \frac{1}{N_{30}^2} \sum_{N=2,4} \overline{S}_{30}^{N0} \left( \sum_{k=0}^{j-1} E_0^{(k+1)} D_{N0}^{(j-k)} \right), \qquad (9.91)$$

$$s_{10} = \frac{1}{N_{32}^2} \sum_{N=2,4} \overline{S}_{32}^{N2} \left( \sum_{k=0}^{j-1} E_0^{(k+1)} D_{N2}^{(j-k)} \right), \qquad (9.92)$$

$$s_{11} = S_{1'1'} \sum_{k=1}^{j-1} E_0^{(k+1)} \beta^{(j-k)}. \qquad (9.93)$$

### 9.4.3  Procedure *zee3s3d* for the 3s and $3d_0$ states

The following Maple procedure implements the iteration scheme (9.78) to (9.93) for the energy corrections. The coefficients $D_{00}$, $D_{10}$, $D_{01}$, $D_{11}$ and $D_{NL}^{(j)}$ are denoted in the program by *D00*, *D10*, *D01*, *D11* and *DD[j,N,L]*. The energy correction $E_0^{(j)}$ is denoted by *DE[j]* and the first order energy corrections $E_0^{(1)}$ and $E_1^{(1)}$ are denoted by *DE10* and *DE11*. Also $\beta^{(j)}$ is denoted by *B[j]*.

```
01    # zee3s3d
02    # Degenerate PT for the doubly degenerate 3s-3d0 states
03    # This version uses unnormalized coefficients and matrix elements
04    zee3s3d := proc(jmax)
05
06       local fW, fS, N0sq, N1sq, S00, S11, S0P0P, S1P1P,
07          j, k, N, L, NP, LP, s, s1, s2, s3, s4, s5, s6,
08          s6a, s7, s8, s9, s9a, s10, s10a, s11, x;
09
10       # global vars (input): D00, D01, D10, D11, DE10, DE11
11       # global vars (input): Norm2, R, W
12       # global vars (output): B, DD, DE
13
14       fW := 3^4;
15       fS := 3^2;
16       N0sq := subs(n=3,l=0,Norm2);
17       N1sq := subs(n=3,l=2,Norm2);
18
19       S00 := fS * subs(n=3,l=0,R[0]);
20       S11 := fS * subs(n=3,l=2,R[0]);
21
22       S0P0P := S00; # C00^2 + C10^2 = 1
23       S1P1P := S00; # C01^2 + C11^2 = 1
24
25    for j from 1 to jmax do
26        lprint(j);
27
28        if j = 1 then
29            DE[j] := DE10;
30            for N from 1 to 3*j + 3 do
31                for L from 0 by 2 to 2*j + 2 do
32                    if (L < N) and ( N <> 3) then
33
34                        s := 0;
35                        if abs(N-3) <= 3 and abs(L) <= 2 then
36                            s := s + D00 * fW
                                 * subs(n=N,l=L,W[3-N,0-L]);
37                        fi;
38                        if abs(N-3) <= 3 and abs(L-2) <= 2 then
39                            s := s + D10 * fW
                                 * subs(n=N,l=L,W[3-N,2-L]);
40                        fi;
41                        if abs(N-3) <= 1 and L = 0 then
42                            s := s - DE[1] * D00 * fS
                                 * subs(n=N,l=L,R[3-N]);
43                        fi;
44                        if abs(N-3) <= 1 and L = 2 then
```

```
45                              s := s - DE[1] * D10 * fS
                                   * subs(n=N,l=L,R[3-N]);
46                      fi;
47
48                      DD[1,N,L] := simplify(s / (3-N));
49
50              fi; # L < N
51           od; # L
52         od; # N
53
54        else # calculate DD[j,N,L] for j > 1
55
56          for N from 1 to 3*j + 3 do
57             for L from 0 by 2 to 2*j + 2 do
58                if (L < N) and (N <> 3) then
59
60                   s1 := 0;
61                   if abs(N-3) <= 3 and abs(L) <= 2 then
62                      s1 := s1 + D01 * fW
                            * subs(n=N,l=L,W[3-N,0-L]);
63                   fi;
64                   if abs(N-3) <= 3 and abs(L-2) <= 2 then
65                      s1 := s1 + D11 * fW
                            * subs(n=N,l=L,W[3-N,2-L]);
66                   fi;
67
68                   s2 := 0;
69                   s3 := 0;
70                   if abs(N-3) <= 1 and L = 0 then
71                      s2 := s2 + D00 * fS * subs(n=N,l=L,R[3-N]);
72                      s3 := s3 + D01 * fS * subs(n=N,l=L,R[3-N]);
73                   fi;
74                   if abs(N-3) <= 1 and L = 2 then
75                      s2 := s2 + D10 * fS * subs(n=N,l=L,R[3-N]);
76                      s3 := s3 + D11 * fS * subs(n=N,l=L,R[3-N]);
77                   fi;
78
79                   s4 := 0;
80                   for k from 1 to j-1 do
81                      s4 := s4 + DE[k] * B[j-k];
82                   od;
83
84                   s := s1 * B[j-1] - s2 * DE[j] - s3 * s4;
85
86                   s5 := 0;
87                   for NP from N-3 to N+3 do
88                      for LP from L-2 by 2 to L+2 do
```

```
 89                                x := DD[j-1,NP,LP];
 90                                if not type(x,name) then
 91                                    s5 := s5 + fW * subs(
                                           n=N,l=L,W[NP-N,LP-L]) * x;
 92                                fi;
 93                            od; # LP
 94                        od; # NP
 95
 96                    s6 := 0;
 97                    for NP from N-1 to N+1 do
 98                        s6a := 0;
 99                        for k from 1 to j-1 do
100                            x := DD[j-k,NP,L];
101                            if not type(x,name) then
102                                s6a := s6a + DE[k] * x;
103                            fi;
104                        od;
105                        s6 := s6 + fS
                                * subs(n=N,l=L,R[NP-N]) * s6a;
106                    od;
107
108                    DD[j,N,L] := simplify((s + s5 - s6)/(3-N));
109
110                fi; # L < N
111            od; # L
112        od; # N
113
114    fi; # j > 1 (else part of DD[j,N,L] calculation)
115
116    # Now calculate B[j] and DE[j+1]
117
118    s7 := 0;
119    for N from 1 to 6 do
120        for L from 0 by 2 to 2 do
121            x := DD[j,N,L];
122            if not type(x, name) then
123                s7 := s7 + subs(n=3,l=0,W[N-3,L]) * x;
124            fi;
125        od;
126    od;
127    s7 := s7 * fW / N0sq;
128
129    s8 := 0;
130    for N from 1 to 6 do
131        for L from 0 by 2 to 4 do
132            x := DD[j,N,L];
133            if not type(x, name) then
```

```
134                   s8 := s8 + subs(n=3,l=2,W[N-3,L-2]) * x;
135             fi;
136          od;
137       od;
138       s8 := s8 * fW / N1sq;
139
140       s9 := 0;
141       for N from 2 to 4 do
142          s9a := 0;
143          for k from 0 to j-1 do
144             x := DD[j-k,N,0];
145             if not type(x,name) then
146                s9a := s9a + DE[k+1]*x;
147             fi;
148          od;
149          s9 := s9 + subs(n=3,l=0,R[N-3]) * s9a;
150       od;
151       s9 := s9 * fS / N0sq;
152
153       s10 := 0;
154       for N from 2 to 4 do # only N=4 contributes
155          s10a := 0;
156          for k from 0 to j-1 do
157             x := DD[j-k,N,2];
158             if not type(x,name) then
159                s10a := s10a + DE[k+1] * x;
160             fi;
161          od;
162          s10 := s10 + subs(n=3,l=2,R[N-3]) * s10a;
163       od;
164       s10 := s10 * fS / N1sq;
165
166       s11 := 0;
167       for k from 1 to j-1 do
168          s11 := s11 + DE[k+1] * B[j-k];
169       od;
170       s11 := S1P1P * s11;
171
172       B[j] := simplify((D01*s7 + D11*s8 - D01*s9
                 - D11*s10 - s11)/S1P1P/(DE[1] - DE11));
173
174       DE[j+1] := simplify((D00*s7 + D10*s8 - D00*s9
                 - D10*s10) / S0P0P);
175
176    od; # j
177 end:
```

Program *zee3s3d* can be used to obtain the wavefunction corrections $\Psi_0^{(j)}$, $\Psi_1^{(j)}$ and the energy corrections $E_0^{(j)}$, $E_1^{(j)}$ for the two degenerate states. For example, the following program uses *zee3s3d* to produce results to order 5 in the wavefunction and to order 6 in the energy for reference state 0.

```
01    # test3-0 : Calculation for reference state 0
02    # Directly substitute the symbolic numeric values
03
04    jmax := 5:
05    read `zrenormdata.m`:
06    read zee3s3d;
07
08    N0sq := subs(n=3,l=0,Norm2):
09    N1sq := subs(n=3,l=2,Norm2):
10    N0 := sqrt(N0sq);
11    N1 := sqrt(N1sq);
12
13    # first order energies
14
15    DE1[0] := 99 + 9 * sqrt(41);
16    DE1[1] := 99 - 9 * sqrt(41);
17
18    # Expansion coefficients for reference states 0 and 1
19
20    ccc := 1/2 + (13/246) * sqrt(41);
21    C0[0] := sqrt(ccc);
22    C1[0] := -sqrt(50/369)/sqrt(ccc);
23    C0[1] := -C1[0];
24    C1[1] := C0[0];
25
26    # Initialize the 6 parameters for the reference state 0
27
28    D00 := N0 * C0[0]:
29    D10 := N1 * C1[0]:
30    DE10 := DE1[0]:
31
32    D01 := N0 * C0[1]:
33    D11 := N1 * C1[1]:
34    DE11 := DE1[1]:
35
36    # Do perturbation theory to energy order jmax
37
38    zee3s3d(jmax):
39
40    # simplify radicals in the energy expressions to form a + b*sqrt(41)
41
42    for j from 1 to jmax + 1 do
```

```
43      radsimp(DE[j], 'e');
44      e0[j] := expand(e);
45   od:
46
47   # Save energy results in a file
48
49   save e0, `e0-`.(jmax+1);
50
51   # Display results in symbolic and floating point form
52
53   op(e0);
54   Digits := 64;
55   map(evalf,op(e0));
56   quit;
```

Similarly, by interchanging the roles of the two reference states we can obtain the wavefunction and energy corrections for reference state 0.

```
01   # test3-1 : Calculation for reference state 1
02   # Directly substitute the symbolic numeric values
03
04   jmax := 5:
05   read `zrenormdata.m`:
06   read zee3s3d;
07
08   N0sq := subs(n=3,l=0,Norm2):
09   N1sq := subs(n=3,l=2,Norm2):
10   N0 := sqrt(N0sq);
11   N1 := sqrt(N1sq);
12
13   # first order energies
14
15   DE1[0] := 99 + 9 * sqrt(41);
16   DE1[1] := 99 - 9 * sqrt(41);
17
18   # Expansion coefficients for reference states 0 and 1
19
20   ccc := 1/2 + (13/246) * sqrt(41);
21   C0[0] := sqrt(ccc);
22   C1[0] := -sqrt(50/369)/sqrt(ccc);
23   C0[1] := -C1[0];
24   C1[1] := C0[0];
25
26   # Initialize the 6 parameters for the reference state 1
27
28   D01 := N0 * C0[0]:
29   D11 := N1 * C1[0]:
30   DE11 := DE1[0]:
```

```
31
32    D00 := N0 * C0[1]:
33    D10 := N1 * C1[1]:
34    DE10 := DE1[1]:
35
36    # Do perturbation theory to energy order jmax
37
38    zee3s3d(jmax):
39
40    # simplify radicals in the energy expressions to form a + b*sqrt(41)
41
42    for j from 1 to jmax + 1 do
43        radsimp(DE[j], 'e');
44        e1[j] := expand(e);
45    od:
46
47    # Save energy results in a file
48
49    save e1, `e1-`.(jmax+1);
50
51    # Display results in symbolic and floating point form
52
53    op(e1);
54    Digits := 64;
55    map(evalf,op(e1));
56    quit;
```

### 9.4.4   Energy corrections for the $3s$ and $3d_0$ states

The energy corrections for the $3s$ and $3d_0$ states to order 8 can be expressed
in the form

$$E_{3s} = \sum_{j=0}^{8} E_0^{(j)} \lambda^j = -\frac{1}{18} + \sum_{j=1}^{8} \left(a_j + b_j \sqrt{41}\right) \lambda^j, \tag{9.94}$$

$$E_{3d_0} = \sum_{j=0}^{8} E_1^{(j)} \lambda^j = -\frac{1}{18} + \sum_{j=1}^{8} \left(a_j - b_j \sqrt{41}\right) \lambda^j, \tag{9.95}$$

where

$$a_1 = 99$$

$$a_2 = -\frac{576639}{2}$$

$$a_3 = \frac{5472956565}{2}$$

$$a_4 = -\frac{1790372774796399}{40}$$

$$a_5 = \frac{198262158135884253459}{200}$$

$$a_6 = -\frac{3810264926043211725335080 41}{14000}$$

$$a_7 = \frac{4338683397293407440555596 03380917}{490000}$$

$$a_8 = -\frac{457102480970569647254932065 8872479690657}{137200000} \tag{9.96}$$

$$b_1 = 9$$

$$b_2 = -\frac{2858409}{82}$$

$$b_3 = \frac{1238374860363}{3362}$$

$$b_4 = -\frac{17642206568880035001}{2756840}$$

$$b_5 = \frac{8264222682769538396 3288781}{565152200}$$

$$b_6 = -\frac{662014263136876914495328064 7896079}{1621986814000}$$

$$b_7 = \frac{44548152957891664885086089154271314385149}{332507296870000}$$

$$b_8 = -\frac{1353725712807351185328350666378529185723599961188023}{26720286376473200000} . \tag{9.97}$$

Due to the symmetry of the energy coefficients expressed by (9.95) and (9.96) for the two states we see that it is only necessary to use *test3-0* to obtain the $a_j$ and $b_j$ for reference state 0. A symmetric energy formula having the same basic form as the non-degenerate formula used in the Stark and Zeeman ground state calculations can also be developed for the degenerate perturbation theory [SI81].

These exact results agree with the floating point results to order 87 obtained by Silverstone and Moats [SI81] using conventional degenerate perturbation theory based on the so(4,2) algebraic approach (perturbed principal quantum number) and also with the unpublished floating point results to order 53 obtained using the modified perturbation formalism of this section [AD80].

It has also been conjectured that the leading asymptotic behaviour of the energy coefficients for the two states (see (9.36), (9.37), (9.67) and (9.72) to (9.74)) is given by

$$E_a^{(n)} = C_a^{(n)}[1 + O(1/n)], \, a = 0, 1, \tag{9.98}$$

where

$$C_a^{(n)} = (-1)^{n+1} \frac{2^{15}}{3^4 \pi^{13/2}} F_a \left(\frac{648}{\pi^2}\right)^n \Gamma\left(2n + \frac{11}{2}\right), \tag{9.99}$$

and

$$F_0 = \left(C_{00} + \frac{C_{10}}{2^{3/2}}\right)^2, \quad F_1 = \left(C_{10} - \frac{C_{00}}{2^{3/2}}\right)^2. \tag{9.100}$$

This leading behaviour was verified numerically using a modified form of the Neville-Richardson extrapolation method [AD80], [CI82].

## 9.5    Method of Moments

Fernández and Castro have developed an alternate perturbation scheme for obtaining the energy series for the Zeeman effect which does not require the wavefunction [FE84], [AR90]. Instead, a compact iteration scheme is developed for the moments of the wavefunction with respect to a suitable class of functions and the energy corrections are simply expressed in terms of these moments.

The method has subsequently been generalized and applied to other problems such as the Stark effect and the hydrogen molecular ion [AR84], [FE85], [FE87a], [FE92]. Here we illustrate the method of moments for the Zeeman effect.

Consider the unperturbed Schrödinger equation

$$H_0\psi_0 = E_0\psi_0, \quad H_0 = -\tfrac{1}{2}\nabla^2 + V_0. \tag{9.101}$$

Express the wavefunction in the exponential form

$$\psi_0(\boldsymbol{r}) = F(\boldsymbol{r})\Phi(\boldsymbol{r}), \quad \Phi(\boldsymbol{r}) = e^{-S(\boldsymbol{r})} \tag{9.102}$$

and substitute into (9.101) to obtain

$$\tfrac{1}{2}\nabla^2 F - \nabla F \cdot \nabla S + \tfrac{1}{2}[(\nabla S)^2 - \nabla^2 S]F = (V_0 - E_0)F. \tag{9.103}$$

The perturbed Schrödinger equation is

$$H\psi = E\psi, \quad H = H_0 + \lambda V, \quad E = E_0 + \Delta E, \tag{9.104}$$

which implies that

$$\langle\psi|H - E|f\Phi\rangle = 0 \tag{9.105}$$

for any reasonable function $f$ and from (9.101) and (9.104) this can be expressed as

$$-\tfrac{1}{2}\langle\psi|(\nabla^2 f)\Phi\rangle + \langle\psi|(\nabla f \cdot \nabla S)\Phi\rangle + \tfrac{1}{2}\langle\psi|(\nabla^2 S - (\nabla S)^2)f\Phi\rangle$$
$$+ \langle\psi|V_0 f\Phi\rangle + \lambda\langle\psi|V f\Phi\rangle - E\langle\psi|f\Phi\rangle = 0. \tag{9.106}$$

In case $f = F$, corresponding to the unperturbed wavefunction (9.102), we can simplify (9.106) using (9.103) to obtain

$$\Delta E\langle\psi|F\Phi\rangle = \lambda\langle\psi|V F\Phi\rangle. \tag{9.107}$$

If we impose the intermediate normalization condition on $\psi$ then

$$\langle \psi | \psi_0 \rangle = \langle \psi | F\Phi \rangle = 1. \tag{9.108}$$

The results (9.106), (9.107) and (9.108) involve only moments of the wavefunction $\psi$ and are the starting point for the method of moments.

For the Zeeman effect in hydrogen we have from (9.1)

$$V_0 = -\frac{1}{r}, \quad V = r^2 - z^2 = r^2 \sin^2 \vartheta, \quad S(r) = \frac{r}{n}, \tag{9.109}$$

where $n$ is the principal quantum number. We choose a set of functions $f(\boldsymbol{r})$ of the form

$$f(\boldsymbol{r}) = r^M \sin^N \vartheta \cos^t \vartheta e^{im\varphi}, \tag{9.110}$$

where $M = N = 0, 1, 2, \ldots$, $m = 0, \pm 1, \pm 2, \ldots$, $t = 0, 1$. It is not necessary to consider functions with $t > 1$ since the identity $\cos^2 \vartheta = 1 - \sin^2 \vartheta$ can be used to obtain functions in the set (9.110) with $t = 0$ or $t = 1$.

With these choices of $f$ and $S$ we have

$$\nabla^2 S = \frac{2}{nr},$$

$$(\nabla S)^2 = \frac{1}{n^2},$$

$$\nabla F \cdot \nabla S = \frac{M}{n} r^{M-1} \sin^N \vartheta \cos^t \vartheta e^{im\varphi},$$

$$\nabla^2 f = \frac{1}{r^2} \frac{\partial}{\partial r} \left( r^2 \frac{\partial f}{\partial r} \right) + \frac{1}{r^2 \sin \vartheta} \frac{\partial}{\partial \vartheta} \left( \sin \vartheta \frac{\partial f}{\partial \vartheta} \right) + \frac{1}{r^2 \sin^2 \vartheta} \frac{\partial^2 f}{\partial \varphi^2}$$

$$= [M(M+1)r^{M-2} \sin^N \vartheta + N^2 r^{M-2} \sin^{N-2} \vartheta (1 - \sin^2 \vartheta)$$

$$- N(t+1)r^{M-2} \sin^N \vartheta - t(N+2)r^{M-2} \sin^N \vartheta$$

$$- m^2 r^{M-2} \sin^{N-2} \vartheta] \cos^t \vartheta e^{im\varphi}.$$

Now define the moments

$$G_{M,N} = \langle \psi | r^M \sin^N \vartheta \cos^t \vartheta e^{im\varphi} \Phi \rangle. \tag{9.111}$$

The dependence on $t$ and $m$ is not indicated since $t$ and $m$ are fixed for any state. Substitution into (9.106) gives the iteration formulas for the moments

$$\frac{M - n + 1}{n} G_{M-1,N} = \tfrac{1}{2}[M(M+1) - (N+1)(N+2t)]G_{M-2,N}$$

$$+ \tfrac{1}{2}(N^2 - m^2)G_{M-2,N-2} + \Delta E\, G_{M,N}$$

$$- \lambda G_{M+2,N+2}. \tag{9.112}$$

For the ground state we have $n = 1$, $m = 0$, $t = 0$ and $\psi_0 = e^{-r}$. Therefore in (9.108) we have $F = 1$, corresponding to $M = N = 0$ in (9.110) and

$$G_{0,0} = 1. \tag{9.113}$$

The energy is given by (9.107) with $F = 1$, $V = r^2 \sin^2 \vartheta$. The moment on the right side corresponds to $M = 2$ and $N = 2$ so

$$\Delta E = \lambda G_{2,2}. \tag{9.114}$$

We can assume the perturbation expansions

$$\Delta E = \sum_{j=1}^{\infty} E^{(j)} \lambda^j, \quad G_{M,N} = \sum_{j=0}^{\infty} G_{M,N}^{(j)} \lambda^j \tag{9.115}$$

and substitute into (9.112), with $M$ replaced by $M + 1$, (9.113), and (9.114) to obtain the iteration scheme for $p = 0, 1, \ldots, p_{\max}$

$$G_{0,0}^{(0)} = 1, \quad G_{0,0}^{(p)} = 0, \ p > 0, \tag{9.116}$$

$$
\begin{aligned}
G_{M,N}^{(p)} = \frac{1}{M+1} \Big[ &\tfrac{1}{2}[(M+1)(M+2) - N(N+1)] G_{M-1,N}^{(p)} \\
&+ \tfrac{1}{2} N^2 G_{M-1,N-2}^{(p)} + \sum_{q=0}^{p-1} G_{2,2}^{(q)} G_{M+1,N}^{(p-q-1)} \\
&- G_{M+3,N+2}^{(p-1)} \Big],
\end{aligned}
\tag{9.117}
$$

$$E^{(p+1)} = G_{2,2}^{(p)}. \tag{9.118}$$

It is clear that $N$ must be even since (9.117) changes $N$ only by 0 or $\pm 2$. Also in order to calculate $G_{2,2}^{(p)}$ to some maximum order $p_{\max}$ it is necessary to calculate the moments $G_{M,N}^{(p)}$ for $p = 0, 1, \ldots, p_{\max}$, $M = 1, 2, \ldots, 2 + 3(p_{\max} - p)$ and $N = 0, 2, \ldots, 2 + 2(p_{\max} - p)$. It is also convenient to define $G_{M,N}^{(p)} = 0$ for $p = -1$.

This iteration scheme is easily implemented symbolically with the following simple Maple program called *moments*. In the program the energy correction $E^{(p)}$ is denoted by *DE[p]* and the moment $G_{M,N}^{(p)}$ is denoted by *G[p,M,N]*.

```
01   # moments
02   # Fernandez and Castro method for Zeeman effect
03   moments := proc(maxorder)
04      local pmax, p, q, M, N, G, s1;
05      # global: DE
06
07      pmax := maxorder-1;
08      for p from -1 to pmax do
09         for M from 0 to 2+3*(pmax-p) do
10            for N from 0 by 2 to 2+2*(pmax-p) do
11               G[p,M,N] := 0;
12            od;
13         od;
14      od;
15      G[0,0,0] := 1;
```

```
16
17        for p from 0 to pmax do
18           for M from 1 to 2 + 3*(pmax-p) do
19              for N from 0 by 2 to 2 + 2*(pmax-p) do
20                 s1 := 0;
21                 for q from 0 to p-1 do
22                    s1 := s1 + G[q,2,2]*G[p-q-1,M+1,N];
23                 od;
24                 s1 := s1 + ( (M+1)*(M+2) - N*(N+1) )*G[p,M-1,N]/2
25                    + N^2*G[p,M-1,N-2]/2 - G[p-1,M+3,N+2];
26                 G[p,M,N] := s1 / (M+1);
27              od;
28           od;
29        DE[p+1] := G[p,2,2];
30        print(p+1, DE[p+1]);
31     od:
32  end:
33
34  maxorder := 20;
35  moments(maxorder);
36  save DE, DE.maxorder;
37  quit;
```

It is interesting to compare program *moments* with program *gszeeman* of Section 9.3.3 based on the algebraic RSPT. It is useful to have two different methods for solving the same problem as a check on the results. Program *moments* is 4 to 5 times faster than *gszeeman* in calculating the energy corrections to a given order and has slightly less memory requirements: 8.1 million bytes compared to 10 million bytes for maximum order 30. On the other hand if we combine *gszeeman* and *symzeeman* which uses results to order $j$ to compute further energy corrections to order $2j + 1$ then results to order 60 can be computed with no additional memory requirements (in fact *symzeeman* runs with less memory requirements than *gszeeman*), whereas the use of *moments* with *maxorder=60* would require considerably more memory.


## 9.6   Exercises

◈ **Exercise 9.1** Show how to obtain the energy expansions for the Zeeman effect for nuclear charge $Z$ from those with $Z = 1$.

◈ **Exercise 9.2** Write a Maple program to diagonalize the $2 \times 2$ secular equation (9.38) to obtain the first order energy corrections (9.71) and coefficients (9.72) to (9.74).

# Chapter 10

# Spherically Symmetric Systems

## 10.1    Introduction

In this chapter we apply algebraic RSPT to spherically symmetric hydrogenic systems. In this case the perturbation is a function only of the radial variable $r$. We consider the most important cases: power potentials of the form $r^d$ where $d > 0$, and the screened Coulomb potentials of which the Yukawa potential is a special case.

First we show how the algebraic RSPT can be applied to obtain both wavefunction and energy corrections in the charmonium and harmonium cases for a general state and for specific states. The general results to order 14 for charmonium and to order 10 for harmonium are reported in Appendix G and Appendix H.

The specific results for the 3-dimensional ground state to order 100 in both rational and floating point form are also reported in these appendices. The charmonium results were first obtained for the first few levels using algebraic RSPT and floating point arithmetic by Vrscay [VR85] who also obtained the leading term asymptotics of the energy coefficients and continued fraction representation.

We also consider the screened Coulomb potential and show that the application of the algebraic RSPT method is more complicated if high order calculations for the energy series are desired. This leads to two alternate approaches to obtaining the energy series. In both cases the wavefunction corrections are not required so these methods are often referred to as "perturbation theory without wavefunctions".

The first method is a power series or difference equation approach first used by Bender and Wu in their study of the ground state energy series of the quartic anharmonic oscillator [BE71], [BE73] and later applied to hydrogenic systems [CI82], [VR85], [VR86]. The disadvantage of this method is that a different set of equations must be formulated for each state.

The second method is referred to as the HVHF method and is based on the application of Hirschfelder's hypervirial theorems [HI60] and the Hellmann-

Feynman theorem [HE37], [FE39] first applied to the energy series for anharmonic oscillators by Swenson and Danforth [SW72] and later applied to the Stark effect by Austin and Lai [AU80], [LA81], to charmonium by Killingbeck [KI78], to a general state of the screened Coulomb potential by Grant and Lai [GR79] and to general power potentials by Vinette and Čížek [VI88], McRae and Vrscay [MC92]. We illustrate the method for power potentials and screened Coulomb potentials using symbolic computation.

While the HVHF method gives the energy series for a general state, some disadvantages are that the wavefunction corrections are not obtained and the method is not applicable to non-separable systems such as the Zeeman effect. Symbolic results for general states and rational and floating point results for the 3-dimensional case are given in Appendix G, Appendix H and Appendix I.

## 10.2   RSPT for Charmonium and Harmonium

In atomic units the Schrödinger equation for a hydrogenic atom in an external spherically symmetric field of the form

$$V(r) = r^d \tag{10.1}$$

is

$$\left[ \frac{1}{2}p^2 - \frac{Z}{r} + \lambda r^d \right] \psi = E\psi. \tag{10.2}$$

From (7.53) to (7.55) we have the $T_3$ eigenvalue problem

$$[T_3 - q + \lambda W - \Delta E\, S]\, \Psi(R) = 0, \tag{10.3}$$

where

$$W = \gamma^2 R V(\gamma R) = \gamma^{d+2} R^{d+1} = \left( \frac{q}{Z} \right)^{d+2} R^{d+1}, \tag{10.4}$$

$$S = \gamma^2 R = \left( \frac{q}{Z} \right)^2 R. \tag{10.5}$$

The Lie algebraic perturbation theory requires only the so(2,1) subalgebra of so(4,2) since $R = T_3 - T_1$. The most important cases are charmonium, corresponding to $d = 1$, and harmonium, corresponding to $d = 2$.

We can treat any state using non-degenerate RSPT since the perturbation is spherically symmetric and we can also use the general $D$-dimensional realization of the so(2,1) generators given in Section 4.8 for a hydrogenic atom [CI87]. Therefore the unperturbed eigenstates of (10.3) can be denoted by $|qk\rangle$ (see (4.18) to (4.21)) where

$$k = \ell + \frac{1}{2}(D - 3),\ \ell = 0, 1, 2, \ldots,\ k \geq -1, \tag{10.6}$$

$$q = k + 1 + n_r = \ell + \frac{1}{2}(D - 1) + n_r,\ n_r = 0, 1, 2, \ldots, \tag{10.7}$$

and the eigenvalues of the so(2,1) Casimir operator are

$$\tau = k(k+1) = \frac{1}{4}(D-1)(D-3) + \ell(\ell + D - 2). \tag{10.8}$$

In the 3-dimensional case this gives the usual interpretation of $q$ and $k$ as the principal and orbital angular momentum quantum numbers $n$ and $\ell$.

As in the case of the Stark effect in parabolic coordinates (see Section 8.4.4) we can denote a general unperturbed state using a single index by

$$\Phi_a = |a\rangle = |q + a, k\rangle, \; a = -n_r, -n_r + 1, \ldots, \infty, \tag{10.9}$$

where the fixed values of $q$ and $k$ define the unperturbed reference state corresponding to index 0 as

$$\Phi_0 = |0\rangle = |qk\rangle. \tag{10.10}$$

The corresponding unperturbed energy is

$$E^{(0)} = E_0 = -\frac{Z^2}{2q^2}. \tag{10.11}$$

The matrix elements of $R$, $R^2$ and $R^3$ needed in the charmonium and harmonium cases are given in Appendix D (see Section D.1.2 and Table D.7).

It is convenient to introduce the notation

$$(R^d)_{ba} = (R^d)_{q+a,k}^{q+b,k} = \langle q + b, k| R^d |q + a, k\rangle, \tag{10.12}$$

$$S_{ba} = \langle b|S|a\rangle = \left(\frac{q}{Z}\right)^2 R_{ba}, \tag{10.13}$$

$$W_{ba} = \langle b|W|a\rangle = \left(\frac{q}{Z}\right)^{d+2} (R^{d+1})_{ba}, \tag{10.14}$$

for matrix elements. The resolvent operator defined in (7.71) can be expressed as

$$G = \sum_{a \neq 0} \frac{|q + a, k\rangle\langle q + a, k|}{q - (q + a)} = \sum_{a \neq 0} \frac{|a\rangle\langle a|}{(-a)}. \tag{10.15}$$

### 10.2.1  Low order calculations

We can let $Z = 1$ in our calculations since the energy corrections as functions of the nuclear charge satisfy (see Exercise 10.1)

$$E^{(n)}(Z) = Z^{2-(d+2)n} E^{(n)}(1), \tag{10.16}$$

so we use $E^{(n)}$ to denote $E^{(n)}(1)$. From (7.78) the first order correction to the energy is given by

$$E^{(1)} = \frac{\langle qk|W|qk\rangle}{\langle qk|S|qk\rangle} = q^d \frac{\langle qk|R^{d+1}|qk\rangle}{\langle qk|R|qk\rangle}. \tag{10.17}$$

For charmonium and harmonium we have

$$E^{(1)}_{charm} = \frac{1}{2}\left[3q^2 - k(k+1)\right] = \frac{1}{2}(3q^2 - \tau), \tag{10.18}$$

$$E^{(1)}_{harm} = \frac{1}{2}q^2\left[5q^2 + 1 - 3k(k+1)\right] = \frac{1}{2}q^2(5q^2 + 1 - 3\tau). \tag{10.19}$$

From (7.80) the second order energy correction is

$$\begin{aligned}
S_{00}E^{(2)} = &\frac{1}{2}\left(W_{-2,0}\right)^2 - \frac{1}{2}\left(W_{2,0}\right)^2 \\
&+ \left[(W_{-1,0})^2 - 2E^{(1)}S_{-1,0}W_{-1,0} + (E^{(1)})^2(S_{-1,0})^2\right] \\
&- \left[(W_{1,0})^2 - 2E^{(1)}S_{1,0}W_{1,0} + (E^{(1)})^2(S_{1,0})^2\right],
\end{aligned}$$

and for charmonium and harmonium, using Table D.7 in Appendix D,

$$\begin{aligned}
E^{(2)}_{charm} &= -\frac{1}{8}q^2(5q^2 - 3k^2 - 6k^3 - 3k^4 + 7q^4) \\
&= -\frac{1}{8}q^2(7q^4 + 5q^2 - 3\tau^2), \tag{10.20} \\
E^{(2)}_{harm} &= -\frac{1}{16}(-126k - 147k^2 + 345q^2 + 28 + 143q^4 - 90kq^2 \\
&\qquad\quad - 42k^3 - 90k^2q^2 - 21k^4) \\
&= -\frac{1}{16}q^6(143q^4 + 345q^2 + 28 - 90q^2\tau - 21\tau^2 - 126\tau).
\end{aligned}$$

$$\tag{10.21}$$

As a special case the 3-dimensional ground state corresponds to $D = 3$, $\ell = 0$, $k = 0$, $q = 1$ and we have

$$E_{charm} = -\frac{1}{2} + \frac{3}{2}\lambda - \frac{3}{2}\lambda^2 + \cdots, \tag{10.22}$$

$$E_{harm} = -\frac{1}{2} + 3\lambda - \frac{129}{4}\lambda^2 + \cdots. \tag{10.23}$$

### 10.2.2   Symbolic calculations

The symbolic calculation of the energy and wavefunction corrections for an arbitrary state can be accomplished using the iteration scheme (7.82) to (7.86) corresponding to the expansion

$$\Psi^{(j)} = \sum_{a=-n_r}^{\infty} C_a^{(j)}\Phi_a \tag{10.24}$$

of the order $j$ wavefunction correction in terms of the unperturbed states (10.9) and the $T_3$ eigenvalues $\kappa_0 = q$, $\kappa_a = q + a$.

As in the case of the Stark effect in Chapter 8 and the Zeeman effect in Chapter 9 it is more convenient and computationally more efficient to define new expansion coefficients $D_a^{(j)}$ by

$$D_a^{(j)} = N_a C_a^{(j)}. \tag{10.25}$$

Here $N_a$ is a suitable "renormalization" factor chosen to remove the square root factors from the matrix elements of $S$ and $W$. Thus, if we define

$$N_a = \left[ \frac{(q + a + k)!}{(q + a - k - 1)!} \right]^{1/2}, \tag{10.26}$$

then the new matrix elements of powers of $R$ are defined by (see (8.23), (8.24) and (8.97))

$$(\overline{R}^d)_{ba} = \frac{N_a}{N_b} (R^d)_{ba}. \tag{10.27}$$

For charmonium and harmonium, matrix elements of the first three powers of $R$ are needed. The results can easily be obtained using simple Maple programs similar to those of Chapter 8 and Chapter 9 (see Section 8.2.2 and Section 9.3.2):

$$\overline{R}_{a,a} = q + a, \tag{10.28}$$

$$\overline{R}_{a+1,a} = -\tfrac{1}{2}(q + a - k), \tag{10.29}$$

$$\overline{R}_{a-1,a} = -\tfrac{1}{2}(q + a + k), \tag{10.30}$$

$$(\overline{R}^2)_{a,a} = \tfrac{1}{2}\left(3(q + a)^2 - k(k + 1)\right), \tag{10.31}$$

$$(\overline{R}^2)_{a+1,a} = -\tfrac{1}{2}(2q + 2a + 1)(q + a - k), \tag{10.32}$$

$$(\overline{R}^2)_{a-1,a} = -\tfrac{1}{2}(2q + 2a - 1)(q + a + k), \tag{10.33}$$

$$(\overline{R}^2)_{a+2,a} = \tfrac{1}{4}(q + a - k)(q + a - k + 1), \tag{10.34}$$

$$(\overline{R}^2)_{a-2,a} = \tfrac{1}{4}(q + a + k - 1)(q + a + k), \tag{10.35}$$

$$(\overline{R}^3)_{a,a} = \tfrac{1}{2}(q + a)[5(q + a)^2 + 1 - 3k(k + 1)], \tag{10.36}$$

$$(\overline{R}^3)_{a+1,a} = -\tfrac{3}{8}[5(q + a)^2 + 5(q + a) + 2 - k(k + 1)](q + a - k), \tag{10.37}$$

$$(\overline{R}^3)_{a-1,a} = -\tfrac{3}{8}[5(q + a)^2 - 5(q + a) + 2 - k(k + 1)](q + a + k), \tag{10.38}$$

$$(\overline{R}^3)_{a+2,a} = \tfrac{3}{4}(q + a + 1)(q + a - k)(q + a - k + 1), \tag{10.39}$$

$$(\overline{R}^3)_{a-2,a} = \tfrac{3}{4}(q + a - 1)(q + a + k - 1)(q + a + k), \tag{10.40}$$

$$(\overline{R}^3)_{a+3,a} = -\tfrac{1}{8}(q + a - k)(q + a - k + 1)(q + a - k + 2), \tag{10.41}$$

$$(\overline{R}^3)_{a-3,a} = -\tfrac{1}{8}(q + a + k - 2)(q + a + k - 1)(q + a + k). \tag{10.42}$$

Then the matrix elements $\bar{S}_{ba}$ and $\bar{W}_{ba}$ are easily obtained from (10.13) and (10.14):

$$\bar{S}_{ba} = q^2 \bar{R}_{ba}, \tag{10.43}$$

$$\bar{W}_{ba} = q^{d+2}(\bar{R}^{d+1})_{ba}. \tag{10.44}$$

The iterative equations (7.82) to (7.85) can now be expressed as

$$D_0^{(0)} = 1, \quad D_A^{(0)} = 0, \, A > 0, \tag{10.45}$$

$$E^{(j)} = \frac{1}{\bar{S}_{00}} \left[ \sum_{A=-(d+1)}^{d+1} \bar{W}_{A0} D_0^{(j-1)} - \sum_{A=\pm 1} \bar{S}_{A0} \sum_{i=1}^{j-1} E^{(i)} D_A^{(j-i)} \right], \tag{10.46}$$

$$D_0^{(j)} = 0, \, j > 0, \tag{10.47}$$

$$D_A^{(j)} = \frac{1}{(-A)} \left[ \sum_{A'=A-(d+1)}^{A+(d+1)} \bar{W}_{A'A} D_{A'}^{(j-1)} - \sum_{A'=A-1}^{A+1} \bar{S}_{A'A} \sum_{i=1}^{j} E^{(i)} D_{A'}^{(j-i)} \right],$$

$$\tag{10.48}$$

where $-(d+1)j \le A \le (d+1)j$, which follows since the perturbation $R^{d+1}$ can change $q$ up or down by at most $d+1$ in each order of perturbation.

### 10.2.3   Maple programs for charmonium and harmonium

The following program, *rpower*, implements the iterative equations (10.45) to (10.48). The wavefunction expansion coefficients $D_A^{(j)}$ are denoted by $DD[j,A]$ and the energy corrections $E^{(j)}$ are denoted by $DE[j]$. They are output by the program.  In the determination of the expansion coefficients the lower loop index over $A$ can be outside the valid range, if the program is used for particular values of $q$ and $k$ rather than symbolic ones.  The corresponding expansion coefficients will have the value 0 so it is necessary in line *100* to unevaluate them.

```
01    # rpower
02    # r^d perturbation for any reference state
03    rpower := proc(jmax, d, q, k)
04        local R, W, fS, fW, S00, i, j, a, A, AP, x, s1, s2, s3;
05        # global DD, DE
06
07        R := table([
08            (1) = -(q+a-k)/2,
09            (0) = q+a,
10            (-1) = -(q+a+k)/2
11        ]);
12
13        if d = 1 then
14
```

```
15                # use matrix elements of R^2 for charmonium
16
17            W := table([
18                (2) = (q+a-k)*(q+a-k+1)/4,
19                (1) = -(2*q+2*a+1)*(q+a-k)/2,
20                (0) = (3*(q+a)^2-k*(k+1))/2,
21                (-1) = -(2*q+2*a-1)*(q+a+k)/2,
22                (-2) = (q+a+k-1)*(q+a+k)/4
23            ]):
24
25        elif d = 2 then # use matrix elements of R^3 for harmonium
26
27            W := table([
28                (0) = 1/2*(q+a)*(5*(q+a)^2+1-3*k-3*k^2),
29                (-1) = -3/8*(q+a+k)*(5*(q+a)^2-5*(q+a)+2-k-k^2),
30                (1) = -3/8*(q+a-k)*(5*(q+a)^2-k-k**2+5*(q+a)+2),
31                (-2) = 3/4*(q+a-1+k)*(q+a+k)*(q+a-1),
32                (2) = 3/4*(q+a-k)*(q+a+1-k)*(q+a+1),
33                (-3) = -1/8*(q+a-2+k)*(q+a-1+k)*(q+a+k),
34                (3) = -1/8*(q+a-k)*(q+a+1-k)*(q+a+2-k)
35            ]):
36
37        else
38            # do other cases here if desired.
39            ERROR(`only d=1 and d=2 are implemented`);
40        fi;
41
42        fS := q^2;
43        fW := q^(d+2);
44
45        DD[0,0] := 1;
46        S00 := fS*subs(a=0, R[0]);
47
48        for j from 1 to jmax do
49            lprint(j);
50
51            # Calculate energy correction DE[j]
52
53            s1 := 0;
54            for A from -(d+1) to (d+1) do
55                x := DD[j-1, A];
56                if not type(x, name) then
57                    s1 := s1 + subs(a=0, W[A])*x;
58                fi;
59            od;
60
61            s2 := 0;
```

```
62              for A in [-1,1] do
63                  s3 := 0;
64                  for i from 1 to j-1 do
65                      x := DD[j-i,A];
66                      if not type(x,name) then
67                          s3 := s3 + DE[i]*x;
68                      fi;
69                  od;
70                  s2 := s2 + subs(a=0, R[A]) * s3
71              od;
72              DE[j] := simplify((fW*s1 - fS*s2)/S00);
73
74              # calculate jth order expansion coefficients
75
76              for A from -(d+1)*j to (d+1)*j do
77                if A <> 0 then
78                  s1 := 0;
79                  for AP from A-(d+1) to A+(d+1) do
80                      x := DD[j-1, AP];
81                      if not type(x, name) then
82                          s1 := s1 + subs(a=A, W[AP-A])*x;
83                      fi;
84                  od;
85
86                  s2 := 0;
87                  for AP from A-1 to A+1 do
88                      s3 := 0;
89                      for i from 1 to j do
90                          x := DD[j-i, AP];
91                          if not type(x,name) then
92                              s3 := s3 + DE[i]*x;
93                          fi;
94                      od;
95                      s2 := s2 + subs(a=A, R[AP-A])*s3;
96                  od;
97
98                  DD[j,A] := simplify((fW*s1 - fS*s2)/(-A));
99                  if DD[j,A] = 0 then # unassign DD[j,A]
100                     DD[j,A] := `DD[`.j.`,`.A.`]`;
101                 fi;
102               fi; # A <> 0
103           od; # A
104     od; # j
105     DE[0] := -1/(2*q^2);
106 end:
```

The following program uses *rpower* to compute the wavefunction and energy corrections for charmonium. The energy corrections are produced as functions of $q$ and $k$ which are then converted to functions of $q$ and $\tau = k(k+1)$. Finally the program produces the ground state energy corrections for the 3-dimensional case by substituting the particular values $q = 1$ and $\tau = 0$ into the general results. The output files in the case shown for *jmax :=4* are *DE1qk4.m* and *DD1qk4.m* containing the energy and wavefunction corrections as functions of $q$ and $k$ and *DE1qt4.m* containing the energy corrections as functions of $q$ and $\tau$.

```
01    # charm
02    # for charmonium the perturbation is r
03
04    read rpower;
05    jmax := 4;
06    rpower(jmax, 1, q, k):
07
08    # Display and save results as functions of q and k
09
10    for j from 1 to jmax do DE[j] := DE[j]; od;
11    save DE, `DE1qk`.jmax.`.m `;
12    save DD, `DD1qk`.jmax.`.m `;
13
14    # Convert from q, k to q, tau
15
16    r1 := (−1 + sqrt(1+4*tau))/2:
17    for j from 1 to jmax do
18       DE[j] := simplify(subs(k=r1,DE[j]));
19       sort(DE[j], [q,tau], plex);
20    od:
21
22    # Display and save energy as functions of q and tau
23
24    for j from 1 to jmax do DE[j] := DE[j]; od;
25    save DE, `DE1qt`.jmax.`.m `;
26
27    # Display 3-dimensional ground state results
28
29    for j from 1 to jmax do subs(q=1,tau=0,DE[j]); od;
30    quit;
```

In a similar manner the results for harmonium can be produced by the following program.

```
01    # harm
02    # for harmonium the perturbation is r^2
03
04    read rpower;
05    jmax := 4;
```

```
06    rpower(jmax, 2, q, k);
07
08    # Display and save results as functions of q and k
09
10    for j from 1 to jmax do DE[j] := DE[j]; od;
11    save DE, `DE2qk`.jmax.`.m`;
12    save DD, `DD2qk`.jmax.`.m`;
13
14    # Convert from q, k to q, tau
15
16    r1 := (−1 + sqrt(1+4*tau))/2:
17    for j from 1 to jmax do
18        DE[j] := simplify(subs(k=r1,DE[j]));
19        sort(DE[j], [q,tau], plex);
20    od:
21
22    # Display and save energy as functions of q and tau
23
24    for j from 1 to jmax do DE[j] := DE[j]; od;
25    save DE, `DE2qt`.jmax.`.m`;
26
27    # Display 3-dimensional ground state results
28    for j from 1 to jmax do subs(q=1,tau=0, DE[j]); od;
29    quit;
```

The general results to order 4 produced by *charm* and *harm* are

$$E^{(1)}_{charm} = \tfrac{3}{2}q^2 - \tfrac{1}{2}\tau, \tag{10.49}$$

$$E^{(2)}_{charm} = -\tfrac{1}{8}q^2(7\,q^4 + 5\,q^2 - 3\,\tau^2), \tag{10.50}$$

$$E^{(3)}_{charm} = \tfrac{1}{16}q^4(33\,q^6 + 75\,q^4 - 7\,q^2\tau^2 - 10\,\tau^3), \tag{10.51}$$

$$\begin{aligned}
E^{(4)}_{charm} = -\tfrac{1}{64}q^6(465\,q^8 + 2275\,q^6 - 99\,q^4\tau^2 + 440\,q^4 \\
- 90\,q^2\tau^3 - 180\,q^2\tau^2 - 84\,\tau^4),
\end{aligned} \tag{10.52}$$

$$E^{(1)}_{harm} = \tfrac{1}{2}q^2(5\,q^2 - 3\,\tau + 1), \tag{10.53}$$

$$E^{(2)}_{harm} = -\tfrac{1}{16}q^6(143\,q^4 - 90\,q^2\tau + 345\,q^2 - 21\,\tau^2 - 126\,\tau + 28), \tag{10.54}$$

$$\begin{aligned}
E^{(3)}_{harm} = \tfrac{1}{16}q^{10}(1530\,q^6 - 1305\,q^4\tau + 11145\,q^4 - 6825\,q^2\tau \\
+ 8645\,q^2 - 33\,\tau^3 + 33\,\tau^2 - 2706\,\tau + 484),
\end{aligned} \tag{10.55}$$

$$\begin{aligned}
E^{(4)}_{harm} = -\tfrac{1}{1024}q^{14}(1502291\,q^8 - 1640100\,q^6\tau + 22937530\,q^6 \\
+ 251370\,q^4\tau^2 - 19742520\,q^4\tau + 54811295\,q^4 \\
- 3060\,q^2\tau^3 + 2184330\,q^2\tau^2 - 31859700\,q^2\tau \\
+ 25371140\,q^2 - 4005\,\tau^4 - 7260\,\tau^3 + 1425540\,\tau^2 \\
- 7286640\,\tau + 1137344).
\end{aligned} \tag{10.56}$$

More complete results to order 14 for charmonium and to order 10 for harmonium are given in Appendix G. The harmonium results in the 2-dimensional case have been used for the application of 2-point Padé approximants to the ground state of the 2-dimensional hydrogen atom in an external magnetic field [AD88a].

Results for specific states can be obtained from these symbolic results by substituting particular values of $q$ and $\tau$. Thus, for the 3-dimensional ground state ($q = 1$, $\tau = 0$) the energy to order 4 is

$$E_{charm} = -\tfrac{1}{2} + \tfrac{3}{2}\lambda - \tfrac{3}{2}\lambda^2 + \tfrac{27}{4}\lambda^3 - \tfrac{795}{16}\lambda^4 + \cdots, \tag{10.57}$$

$$E_{harm} = -\tfrac{1}{2} + 3\lambda - \tfrac{129}{4}\lambda^2 + \tfrac{5451}{4}\lambda^3 - \tfrac{6609975}{64}\lambda^4 + \cdots. \tag{10.58}$$

A more efficient way to obtain high order results for specific states is to substitute particular values of $q$ and $k$ from the beginning so that the symbolic calculation involves only rational numbers. The following program, *charmgs*, shows how to obtain the energy corrections for the 3-dimensional ground state of charmonium. The energy and wavefunction corrections in the case shown for *jmax := 20* are output to the files *DE1GS20.m* and *DD1GS20.m*.

```
01    # charmgs
02    # for charmonium perturbation is r
03    # Do ground state 3-dim case directly
04
05    read rpower;
06    jmax := 20;
07    rpower(jmax, 1, 1, 0):
08    for j from 1 to jmax do DE[j] := DE[j]; od;
09    save DE, DE1GS.jmax.`.m `;
10    save DD, DD1GS.jmax.`.m `;
11    quit;
```

The resulting energy series to order 20 is

$$
\begin{aligned}
E_{charm} = {}& -\frac{1}{2} + \frac{3}{2}\lambda - \frac{3}{2}\lambda^2 + \frac{27}{4}\lambda^3 \\
& - \frac{795}{16}\lambda^4 + \frac{3843}{8}\lambda^5 - 5583\,\lambda^6 \\
& + \frac{9543339}{2^7}\lambda^7 - \frac{1141062999}{2^{10}}\lambda^8 + \frac{18769071555}{2^{10}}\lambda^9 \\
& - \frac{1343699301873}{2^{12}}\lambda^{10} + \frac{51910283674773}{2^{13}}\lambda^{11} \\
& - \frac{4302261498085317}{2^{15}}\lambda^{12} + \frac{23790217856283351}{2^{13}}\lambda^{13} \\
& - \frac{2238202979629389225}{2^{15}}\lambda^{14} + \frac{446349518125914065265}{2^{18}}\lambda^{15} \\
& - \frac{188137797615156926767995}{2^{22}}\lambda^{16}
\end{aligned}
$$

$$+ \frac{522442559730883232324993495}{2^{22}} \lambda^{17}$$
$$- \frac{610283093633587591852031535}{2^{24}} \lambda^{18}$$
$$+ \frac{37405459833681441600909796725}{2^{25}} \lambda^{19}$$
$$- \frac{480228806704592088987259370740 5}{2^{27}} \lambda^{20}. \tag{10.59}$$

More complete results to order 100 are given in Appendix G.

Similarly the following program, *harmgs*, directly calculates the energy corrections for the 3-dimensional ground state of harmonium.

```
01    # harmgs
02    # for harmonium perturbation is r^2
03    # Do ground state 3-dim case directly
04
05    read rpower;
06    jmax := 20;
07    rpower(jmax, 2, 1, 0):
08    for j from 1 to jmax do DE[j] := DE[j]; od;
09    save DE, DE2GS.jmax.`.m`;
10    save DD, DD2GS.jmax.`.m`;
11    quit;
```

The resulting energy series to order 18 is

$$E_{harm} = -\frac{1}{2} + 3\lambda - \frac{129}{4}\lambda^2 + \frac{5451}{4}\lambda^3$$
$$- \frac{6609975}{2^6}\lambda^4 + \frac{734589303}{2^6}\lambda^5$$
$$- \frac{880224055389}{2^9}\lambda^6 + \frac{169960252839003}{2^9}\lambda^7$$
$$- \frac{13164581517 45974019}{2^{14}}\lambda^8$$
$$+ \frac{39139889611321886653 5}{2^{14}}\lambda^9$$
$$- \frac{1124700335952727250306379}{2^{17}}\lambda^{10}$$
$$+ \frac{48115566529292698500918252 9}{2^{17}}\lambda^{11}$$
$$- \frac{3873547640308281898602688756323}{2^{21}}\lambda^{12}$$
$$+ \frac{226677256518268483127528120603594 7}{2^{21}}\lambda^{13}$$
$$- \frac{1221997346246102747455824893300126068 5}{2^{24}}\lambda^{14}$$

$$+ \frac{93987176784026337318551785089606242358 75}{2^{24}} \lambda^{15}$$

$$- \frac{5238932756293139237468652943166508304163 71515}{2^{30}} \lambda^{16}$$

$$+ \frac{513057352091621880993517596730198301652201013575}{2^{30}} \lambda^{17}$$

$$- \frac{44911262829695580863651801759650521251871692212 15455}{2^{33}} \lambda^{18}$$

$$(10.60)$$

More complete results to order 100 are given in Appendix G.

The symmetric energy formula (7.86) can be used to calculate energy corrections to order $2j+1$ in terms of the wavefunction and energy corrections to order $j$. In the case of charmonium or harmonium (7.86) is given by

$$\bar{S}_{00} E^{(j)} = \sum_{A=-(d+1)j}^{(d+1)j} \frac{D_A^{(P)}}{N_A^2} \left( \sum_{A'=A-(d+1)}^{A+(d+1)} \bar{W}_{A'A} D_{A'}^{(Q)} \right)$$

$$- 2 \sum_{A=\pm 1} \bar{S}_{A0} \left( \sum_{r=1}^{Q} E^{(j-r)} D_A^{(r)} \right) - \sum_{A=\pm 1} \bar{S}_{A0} \left( \sum_{r=Q+1}^{P} E^{(j-r)} D_A^{(r)} \right)$$

$$- \sum_{r=1}^{P} \left( \sum_{A=-(d+1)r}^{(d+1)r} \frac{D_A^{(r)}}{N_A^2} \sum_{A'=A-1}^{A+1} \bar{S}_{A'A} \left( \sum_{s=1}^{Q} E^{(j-r-s)} D_{A'}^{(s)} \right) \right), \qquad (10.61)$$

where $j = P + Q + 1$ and $P = \lfloor j/2 \rfloor$, $Q = \lfloor (j-1)/2 \rfloor$.

The following program, *sympower*, implements this symmetric energy formula.

```
01    # sympower
02    # symmetric energy formula for charmonium and harmonium
03
04    symEnergy := proc(jmin, jmax, d, q, k)
05        local R, W, a, A, AP, j, P, Q, r, s, x, z, Na2, Norm2, fS, fW,
06            S00, s1, s2, s21, s3, s31, s4, s41, s42, s43;
07        # global DD, DE
08
09        R := table([
10            (1) = -(q+a-k)/2,
11            (0) = q+a,
12            (-1) = -(q+a+k)/2
13        ]);
14
15        if d = 1 then
16
17            # use matrix elements of R^2 for charmonium
18
19            W := table([
```

```
20           (2) = (q+a-k)*(q+a-k+1)/4,
21           (1) = -(2*q+2*a+1)*(q+a-k)/2,
22           (0) = (3*(q+a)^2-k*(k+1))/2,
23           (-1) = -(2*q+2*a-1)*(q+a+k)/2,
24           (-2) = (q+a+k-1)*(q+a+k)/4
25        ]):
26

27     elif d = 2 then # use matrix elements of R^3 for harmonium
28

29        W := table([
30           (0) = 1/2*(q+a)*(5*(q+a)^2+1-3*k-3*k^2),
31           (-1) = -3/8*(q+a+k)*(5*(q+a)^2-5*(q+a)+2-k-k^2),
32           (1) = -3/8*(q+a-k)*(5*(q+a)^2-k-k**2+5*(q+a)+2),
33           (-2) = 3/4*(q+a-1+k)*(q+a+k)*(q+a-1),
34           (2) = 3/4*(q+a-k)*(q+a+1-k)*(q+a+1),
35           (-3) = -1/8*(q+a-2+k)*(q+a-1+k)*(q+a+k),
36           (3) = -1/8*(q+a-k)*(q+a+1-k)*(q+a+2-k)
37        ]):
38

39     else
40        # do other cases here if desired.
41        ERROR(`only d=1 and d=2 are implemented`);
42     fi;
43

44     Norm2 := (q+a+k)! / (q+a-k-1)!;
45

46     fS := q^2;
47     fW := q^(d+2);
48     S00 := fS*subs(a=0, R[0]);
49

50     for j from jmin to jmax do
51        P := iquo(j,2);
52        Q := iquo(j-1,2);
53        lprint(j,P,Q);
54

55        # do terms involving the perturbation matrix
56

57        s1 := 0;
58        for A from -(d+1)*j to (d+1)*j do
59           s2 := 0;
60           z := DD[P,A];
61           if not type(z,name) then
62              for AP from A-(d+1) to A+(d+1) do
63                 x := DD[Q,AP];
64                 if not type(x,name) then
65                    s2 := s2 + subs(a=A, W[AP-A])*x;
66                 fi;
```

```
67                    od;
68                    Na2 := subs(a=A, Norm2);
69                    s1 := s1 + z*s2/Na2;
70                fi;
71            od; # A
72
73            # do terms involving lower order corrections
74
75            s2 := 0;
76            for A in [-1,1] do
77                s21 := 0;
78                for r from 1 to Q do
79                    x := DD[r,A];
80                    if not type(x,name) then
81                        s21 := s21 + DE[P+Q+1-r]*x;
82                    fi;
83                od;
84                s2 := s2 + subs(a=0, R[A])*s21;
85            od;
86
87            s3 := 0;
88            for A in [-1,1] do
89                s31 := 0;
90                for r from Q+1 to P do
91                    x := DD[r,A];
92                    if not type(x,name) then
93                        s31 := s31 + DE[P+Q+1-r]*x;
94                    fi;
95                od;
96                s3 := s3 + subs(a=0, R[A])*s31;
97            od;
98
99            s4 := 0;
100           for r from 1 to P do
101               s41 := 0;
102               for A from -(d+1)*r to (d+1)*r do
103                   s42 := 0;
104                   z := DD[r,A];
105                   if not type(z,name) then
106                       for AP from A-1 to A+1 do
107                           s43 := 0;
108                           for s from 1 to Q do
109                               x := DD[s,AP];
110                               if not type(x,name) then
111                                   s43 := s43 + DE[P+Q+1-r-s]*x;
112                               fi;
113                           od;
```

```
114                          s42 := s42 + subs(a=A,R[AP-A])*s43;
115                    od; # AP
116                    Na2 := subs(a=A, Norm2);
117                    s41 := s41 + z*s42/Na2;
118              fi;
119           od; # A
120              s4 := s4 + s41;
121        od; # r
122        DE[j] := simplify( (fW*(s1-s4) - fS*(2*s2+s3))/S00 );
123     od; # j
124  end:
```

The following program shows how *sympower* can be used in the charmonium case ($d = 1$) to produce energy corrections to order 40 from wavefunction and energy corrections to order 20. The file *DD1GS20.m* contains the wavefunction corrections to order 20. The file *DE1GS20.m* contains the energy corrections to order 20. The energy corrections to order 40 are output to the file *DE1GS40.m*.

```
01    # symcharmgs
02    read `DD1GS20.m`;
03    read `DE1GS20.m`;
04    read sympower;
05    jmin := 21;
06    jmax := 40;
07    symEnergy(jmin,jmax,1,1,0):
08    save DE, `DE1GS`.jmax.`.m`;
09    save DE, `DE1GS`.jmax;
10    quit;
```

A similar program with $d = 2$ could be used for harmonium.

```
01    # symharmgs
02    read `DD2GS20.m`;
03    read `DE2GS20.m`;
04    read sympower;
05    jmin := 21;
06    jmax := 40;
07    symEnergy(jmin,jmax,2,1,0):
08    save DE, `DE2GS`.jmax.`.m`;
09    save DE, `DE2GS`.jmax;
10    quit;
```

The results of these two programs are given in Appendix G to order 100.

## 10.3    RSPT for Screened Coulomb Potential

The hamiltonian for a hydrogenic atom in a screened Coulomb potential is

$$H = \frac{1}{2}p^2 + U(r,\lambda), \tag{10.62}$$

where

$$U(r,\lambda) = -\frac{Z}{r} + V(r,\lambda), \tag{10.63}$$

$$V(r,\lambda) = -\frac{Z}{r}\sum_{j=1}^{\infty} V_j\lambda^j r^j. \tag{10.64}$$

An important special case is the Yukawa potential defined by

$$U(r,\lambda) = -\frac{Z}{r}e^{-\lambda r}, \tag{10.65}$$

$$V(r,\lambda) = -\frac{Z}{r}(e^{-\lambda r} - 1), \tag{10.66}$$

$$V_j = \frac{(-1)^j}{j!}. \tag{10.67}$$

The Schrödinger equation is

$$\left[\frac{1}{2}p^2 - \frac{Z}{r} + V(r,\lambda) - E\right]\psi(r) = 0. \tag{10.68}$$

We can now multiply by $r$ and apply the scaling transformation (7.49) to obtain
the scaled hydrogenic equation

$$\left[\frac{1}{2}RP^2 + \frac{1}{2}R - q + \gamma^2 RV(\gamma R,\lambda) - \Delta E\gamma^2 R\right]\Psi(R) = 0, \tag{10.69}$$

where

$$E = E^{(0)} + \Delta E, \quad E^{(0)} = E_0 = -\frac{1}{2\gamma^2}, \quad \gamma = \frac{q}{Z}, \tag{10.70}$$

so we obtain the perturbed $T_3$ eigenvalue problem

$$[T_3 - q + W(R,\lambda) - \Delta E\, S]\,\Psi(R) = 0, \tag{10.71}$$

where $S = \gamma^2 R$ and

$$W(R,\lambda) = \gamma^2 RV(\gamma R,\lambda) = \sum_{j=1}^{\infty} W_j(R)\lambda^j, \tag{10.72}$$

$$W_j(R) = -\gamma Z V_j\gamma^j R^j. \tag{10.73}$$

This is a simple generalization of (7.52) and the modified algebraic RSPT of
Section 7.4 can be adapted to (10.71).

The resulting equations are

$$\Psi^{(j)} = \sum_{k=1}^{j} G(W_k - E^{(k)}S)\Psi^{(j-k)}, \tag{10.74}$$

which is the appropriate generalization of (7.73) for the wavefunction corrections and

$$\langle\Psi_0|S|\Psi_0\rangle E^{(j)} = \sum_{k=1}^{j}\langle\Psi_0|W_k|\Psi^{(j-k)}\rangle - \sum_{k=1}^{j-1}\langle\Psi_0|S|\Psi^{(j-k)}\rangle E^{(k)}, \qquad (10.75)$$

which is the appropriate generalization of (7.74) for the energy corrections.

We can let $\mathcal{Z} = 1$ since the energy corrections $E^{(j)}(\mathcal{Z})$, defined as a function of the nuclear charge $\mathcal{Z}$, can be expressed in terms of those for $\mathcal{Z} = 1$ by (see Exercise 10.2)

$$E^{(j)}(\mathcal{Z}) = \mathcal{Z}^{2-j}E^{(j)}(1). \qquad (10.76)$$

The first two energy corrections can easily be obtained by hand using the definitions (10.9) to (10.15) of Section 10.2 and the formulas in Section 7.4 (see Exercise 10.3):

$$E^{(1)} = -V_1, \qquad (10.77)$$

$$E^{(2)} = -\frac{1}{2}V_2\left[3q^2 - k(k+1)\right]. \qquad (10.78)$$

The symbolic computation of high order energy corrections for a general state is complicated by the presence of high powers of $R$. To calculate $\Psi^{(j)}$ and $E^{(j)}$ it is necessary to have matrix elements of $R^d$ for $1 \le d \le j$. In principle this is not difficult since the "renormalized" matrix elements (10.27) of $R^d$ can be obtained iteratively using the matrix multiplication formula

$$(\overline{R}^d)_{ba} = \sum_{c=b-1}^{b+1} \overline{R}_{bc}(\overline{R}^{d-1})_{ca}, \quad |b - a| \le d. \qquad (10.79)$$

but the program storage requirements for the wavefunction corrections is much larger than is the case for charmonium and harmonium so we shall illustrate two alternate approaches to algebraic RSPT for obtaining the energy corrections. One is based on the direct series solution of the differential equation and the other is based on the hypervirial and Hellmann-Feynman theorems.

## 10.4   Power Series Method

It is possible to obtain the wavefunction and energy corrections for any particular state of (10.69) by directly substituting a trial solution which is both a perturbation expansion in powers of $\lambda$ and an expansion in powers of $R$. This approach was first used by Bender and Wu in their study of the large order perturbation theory for the quartic anharmonic oscillator [BE71], [BE73]. It was also used by Vrscay to obtain the high order perturbation theory for the Yukawa potential for the 3-dimensional 1s, 2s and 2p states [VR86], and for the ground state of charmonium [VR85]. The method can also be used for the ground state Zeeman effect [CI82]. A disadvantage of this approach is that a different set of difference equations must be derived for each state.

To illustrate the method we consider only the ground state of (10.69) in the D-dimensional case. From Chapter 4

$$P^2 = P_R^2 + \frac{(D-1)(D-3)}{4R^2} + \frac{L^2}{R^2}, \tag{10.80}$$

$$P_R^2 = -\frac{d^2}{dR^2} - \frac{D-1}{R}\frac{d}{dR} - \frac{(D-1)(D-3)}{4R^2}, \tag{10.81}$$

so (10.69) becomes

$$\left[-\frac{R}{2}\frac{d^2}{dR^2} - \frac{D-1}{2}\frac{d}{dR} + \frac{L^2}{2R} + \frac{R}{2} - q + W - \Delta E\, S\right]\Psi(R) = 0. \tag{10.82}$$

For the ground state

$$q = k + 1 = \frac{D-1}{2}, \quad L^2\Psi = 0, \tag{10.83}$$

so

$$\left[-\frac{R}{2}\frac{d^2}{dR^2} - q\frac{d}{dR} + \frac{R}{2} - q + W - q^2\Delta E\, R\right]\Psi(R) = 0, \tag{10.84}$$

where from (10.73) the components of $W$ are

$$W_j(R) = -q^{j+1}V_j R^j. \tag{10.85}$$

In the general case a trial solution would have the form

$$\Psi_{qk}(R,\lambda) = F_{qk}(R,\lambda)Y(\Omega), \tag{10.86}$$

where $F_{qk}(R,0)$ is defined by (4.150) in terms of the associated Laguerre polynomials and $Y(\Omega)$ is a generalized spherical harmonic function in $D$ dimensions. For the ground state, neglecting normalization, we can choose the simple functional form

$$\Psi_0(R,\lambda) = e^{-R}B(R,\lambda), \quad B(R,0) = 1. \tag{10.87}$$

Substitute into (10.84) to obtain a differential equation for $B(R,\lambda)$:

$$RB''(R,\lambda) + 2(q-R)B'(R,\lambda) - 2(W - q^2\Delta E\, R)B(R,\lambda) = 0. \tag{10.88}$$

Next substitute (10.72) and the perturbation expansions

$$B(R,\lambda) = \sum_{n=0}^{\infty} B_n(R)\lambda^n, \quad \Delta E = \sum_{n=1}^{\infty} E^{(n)}\lambda^n, \tag{10.89}$$

to obtain a differential equation for $B_n(R)$:

$$RB_n''(R) + 2(q - R)B_n'(R) + 2R\sum_{k=1}^{\infty} q^2 E^{(k)} B_{n-k}(R)$$

$$- 2\sum_{k=1}^{\infty} W_k(R)B_{n-k}(R) = 0. \tag{10.90}$$

Now assume a polynomial form for $B_n(R)$:

$$B_n(R) = \sum_{j=0}^{n} b_{jn}R^j. \tag{10.91}$$

For $n = 0$ it follows from (10.87) that

$$B_0(R) = 1, \quad b_{00} = 1. \tag{10.92}$$

For $n = 1$ (10.90) reduces to

$$qb_{11} + (-b_{11} + q^2 E^{(1)} + q^2 V_1)R = 0, \tag{10.93}$$

so

$$b_{01} = 0, \; b_{11} = 0, \; E^{(1)} = -V_1, \; B_1(R) = 0, \tag{10.94}$$

where we observe that the coefficients $b_{0n}$ for $n > 0$ are not determined by (10.90) so we are free to choose them to be zero. This is analogous to the intermediate normalization condition chosen for the algebraic RSPT. Similarly for $n = 2$ (10.90) reduces to

$$qb_{12} + ((2q + 1)b_{22} - b_{12} + q^2 E^{(2)})R + (q^3 V_2 - 2b_{22})R^2 = 0, \tag{10.95}$$

so

$$b_{02} = 0, \; b_{12} = 0, \; b_{22} = \frac{1}{2}q^3 V_2, \; E^{(2)} = -\frac{1}{2}(2q + 1)qV_2. \tag{10.96}$$

For the general case substitute (10.91) into (10.90) to obtain

$$qb_{1n} + \frac{1}{2}\sum_{j=1}^{n-1}(j + 1)(j + 2q)b_{j+1,n}R^j - \sum_{j=1}^{n} jb_{jn}R^j$$

$$+ q^2 R\sum_{k=1}^{n}\sum_{j=0}^{n-k} E^{(k)}b_{j,n-k}R^j + \sum_{k=1}^{n}\sum_{j=0}^{n-k} q^{k+1}V_k b_{j,n-k}R^{j+k} = 0.$$

We need to rearrange the double sums:

$$q^2 R\sum_{k=1}^{n}\sum_{j=0}^{n-k} E^{(k)}b_{j,n-k}R^j = q^2 \sum_{j=1}^{n}\left(\sum_{k=1}^{n-j+1} E^{(k)}b_{j-1,n-k}\right)R^j,$$

$$\sum_{k=1}^{n}\sum_{j=0}^{n-k} q^{k+1}V_k b_{j,n-k}R^{j+k} = \sum_{j=1}^{n}\left(\sum_{k=1}^{j} q^{k+1}V_k b_{j-k,n-k}\right)R^j.$$

Therefore

$$q b_{1n} + \frac{1}{2} \sum_{j=1}^{n-1} (j+1)(j+2q) b_{j+1,n} R^j - \sum_{j=1}^{n} j b_{jn} R^j$$

$$+ q^2 \sum_{j=1}^{n} \left( \sum_{k=1}^{n-j+1} E^{(k)} b_{j-1,n-k} \right) R^j$$

$$+ \sum_{j=1}^{n} \left( \sum_{k=1}^{j} q^{k+1} V_k b_{j-k,n-k} \right) R^j = 0.$$

$$(10.97)$$

The first sum can be made to go from $j = 1$ to $j = n$ since the added term is zero from (10.91).

Using $b_{0n} = b_{1n} = 0$ for $n > 0$, the $j = 0$ term in (10.97) gives the formula for the energy corrections

$$q^2 E^{(n)} = -(2q+1) b_{2n}, \qquad (10.98)$$

and the other terms give the iteration scheme for $2 \le j \le n$ and $n \ge 2$:

$$b_{jn} = \frac{1}{j} \left( \frac{1}{2}(j+1)(j+2q) b_{j+1,n} + \sum_{k=1}^{j} q^{k+1} V_k b_{j-k,n-k} \right.$$

$$\left. + q^2 \sum_{k=1}^{n-j+1} E^{(k)} b_{j-1,n-k} \right). \qquad (10.99)$$

For each $n \ge 2$ the computational order is

$$b_{nn} \to b_{n-1,n} \to b_{n-2,n} \to \cdots \to b_{2n} \to E^{(n)},$$

subject to the boundary conditions

$$b_{00} = 1, \quad E^{(1)} = -V_1, \qquad (10.100)$$

$$b_{0n} = b_{1n} = 0,\, n > 0, \quad b_{jn} = 0,\, j > n. \qquad (10.101)$$

The iteration scheme (10.98) to (10.101) is easily programmed in Maple to give energy corrections to high order as polynomial functions of $q$ related to the dimensionality by (10.82) or in rational form for particular values of $q$ ($q = 1$ corresponds to $D = 3$ and $q = 1/2$ corresponds to $D = 2$).  The following program, *yukawa*, implements the iteration scheme for the Yukawa potential defined by the coefficients $V_k = (-1)^k / k!$.

```
01    # yukawa
02    # energy corrections for yukawa potential
03    # D-dimensional space: q = (D-1)/2
04
05    yukawa := proc(nmax, q)
```

```
06        local b, n, j, k, x, s1, s2;
07        # global DEq
08
09        b[0,0] := 1;
10        DEq[1] := 1;
11
12        for n from 1 to nmax do b[0,n] := 0; od;
13        for n from 1 to nmax do b[1,n] := 0; od;
14        for n from 2 to nmax do b[n+1,n] := 0; od;
15
16        for n from 2 to nmax do
17           for j from n by -1 to 2 do
18              s1 := sum('q^(k+1)*(-1)^k*b[j-k,n-k]/k!', 'k'=1..j);
19              s2 := sum('q^2*DEq[k]*b[j-1,n-k]', 'k'=1..n-j+1);
20              b[j,n] := simplify(( (j+1)*(j+2*q)*b[j+1,n]/2
21                    + s1 + s2 )/j);
22           od;
23           DEq[n] := simplify(-(2*q+1)*b[2,n]/q^2);
24        od;
25        DEq[0] := -1/(2*q^2);
26     end:
27
28     nmax := 20;
29     yukawa(nmax,q):
30     for n from 1 to nmax do DEq[n] := DEq[n]; od;
31     save DEq, DEq.nmax.`.m `;
32     save DEq, DEq.nmax;
33     quit;
```

The results to order 6 are

$$E^{(1)} = 1,$$

$$E^{(2)} = -\frac{1}{4}(2q+1)q,$$

$$E^{(3)} = \frac{1}{12}(2q+1)(q+1)q^2,$$

$$E^{(4)} = -\frac{1}{96}(2q+1)(q+1)q^3(8q+3),$$

$$E^{(5)} = \frac{1}{160}(2q+1)(q+1)q^4(14q^2+19q+2),$$

$$E^{(6)} = -\frac{1}{2880}(2q+1)(q+1)q^5(352q^3+717q^2+366q+15).$$

$$(10.102)$$

More complete results to order 20 are given in Appendix I. These results are
special cases of the general results for an arbitrary state obtained using the

hypervirial and Hellmann-Feynman theorems in Section 10.7 and also reported
to order 14 in Appendix I.

The program can also be used for a particular value of $q$.  For the 3-
dimensional ground state case the procedure call *yukawa(100,1)* can be used
to obtain the energy coefficients in rational form to order 100.  The results to
order 20 are

$$
\begin{aligned}
E = {} & -\frac{1}{2} + \lambda - \frac{3}{4}\lambda^2 + \frac{1}{2}\lambda^3 - \frac{11}{16}\lambda^4 + \frac{21}{16}\lambda^5 - \frac{145}{48}\lambda^6 + \frac{757}{96}\lambda^7 \\
& - \frac{69433}{3072}\lambda^8 + \frac{321449}{4608}\lambda^9 - \frac{2343967}{10240}\lambda^{10} + \frac{24316577}{30720}\lambda^{11} \\
& - \frac{2536041607}{884736}\lambda^{12} + \frac{47860811537}{4423680}\lambda^{13} - \frac{1459923785051}{3440640}\lambda^{14} \\
& + \frac{1599957248809633}{928972800}\lambda^{15} - \frac{42949294634584421}{59454259200}\lambda^{16} \\
& + \frac{46466864975430973}{14863564800}\lambda^{17} - \frac{236707218502703471}{16986931200}\lambda^{18} \\
& + \frac{25325036123322014333}{39636172800}\lambda^{19} - \frac{572761545510214991993}{1902536294400}\lambda^{20}.
\end{aligned}
$$

$$(10.103)$$

More complete results in rational and floating point form are given in Appendix I to order 100.

## 10.5  HVHF Perturbation Method

The classical and quantum mechanical hypervirial (HV) theorems were first
introduced by Hirschfelder [HI60]. Together with the Hellmann-Feynman (HF)
theorems [HE37], [FE39] the HV theorems provide a powerful perturbation
method (HVHF) for separable quantum mechanical systems. This method is
often referred to as "perturbation theory without wavefunctions" since it provides iteration formulas for the energy corrections for a general state which do
not require the wavefunction corrections. Recent reviews of the HV theorems
and their applications have been given by Marc and McMillan [MA85] and by
Fernández and Castro [FE87b].

The HVHF method was first applied to anharmonic oscillators by Swenson and Danforth [SW72], to the charmonium ground state by Killingbeck
[KI78] and Austin [AU80] and to a general state of the screened Coulomb potential by Grant and Lai [GR79]. The method was also applied to the Stark
effect by Austin [AU80] and Lai [LA81]. This is possible since the Schrödinger
equation for the Stark effect is separable in parabolic coordinates into two
one-dimensional systems.

More recently the symbolic computation for the $D$-dimensional hydrogenic
atom with power potentials of the form $r^d$, $d > 0$, has been discussed by Vinette

and Čížek [VI88] and the HVHF theorems have been used by McRae and Vrscay [MC92] to obtain energy expansions for both the classical and quantum mechanical anharmonic oscillators and Kepler problems with power potentials, including the cases $d < 0$.

We shall illustrate the HVHF method for hydrogenic systems with spherically symmetric potentials of the power and screened Coulomb types.

### 10.5.1   Hypervirial and Hellmann-Feynman theorems

Let $H$ be a time-independent and hermitian hamiltonian and let $A$ be any time-independent linear operator. If $\psi$ is any eigenstate of $H$ such that $H$ is hermitian with respect to both $\psi$ and $A\psi$ then

$$\langle\psi|[A,H]|\psi\rangle = 0. \tag{10.104}$$

This result is called the hypervirial theorem and easily follows from $H\psi = E\psi$ and the assumed hermitian conditions on $H$ [EP62].

If $H = H(\lambda)$ depends on a parameter $\lambda$, such as a perturbation parameter, and $H(\lambda)\psi = E(\lambda)\psi$ for some range of values of $\lambda$ and $\psi$ is normalized, $\langle\psi|\psi\rangle = 1$, then

$$\frac{\partial E}{\partial \lambda} = \left\langle \psi \left| \frac{\partial H(\lambda)}{\partial \lambda} \right| \psi \right\rangle. \tag{10.105}$$

This result is called the Hellmann-Feynman (HF) theorem and can be proved as follows. The energy is given by $E(\lambda) = \langle\psi|H(\lambda)|\psi\rangle$. Differentiating and assuming that $H$ is hermitian with respect to both $\psi$ and $\partial\psi/\partial\lambda$,

$$\frac{\partial E}{\partial \lambda} = \left\langle \frac{\partial \psi}{\partial \lambda} \middle| H \middle| \psi \right\rangle + \left\langle \psi \middle| H \middle| \frac{\partial \psi}{\partial \lambda} \right\rangle + \left\langle \psi \middle| \frac{\partial H}{\partial \lambda} \middle| \psi \right\rangle$$

$$= E\frac{\partial}{\partial \lambda}\langle\psi|\psi\rangle + \left\langle \psi \middle| \frac{\partial H}{\partial \lambda} \middle| \psi \right\rangle.$$

Then (10.105) follows since $\langle\psi|\psi\rangle = 1$. Together (10.104) and (10.105), with the appropriate choices of the operator $A$, give the HVHF method.

### 10.5.2   HVHF method for hydrogenic systems

For a $D$-dimensional hydrogenic system with a spherically symmetric perturbation the hamiltonian is given by (see Chapter 4)

$$H = \frac{1}{2}p^2 - \frac{Z}{r} + V(r,\lambda), \tag{10.106}$$

and the Schrödinger equation is

$$H\psi_{qk}(r) = E\psi_{qk}(r), \tag{10.107}$$

where

$$p^2 = p_r^2 + \frac{\tau}{r^2}, \tag{10.108}$$

$$\tau = k(k+1) = \frac{1}{4}(D-1)(D-3) + \ell(\ell + D - 2), \tag{10.109}$$

$$k = \ell + \frac{1}{2}(D-3), \quad \ell \geq 0, \tag{10.110}$$

$$q = k + 1 + n_r, \quad n_r \geq 0. \tag{10.111}$$

The key idea in the derivation of the HVHF method is to consider the set of operators of the form

$$A = r^{j+1}p_r \tag{10.112}$$

and evaluate the commutators $[A, H]$ using the commutator identities (see Chapter 4)

$$[p_r, r^n] = -inr^{n-1}, \tag{10.113}$$

$$[p_r, r^{-n}] = inr^{-(n+1)}, \tag{10.114}$$

$$[p_r, V(r)] = -i\frac{dV}{dr} = -iV'(r), \tag{10.115}$$

which are valid in $D$-dimensional space. Therefore

$$[r^{j+1}p_r, H] = r^{j+1}[p_r, H] + [r^{j+1}, H]p_r, \tag{10.116}$$

where

$$\begin{aligned}
[p_r, H] &= [p_r, \tau r^{-2}/2 - \mathcal{Z}r^{-1} + V] \\
&= \frac{\tau}{2}[p_r, r^{-2}] - \mathcal{Z}[p_r, r^{-1}] + [p_r, V] \\
&= i(\tau r^{-3} - \mathcal{Z}r^{-2} - V'), \tag{10.117}
\end{aligned}$$

$$\begin{aligned}
[r^{j+1}, H] &= -\frac{1}{2}[p_r^2, r^{j+1}] \\
&= -\frac{1}{2}p_r[p_r, r^{j+1}] - \frac{1}{2}[p_r, r^{j+1}]p_r \\
&= \frac{1}{2}i(j+1)p_r r^j + \frac{1}{2}i(j+1)r^j p_r \\
&= \frac{1}{2}i(j+1)(r^j p_r - ijr^{j-1}) + \frac{1}{2}i(j+1)r^j p_r \\
&= i(j+1)r^j p_r + \frac{1}{2}j(j+1)r^{j-1}. \tag{10.118}
\end{aligned}$$

Substitute into (10.116) to get

$$\begin{aligned}
[r^{j+1}p_r, H] &= i\tau r^{j-2} - i\mathcal{Z}r^{j-1} - ir^{j+1}V' \\
&\quad + i(j+1)r^j p_r^2 + \frac{1}{2}j(j+1)r^{j-1}p_r.
\end{aligned}$$

Now substitute for $p_r^2$ using (10.106) and (10.108) to obtain

$$[r^{j+1}p_r, H] = -ij\tau r^{j-2} + i\mathcal{Z}(2j+1)r^{j-1} + 2i(j+1)r^j(H-V)$$
$$- ir^{j+1}V' + \frac{1}{2}j(j+1)r^{j-1}p_r.$$

Now take expectation values of both sides and use the HV theorem (10.104) to obtain

$$2(j+1)E\langle r^j\rangle = j\tau\langle r^{j-2}\rangle - \mathcal{Z}(2j+1)\langle r^{j-1}\rangle + 2(j+1)\langle r^jV\rangle$$
$$+ \langle r^{j+1}V'\rangle + \frac{1}{2}ij(j+1)\langle r^{j-1}p_r\rangle, \tag{10.119}$$

where we have introduced the shorthand notation

$$\langle A\rangle = \langle\psi|A|\psi\rangle \tag{10.120}$$

for the expectation value of an operator $A$ with respect to an eigenstate $\psi$ of $H$ having eigenvalue $E$.

To eliminate $p_r$ from (10.119) we can again apply the HV theorem but now with $A = r^j$ to obtain

$$\langle[r^j, H]\rangle = ij\langle r^{j-1}p_r\rangle + \frac{1}{2}j(j-1)\langle r^{j-2}\rangle = 0. \tag{10.121}$$

Thus the last term in (10.119) can be eliminated and we obtain the following result involving only expectation values of powers of $r$:

$$2(j+1)E\langle r^j\rangle = 2(j+1)\langle r^jV\rangle + \langle r^{j+1}V'\rangle - (2j+1)\mathcal{Z}\langle r^{j-1}\rangle$$
$$+ \frac{1}{4}j[4\tau - (j^2-1)]\langle r^{j-2}\rangle. \tag{10.122}$$

The HF theorem can be expressed as

$$\frac{\partial E}{\partial\lambda} = \left\langle\frac{\partial V(r,\lambda)}{\partial\lambda}\right\rangle. \tag{10.123}$$

For any spherically symmetric potential $V$, we can now use these two results to obtain an efficient iteration scheme for the energy corrections.

## 10.6   HVHF Method for Power Potentials

For $V(r,\lambda) = \lambda r^d$ (10.122) and (10.123) become

$$2(j+1)E\langle r^j\rangle = [2(j+1)+d]\lambda\langle r^{j+d}\rangle - (2j+1)\mathcal{Z}\langle r^{j-1}\rangle$$
$$+ \frac{1}{4}j[4\tau - (j^2-1)]\langle r^{j-2}\rangle, \tag{10.124}$$

$$\frac{\partial E}{\partial\lambda} = \langle r^d\rangle. \tag{10.125}$$

Now substitute the perturbation expansions

$$E = \sum_{n=0}^{\infty} E^{(n)}\lambda^n, \quad \langle r^j \rangle = \sum_{n=0}^{\infty} C_j^{(n)}\lambda^n, \tag{10.126}$$

into (10.124) to obtain

$$2(j+1)\sum_{n=0}^{\infty}\left(\sum_{m=0}^{n} E^{(n-m)}C_j^{(m)}\right)\lambda^n = [2(j+1)+d]\sum_{n=1}^{\infty} C_{j+d}^{(n-1)}\lambda^n$$

$$- (2j+1)\mathcal{Z}\sum_{n=0}^{\infty} C_{j-1}^{(n)}\lambda^n + \frac{1}{4}j[4\tau - (j^2-1)]\sum_{n=0}^{\infty} C_{j-2}^{(n)}\lambda^n.$$

We can make the second sum begin at $n = 0$ if we define

$$C_j^{(-1)} = 0, \quad j \geq 0. \tag{10.127}$$

Therefore, equating coefficients of $\lambda^n$ we obtain

$$2(j+1)\sum_{m=0}^{n} E^{(n-m)}C_j^{(m)} = [2(j+1)+d]C_{j+d}^{(n-1)}$$

$$- (2j+1)\mathcal{Z}C_{j-1}^{(n)} + \frac{1}{4}j[4\tau - (j^2-1)]C_{j-2}^{(n)}.$$

Take out the $m = n$ term to get

$$2(j+1)E^{(0)}C_j^{(n)} = -2(j+1)\sum_{m=0}^{n-1} E^{(n-m)}C_j^{(m)} - (2j+1)\mathcal{Z}C_{j-1}^{(n)}$$

$$+ [2(j+1)+d]C_{j+d}^{(n-1)} + \frac{1}{4}j[4\tau - (j^2-1)]C_{j-2}^{(n)}. \tag{10.128}$$

The HF result (10.125) gives

$$\frac{\partial}{\partial\lambda}\sum_{n=0}^{\infty} E^{(n)}\lambda^n = \sum_{n=0}^{\infty}(n+1)E^{(n+1)}\lambda^n = \sum_{n=0}^{\infty} C_d^{(n)}\lambda^n.$$

Therefore the energy corrections are given by

$$E^{(n+1)} = \frac{1}{n+1}C_d^{(n)}, \, n \geq 0. \tag{10.129}$$

Since $\langle r^0 \rangle = 1$ it follows from (10.126) that

$$C_0^{(0)} = 1, \quad C_0^{(n)} = 0, \, n > 0. \tag{10.130}$$

To initialize the iteration scheme (10.128) for each $n$ we also need values for $C_{-1}^{(n)}$. To obtain them substitute $j = 0$ into (10.128) to obtain

$$2E^{(0)}C_0^{(n)} = -2\sum_{m=0}^{n-1} E^{(n-m)}C_0^{(m)} - \mathcal{Z}C_{-1}^{(n)} + (d+2)C_d^{(n-1)}.$$

For $n = 0$ this gives $C_{-1}^{(0)} = -2E^{(0)}/\mathcal{Z}$ and for $n > 0$

$$C_{-1}^{(n)} = \frac{1}{\mathcal{Z}}[-2E^{(n)} + (d+2)C_d^{(n-1)}], \tag{10.131}$$

which is also valid if $n = 0$ from (10.127). In case $d > 0$ the calculation of the $C_j^{(n)}$ and $E^{(n+1)}$ proceeds according to the scheme

$$n = 0 : C_{-1}^{(0)} \to C_0^{(0)} \to C_1^{(0)} \to \cdots\cdots\cdots\cdots \to C_{n_0}^{(0)} \to E^{(1)}$$
$$n = 1 : C_{-1}^{(1)} \to C_0^{(1)} \to C_1^{(1)} \to \cdots\cdots\cdots \to C_{n_1}^{(1)} \to E^{(2)}$$

$$\cdot$$
$$\cdot$$
$$\cdot$$

$$n = k : C_{-1}^{(k)} \to C_0^{(k)} \to C_1^{(k)} \to \cdots \to C_{n_k}^{(k)} \to E^{(k+1)}$$

It is clear from (10.128) that the calculation of $C_{n_k}^{(k)}$ in row $k$ requires entries up to $C_{n_k+d}^{(k-1)}$ in row $k - 1$. Therefore if we want to calculate energy corrections to order $n_{max}$ then we calculate rows $n = 0$ to $n = n_{max} - 1$ so we have $n_k = (n_{max} - k)d$ as the maximum subscript in each row $k$.

The entire iteration scheme for $d > 0$ can now be obtained starting only with $C_0^{(0)} = 1$ and $E^{(0)} = -\mathcal{Z}^2/(2q^2)$ using (10.130) and (10.131) to obtain the first two entries in each row and (10.128) for the remaining entries.

The following simple Maple procedure, *hvhfpower*, can be used to obtain the energy corrections as functions of $q$, $\tau$ and $\mathcal{Z}$ for any power potential $r^d$, $d > 0$. In the program the energy corrections $E^{(n)}$ are denoted by *DE[n]*, the coefficients $C_j^{(n)}$ are denoted by *C[n,j]* and the nuclear charge $\mathcal{Z}$ is denoted by *Z*. Of course we can do all calculations with $\mathcal{Z} = 1$ and use (10.16) to obtain the energy corrections for any other value of $\mathcal{Z}$.

```
01    # hvhfpower
02    # HVHF method for power potential r^d, d > 0
03    hvhfpower := proc(nmax, d, q, tau, Z)
04        local j, n, s1, m;
05        DE[0] := - Z^2/(2*q^2):
06        C[0,0] := 1:
07        for j from 0 to (nmax+1)*d do C[-1,j] := 0; od:
08        for n from 1 to nmax-1 do C[n,0] := 0; od:
09
10        for n from 0 to nmax-1 do
11            C[n,-1] := simplify((-2*DE[n] + (d+2)*C[n-1,d]) / Z);
12            for j from 1 to (nmax-n)*d do
13                s1 := sum('DE[n-m]*C[m,j]', 'm'=0..n-1);
14                C[n,j] := simplify( ( -2*(j+1)*s1 - (2*j+1)*Z*C[n,j-1]
15                    + (2*(j+1)+d)*C[n-1,j+d]
16                    + j*(tau-(j^2-1)/4)*C[n,j-2] ) / (2*(j+1)*DE[0]) );
17            od;
```

```
18          DE[n+1] := C[n,d]/(n+1);
19      od:
20  end:
```

```
01   # hvhfcharm
02   # test for charmonium using hvhfpower
03   read hvhfpower;
04   nmax := 6;
05   d := 1;
06   hvhfpower(nmax, d, q, tau, 1):
07   DEqt := op(DE); # make a new name for energy coefficients
08   for n from 0 to nmax do DEqt[n] := DEqt[n]; od;
09   save DEqt, DE.d.`qt `.nmax.`.m`;
10   save DEqt, DE.d.`qt `.nmax;
11   quit;
```

```
01   # hvhfharm
02   # test for harmonium using hvhfpower
03   read hvhfpower;
04   nmax := 6;
05   d := 2;
06   hvhfpower(nmax, d, q, tau, 1):
07   DEqt := op(DE); # make a new name for energy coefficients
08   for n from 0 to nmax do DEqt[n] := DEqt[n]; od;
09   save DEqt, DE.d.`qt `.nmax.`.m`;
10   save DEqt, DE.d.`qt `.nmax;
11   quit;
```

Programs *hvhfcharm* and *hvhfharm* use *hvhfpower* to efficiently produce
the charmonium and harmonium energy corrections for a general state as previously obtained by programs *charm* and *harm* in Section 10.2 and given to
order 14 for charmonium in Appendix G and to order 10 for harmonium in
Appendix H. With *hvhfpower* it is possible to go to much higher orders for a
general state.

Procedure *hvhfpower* can also be used to produce the energy corrections
to high order for any particular state. For example, the 3-dimensional ground
state results for charmonium given in Appendix G can easily be obtained using the procedure call *hvhfpower(100,1,1,0,1)* and the similar results in Appendix H for harmonium can be obtained using *hvhfpower(100,2,1,0,1)*.

## 10.7   HVHF Method: Screened Coulomb Case

For $V(r, \lambda)$ given by (10.64) the HVHF equations (10.122) and (10.123) become

$$2(j+1)(E + \mathcal{Z}V_1\lambda)\langle r^j \rangle = -(2j+1)\mathcal{Z}\langle r^{j-1} \rangle$$
$$+ \frac{1}{4}j[4\tau - (j^2 - 1)]\langle r^{j-2} \rangle$$
$$- \mathcal{Z}\sum_{m=2}^{\infty}(2j+m+1)V_m\lambda^m \langle r^{j+m-1} \rangle, \qquad (10.132)$$

$$\frac{\partial E}{\partial \lambda} = -\mathcal{Z}\sum_{m=1}^{\infty} mV_m\lambda^{m-1}\langle r^{m-1} \rangle. \qquad (10.133)$$

Substitute the perturbation expansions (10.126) into (10.132) to obtain

$$-2\sum_{n=0}^{\infty}\left(\sum_{m=0}^{n} E^{(n-m)}C_j^{(m)}\right)\lambda^n - 2\mathcal{Z}V_1\sum_{n=1}^{\infty}C_j^{(n-1)}\lambda^n$$
$$= \frac{2j+1}{j+1}\mathcal{Z}\sum_{n=0}^{\infty}C_{j-1}^{(n)}\lambda^n + j\left(-\frac{\tau}{j+1} + \frac{j-1}{4}\right)\sum_{n=0}^{\infty}C_{j-2}^{(n)}\lambda^n$$
$$+ \mathcal{Z}\sum_{m=2}^{\infty}\frac{2j+m+1}{j+1}V_m\lambda^m\sum_{n=0}^{\infty}C_{j+m-1}^{(n)}\lambda^n.$$

Then using the definition (10.127) and writing the double sum on the right side as

$$\sum_{n=0}^{\infty}\left(\sum_{m=2}^{n}\frac{2j+m+1}{j+1}V_mC_{j+m-1}^{(n-m)}\right)\lambda^n,$$

we can equate coefficients of $\lambda^n$ on both sides to obtain

$$-2\sum_{m=0}^{n} E^{(n-m)}C_j^{(m)} - 2\mathcal{Z}V_1C_j^{(n-1)} = \frac{2j+1}{j+1}\mathcal{Z}C_{j-1}^{(n)}$$
$$+ j\left(-\frac{\tau}{j+1} + \frac{j-1}{4}\right)C_{j-2}^{(n)} + \mathcal{Z}\sum_{m=2}^{n}\frac{2j+m+1}{j+1}V_mC_{j+m-1}^{(n-m)}.$$

Take out the $m = n$ term and solve for $C_j^{(n)}$ to obtain the iteration formula

$$C_j^{(n)} = -\frac{1}{2E^{(0)}}\left[2\sum_{m=0}^{n-1} E^{(n-m)}C_j^{(m)} + 2\mathcal{Z}V_1C_j^{(n-1)}\right.$$
$$+ \frac{2j+1}{j+1}\mathcal{Z}C_{j-1}^{(n)} + j\left(-\frac{\tau}{j+1} + \frac{j-1}{4}\right)C_{j-2}^{(n)}$$
$$\left. + \mathcal{Z}\sum_{m=2}^{n}\frac{2j+m+1}{j+1}V_mC_{j+m-1}^{(n-m)}\right]. \qquad (10.134)$$

For $j = 0$ we can solve for the starting value $C_{-1}^{(n)}$:

$$C_{-1}^{(n)} = -\frac{1}{\mathcal{Z}}\left[2E^{(n)} + 2\mathcal{Z}V_1 C_0^{(n-1)} + \mathcal{Z}\sum_{m=2}^{n}(m+1)V_m C_{m-1}^{(n-m)}\right],$$

$$(10.135)$$

which is valid for $n \geq 0$ under the assumption (10.127). Therefore the starting values $C_{-1}^{(n)}$ and $C_0^{(n)}$ can be used in (10.134) to calculate the coefficients $C_j^{(n)}$ for $j \geq 1$.

To obtain the energy formula substitute perturbation expansions (10.126) into (10.133) to obtain

$$\sum_{n=1}^{\infty} nE^{(n)}\lambda^{n-1} = -\mathcal{Z}\sum_{n=1}^{\infty} nV_n\lambda^{n-1}\sum_{m=0}^{\infty} C_{n-1}^{(m)}\lambda^m$$

$$= -\mathcal{Z}\sum_{n=0}^{\infty}\left(\sum_{m=1}^{n+1} mV_m C_{m-1}^{(n-m+1)}\right)\lambda^n.$$

Therefore for $n \geq 0$

$$E^{(n+1)} = -\frac{\mathcal{Z}}{n+1}\sum_{m=1}^{n+1} mV_m C_{m-1}^{(n-m+1)}.$$

$$(10.136)$$

To calculate $E^{(n+1)}$ we need to calculate the coefficients in row 0 to $C_n^{(0)}$, the coefficients in row 1 to $C_{n-1}^{(1)}$ and finally in row $n$ the last coefficient needed is $C_0^{(n)}$. As in the power potential case of the preceding section the entire iteration scheme can be obtained starting only with $C_0^{(0)} = 1$ and $E^{(0)} = -\mathcal{Z}^2/(2q^2)$ using (10.135) and (10.130) to obtain the first two entries in each row and (10.134) for the remaining entries.

The following simple Maple procedure, *hvhfscr*, implements this iteration scheme. The notation is the same as in procedure *hvhfpower* of the preceding section. The program produces symbolic energy corrections as functions of $q$, $\tau$, $\mathcal{Z}$, and the coefficients $V_j$ defining the screened Coulomb potential. The specific form $V_j = (-1)^j/j!$ can be substituted, as shown in line *15* of program *scrtest*, to obtain energy corrections for a general state for the Yukawa potential.

```
01    # hvhfscr
02    # HVHF method for screened Coulomb potential
03    hvhfscr := proc(nmax, q, tau, Z)
04        local j, n, s1, s2, s3, m;
05        DE[0] := - Z^2/(2*q^2):
06        C[0,0] := 1:
07        for j from 0 to nmax-1 do C[-1,j] := 0; od:
08        for n from 1 to nmax-1 do C[n,0] := 0; od:
09        for n from 0 to nmax-1 do
10            s1 := sum('(m+1)*Z*V[m]*C[n-m,m-1]', 'm'=2..n);
11            C[n,-1] := simplify(
```

```
12                -(2*DE[n] + 2*Z*V[1]*C[n-1,0] + s1)/Z ):
13            for j from 1 to nmax-n-1 do
14                s2 := sum('DE[n-m]*C[m,j]', 'm'=0..n-1);
15                s3 := sum('((2*j+m+1)/(j+1))*Z*V[m]*C[n-m,j+m-1]',
16                    'm'=2..n);
17                C[n,j] := simplify( -( 2*s2 + 2*Z*V[1]*C[n-1,j]
18                    + ((2*j+1)/(j+1))*Z*C[n,j-1]
19                    + j*( -tau/(j+1) + (j-1)/4 )*C[n,j-2] + s3 )
20                    / (2*DE[0]) );
21            od:
22            DE[n+1] := simplify(-Z*sum('m*V[m]*C[n-m+1,m-1]',
23                'm'=1..n+1)/(n+1));
24        od:
25    end:
```

```
01    # scrtest
02    read hvhfscr;
03
04    # Do the screened coulomb potential
05
06    nmax := 6;
07    hvhfscr(nmax, q, tau, 1):
08    DEqt := op(DE); # make a new name
09    for n from 0 to nmax do DEqt[n] := factor(DEqt[n]); od;
10    save DEqt, DEsqt.nmax.`.m`;
11    save DEqt, DEsqt.nmax;
12
13    # Do the yukawa potential
14
15    for n from 1 to nmax do V[n] := (-1)^n/n!; od:
16    for n from 0 to nmax do DEqt[n] := factor(DEqt[n]); od;
17    save DEqt, DEyqt.nmax.`.m`;
18    save DEqt, DEyqt.nmax;
19    quit;
```

The general screened Coulomb results to order 5 are given by

$$E^{(1)} = -V_1,$$

$$E^{(2)} = -\frac{1}{2}V_2(3\,q^2 - \tau),$$

$$E^{(3)} = -\frac{1}{2}V_3 q^2(5\,q^2 - 3\,\tau + 1),$$

$$E^{(4)} = \frac{1}{8}q^2(-7\,V_2^2 q^4 - 35\,V_4 q^4 + 30\,V_4 q^2\tau - 25\,V_4 q^2 - 5\,V_2^2 q^2$$
$$+ 3\,V_2^2\tau^2 - 3\,V_4\tau^2 + 6\,V_4\tau),$$

$$E^{(5)} = \frac{1}{8}q^4(-45\,V_2V_3q^4 - 63\,V_5q^4 + 70\,V_5q^2\tau + 14\,V_2V_3q^2\tau$$
$$- 105\,V_5q^2 - 63\,V_2V_3q^2 - 15\,V_5\tau^2 + 15\,V_2V_3\tau^2$$
$$+ 10\,V_2V_3\tau + 50\,V_5\tau - 12\,V_5). \tag{10.137}$$

More complete results to order 10 are given in Appendix I.

In the special case of the Yukawa potential the results to order 6 are

$$E^{(1)} = 1,$$

$$E^{(2)} = -\frac{1}{4}(3\,q^2 - \tau),$$

$$E^{(3)} = \frac{1}{12}q^2(5\,q^2 - 3\,\tau + 1),$$

$$E^{(4)} = -\frac{1}{192}q^2(77\,q^4 - 30\,q^2\tau + 55\,q^2 - 15\,\tau^2 - 6\,\tau),$$

$$E^{(5)} = \frac{1}{320}q^4(171\,q^4 - 70\,q^2\tau + 245\,q^2 - 45\,\tau^2 - 50\,\tau + 4),$$

$$E^{(6)} = -\frac{1}{5760}q^4(4763\,q^6 - 2070\,q^4\tau + 11580\,q^4 - 945\,q^2\tau^2$$
$$- 2940\,q^2\tau + 1057\,q^2 - 340\,\tau^3 - 205\,\tau^2 - 30\,\tau), \tag{10.138}$$

and more complete results to order 14 are given in Appendix I. Results for a specific state can be easily obtained directly. For example the 3-dimensional ground state results for the Yukawa potential are easily obtained to order 100 using procedure *hvhfscr* by first initializing the coefficients $V[j]$ with a statement such as

**for** *n* **from** *1* **to** *100* **do** *V[n] := (−1)ˆn/n!;* **od:**

and then using the procedure call

*hvhfscr(100,1,0,1);*

The results are given in Appendix I.

We also note that the results obtained in Section 10.4 for the D-dimensional ground state by program *yukawa* using the power series method are special cases of the general results produced by program *hvhfscr*. In fact the results (10.102) are easily obtained by substituting $\tau = k(k+1) = q(q-1)$ into the results (10.138) since for the ground state we have from (10.83) $k = q - 1$.

## 10.8   Exercises

◈ **Exercise 10.1** Derive (10.16) for the $\mathcal{Z}$ dependence of the energy corrections for power potentials.

◈ **Exercise 10.2** Derive (10.76) for the $\mathcal{Z}$ dependence of the energy corrections for the Yukawa potential.

◈ **Exercise 10.3** Derive the formulas (10.77) and (10.78) for the energy corrections $E^{(1)}$ and $E^{(2)}$ for the screened Coulomb potential.

# Appendix A

# The Levi-Civita Symbol

## A.1  Basic Properties

The Levi-Civita symbol is denoted by $\epsilon_{ijk}$. It is antisymmetric in all three indices and defined by

$$\epsilon_{ijk} = \begin{cases} 1 & \text{if } ijk \text{ is an even permutation of 123,} \\ -1 & \text{if } ijk \text{ is an odd permutation of 123,} \\ 0 & \text{if two or more of } i,j,k \text{ are equal.} \end{cases} \tag{A.1}$$

It is very useful for the manipulation of various commutation relations and for expressing them in compact form and arises most naturally in the discussion of the vector product in the three-dimensional vector space $\mathbb{R}^3$.

If an orthonormal basis $\{e_1, e_2, e_3\}$ is chosen then any vector $\boldsymbol{x}$ can be expressed as $\boldsymbol{x} = (x_1, x_2, x_3) = \sum_j x_j e_j$ and the dot and cross products can be expressed as

$$e_i \cdot e_j = \delta_{ij}, \tag{A.2}$$

$$e_i \times e_j = \sum_k \epsilon_{ijk} e_k. \tag{A.3}$$

Because of the properties of $\epsilon_{ijk}$ given in (A.1) the sum over $k$ reduces to a single term and provides a compact way of writing the three identities $e_1 \times e_2 = e_3$, $e_2 \times e_3 = e_1$ and $e_3 \times e_1 = e_2$ and the antisymmetry property $e_i \times e_j = -e_j \times e_i$ of the vector cross product.

From (A.2) and (A.3) we obtain the following definition of the Levi-Civita symbol as the signed volume of the unit cube:

$$\epsilon_{ijk} = (e_i \times e_j) \cdot e_k. \tag{A.4}$$

## A.2  Important Identities

Important identities involving the Levi-Civita symbol can be obtained from the following well known vector identities

$$a \times (b \times c) = (a \cdot c)b - (a \cdot b)c, \tag{A.5}$$

$$a \times (b \times c) + b \times (c \times a) + c \times (a \times b) = 0, \qquad (A.6)$$

$$(a \times b) \cdot (c \times d) = (a \cdot c)(b \cdot d) - (a \cdot d)(b \cdot c), \qquad (A.7)$$

where $a$, $b$, $c \in \mathbb{R}^3$.  Equation (A.6) is a consequence of (A.5).  From these identities it can be shown that

$$\sum_m (\epsilon_{ijm}\epsilon_{k\ell m} + \epsilon_{jkm}\epsilon_{i\ell m} + \epsilon_{kim}\epsilon_{j\ell m}) = 0, \qquad (A.8)$$

$$\sum_m \epsilon_{ijm}\epsilon_{k\ell m} = \delta_{ik}\delta_{j\ell} - \delta_{i\ell}\delta_{jk}, \qquad (A.9)$$

$$\sum_{\ell m} \epsilon_{i\ell m}\epsilon_{k\ell m} = 2\delta_{ik}, \qquad (A.10)$$

$$\sum_{i\ell m} \epsilon_{i\ell m}\epsilon_{i\ell m} = 6. \qquad (A.11)$$

Note how the subscripts are obtained in (A.8): cyclic permutations of $i, j, k$ but $\ell, m$ are in fixed positions.

## A.3  Exercises

◈ **Exercise A.1**  Derive (A.8) to (A.11) using vector identities (A.5) to (A.8).

# Appendix B

# Lie Groups and Lie Algebras

In order to make the book as self-contained as possible we briefly outline the connection between Lie groups and Lie algebras using the familiar concepts of matrices and linear algebra. The special orthogonal groups in 2 and 3 dimensions, SO(2) and SO(3), are used as familiar examples. It is also shown how the infinitesimal group transformations corresponding to one-parameter subgroups give rise to the generators of the associated Lie algebra. Also we discuss the special unitary group SU(2), which is important for spin in quantum mechanics, and show its connection with SO(3).

The Lie groups $SO(n)$ and $SO(p,q)$ and their corresponding Lie algebras $so(n)$ and $so(p,q)$ are also discussed in general terms as transformation groups since realizations and representations of so(4) and so(4,2) are the subject of Chapter 5 and Chapter 6. We do not present a mathematically rigorous or advanced approach here (see [MI72], [SA86], [HU72], [CO84a,b] for accessible and more comprehensive treatments). A pedagogical discussion of the unitary Lie groups and algebras and their application to the many-electron correlation problem in quantum chemistry has been given by Paldus [PA76].

## B.1   Some Definitions

A group $\mathcal{G}$ is a set of elements with a binary operation (denoted by juxtaposition) such that for all $a, b, c \in \mathcal{G}$

    **(a)**  $ab \in \mathcal{G}$, (closure),

    **(b)**  $(ab)c = a(bc)$, (associativity),

    **(c)**  there exists an identity element $e \in \mathcal{G}$ such that $ae = ea = a$,

    **(d)**  there exists an inverse element $a^{-1} \in \mathcal{G}$ such that $aa^{-1} = e$ and $a^{-1}a = e$.

It can be shown that the identity element is unique and also that the inverse of each element is unique. A *subgroup* is a subset of the elements which also forms a group.

Groups can be broadly classified as discrete (finite or countably infinite number of elements) or continuous. For a continuous group each element depends continuously on one or more parameters and a Lie group is a special kind of continuous group. A rigorous definition will not be attempted here since it would require a knowledge of topology and differentiable manifolds.

For our purposes an $n$-parameter Lie group is a group whose elements can be expressed as analytic functions $g(\alpha_1, \ldots, \alpha_n)$ of $n$ independent parameters $\alpha_j$ in such a way that

(a) the identity element $e = e(0, \ldots, 0)$ corresponds to parameter values $\alpha_j = 0$, $j = 1, \ldots, n$,

(b) the group multiplication and inverse operations are analytic functions of the parameters in some neighborhood of the identity element.

Using the vector notation $\boldsymbol{\alpha} = (\alpha_1, \ldots, \alpha_n)$, property (b) implies that if $g(\boldsymbol{\alpha})$ and $g(\boldsymbol{\beta})$ are two group elements defined in some suitable neighborhood of the identity element then there exist analytic functions $\gamma_j = \gamma_j(\boldsymbol{\alpha}, \boldsymbol{\beta})$ and $\delta_j = \delta_j(\boldsymbol{\alpha})$, $j = 1, \ldots, n$ such that in some neighborhood of the identity element we have $g(\boldsymbol{\gamma}) = g(\boldsymbol{\alpha})g(\boldsymbol{\beta})$ and $g^{-1}(\boldsymbol{\alpha}) = g(\boldsymbol{\delta})$.

The simplest examples of Lie groups are the matrix Lie groups for which each group element $g$ is an $m \times m$ matrix and each matrix element $g_{ij} = g_{ij}(\boldsymbol{\alpha})$ is an analytic function of the $n$ parameters. In Section B.3 we shall consider two important examples: the special orthogonal group SO(3) of proper rotations in 3 dimensions and the special unitary group SU(2) of unitary, unimodular $2 \times 2$ matrices. In each case we obtain the Lie algebra associated with the group and show how the Lie algebra generates the group elements in some neighborhood of the identity via exponentiation. In fact each basis vector of the Lie algebra generates a one-parameter subgroup of the Lie group.

## B.2    One-parameter Subgroups

Given an $n$-parameter Lie group we can construct $n$ one-parameter subgroups in many ways. We shall not give the details here but it is possible to choose a canonical parameterization of the group elements such that the elements of subgroup $j$ have the form $g(t) = g(0, \ldots, 0, t, 0, \ldots, 0)$ for $t$ in position $j$ and for $t$ in some neighborhood of $t = 0$ corresponding to the identity element $e = g(0)$. Moreover this parameterization can be chosen so that the group multiplication has the particularly simple additive form

$$g(t_1 + t_2) = g(t_1)g(t_2). \tag{B.1}$$

Letting $t_1 = t$ and $t_2 = -t$ and using $g(0) = e$, inverses are given by

$$g^{-1}(t) = g(-t). \tag{B.2}$$

If we consider only matrix Lie groups ($m \times m$ matrices) it is possible to show that the matrices $g(t)$ satisfy a first-order linear matrix differential equation of the form

$$\frac{dg}{dt} = Ag, \tag{B.3}$$

where $A$ is a constant matrix. The solution satisfying $g(0) = I$, the $m \times m$ identity matrix, is

$$g(t) = \exp(tA) = e^{tA}. \tag{B.4}$$

Since there are $n$ such independent one-parameter subgroups we can use a subscript to distinguish them. Thus, the elements of subgroup $j$ have the form

$$g_j(\alpha_j) = e^{\alpha_j A_j}, \tag{B.5}$$

where $\alpha_j$ is the parameter and $A_j$ is an $m \times m$ constant matrix.

The important results, whose proofs are beyond the scope of this book, are that the $n$ matrices $A_j$, $j = 1,\ldots,n$ satisfy the commutation relations

$$[A_j, A_k] = \sum_\ell c_{jk\ell} A_\ell, \tag{B.6}$$

and thus form a basis for a Lie algebra, and that any group element sufficiently close to the identity element can be expressed in the form

$$g(\alpha_1,\ldots,\alpha_n) = e^{\sum_j \alpha_j A_j}, \tag{B.7}$$

or the form

$$g(\alpha_1',\ldots,\alpha_n') = e^{\alpha_1' A_1} \cdots e^{\alpha_n' A_n}, \tag{B.8}$$

for some other set of parameters $\alpha_j'$.

If we know the exponential form of the one-parameter subgroups then the basis vectors of the Lie algebra can easily be obtained as

$$A_j = \left.\frac{dg(\alpha_j)}{d\alpha_j}\right|_{\alpha_j=0}, \quad j = 1,\ldots,n. \tag{B.9}$$

Each $A_j$ is often called an infinitesimal generator of a one-parameter subgroup since for small values ($\alpha_j \approx 0$) of the parameter the group element $g(\alpha_j)$ can be approximated by the matrix

$$g(\alpha_j) \approx I_m + \alpha_j A_j, \tag{B.10}$$

where $I_m$ is the $m \times m$ identity matrix.

Finally we note without proof that the correspondence between Lie groups and Lie algebras is many-to-one in the sense that several non-isomorphic Lie groups may have the same Lie algebra. This is not surprising since a Lie algebra determines a Lie group only locally within a neighborhood of the identity.

## B.3   Rotation Group SO(2)

Consider the one-parameter group of $2 \times 2$ real matrices

$$R(\theta) = \begin{pmatrix} \cos\theta & \sin\theta \\ -\sin\theta & \cos\theta \end{pmatrix}, \quad 0 \le \theta \le 2\pi. \tag{B.11}$$

These matrices have the following properties

$$R^T(\theta)R(\theta) = I_2, \tag{B.12}$$
$$R(\theta_1 + \theta_2) = R(\theta_1)R(\theta_2), \tag{B.13}$$
$$R^{-1}(\theta) = R(-\theta) = R^T(\theta), \tag{B.14}$$
$$\det(R(\theta)) = 1, \tag{B.15}$$

where $R^T$ denotes matrix transposition and $I_2$ denotes the $2\times 2$ identity matrix. These matrices form a one-parameter Lie group called the *special orthogonal group* in two dimensions and denoted by SO(2). The matrices are orthogonal, property (B.12), and "special", property (B.15). Properties (B.13) and (B.14) show that the group multiplication and inverse operations are analytic functions of parameter $\theta$.

Considered as a transformation group these matrices rotate points $X = (x_1, x_2)^T$ about the origin in $\mathbb{R}^2$ to their new positions $X' = R(\theta)X$. The group is often referred to as the proper rotation group in 2 dimensions.

Since every orthogonal matrix $R$ satisfies $\det(R) = \pm 1$ we can define the more general group O(2) consisting of all real orthogonal $2 \times 2$ matrices. O(2) contains SO(2) as a subgroup and also contains the improper rotation matrices of the form $SR(\theta)$ where $S = \mathrm{diag}(1, -1)$ is the diagonal reflection matrix.

From (B.9) the infinitesimal generator and basis vector for the 1-dimensional Lie algebra so(2) is given by

$$A = \left.\frac{dR(\theta)}{d\theta}\right|_{\theta=0} = \begin{pmatrix} 0 & 1 \\ -1 & 0 \end{pmatrix}. \tag{B.16}$$

Since $A^{2n} = (-1)^n I_2$ and $A^{2n+1} = (-1)^n A$ it follows that each element of the one-parameter group can be expressed in the form

$$R(\theta) = e^{\theta A}. \tag{B.17}$$

These results can easily be generalized to 3 dimensions.

## B.4   Rotation Group SO(3)

In 3 dimensions every proper rotation can be expressed in terms of rotations about the $x_1$, $x_2$ and $x_3$ axes. There are now three one-parameter subgroups whose matrices are

$$R_1(\alpha) = \begin{pmatrix} 1 & 0 & 0 \\ 0 & \cos\alpha & -\sin\alpha \\ 0 & \sin\alpha & \cos\alpha \end{pmatrix}, \tag{B.18}$$

$$R_2(\beta) = \begin{pmatrix} \cos\beta & 0 & \sin\beta \\ 0 & 1 & 0 \\ -\sin\beta & 0 & \cos\beta \end{pmatrix}, \tag{B.19}$$

$$R_3(\gamma) = \begin{pmatrix} \cos\gamma & -\sin\gamma & 0 \\ \sin\gamma & \cos\gamma & 0 \\ 0 & 0 & 1 \end{pmatrix}. \tag{B.20}$$

The matrices of each of these subgroups also have properties (B.12) to (B.15). From (B.9) we obtain the infinitesimal generators

$$A_1 = \left.\frac{dR_1(\alpha)}{d\alpha}\right|_{\alpha=0} = \begin{pmatrix} 0 & 0 & 0 \\ 0 & 0 & -1 \\ 0 & 1 & 0 \end{pmatrix}, \tag{B.21}$$

$$A_2 = \left.\frac{dR_2(\beta)}{d\beta}\right|_{\beta=0} = \begin{pmatrix} 0 & 0 & 1 \\ 0 & 0 & 0 \\ -1 & 0 & 0 \end{pmatrix}, \tag{B.22}$$

$$A_3 = \left.\frac{dR_3(\gamma)}{d\gamma}\right|_{\gamma=0} = \begin{pmatrix} 0 & -1 & 0 \\ 1 & 0 & 0 \\ 0 & 0 & 0 \end{pmatrix}. \tag{B.23}$$

These matrices satisfy the defining commutation relations

$$[A_1, A_2] = A_3, \quad [A_2, A_3] = A_1, \quad [A_3, A_1] = A_2 \tag{B.24}$$

of the Lie algebra so(3). It follows from (B.8) that in some neighborhood of the identity ($\alpha = \beta = \gamma = 0$) a general rotation can be expressed as

$$R(\alpha, \beta, \gamma) = e^{\alpha A_1 + \beta A_2 + \gamma A_3} = e^{\alpha' A_1} e^{\beta' A_2} e^{\gamma' A_3}. \tag{B.25}$$

The three one-parameter subgroups can also be expressed as $R_1(\alpha) = e^{-i\alpha L_1}$, $R_2(\beta) = e^{-i\beta L_2}$, $R_3(\gamma) = e^{-i\gamma L_3}$ in terms of the matrix representation

$$L_1 = \begin{pmatrix} 0 & 0 & 0 \\ 0 & 0 & -i \\ 0 & i & 0 \end{pmatrix}, \ L_2 = \begin{pmatrix} 0 & 0 & i \\ 0 & 0 & 0 \\ -i & 0 & 0 \end{pmatrix}, \ L_3 = \begin{pmatrix} 0 & -i & 0 \\ i & 0 & 0 \\ 0 & 0 & 0 \end{pmatrix}, \tag{B.26}$$

of the orbital angular momentum components:

$$[L_1, L_2] = iL_3, \quad [L_2, L_3] = iL_1, \quad [L_3, L_1] = iL_2. \tag{B.27}$$

## B.5   Special Unitary Group SU(2)

The special unitary group SU(2) consists of all unitary, unimodular $2 \times 2$ matrices. Every $g \in$ SU(2) has the form

$$g = \begin{pmatrix} a & b \\ -b^* & a^* \end{pmatrix} = \begin{pmatrix} a_0 + ia_3 & a_2 + ia_1 \\ -a_2 + ia_1 & a_0 - ia_3 \end{pmatrix}, \tag{B.28}$$

$$\det(g) = |a|^2 + |b|^2 = a_0^2 + a_1^2 + a_2^2 + a_3^2 = 1, \tag{B.29}$$

where $a_0$, $a_1$, $a_2$ and $a_3$ are real. It follows that SU(2) is a real 3-parameter Lie group whose matrices have the form

$$g = a_0 \begin{pmatrix} 1 & 0 \\ 0 & 1 \end{pmatrix} + a_1 \begin{pmatrix} 0 & i \\ i & 0 \end{pmatrix} + a_2 \begin{pmatrix} 0 & 1 \\ -1 & 0 \end{pmatrix} + a_3 \begin{pmatrix} i & 0 \\ 0 & -i \end{pmatrix}. \tag{B.30}$$

The last three matrices are related to the Pauli spin matrices so we obtain the compact expression

$$g = a_0 I_2 + i a \cdot \sigma, \tag{B.31}$$

where $a = (a_1, a_2, a_3)$ and $\sigma = (\sigma_1, \sigma_2, \sigma_3)$ is the vector of Pauli spin matrices. Defining

$$a_0 = \cos \frac{\omega}{2}, \tag{B.32}$$

$$a = \sqrt{a \cdot a} = \sin \frac{\omega}{2}, \tag{B.33}$$

the group elements can be expressed as

$$g(\omega, \hat{a}) = \left( \cos \frac{\omega}{2} \right) I_2 + i \left( \sin \frac{\omega}{2} \right) \hat{a} \cdot \sigma, \tag{B.34}$$

where $\hat{a} = (a_1/a, a_2/a, a_3/a)$ is the unit vector in the direction of $a$.

We can now relate the elements of SU(2) to rotations in $\mathbb{R}^3$ where $\hat{a}$ specifies the direction of the rotation axis and $\omega$ is the rotation angle. Since $(\hat{a} \cdot \sigma)^{2n} = I_2$ and $(\hat{a} \cdot \sigma)^{2n+1} = \hat{a} \cdot \sigma$, the elements of SU(2) can be expressed in the exponential form

$$g(\omega, \hat{a}) = \exp \left( i \frac{\omega}{2} \hat{a} \cdot \sigma \right). \tag{B.35}$$

Therefore there are three 1-parameter subgroups given by $g_1(\alpha) = g(\alpha, 1, 0, 0)$, $g_2(\beta) = g(\beta, 0, 1, 0)$ and $g_3(\gamma) = g(\gamma, 0, 0, 1)$ whose infinitesimal generators are $S_j = \frac{1}{2}\sigma_j$, $j = 1, 2, 3$ and satisfy the su(2) commutation relations (also satisfied by the $L_j$ in (B.26))

$$[S_1, S_2] = iS_3, \quad [S_2, S_3] = iS_1, \quad [S_3, S_1] = iS_2. \tag{B.36}$$

With respect to these commutation relations the Lie algebra su(2) consists of all real linear combinations of the Pauli spin matrices and can also be characterized as the set of traceless $2 \times 2$ complex Hermitian matrices: every $A \in$ su(2) can be expressed as

$$A = x \cdot \sigma, \quad x \in \mathbb{R}^3. \tag{B.37}$$

Even though the Lie algebras so(3) and su(2) are the same, the corresponding Lie groups are not isomorphic as will be shown in the next section.

## B.6  Mapping from SU(2) to SO(3)

From (B.34) it follows that the correspondence between SU(2) and SO(3) is 2-to-1 since the angles $\omega_0$ and $\omega_0 + 2\pi$, $0 \le \omega_0 \le 2\pi$, correspond to distinct elements of SU(2) but to the same element of SO(3). This relationship can be made more explicit.

Each $G \in$ SU(2) defines a Lie algebra homomorphism denoted by $\boldsymbol{T}_G$ : su(2) $\rightarrow$ su(2) and defined by

$$\boldsymbol{T}_G(A) = B \equiv GAG^{-1}, \quad A, B \in \text{su}(2). \tag{B.38}$$

Defining $A = \boldsymbol{x} \cdot \boldsymbol{\sigma}$ and $B = \boldsymbol{y} \cdot \boldsymbol{\sigma}$ for $\boldsymbol{x}, \boldsymbol{y} \in \mathbb{R}^3$ we can associate with each $G$ the linear transformation on $\mathbb{R}^3$ defined by

$$\boldsymbol{y} = R_G \boldsymbol{x}. \tag{B.39}$$

Letting $G = a_0 I_2 + i\boldsymbol{a} \cdot \boldsymbol{\sigma}$, then $G^{-1} = a_0 I_2 - i\boldsymbol{a} \cdot \boldsymbol{\sigma}$. Substituting into (B.38) we obtain

$$\boldsymbol{y} = (1 - 2a^2)\boldsymbol{x} + 2(\boldsymbol{a} \cdot \boldsymbol{x})\boldsymbol{a} - 2a_0(\boldsymbol{a} \times \boldsymbol{x}), \tag{B.40}$$

$$y_j = \sum_k (R_G)_{jk} x_k, \ j, k = 1, 2, 3, \text{ where} \tag{B.41}$$

$$(R_G)_{jk} = (1 - 2a^2)\delta_{jk} + 2a_j a_k + 2a_0 \sum_\ell \epsilon_{jk\ell} a_\ell. \tag{B.42}$$

From these results it can be shown that $R_G$ is orthogonal $(R_G R_G^T = I_3)$ and that $\det(R_G) = 1$. Therefore $R_G \in$ SO(3). Finally we note that $-G \in$ SU(2) and $R_{-G} = R_G$ so the correspondence is 2-to-1.

## B.7  SO($n$) and its Lie Algebra so($n$)

Rotations in 2 and 3 dimensions can be defined as matrix transformation groups which leave the quadratic forms $x_1^2 + x_2^2$ and $x_1^2 + x_2^2 + x_3^2$ invariant, where $(x_1, x_2) \in \mathbb{R}^2$ and $(x_1, x_2, x_3) \in \mathbb{R}^3$. This definition is easily generalized to rotations in $n$-dimensional Euclidean space $\mathbb{R}^n$. Consider the set of $n \times n$ matrices which preserve the quadratic form

$$Q(x) = x_1^2 + x_2^2 + \cdots + x_n^2 = x^T x, \tag{B.43}$$

where $x$ denotes the column vector $(x_1, x_2, \ldots, x_n)$ and $x^T$ denotes its transpose, a row vector. The invariance of $Q(x)$ under a transformation by a matrix $A$ can be expressed using matrix multiplication as

$$Q(Ax) = (Ax)^T (Ax) = x^T A^T A x = x^T x = Q(x). \tag{B.44}$$

Therefore $Q$ is invariant if $A$ is an orthogonal matrix:

$$A^T A = I_n, \tag{B.45}$$

where $I_n$ is the $n \times n$ identity matrix (the converse is also true) and we define the orthogonal Lie group $O(n)$ as

$$O(n) = \{A \in GL(n,\mathbb{R}): A^T A = I_n\}, \qquad (B.46)$$

where $GL(n,\mathbb{R})$ is the general linear group of all invertible $n \times n$ real matrices. The special orthogonal group $SO(n)$ is defined as the subgroup of matrices having unit determinant:

$$SO(n) = \{A \in O(n): \det(A) = 1\}. \qquad (B.47)$$

Thus the reflections are not included in $SO(n)$. We note that $O(n)$ and $SO(n)$ are indeed subgroups of $GL(n,\mathbb{R})$ (see Exercise B.6).

To characterize the Lie algebra $so(n)$ write

$$A = e^{\mathcal{L}}. \qquad (B.48)$$

Then from (B.45) it follows that

$$A^T = (e^{\mathcal{L}})^T = e^{\mathcal{L}^T} = (e^{\mathcal{L}})^{-1} = e^{-\mathcal{L}},$$

so $so(n)$ is the set of all $n \times n$ antisymmetric matrices:

$$so(n) = \{\mathcal{L} \in gl(n,\mathbb{R}): \mathcal{L}^T = -\mathcal{L}\}, \qquad (B.49)$$

where $gl(n,\mathbb{R})$ is the general linear Lie algebra (see Chapter 1).

The Lie algebra $so(n)$ has $n(n-1)/2$ independent generators (see Exercise B.1) which can be expressed in terms of the elementary $n \times n$ matrices $E_{jk}$, having 1 in position $(jk)$ and 0 elsewhere (see (1.5)), as

$$\mathcal{L}_{jk} = E_{jk} - E_{kj}, \quad 1 \le j < k \le n. \qquad (B.50)$$

We can extend this definition to all indices $j$ and $k$ by defining

$$\mathcal{L}_{kj} = -\mathcal{L}_{jk}, \quad 1 \le j,k \le n. \qquad (B.51)$$

The defining commutation relations can be obtained from the defining commutation relations (1.6) for $gl(n,\mathbb{R})$. The result is (see Exercise B.2)

$$[\mathcal{L}_{jk}, \mathcal{L}_{\ell m}] = -(\delta_{j\ell}\mathcal{L}_{km} + \delta_{km}\mathcal{L}_{j\ell} - \delta_{jm}\mathcal{L}_{k\ell} - \delta_{k\ell}\mathcal{L}_{jm}). \qquad (B.52)$$

As mentioned in Chapter 1 it is usually more convenient in quantum mechanics to use the generators

$$L_{jk} = -i\mathcal{L}_{jk}. \qquad (B.53)$$

This corresponds to writing $A = e^{iL}$ instead of (B.48) and makes the generators $L_{jk}$ antihermitian ($L_{jk}^\dagger = -L_{kj}$) rather than antisymmetric. Then $A$ is a unitary matrix. Of course this does not change the fact that so($n$) is a real Lie algebra (see Chapter 1).

The defining commutation relations are now expressed as

$$[L_{jk}, L_{\ell m}] = i(\delta_{j\ell}L_{km} + \delta_{km}L_{j\ell} - \delta_{jm}L_{k\ell} - \delta_{k\ell}L_{jm}). \qquad (B.54)$$

If all indices $j$, $k$, $\ell$ amd $m$ are unequal then $[L_{jk}, L_{\ell m}] = 0$. Otherwise, from (B.51), we can assume that $\ell = j$ to obtain (no sum over j)

$$[L_{jk}, L_{jm}] = i(L_{km} - \delta_{jm}L_{kj} - \delta_{kj}L_{jm}).$$

Also if $k = m$ then these commutators are zero so all nonzero commutators can be obtained from

$$[L_{jk}, L_{jm}] = iL_{km}, \quad j \neq k \neq m, \text{ (no sum on } j) \qquad (B.55)$$

and the antisymmetry property $L_{jk} = -L_{kj}$.

We have used $n \times n$ transformation matrices to obtain the defining representation and commutation relations of so($n$). Other realizations are possible. In particular the orbital angular momentum realization in terms of position and momentum coordinates given by

$$L_{jk} = x_j p_k - x_k p_j \qquad (B.56)$$

is important in quantum mechanics and is discussed in Chapter 4 (compare (B.54) and (4.109)). Since $\boldsymbol{p} = -i\nabla$, (B.56) corresponds to the original antisymmetric generators

$$\mathcal{L}_{jk} = x_j \frac{\partial}{\partial x_k} - x_k \frac{\partial}{\partial x_j}, \qquad (B.57)$$

which satisfy the commutation relations (B.52). Another important realization, in case $n = 4$, defined in terms of the orbital angular momentum vector $\boldsymbol{L}$ and the scaled Laplace-Runge-Lenz vector $\boldsymbol{A}$, is used in Chapter 5 and Chapter 6.

## B.7.1  Special case: so(3)

The general results correspond to the familiar angular momentum case if we define

$$L_1 = L_{23}, \quad L_2 = L_{31}, \quad L_3 = L_{12} \qquad (B.58)$$

to obtain the defining commutation relations (1.14). The Casimir operator for so(3) is given by

$$\mathcal{C}_1 = \frac{1}{2} \sum_{j,k=1}^{3} L_{jk} L_{jk} = \sum_{j<k} L_{jk} L_{jk}$$

$$= L_{12}^2 + L_{13}^2 + L_{23}^2 = L_1^2 + L_2^2 + L_3^2 = L^2. \qquad (B.59)$$

In fact it can easily be shown that (see Exercise B.3)

$$\mathcal{C}_1 = \frac{1}{2} \sum_{j,k=1}^{n} L_{jk} L_{jk} \qquad (B.60)$$

is one of the Casimir operators for so($n$) for any $n$ ($[\mathcal{C}_1, L_{jk}] = 0$).

### B.7.2  Special case: so(4)

In this case there are $(4)(3)/2 = 6$ independent generators for the Lie algebra
so(4). If we define

$$L_1 = L_{23}, \quad L_2 = L_{31}, \quad L_3 = L_{12},$$
$$A_1 = L_{14}, \quad A_2 = L_{24}, \quad A_3 = L_{34}, \tag{B.61}$$

then these six generators satisfy the defining commutation relations (5.129) to
(5.131) (see Exercise B.4).  The correspondence (B.61) can also be conveniently
displayed as

$$L_{jk} \iff \begin{bmatrix} 0 & L_3 & -L_2 & A_1 \\ & 0 & L_1 & A_2 \\ & & 0 & A_3 \\ & & & 0 \end{bmatrix}, \tag{B.62}$$

extended to the lower half using antisymmetry.

For so(4) there are two independent Casimir operators given by (B.60) and

$$C_2 = \frac{1}{8} \sum_{j,k,\ell,m=1}^{4} \epsilon_{jk\ell m} L_{jk} L_{\ell m}, \tag{B.63}$$

where $\epsilon_{jk\ell m}$ is the 4-index Levi-Civita symbol defined so that $\epsilon_{jk\ell m} = 1$ if
$(jk\ell m)$ is an even permutation of (1234), and $\epsilon_{jk\ell m} = -1$ if $(jk\ell m)$ is an odd
permutation of (1234).  In terms of the correspondence (B.62) and the vector
notation $\boldsymbol{L} = (L_1, L_2, L_3)$ and $\boldsymbol{A} = (A_1, A_2, A_3)$ they can be expressed as (see
Exercise B.5)

$$C_1 = \boldsymbol{L} \cdot \boldsymbol{L} + \boldsymbol{A} \cdot \boldsymbol{A} = L^2 + A^2, \tag{B.64}$$
$$C_2 = \tfrac{1}{2}(\boldsymbol{L} \cdot \boldsymbol{A} + \boldsymbol{A} \cdot \boldsymbol{L}). \tag{B.65}$$

## B.8  SO($p,q$) and its Lie Algebra so($p,q$)

If $n = p+q$ then the real Lie group SO($p,q$) can be defined as the transformation
group of $n \times n$ matrices with unit determinant which preserve the quadratic
form[1]

$$Q(x) = \sum_{j=1}^{p} x_j^2 - \sum_{j=p+1}^{n} x_j^2 = x^T G x, \tag{B.66}$$

where $x$ and $x^T$ are defined in the preceding section and $G$ is the diagonal
metric matrix

$$G = \begin{pmatrix} I_p & 0 \\ 0 & -I_q \end{pmatrix} = \mathrm{diag}(1, \ldots, 1, -1, \ldots, -1), \tag{B.67}$$

---

[1]Some authors use the opposite signs in (B.66) which reverses all the signs in $G$.

defined in terms of the $p \times p$ and $q \times q$ identity matrices $I_p$ and $I_q$.

If $A \in \mathrm{SO}(p,q)$ then the invariance of $\mathcal{Q}$ can be expressed as

$$\mathcal{Q}(Ax) = (Ax)^T G(Ax) = x^T(A^T G A)x, \tag{B.68}$$

so we define (compare with (B.46) and (B.47))

$$\mathrm{O}(p,q) = \{A \in \mathrm{GL}(n,\mathbb{R}): A^T G A = G\}, \tag{B.69}$$

$$\mathrm{SO}(p,q) = \{A \in \mathrm{O}(p,q): \det A = 1\}. \tag{B.70}$$

We note that the matrices in $\mathrm{O}(p,q)$ and $\mathrm{SO}(p,q)$ are indeed subgroups of $\mathrm{GL}(n,\mathbb{R})$ (see Exercise B.6).

As in the preceding section we can write $A = e^{\mathcal{L}}$ to obtain (using $G^2 = I_n$ and $G^{-1} = G$)

$$A^T = e^{\mathcal{L}^T} = GA^{-1}G = Ge^{-\mathcal{L}}G = Ge^{-\mathcal{L}}G^{-1}$$
$$= e^{-G\mathcal{L}G^{-1}} = e^{-G\mathcal{L}G}.$$

Therefore $\mathcal{L}^T = -G\mathcal{L}G$ and we can define the real Lie algebra so($p,q$) by

$$\mathrm{so}(p,q) = \{\mathcal{L} \in \mathrm{gl}(n,\mathbb{R}): \mathcal{L}^T = -G\mathcal{L}G\}. \tag{B.71}$$

To obtain generators for so($p,q$) write $\mathcal{L}$ in the block form

$$\mathcal{L} = \begin{pmatrix} A & C \\ D & B \end{pmatrix},$$

where $A$ is a $p \times p$ matrix and $B$ is a $q \times q$ matrix.  Then the condition $\mathcal{L}^T = -G\mathcal{L}G$ implies that

$$\mathcal{L} = \begin{pmatrix} A & C \\ C^T & B \end{pmatrix}, \tag{B.72}$$

where $A$ and $B$ are antisymmetric matrices and $C$ is an arbitrary $p \times q$ matrix. It then follows that there are $n(n-1)/2$ independent generators for so($p,q$) which can be expressed in terms of the elementary matrices $E_{jk}$ by (see Exercise B.7)

$$\mathcal{L}_{jk} = \begin{cases} E_{jk} - E_{kj}, & 1 \leq j < k \leq p, \\ E_{kj} - E_{jk}, & p+1 \leq j < k \leq n, \\ E_{jk} + E_{kj}, & 1 \leq j \leq p, p+1 \leq k \leq n \end{cases} \tag{B.73}$$

As in (B.51) we also define $\mathcal{L}_{kj} = -\mathcal{L}_{jk}$ so that $\mathcal{L}_{jk}$ is defined for all index pairs $(jk)$. All cases can be conveniently expressed in terms of the matrix elements $g_{jk}$ of the diagonal metric matrix $G$ by

$$\mathcal{L}_{jk} = g_{j\ell}E_{\ell k} - g_{k\ell}E_{\ell j}. \tag{B.74}$$

We can now obtain the defining so($p,q$) commutation relations (see Exercise B.8)

$$[\mathcal{L}_{jk}, \mathcal{L}_{\ell m}] = -(g_{j\ell}\mathcal{L}_{km} + g_{km}\mathcal{L}_{j\ell} - g_{jm}\mathcal{L}_{k\ell} - g_{k\ell}\mathcal{L}_{jm}), \qquad \text{(B.75)}$$

so the metric matrix elements $g_{jk}$ play the same role for so($p,q$) as the identity matrix elements $\delta_{jk}$ do for so($n$). Also, from (B.53), the alternate form is given by

$$[L_{jk}, L_{\ell m}] = i(g_{j\ell}L_{km} + g_{km}L_{j\ell} - g_{jm}L_{k\ell} - g_{k\ell}L_{jm}). \qquad \text{(B.76)}$$

The nonzero commutators can be expressed in the simple form (compare with (B.55))

$$[L_{jk}, L_{jm}] = ig_{jj}L_{km}, \quad j \neq k \neq m, \text{ (no sum on } j). \qquad \text{(B.77)}$$

It can also be shown that the second order Casimir operator for so($p,q$) is given by (see Exercise B.9)

$$Q_2 = \tfrac{1}{2}L_{ab}L^{ab}, \qquad \text{(B.78)}$$

where $L^{ab}$ is related to $L_{ab}$ by

$$L^{ab} = g_{ac}g_{bd}L_{cd}, \qquad \text{(B.79)}$$

and summation is implied over all indices in (B.78) and (B.79). We now consider three important cases

### B.8.1   Special case: so(2,1)

If we define

$$T_1 = L_{23}, \quad T_2 = L_{31}, \quad T_3 = L_{12},$$

and use (B.77) and $L_{kj} = -L_{jk}$ then the standard so(2,1) commutation relations (4.1) are obtained. There is only one Casimir operator for so(2,1) given by (B.78) (compare with (4.2))

$$\begin{aligned}
Q_2 &= \sum_{a<b}^{3} g_{aa}g_{bb}L_{ab}L_{ab} \\
&= g_{11}g_{22}L_{12}L_{12} + g_{11}g_{33}L_{13}L_{13} + g_{22}g_{33}L_{23}L_{23} \\
&= L_{12}^2 - L_{13}^2 - L_{23}^2 = T_3^2 - T_1^2 - T_2^2.
\end{aligned}$$

### B.8.2   Special case: so(4,1)

Here there are $5(4)/2 = 10$ generators. In Chapter 6 the scaled hydrogenic re-
alization of these ten generators is derived using the operator set $\{\boldsymbol{L}, \boldsymbol{A}, \boldsymbol{B}, T_2\}$
satisfying the commutation relations (6.3) to (6.11). The correspondence with
the generators $L_{jk}$ can be conveniently displayed as

$$
L_{jk} \Longleftrightarrow
\begin{bmatrix}
0 & L_3 & -L_2 & A_1 & B_1 \\
  & 0 & L_1 & A_2 & B_2 \\
  &   & 0 & A_3 & B_3 \\
  &   &   & 0 & T_2 \\
  &   &   &   & 0
\end{bmatrix},
\tag{B.80}
$$

extended to the lower half using antisymmetry. This notation makes it clear
that so(4) is a subalgebra (compare with (B.62)).

The Lie algebra so(4,1) has two independent Casimir operators. The second
order one is given by (B.78) and there is also a fourth order one given by the
more complicated expression [BO66], [ST65]

$$
Q_4 = \sum_{j,k=1}^{5} g_{jk} w_j w_k,
\tag{B.81}
$$

where

$$
w_j = \frac{1}{8} \sum_{a,b,c,d=1}^{5} \epsilon_{jabcd} L_{ab} L_{cd}.
\tag{B.82}
$$

Under the correspondence (B.80) $Q_2$ and $Q_4$ reduce to the expressions $Q$ and
$W$ defined in (6.13) and (6.14) In fact (see Exercise B.10)

$$
\begin{aligned}
w_j &= (T_2 \boldsymbol{L} - \boldsymbol{A} \times \boldsymbol{B})_j, \quad j = 1, 2, 3, \\
w_4 &= -\boldsymbol{L} \cdot \boldsymbol{B}, \quad w_5 = \boldsymbol{L} \cdot \boldsymbol{A}.
\end{aligned}
\tag{B.83}
$$

### B.8.3   Special case: so(4,2)

Here there are $6(5)/2 = 15$ generators. In Chapter 6 the scaled hydrogenic real-
ization of these 15 generators is derived using the operator set $\{\boldsymbol{L}, \boldsymbol{A}, \boldsymbol{B}, \boldsymbol{\Gamma}, T_1,$
$T_2, T_3\}$ satisfying the commutation relations (6.25) to (6.37). As in the so(4,1)
case the correspondence with the generators $L_{jk}$ is

$$
L_{jk} \Longleftrightarrow
\begin{bmatrix}
0 & L_3 & -L_2 & A_1 & B_1 & \Gamma_1 \\
  & 0 & L_1 & A_2 & B_2 & \Gamma_2 \\
  &   & 0 & A_3 & B_3 & \Gamma_3 \\
  &   &   & 0 & T_2 & T_1 \\
  &   &   &   & 0 & T_3 \\
  &   &   &   &   & 0
\end{bmatrix},
\tag{B.84}
$$

extended to the lower half using antisymmetry.  The subalgebra structure
so(4,2) $\supset$ so(4,1) $\supset$ so(4) $\supset$ so(3) is clearly evident.

Again the second order Casimir operator is given by (B.78) for $n = 6$ and
metric matrix $G = \mathrm{diag}(1,1,1,1,-1,-1)$.  There are two additional independent Casimir operators, a third order one and a fourth order one defined by
[BA71a,b]

$$Q_3 = \tfrac{1}{48}\epsilon_{abcdef}L^{ab}L^{cd}L^{ef}, \qquad\qquad (B.85)$$

$$Q_4 = L_{ab}L^{bc}L_{cd}L^{da}, \qquad\qquad (B.86)$$

where $\epsilon_{abcdef}$ is fully antisymmetric in all indices with $\epsilon_{123456} = 1$, summation
over all indices is implied and $L^{ab}$ is related to $L_{ab}$ using (B.79).

## B.9    Exercises

$\otimes$ **Exercise B.1** Show that the Lie algebra so($n$) has $n(n-1)/2$ independent
generators given in terms of the $n \times n$ elementary matrices $E_{jk}$ (see (1.5) and
(1.6)) by $\mathcal{L}_{jk} = E_{jk} - E_{kj}$, $1 \le j < k \le n$.

$\otimes$ **Exercise B.2** Derive the defining commutation relations (B.52) for so($n$)
using the commutation relations (1.6) for the elementary matrices $E_{jk}$.

$\otimes$ **Exercise B.3** Show that (B.60) is one of the Casimir operators of so($n$).

$\otimes$ **Exercise B.4** In case $n = 4$ show that the definitions (B.61) give the
commutation relations (5.129) to (5.131).

$\otimes$ **Exercise B.5** In case $n = 4$ show that the operator $C_2$ defined by (B.63)
is a Casimir operator.

$\otimes$ **Exercise B.6** Show that O($n$), SO($n$), O($p,q$) and SO($p,q$) satisfy the
group properties (a) to (d) of Section B.1.

$\otimes$ **Exercise B.7** Show that the matrices (B.73) form a basis for the Lie algebra so($p,q$).

$\otimes$ **Exercise B.8** Derive the defining commutation relations (B.75), or equivalently (B.76), for so($p,q$) using the commutation relations (1.6) for the elementary matrices $E_{jk}$ and the definition (B.74).

$\otimes$ **Exercise B.9** Show that $Q_2$ defined by (B.78) is a Casimir operator for
so($p,q$).

$\otimes$ **Exercise B.10** Show that the fourth order so(4,1) Casimir operator $Q_4$
defined by (B.80) and (B.81) reduces to (6.14) under the correspondence (B.79).

# Appendix C

# The Tilting Transformation

## C.1   Transformation of Functions

The scaling transformation

$$r = \gamma R, \quad p_r = \frac{1}{\gamma} P_R, \tag{C.1}$$

or more generally, in vector form

$$\boldsymbol{r} = \gamma \boldsymbol{R}, \quad \boldsymbol{p} = \frac{1}{\gamma} \boldsymbol{P}, \tag{C.2}$$

is used in Chapter 4 to convert a conventional hamiltonian eigenvalue problem
into a $T_3$ eigenvalue problem. It is also used in Chapter 5 to obtain the scaled
Laplace-Runge-Lenz (LRL) vector from the modified LRL vector. This simple
approach from the passive viewpoint uses two coordinate systems and has the
advantage of being simpler to introduce and apply than the so called "tilting
transformation" commonly used in the literature [AD82], [AD88b], [BA71a],
[WY74]. The tilting transformation is just an active approach to scaling in
operator form.

An operator $T_S$ is a scaling transformation if

$$T_S f(r) = c f(e^{\alpha} r), \tag{C.3}$$

where $\gamma = e^{\alpha}$ is the scaling parameter and $c$ is some proportionality constant.
Since (see Exercise C.1)

$$f(e^{\alpha} r) = e^{\alpha r \frac{d}{dr}} f(r), \tag{C.4}$$

we can choose $T_S$ to be the exponential operator

$$T_S = c e^{\alpha r \frac{d}{dr}}. \tag{C.5}$$

Using the realization (4.53) to (4.55) of the so(2,1) generators with $\{r, p_r\}$
rather than $\{R, P_R\}$ we can express $T_S$ in terms of the unitary operator $e^{i a \alpha T_2}$.

In $D$-dimensional space $p_r$ is defined by (4.105) and

$$T_2 = \frac{1}{2}\left[ rp_r - \frac{i}{a}(a-1)\right]$$
$$= -\frac{i}{a}r\frac{\partial}{\partial r} - i\left[\frac{D+a-2}{2a}\right]$$
$$= -\frac{i}{a}r\frac{\partial}{\partial r} - iG, \qquad (C.6)$$

using the definition (4.136). Therefore, as an operator acting on functions of $r$,

$$T_S = e^{ia\alpha T_2} = e^{a\alpha G}e^{\alpha r\frac{d}{dr}}, \qquad (C.7)$$

and

$$\widetilde{f(r)} = T_S f(r) = e^{a\alpha G}f(e^{\alpha}r). \qquad (C.8)$$

## C.2  Transformation of Operators

The scaling transformation of an operator $X$ can now be defined as

$$\widetilde{X} = e^{ia\alpha T_2} X e^{-ia\alpha T_2}. \qquad (C.9)$$

We need the scaled versions of the operators $r$, $p_r$ and $T_3$ in order to apply the scaling transformation to the general radial equation (4.199). It is shown in Exercise C.2 that

$$\widetilde{r} = e^{ia\alpha T_2} r e^{-ia\alpha T_2} = e^{\alpha}r, \qquad (C.10)$$
$$\widetilde{p_r} = e^{ia\alpha T_2} p_r e^{-ia\alpha T_2} = e^{-\alpha}p_r, \qquad (C.11)$$
$$\widetilde{\boldsymbol{r}} = e^{ia\alpha T_2} \boldsymbol{r} e^{-ia\alpha T_2} = e^{\alpha}\boldsymbol{r}, \qquad (C.12)$$
$$\widetilde{\boldsymbol{p}} = e^{ia\alpha T_2} \boldsymbol{p} e^{-ia\alpha T_2} = e^{-\alpha}\boldsymbol{p}, \qquad (C.13)$$
$$\widetilde{T_3} = e^{ia\alpha T_2} T_3 e^{-ia\alpha T_2} = T_3 \cosh a\alpha - T_1 \sinh a\alpha. \qquad (C.14)$$

The first four transformations are the active versions of (C.1) and (C.2).

## C.3  Transformation of the Radial Equation

### C.3.1  General case

Now consider the radial equation (4.202)

$$\left[\frac{1}{2a^2}p_r^2 + \frac{\tau}{2r^2} - \sigma r^{a-2} - \varepsilon r^{2a-2}\right]f(r) = 0. \qquad (C.15)$$

Multiply by $2r^{2-a}$ to obtain

$$\left[\frac{1}{a^2}r^{2-a}p_r^2 + \frac{\tau}{r^a} - \sigma - 2\varepsilon r^a\right]f(r) = 0.$$

From the realization (4.53) to (4.55) the first two terms are just $T_1 + T_3$ and in the last term $r^a = T_3 - T_1$. Therefore

$$[(1 - 2\varepsilon)T_3 + (1 + 2\varepsilon)T_1 - 2\sigma] f(r) = 0. \tag{C.16}$$

From the identity $\cosh^2 x - \sinh^2 x = 1$ it follows that

$$\cosh a\alpha = \frac{1 - 2\varepsilon}{\sqrt{-8\varepsilon}}, \quad \sinh a\alpha = \frac{1 + 2\varepsilon}{\sqrt{-8\varepsilon}}. \tag{C.17}$$

Therefore, from (C.14) with $\alpha$ replaced by $-\alpha$, (C.16) can be expressed as

$$\left[\frac{1 - 2\varepsilon}{\sqrt{-8\varepsilon}}T_3 + \frac{1 + 2\varepsilon}{\sqrt{-8\varepsilon}}T_1 - \frac{2\sigma}{\sqrt{-8\varepsilon}}\right] f(r) = 0,$$

$$\left[e^{-ia\alpha T_2}T_3 e^{ia\alpha T_2} - \frac{2\sigma}{\sqrt{-8\varepsilon}}\right] f(r) = 0,$$

$$e^{-ia\alpha T_2}\left[T_3 - \frac{2\sigma}{\sqrt{-8\varepsilon}}\right] e^{ia\alpha T_2} f(r) = 0, \tag{C.18}$$

which is just a $T_3$ eigenvalue problem of the form $(T_3 - q)\widetilde{f(r)} = 0$ with

$$q = \frac{2\sigma}{\sqrt{-8\varepsilon}} = \frac{\sigma}{\sqrt{-2\varepsilon}}. \tag{C.19}$$

From (C.17) and the definition $\gamma = e^\alpha$ of the scaling parameter,

$$\gamma^{2a} = e^{2a\alpha} = (\cosh a\alpha + \sinh a\alpha)^2 = \frac{1}{-2\varepsilon}. \tag{C.20}$$

These results are identical to (4.201) so the active tilting transformation is equivalent to the simpler passive approach.

### C.3.2  Special case of hydrogenic atom

In the special case of the hydrogenic atom, $a = 1$, $\sigma = Z$ and $\varepsilon = E_n$ (see Table 4.1) we have for the principal quantum number, $q = n = Z\gamma$, and for the energy levels, $E_n = -\gamma^{-2}/2$, and for the scaling parameter, $\gamma = e^\alpha$. Therefore

$$\begin{aligned}
e^{i\alpha T_2}r(H - E_n)e^{-i\alpha T_2} &= e^{i\alpha T_2}r(H - E_n)e^{-i\alpha T_2} \\
&= e^{i\alpha T_2}re^{-i\alpha T_2}e^{i\alpha T_2}(H - E_n)e^{-i\alpha T_2} \\
&= \tilde{r}(\widetilde{H} - E_n) \\
&= e^{-\alpha}\tfrac{1}{2}rp^2 - Z - e^{\alpha}rE_n \\
&= e^{-\alpha}(\tfrac{1}{2}rp^2 - n + \tfrac{1}{2}r) \\
&= e^{-\alpha}(T_3 - n), \tag{C.21}
\end{aligned}$$

which shows that the original eigenvalue problem is transformed into a $T_3$ eigenvalue problem by the tilting transformation.

## C.4  Tilted Laplace-Runge-Lenz Vector

The tilting transformation (C.9) can be used to show how the scaled LRL vector $A$ defined in (5.128) is related to the LRL vector $U$ defined in (5.107) and the modified LRL vector defined in (5.103).

From the commutation relations $[T_2, A] = iB$ and $[T_2, B] = iA$ (see Section 6.2) and the exponential operator identity (2.6) it follows that (see Exercise C.2)

$$e^{-i\alpha T_2} A e^{i\alpha T_2} = A \cosh\alpha + B \sinh\alpha. \tag{C.22}$$

Using the definitions (6.39) and (6.40) of $A$ and $B$, the identity $r = B - A$, and $E_n = e^{-2\alpha}/2$,

$$\begin{aligned}
e^{-i\alpha T_2} A e^{i\alpha T_2} &= \tfrac{1}{2}e^{\alpha}(A + B) - \tfrac{1}{2}e^{-\alpha}(B - A) \\
&= e^{\alpha}(\tfrac{1}{2}rp^2 - p(r \cdot p)) - \tfrac{1}{2}e^{-\alpha}r \\
&= e^{\alpha}(U - rH) - \tfrac{1}{2}e^{-\alpha}r, \text{ from (5.107)} \\
&= e^{\alpha}(U + \tfrac{1}{2}e^{-2\alpha}r) - \tfrac{1}{2}e^{-\alpha}r \\
&= e^{-\alpha}U = V, \text{ see (5.113).}
\end{aligned}$$

Therefore

$$e^{i\alpha T_2} V e^{-i\alpha T_2} = A, \tag{C.23}$$

which shows that $A$ is really the scaled LRL vector.

## C.5  Exercises

◈ **Exercise C.1** Derive the scaling transformation (C.4).

◈ **Exercise C.2** Derive the scaling transformations (C.10) to (C.14) and (C.22) for $r$, $p_r$, $T_3$ and $A$ using the operator identity

$$e^{-B} A e^{B} = A + [A, B] + \frac{1}{2}[[A, B], B] + \cdots = \sum_{n=0}^{\infty} \frac{1}{n!}[A, B]^{(n)}$$

derived in Chapter 2.

# Appendix D

# Perturbation Matrix Elements

In our applications to hydrogenic perturbation theory we need matrix elements
of $R$, $R^2$, $R^3$, in the D-dimensional case, and matrix elements of $Z$, $Z^2$, $RZ$ and
$R(R^2 - Z^2)$ in the 3-dimensional case. They are conveniently collected together
here in tables derived using the Maple computer algebra system [CH91a,b,c].

## D.1   Basic Formulas for Matrix Elements

### D.1.1   3-dimensional case

All of these matrix elements can be obtained from the matrix elements of $R$
and $Z$ by matrix multiplication. The matrix elements of $R$ and $Z$ have been
calculated in Chapter 6 (see (6.85) to (6.90)). The results are given in the
first two rows of Table D.1 and in Table D.2 where we have used the compact
notation

$$R_{n\ell}^{n+\mu,\ell} = \langle n + \mu, \ell m|R|n\ell m\rangle, \tag{D.1}$$

$$Z_{n\ell m}^{n+\mu,\ell+\nu} = \langle n + \mu, \ell + \nu, m|Z|n\ell m\rangle. \tag{D.2}$$

Only two of the three kinds of matrix elements of $R$, $\mu = -1, 0, 1$, are
given in Table D.1 and only three of the six kinds of matrix elements of $Z$,
$\mu = -1, 0, 1$, $\nu = -1, 1$, are given in Table D.2. The other matrix elements are
obtained from the symmetry properties

$$R_{n\ell}^{n+\mu,\ell} = R_{n+\mu,\ell}^{n\ell}, \qquad Z_{n\ell m}^{n+\mu,\ell+\nu} = Z_{n+\mu,\ell+\nu,m}^{n\ell}, \tag{D.3}$$

which follow from the hermiticity of $R$ and $Z$ and the fact that their matrix
elements are real.

Higher powers of $R$ are obtained by matrix multiplication. For example,

$$(R^2)_{n\ell}^{n+\mu,\ell} = (R^2)_{n+\mu,\ell}^{n\ell} = \sum_{\mu'} R_{n+\mu',\ell}^{n+\mu,\ell} R_{n\ell}^{n+\mu',\ell}, \tag{D.4}$$

where $|\mu| \leq 2$, $|\mu'| \leq 1$ and $|\mu - \mu'| \leq 1$, and

$$(R^3)_{n\ell}^{n+\mu,\ell} = (R^3)_{n+\mu,\ell}^{n\ell} = \sum_{\mu'} R_{n+\mu',\ell}^{n+\mu,\ell}(R^2)_{n\ell}^{n+\mu',\ell}, \tag{D.5}$$

where $|\mu| \leq 3$, $|\mu'| \leq 2$, $|\mu - \mu'| \leq 1$. The results are given in Table D.2.

Similarly matrix elements of $Z^2$ are obtained from

$$(Z^2)_{n\ell m}^{n+\mu,\ell+\nu} = (Z^2)_{n+\mu,\ell+\nu,m}^{n\ell} = \sum_{\mu',\nu'} Z_{n+\mu',\ell+\nu',m}^{n+\mu,\ell+\nu} \, Z_{n\ell m}^{n+\mu',\ell+\nu'}, \qquad (D.6)$$

where $|\mu| \leq 2$, $\nu \in \{-2,0,2\}$, $|\mu'| \leq 1$, $|\nu'| = 1$, $|\mu - \mu'| \leq 1$, $|\nu - \nu'| = 1$. They are given in Table D.3.

The matrix elements of $RZ$, needed for the Stark perturbation, are obtained from

$$(RZ)_{n\ell m}^{n+\mu,\ell+\nu} = (RZ)_{n+\mu,\ell+\nu,m}^{n\ell} = \sum_{\mu'} R_{n+\mu',\ell+\nu}^{n+\mu,\ell+\nu} \, Z_{n\ell m}^{n+\mu',\ell+\nu}, \qquad (D.7)$$

where $|\mu| \leq 2$, $|\nu| = 1$, $|\mu'| \leq 1$, $|\mu - \mu'| \leq 1$. They are given in Table D.4.

Finally the matrix elements of $U = R(R^2 - Z^2)$, needed for the Zeeman perturbation, are obtained from

$$U_{n\ell m}^{n+\mu,\ell+\nu} = U_{n+\mu,\ell+\nu,m}^{n\ell} = (R^3)_{n\ell}^{n+\mu,\ell}\delta_{\nu 0} - (RZ^2)_{n\ell m}^{n+\mu,\ell+\nu}, \qquad (D.8)$$

$$(RZ^2)_{n\ell m}^{n+\mu,\ell+\nu} = \sum_{\mu'} R_{n+\mu',\ell+\nu}^{n+\mu,\ell+\nu}(Z^2)_{n\ell m}^{n+\mu',\ell+\nu}, \qquad (D.9)$$

where $|\mu| \leq 3$, $\nu \in \{-2,0,2\}$, $|\mu'| \leq 2$, $|\mu - \mu'| \leq 1$. They are given in Tables D.5 and D.6.


### D.1.2   D-dimensional case

Since $R = T_+ - T_-$ we can use so(2,1) to do perturbation theory based on the $D$-dimensional hydrogenic atom in case the perturbation is central. Denote the matrix elements of $R$ by

$$R_{qk}^{q+\mu,k} = \langle q + \mu, k | R | qk \rangle, \qquad (D.10)$$

where we have used the notation $|qk\rangle$ rather than $|kq\rangle$ (as in Chapter 4) for the basis vectors. From Section 4.8 in the $D$-dimensional hydrogenic case we have

$$k = \ell + \frac{D-3}{2}, \qquad (D.11)$$

$$q = k + 1 + n_r = \ell + \frac{D-1}{2} + n_r, \; n_r = 0,1,2,\ldots, \qquad (D.12)$$

and the matrix elements of the first three powers of $R$ are given in Table D.7 using the symmetry property analogous to (D.3). It is clear that this is just Table D.1 but with $n$ and $\ell$ generalized to $q$ and $k$ as given in (D.11) and (D.12).

## D.2 Symbolic Calculation of Tables

The following Maple program, *rzmatrix*, was used to calculate the matrix elements for Tables D.1 to D.6.

In lines *3* to *19* the matrix elements of $R$ and $Z$ are defined as expressions in terms of $n$ and $\ell$ from the results obtained for the so(2,1) and so(4,2) Lie algebras. The notation *r1[mu]* is used for the matrix elements of $R$ in (D.1) and *z1[mu,nu]* is used for the matrix elements of $Z$ in (D.2) as functions of $n$ and $\ell$. The Maple *subs* function is used in several places to substitute new values of $n$ and $\ell$ into the expressions for the matrix elements of $R$ and $Z$ according to the symmetry properties given in (D.1) and (D.2).

In lines *23* to *31* (D.4) is used to derive formulas for the matrix elements of $R^2$ using matrix multiplication. The *factor* and *simplify* functions are also used to simplify the expressions involving square roots. Similarly the formulas for matrix elements of $R^3$, $Z^2$, $RZ$ and $U$ are obtained from (D.5) to (D.9).

It is convenient to use loops involving **in** to loop over only the specified elements of the lists *[-1,1]* or *[-2,0,2]*.

```
01   # rzmatrix
02
03   # matrix elements of r from so(2,1) Lie algebra
04
05   r1[1] := -(1/2)*((n-l)*(n+l+1))^(1/2):
06   r1[0] := n:
07   r1[-1] := subs(n=n-1, r1[1]):
08
09   # matrix elements of z from so(4,2) Lie algebra
10
11   z1[0,1] := -((2*l+1)*(2*l+3))^(-1/2)
12              * ((l-m+1)*(l+m+1)*(n-l-1)*(n+l+1))^(1/2):
13   z1[1,-1] := (1/2)*((2*l-1)*(2*l+1))^(-1/2)
14              * ((l-m)*(l+m)*(n-l+1)*(n-l))^(1/2):
15   z1[1,1] := (1/2)*((2*l+1)*(2*l+3))^(-1/2)
16              * ((l-m+1)*(l+m+1)*(n+l+1)*(n+l+2))^(1/2):
17   z1[0,-1] := subs(l=l-1, z1[0,1]):
18   z1[-1,1] := subs(n=n-1, l=l+1, z1[1,-1]):
19   z1[-1,-1] := subs(n=n-1, l=l-1, z1[1,1]):
20
21   # Matrix element formulas for r^2 from those of r
22
23   for mu from-2 to 2 do
24       s := 0;
25       for mu1 from -1 to 1 do
26           if abs(mu-mu1) <= 1 then
27               s := s + subs(n=n+mu1, r1[mu-mu1])*r1[mu1];
28           fi;
29       od;
```

```
30       r2[mu] := factor(simplify(s));
31    od:
32
33    # Matrix element formulas for r^3 from those of r^2
34
35    for mu from -3 to 3 do
36       s := 0;
37       for mu1 from -2 to 2 do
38          if abs(mu-mu1) <= 1 then
39             s := s + subs(n=n+mu1, r1[mu-mu1])*r2[mu1]
40          fi;
41       od;
42       r3[mu] := factor(simplify(s));
43    od:
44
45    # Matrix element formulas for z^2 from those of z
46
47    for mu from -2 to 2 do
48       for nu in [-2,0,2] do
49          s := 0;
50          for mu1 from -1 to 1 do
51             for nu1 in [-1,1] do
52                if abs(mu-mu1)<=1 and abs(nu-nu1)=1 then
53                   s := s + subs(n=n+mu1, l=l+nu1,
54                      z1[mu-mu1, nu-nu1]) * z1[mu1,nu1];
55                fi;
56             od;
57          od;
58          z2[mu, nu] := factor(simplify(s));
59       od;
60    od:
61
62    # Matrix element formulas for rz
63
64    for mu from -2 to 2 do
65       for nu in [-1,1] do
66          s := 0;
67          for mu1 from -1 to 1 do
68             if abs(mu-mu1) <= 1 then
69                s := s + subs(n=n+mu1, l=l+nu, r1[mu-mu1])
70                   *z1[mu1,nu];
71             fi;
72          od;
73          rz[mu,nu] := factor(simplify(s));
74       od;
75    od;
76
```

```
77    # Matrix element formulas for u := r^3 - r*z^2
78
79    for mu from -3 to 3 do
80        for nu in [-2,0,2] do
81        s := 0;
82        for mu1 from -2 to 2 do
83            if abs(mu-mu1) <= 1 then
84                s := s + subs(n=n+mu1, l=l+nu, r1[mu-mu1])
85                    * z2[mu1,nu];
86            fi;
87        od;
88        if nu = 0 then
89            u[mu,nu] := factor(simplify(r3[mu] - s));
90        else
91            u[mu,nu] := factor(simplify(-s));
92        fi;
93        od;
94    od:
95
96    save r1, r2, r3, z1, z2, rz, u, `rzdata`;
97    save r1, r2, r3, z1, z2, rz, u, `rzdata.m`;
98    quit;
```

Using the **save** statement the program creates two files for use in other programs. The first, *rzdata*, contains all the tables of matrix elements in a "human readable" form and the second, with the $m$ suffix, indicates that the file should be stored in Maple's internal format so that it can be loaded more quickly using the **read** statement

> **read** `rzdata.m`;

The file *rzdata* contains the following tables

$r1 := \mathbf{table}([$
$\quad (0) \quad = n,$
$\quad (-1) = -1/2*(n-1-\ell)\char`^(1/2)*(n+\ell)\char`^(1/2),$
$\quad (1) \quad = -1/2*(n-\ell)\char`^(1/2)*(n+\ell+1)\char`^(1/2)$
$]);$

$r2 := \mathbf{table}([$
$\quad (0) \quad = 3/2*n\char`^2-1/2*\ell-1/2*\ell\char`^2,$
$\quad (-1) = -1/2*(n-1-\ell)\char`^(1/2)*(n+\ell)\char`^(1/2)*(2*n-1),$
$\quad (1) \quad = -1/2*(n+\ell+1)\char`^(1/2)*(n-\ell)\char`^(1/2)*(2*n+1),$
$\quad (-2) = 1/4*(n-2-\ell)\char`^(1/2)*(n+\ell-1)\char`^(1/2)*(n-1-\ell)\char`^(1/2)$
$\qquad\qquad *(n+\ell)\char`^(1/2),$
$\quad (2) \quad = 1/4*(n-\ell+1)\char`^(1/2)*(n+\ell+2)\char`^(1/2)*(n-\ell)\char`^(1/2)$
$\qquad\qquad *(n+\ell+1)\char`^(1/2)$

```
]);

r3 := table([
   (0)   = 1/2*n*(5*n^2+1-3*l-3*l^2),
  (-1)  = -3/8*(n-1-l)^(1/2)*(n+l)^(1/2)
          *(5*n^2-5*n-l+2-l^2),
   (1)  = -3/8*(n+l+1)^(1/2)*(n-l)^(1/2)
          *(5*n^2-l-l^2+5*n+2),
  (-2)  = 3/4*(n-2-l)^(1/2)*(n+l-1)^(1/2)*(n-1-l)^(1/2)
          *(n+l)^(1/2)*(n-1),
   (2)  = 3/4*(n-l+1)^(1/2)*(n+l+2)^(1/2)*(n+l+1)^(1/2)
          *(n-l)^(1/2)*(n+1),
  (-3)  = -1/8*(n-3-l)^(1/2)*(n-2+l)^(1/2)*(n-2-l)^(1/2)
          *(n+l-1)^(1/2)*(n-1-l)^(1/2)*(n+l)^(1/2),
   (3)  = -1/8*(n+2-l)^(1/2)*(n+3+l)^(1/2)*(n-l+1)^(1/2)
          *(n+l+2)^(1/2)*(n-l)^(1/2)*(n+l+1)^(1/2)
]);


z1 := table([
   (1, 1)   = 1/2/(2*l+1)^(1/2)/(2*l+3)^(1/2)*(l-m+1)^(1/2)
             *(l+m+1)^(1/2)*(n+l+1)^(1/2)*(n+l+2)^(1/2),
  (-1, 1)   = 1/2/(2*l+1)^(1/2)/(2*l+3)^(1/2)*(l-m+1)^(1/2)
             *(l+m+1)^(1/2)*(n-1-l)^(1/2)*(n-2-l)^(1/2),
   (1,-1)   = 1/2/(2*l-1)^(1/2)/(2*l+1)^(1/2)*(l-m)^(1/2)
             *(l+m)^(1/2)*(n-l+1)^(1/2)*(n-l)^(1/2),
   (0,-1)   = -1/(2*l-1)^(1/2)/(2*l+1)^(1/2)*(l-m)^(1/2)
             *(l+m)^(1/2)*(n-l)^(1/2)*(n+l)^(1/2),
   (0, 1)   = -1/(2*l+1)^(1/2)/(2*l+3)^(1/2)*(l-m+1)^(1/2)
             *(l+m+1)^(1/2)*(n-1-l)^(1/2)*(n+l+1)^(1/2),
  (-1,-1)   = 1/2/(2*l-1)^(1/2)/(2*l+1)^(1/2)*(l-m)^(1/2)
             *(l+m)^(1/2)*(n+l-1)^(1/2)*(n+l)^(1/2)
]);


z2 := table([
   (2,-2)   = 1/4/(2*l-3)^(1/2)/(2*l-1)*(l-1-m)^(1/2)
             *(l-1+m)^(1/2)*(n+3-l)^(1/2)*(n+2-l)^(1/2)
             /(2*l+1)^(1/2)*(l-m)^(1/2)*(l+m)^(1/2)
             *(n-l+1)^(1/2)*(n-l)^(1/2),
   (1,-2)   = -1/(2*l-3)^(1/2)/(2*l-1)*(l-1-m)^(1/2)
             *(l-1+m)^(1/2)*(n+2-l)^(1/2)*(n-l+1)^(1/2)
             /(2*l+1)^(1/2)*(l-m)^(1/2)*(l+m)^(1/2)
             *(n-l)^(1/2)*(n+l)^(1/2),
   (1, 2)   = -1/(2*l+3)/(2*l+5)^(1/2)*(l+2-m)^(1/2)
```

$$*(\ell+2+m)\hat{}(1/2)*(n+\ell+2)\hat{}(1/2)*(n+3+\ell)\hat{}(1/2)$$
$$/(2*\ell+1)\hat{}(1/2)*(\ell-m+1)\hat{}(1/2)*(\ell+m+1)\hat{}(1/2)$$
$$*(n-1-\ell)\hat{}(1/2)*(n+\ell+1)\hat{}(1/2),$$

$(-2,\,0)\ \ = 1/4*(2*\ell\hat{}2+2*\ell-1-2*m\hat{}2)*(n-1-\ell)\hat{}(1/2)$
$\qquad\qquad *(n-2-\ell)\hat{}(1/2)*(n+\ell-1)\hat{}(1/2)*(n+\ell)\hat{}(1/2)$
$\qquad\qquad /(2*\ell+3)/(2*\ell-1),$

$(2,\,2)\ \ \ = 1/4/(2*\ell+3)/(2*\ell+5)\hat{}(1/2)*(\ell+2-m)\hat{}(1/2)$
$\qquad\qquad *(\ell+2+m)\hat{}(1/2)*(n+3+\ell)\hat{}(1/2)*(n+4+\ell)\hat{}(1/2)$
$\qquad\qquad /(2*\ell+1)\hat{}(1/2)*(\ell-m+1)\hat{}(1/2)*(\ell+m+1)\hat{}(1/2)$
$\qquad\qquad *(n+\ell+1)\hat{}(1/2)*(n+\ell+2)\hat{}(1/2),$

$(0,\,2)\ \ \ = 3/2/(2*\ell+3)/(2*\ell+5)\hat{}(1/2)*(\ell+2-m)\hat{}(1/2)$
$\qquad\qquad *(\ell+2+m)\hat{}(1/2)*(n+\ell+1)\hat{}(1/2)*(n+\ell+2)\hat{}(1/2)$
$\qquad\qquad /(2*\ell+1)\hat{}(1/2)*(\ell-m+1)\hat{}(1/2)*(\ell+m+1)\hat{}(1/2)$
$\qquad\qquad *(n-1-\ell)\hat{}(1/2)*(n-2-\ell)\hat{}(1/2),$

$(2,\,0)\ \ \ = 1/4*(2*\ell\hat{}2+2*\ell-1-2*m\hat{}2)*(n+\ell+1)\hat{}(1/2)$
$\qquad\qquad *(n+\ell+2)\hat{}(1/2)*(n-\ell+1)\hat{}(1/2)*(n-\ell)\hat{}(1/2)$
$\qquad\qquad /(2*\ell+3)/(2*\ell-1),$

$(-1,\,2)\ \ = -1/(2*\ell+3)/(2*\ell+5)\hat{}(1/2)*(\ell+2-m)\hat{}(1/2)$
$\qquad\qquad *(\ell+2+m)\hat{}(1/2)*(n-3-\ell)\hat{}(1/2)*(n+\ell+1)\hat{}(1/2)$
$\qquad\qquad /(2*\ell+1)\hat{}(1/2)*(\ell-m+1)\hat{}(1/2)*(\ell+m+1)\hat{}(1/2)$
$\qquad\qquad *(n-1-\ell)\hat{}(1/2)*(n-2-\ell)\hat{}(1/2),$

$(-2,\,2)\ \ = 1/4/(2*\ell+3)/(2*\ell+5)\hat{}(1/2)*(\ell+2-m)\hat{}(1/2)$
$\qquad\qquad *(\ell+2+m)\hat{}(1/2)*(n-3-\ell)\hat{}(1/2)*(n-4-\ell)\hat{}(1/2)$
$\qquad\qquad /(2*\ell+1)\hat{}(1/2)*(\ell-m+1)\hat{}(1/2)*(\ell+m+1)\hat{}(1/2)$
$\qquad\qquad *(n-1-\ell)\hat{}(1/2)*(n-2-\ell)\hat{}(1/2),$

$(0,-2)\ \ = 3/2/(2*\ell-3)\hat{}(1/2)/(2*\ell-1)*(\ell-1-m)\hat{}(1/2)$
$\qquad\qquad *(\ell-1+m)\hat{}(1/2)*(n-\ell+1)\hat{}(1/2)*(n-\ell)\hat{}(1/2)$
$\qquad\qquad /(2*\ell+1)\hat{}(1/2)*(\ell-m)\hat{}(1/2)*(\ell+m)\hat{}(1/2)$
$\qquad\qquad *(n+\ell-1)\hat{}(1/2)*(n+\ell)\hat{}(1/2),$

$(-1,-2) = -1/(2*\ell-3)\hat{}(1/2)/(2*\ell-1)*(\ell-1-m)\hat{}(1/2)$
$\qquad\qquad *(\ell-1+m)\hat{}(1/2)*(n-\ell)\hat{}(1/2)*(n-2+\ell)\hat{}(1/2)$
$\qquad\qquad /(2*\ell+1)\hat{}(1/2)*(\ell-m)\hat{}(1/2)*(\ell+m)\hat{}(1/2)$
$\qquad\qquad *(n+\ell-1)\hat{}(1/2)*(n+\ell)\hat{}(1/2),$

$(1,\,0)\ \ \ = -1/2*(2*\ell\hat{}2+2*\ell-1-2*m\hat{}2)*(2*n+1)*(n+\ell+1)\hat{}(1/2)$
$\qquad\qquad *(n-\ell)\hat{}(1/2)/(2*\ell+3)/(2*\ell-1),$

$(-1,\,0)\ \ = -1/2*(2*\ell\hat{}2+2*\ell-1-2*m\hat{}2)*(2*n-1)*(n-1-\ell)\hat{}(1/2)$
$\qquad\qquad *(n+\ell)\hat{}(1/2)/(2*\ell+3)/(2*\ell-1),$

$(0,\,0)\ \ \ = 1/2*(2*\ell\hat{}2+2*\ell-1-2*m\hat{}2)*(-\ell\hat{}2-\ell+3*n\hat{}2)$
$\qquad\qquad /(2*\ell+3)/(2*\ell-1),$

$(-2,-2) = 1/4/(2*\ell-3)\hat{}(1/2)/(2*\ell-1)*(\ell-1-m)\hat{}(1/2)$
$\qquad\qquad *(\ell-1+m)\hat{}(1/2)*(n-3+\ell)\hat{}(1/2)*(n-2+\ell)\hat{}(1/2)$
$\qquad\qquad /(2*\ell+1)\hat{}(1/2)*(\ell-m)\hat{}(1/2)\ *(\ell+m)\hat{}(1/2)$
$\qquad\qquad *(n+\ell-1)\hat{}(1/2)*(n+\ell)\hat{}(1/2)$

$]);$

```
rz := table([
  (1, 1)    = 1/2*(ℓ−m+1)^(1/2)*(ℓ+m+1)^(1/2)
              *(n+ℓ+1)^(1/2)*(n+ℓ+2)^(1/2)*(2*n−ℓ)
              /(2*ℓ+1)^(1/2)/(2*ℓ+3)^(1/2),
  (−2,−1) = −1/4*(n−1−ℓ)^(1/2)*(n−2+ℓ)^(1/2)
              /(2*ℓ−1)^(1/2)/(2*ℓ+1)^(1/2)*(ℓ−m)^(1/2)
              *(ℓ+m)^(1/2)*(n+ℓ−1)^(1/2)*(n+ℓ)^(1/2),
  (−1, 1)  = 1/2*(ℓ−m+1)^(1/2)*(ℓ+m+1)^(1/2)
              *(n−1−ℓ)^(1/2)*(n−2−ℓ)^(1/2)*(2*n+ℓ)
              /(2*ℓ+1)^(1/2)/(2*ℓ+3)^(1/2),
  (1,−1)   = 1/2*(ℓ−m)^(1/2)*(ℓ+m)^(1/2)*(n−ℓ+1)^(1/2)
              *(n−ℓ)^(1/2)*(2*n+ℓ+1)/(2*ℓ−1)^(1/2)
              /(2*ℓ+1)^(1/2),
  (−2, 1)  = −1/4*(n−3−ℓ)^(1/2)*(n+ℓ)^(1/2)
              /(2*ℓ+1)^(1/2)/(2*ℓ+3)^(1/2)*(ℓ−m+1)^(1/2)
              *(ℓ+m+1)^(1/2)*(n−1−ℓ)^(1/2)*(n−2−ℓ)^(1/2),
  (0,−1)   = −3/2*n/(2*ℓ−1)^(1/2)/(2*ℓ+1)^(1/2)
              *(ℓ−m)^(1/2)*(ℓ+m)^(1/2)*(n−ℓ)^(1/2)
              *(n+ℓ)^(1/2),
  (2, 1)   = −1/4*(n−ℓ)^(1/2)*(n+3+ℓ)^(1/2)/(2*ℓ+1)^(1/2)
              /(2*ℓ+3)^(1/2)*(ℓ−m+1)^(1/2)*(ℓ+m+1)^(1/2)
              *(n+ℓ+1)^(1/2)*(n+ℓ+2)^(1/2),
  (2,−1)   = −1/4*(n+2−ℓ)^(1/2)*(n+ℓ+1)^(1/2)/(2*ℓ−1)^(1/2)
              /(2*ℓ+1)^(1/2)*(ℓ−m)^(1/2)*(ℓ+m)^(1/2)
              *(n−ℓ+1)^(1/2)*(n−ℓ)^(1/2),
  (0, 1)   = −3/2*n/(2*ℓ+1)^(1/2)/(2*ℓ+3)^(1/2)
              *(ℓ−m+1)^(1/2)*(ℓ+m+1)^(1/2)*(n−1−ℓ)^(1/2)
              *(n+ℓ+1)^(1/2),
  (−1,−1) = 1/2*(ℓ−m)^(1/2)*(ℓ+m)^(1/2)*(n+ℓ−1)^(1/2)
              *(n+ℓ)^(1/2)*(2*n−1−ℓ)/(2*ℓ−1)^(1/2)
              /(2*ℓ+1)^(1/2)
]);

u := table([
  (2,−2)   = −1/4*(n+3−ℓ)^(1/2)*(ℓ−1−m)^(1/2)
              *(ℓ−1+m)^(1/2)*(n+2−ℓ)^(1/2)*(n−ℓ+1)^(1/2)
              *(ℓ−m)^(1/2)*(ℓ+m)^(1/2)*(n−ℓ)^(1/2)
              *(3*n+2*ℓ+2)/(2*ℓ−3)^(1/2)/(2*ℓ−1)/(2*ℓ+1)^(1/2),
  (1,−2)   = 5/8*(n−ℓ+1)^(1/2)*(ℓ−1−m)^(1/2)
              *(ℓ−1+m)^(1/2)*(n−ℓ)^(1/2)*(ℓ−m)^(1/2)
              *(ℓ+m)^(1/2)*(n+2−ℓ)^(1/2)*(n+ℓ)^(1/2)
              *(3*n+ℓ+1)/(2*ℓ−3)^(1/2)/(2*ℓ−1)/(2*ℓ+1)^(1/2),
  (1, 2)   = 5/8*(n+ℓ+1)^(1/2)*(ℓ+2−m)^(1/2)
```

$$*(\ell+2+m)\hat{}(1/2)*(n+3+\ell)\hat{}(1/2)*(\ell-m+1)\hat{}(1/2)$$
$$*(\ell+m+1)\hat{}(1/2)*(n-1-\ell)\hat{}(1/2)$$
$$*(n+\ell+2)\hat{}(1/2)*(3*n-\ell)/(2*\ell+3)/(2*\ell+5)\hat{}(1/2)$$
$$/(2*\ell+1)\hat{}(1/2),$$

$$(-3,-2) = 1/8*(n-1-\ell)\hat{}(1/2)*(n-4+\ell)\hat{}(1/2)/(2*\ell-3)\hat{}(1/2)$$
$$/(2*\ell-1)*(\ell-1-m)\hat{}(1/2)*(\ell-1+m)\hat{}(1/2)$$
$$*(n-3+\ell)\hat{}(1/2)*(n-2+\ell)\hat{}(1/2)/(2*\ell+1)\hat{}(1/2)$$
$$*(\ell-m)\hat{}(1/2)*(\ell+m)\hat{}(1/2)*(n+\ell-1)\hat{}(1/2)$$
$$*(n+\ell)\hat{}(1/2),$$

$$(-2, 0) = 3/2*(n-2-\ell)\hat{}(1/2)*(n+\ell-1)\hat{}(1/2)*(n-1-\ell)\hat{}(1/2)$$
$$*(n+\ell)\hat{}(1/2)*(\ell\hat{}2+\ell-1+m\hat{}2)*(n-1)/(2*\ell+3)/(2*\ell-1),$$

$$(-3, 2) = 1/8*(n-5-\ell)\hat{}(1/2)*(n+\ell)\hat{}(1/2)/(2*\ell+3)$$
$$/(2*\ell+5)\hat{}(1/2)*(\ell+2-m)\hat{}(1/2)*(\ell+2+m)\hat{}(1/2)$$
$$*(n-3-\ell)\hat{}(1/2)*(n-4-\ell)\hat{}(1/2)/(2*\ell+1)\hat{}(1/2)$$
$$*(\ell-m+1)\hat{}(1/2)*(\ell+m+1)\hat{}(1/2)*(n-1-\ell)\hat{}(1/2)$$
$$*(n-2-\ell)\hat{}(1/2),$$

$$(2, 2) = -1/4*(n+4+\ell)\hat{}(1/2)*(\ell+2-m)\hat{}(1/2)*(\ell+2+m)\hat{}(1/2)$$
$$*(n+\ell+2)\hat{}(1/2)*(n+3+\ell)\hat{}(1/2)*(\ell-m+1)\hat{}(1/2)$$
$$*(\ell+m+1)\hat{}(1/2)*(n+\ell+1)\hat{}(1/2)*(3*n-2*\ell)$$
$$/(2*\ell+3)/(2*\ell+5)\hat{}(1/2)\ /(2*\ell+1)\hat{}(1/2),$$

$$(0, 2) = -5/2*n/(2*\ell+3)/(2*\ell+5)\hat{}(1/2)*(\ell+2-m)\hat{}(1/2)$$
$$*(\ell+2+m)\hat{}(1/2)*(n+\ell+1)\hat{}(1/2)*(n+\ell+2)\hat{}(1/2)$$
$$/(2*\ell+1)\hat{}(1/2)*(\ell-m+1)\hat{}(1/2)*(\ell+m+1)\hat{}(1/2)$$
$$*(n-1-\ell)\hat{}(1/2)*(n-2-\ell)\hat{}(1/2),$$

$$(2, 0) = 3/2*(n-\ell+1)\hat{}(1/2)*(n+\ell+2)\hat{}(1/2)*(n+\ell+1)\hat{}(1/2)$$
$$*(n-\ell)\hat{}(1/2)*(\ell\hat{}2+\ell-1+m\hat{}2)*(n+1)/(2*\ell+3)/(2*\ell-1),$$

$$(3, 2) = 1/8*(n-\ell)\hat{}(1/2)*(n+5+\ell)\hat{}(1/2)/(2*\ell+3)$$
$$/(2*\ell+5)\hat{}(1/2)*(\ell+2-m)\hat{}(1/2)*(\ell+2+m)\hat{}(1/2)$$
$$*(n+3+\ell)\hat{}(1/2)*(n+4+\ell)\hat{}(1/2)/(2*\ell+1)\hat{}(1/2)$$
$$*(\ell-m+1)\hat{}(1/2)*(\ell+m+1)\hat{}(1/2)$$
$$*(n+\ell+1)\hat{}(1/2)*(n+\ell+2)\hat{}(1/2),$$

$$(3,-2) = 1/8*(n+4-\ell)\hat{}(1/2)*(n+\ell+1)\hat{}(1/2)/(2*\ell-3)\hat{}(1/2)$$
$$/(2*\ell-1)*(\ell-1-m)\hat{}(1/2)*(\ell-1+m)\hat{}(1/2)$$
$$*(n+3-\ell)\hat{}(1/2)*(n+2-\ell)\hat{}(1/2)/(2*\ell+1)\hat{}(1/2)$$
$$*(\ell-m)\hat{}(1/2)*(\ell+m)\hat{}(1/2)*(n-\ell+1)\hat{}(1/2)$$
$$*(n-\ell)\hat{}(1/2),$$

$$(-3, 0) = -1/4*(n+\ell-1)\hat{}(1/2)*(n-3-\ell)\hat{}(1/2)*(n-1-\ell)\hat{}(1/2)$$
$$*(n+\ell)\hat{}(1/2)*(n-2-\ell)\hat{}(1/2)*(n-2+\ell)\hat{}(1/2)$$
$$*(\ell\hat{}2+\ell-1+m\hat{}2)/(2*\ell+3)/(2*\ell-1),$$

$$(-1, 2) = 5/8*(n+\ell+1)\hat{}(1/2)*(\ell+2-m)\hat{}(1/2)*(\ell+2+m)\hat{}(1/2)$$
$$*(n-3-\ell)\hat{}(1/2)*(\ell-m+1)\hat{}(1/2)*(\ell+m+1)\hat{}(1/2)$$
$$*(n-1-\ell)\hat{}(1/2)*(n-2-\ell)\hat{}(1/2)*(3*n+\ell)$$
$$/(2*\ell+3)/(2*\ell+5)\hat{}(1/2)/(2*\ell+1)\hat{}(1/2),$$

$$(-2, 2) = -1/4*(\ell+2-m)\hat{}(1/2)*(\ell+2+m)\hat{}(1/2)*(n-3-\ell)\hat{}(1/2)$$

$$*(n-4-\ell)\hat{\ }(1/2)*(\ell-m+1)\hat{\ }(1/2)*(\ell+m+1)\hat{\ }(1/2)$$
$$*(n-1-\ell)\hat{\ }(1/2)*(n-2-\ell)\hat{\ }(1/2)*(3*n+2*\ell)/(2*\ell+3)$$
$$/(2*\ell+5)\hat{\ }(1/2)/(2*\ell+1)\hat{\ }(1/2),$$

$$(0,-2) \quad = -5/2*n/(2*\ell-3)\hat{\ }(1/2)/(2*\ell-1)*(\ell-1-m)\hat{\ }(1/2)$$
$$*(\ell-1+m)\hat{\ }(1/2)*(n-\ell+1)\hat{\ }(1/2)*(n-\ell)\hat{\ }(1/2)$$
$$/(2*\ell+1)\hat{\ }(1/2)*(\ell-m)\hat{\ }(1/2)*(\ell+m)\hat{\ }(1/2)$$
$$*(n+\ell-1)\hat{\ }(1/2)*(n+\ell)\hat{\ }(1/2),$$

$$(3,\,0) \quad = -1/4*(n+2-\ell)\hat{\ }(1/2)*(n+3+\ell)\hat{\ }(1/2)*(n-\ell+1)\hat{\ }(1/2)$$
$$*(n+\ell+2)\hat{\ }(1/2)*(n-\ell)\hat{\ }(1/2)*(n+\ell+1)\hat{\ }(1/2)$$
$$*(\ell\hat{\ }2+\ell-1+m\hat{\ }2)/(2*\ell+3)/(2*\ell-1),$$

$$(-1,-2) = 5/8*(\ell-1-m)\hat{\ }(1/2)*(\ell-1+m)\hat{\ }(1/2)*(n-2+\ell)\hat{\ }(1/2)$$
$$*(n+\ell-1)\hat{\ }(1/2)*(\ell-m)\hat{\ }(1/2)*(\ell+m)\hat{\ }(1/2)$$
$$*(n+\ell)\hat{\ }(1/2)*(n-\ell)\hat{\ }(1/2)*(3*n-1-\ell)/(2*\ell-3)\hat{\ }(1/2)$$
$$/(2*\ell-1)/(2*\ell+1)\hat{\ }(1/2),$$

$$(1,\,0) \quad = -3/4*(n+\ell+1)\hat{\ }(1/2)*(n-\ell)\hat{\ }(1/2)*(\ell\hat{\ }2+\ell-1+m\hat{\ }2)$$
$$*(5*n\hat{\ }2-\ell-\ell\hat{\ }2+5*n+2)/(2*\ell+3)/(2*\ell-1),$$

$$(-1,\,0) \quad = -3/4*(n+\ell)\hat{\ }(1/2)*(n-1-\ell)\hat{\ }(1/2)*(\ell\hat{\ }2+\ell-1+m\hat{\ }2)$$
$$*(5*n\hat{\ }2-5*n-\ell+2-\ell\hat{\ }2)/(2*\ell+3)/(2*\ell-1),$$

$$(0,\,0) \quad = n*(\ell\hat{\ }2+\ell-1+m\hat{\ }2)*(5*n\hat{\ }2+1-3*\ell-3*\ell\hat{\ }2)$$
$$/(2*\ell+3)/(2*\ell-1),$$

$$(-2,-2) = -1/4*(\ell-1-m)\hat{\ }(1/2)*(\ell-1+m)\hat{\ }(1/2)*(n-3+\ell)\hat{\ }(1/2)$$
$$*(n-2+\ell)\hat{\ }(1/2)*(\ell-m)\hat{\ }(1/2)*(\ell+m)\hat{\ }(1/2)$$
$$*(n+\ell-1)\hat{\ }(1/2)*(n+\ell)\hat{\ }(1/2)*(3*n-2-2*\ell)$$
$$/(2*\ell-3)\hat{\ }(1/2)/(2*\ell-1)/(2*\ell+1)\hat{\ }(1/2)$$

$$]);$$

## D.3  Special Values of Matrix Elements

Here we give the special values of the matrix elements of $S = R$ and the scaled
perturbation $W = RZ$ for the ground state Stark effect in the hydrogen atom
(see (7.53) to (7.55)). These results were used to do the "hand calculations"
of the second order energy correction $E^{(2)}$ of Section 7.4.4.

### D.3.1  Matrix elements of $R$ for $n = 1, \ldots, 5$

Using the general values in Table D.1 we can calculate specific values of the
matrix elements of $R$ for the 3-dimensional hydrogen atom. Since the $R$ matrix
is rather sparse and symmetric (at most 3 nonzero elements $R_{n\ell}^{n'\ell'}$ per row if the
rows are labeled by $n\ell$ and the columns are labeled by $n'\ell'$) only the nonzero
elements are listed for principal quantum number $n = 1, \ldots, 5$.

$$R_{10}^{10} = 1 \qquad\qquad R_{20}^{10} = -\sqrt{2}/2$$

$$R_{10}^{20} = -\sqrt{2}/2 \qquad R_{20}^{20} = 2 \qquad\qquad R_{30}^{20} = -\sqrt{6}/2$$
$$R_{21}^{21} = 2 \qquad\qquad R_{31}^{21} = -1$$

$$R^{30}_{20} = -\sqrt{6}/2 \qquad R^{30}_{30} = 3 \qquad R^{30}_{40} = -\sqrt{3}$$
$$R^{31}_{21} = -1 \qquad R^{31}_{31} = 3 \qquad R^{31}_{41} = -\sqrt{10}/2$$
$$R^{32}_{32} = 3 \qquad R^{32}_{42} = -\sqrt{6}/2$$

$$R^{40}_{30} = -\sqrt{3} \qquad R^{40}_{40} = 4 \qquad R^{50}_{50} = -\sqrt{5}$$
$$R^{41}_{31} = -\sqrt{10}/2 \qquad R^{41}_{41} = 4 \qquad R^{41}_{51} = -3\sqrt{2}/2$$
$$R^{42}_{32} = -\sqrt{6}/2 \qquad R^{42}_{42} = 4 \qquad R^{42}_{52} = -\sqrt{14}/2$$
$$R^{43}_{43} = 4 \qquad R^{43}_{53} = -\sqrt{2}$$

$$R^{50}_{40} = -\sqrt{5} \qquad R^{50}_{50} = 5$$
$$R^{51}_{41} = -3\sqrt{2}/2 \qquad R^{51}_{51} = 5$$
$$R^{52}_{42} = -\sqrt{14}/2 \qquad R^{52}_{52} = 5$$
$$R^{53}_{43} = -\sqrt{2} \qquad R^{53}_{53} = 5$$
$$R^{54}_{54} = 5$$

## D.3.2  Matrix elements of $W = RZ$ for $n = 1,\ldots,5$, $m = 0$

Using the general results in Table D.4 the matrix of the scaled Stark perturbation $RZ$ is given here for $n = 1,\ldots,5$, $\ell = 0,\ldots,n-1$ and for the special case $m = 0$ (the subscript $m$ is omitted).

$$W^{10}_{21} = \sqrt{2} \qquad\qquad W^{10}_{31} = -\sqrt{2}/2$$

$$W^{20}_{21} = -3 \qquad\qquad W^{20}_{31} = 4 \qquad\qquad W^{20}_{41} = -\sqrt{10}/2$$

$$W^{21}_{10} = \sqrt{2} \qquad\qquad W^{21}_{20} = -3 \qquad\qquad W^{21}_{30} = \sqrt{6}$$
$$W^{21}_{32} = 2\sqrt{3} \qquad\qquad W^{21}_{40} = -\sqrt{2}/2 \qquad\qquad W^{21}_{42} = -\sqrt{2}$$

$$W^{30}_{21} = \sqrt{6} \qquad\qquad W^{30}_{31} = -3\sqrt{6} \qquad\qquad W^{30}_{41} = 2\sqrt{15}$$
$$W^{30}_{51} = -\sqrt{30}/2$$

$$W^{31}_{10} = -\sqrt{2}/2 \qquad\qquad W^{31}_{20} = 4 \qquad\qquad W^{31}_{30} = -3\sqrt{6}$$
$$W^{31}_{32} = -3\sqrt{3} \qquad\qquad W^{31}_{40} = 4\sqrt{2} \qquad\qquad W^{31}_{42} = 5\sqrt{2}$$
$$W^{31}_{50} = -\sqrt{10}/2 \qquad\qquad W^{31}_{52} = -\sqrt{7}$$

$$W^{32}_{21} = 2\sqrt{3} \qquad\qquad W^{32}_{31} = -3\sqrt{3} \qquad\qquad W^{32}_{41} = 3\sqrt{30}/5$$
$$W^{32}_{43} = 6\sqrt{30}/5 \qquad\qquad W^{32}_{51} = -\sqrt{15}/5 \qquad\qquad W^{32}_{53} = -3\sqrt{15}/5$$

$$W^{40}_{21} = -\sqrt{2}/2 \qquad\qquad W^{40}_{31} = 4\sqrt{2} \qquad\qquad W^{40}_{41} = -6\sqrt{5}$$
$$W^{40}_{51} = 4\sqrt{10}$$

$$W^{41}_{20} = -\sqrt{10}/2 \qquad\qquad W^{41}_{30} = 2\sqrt{15} \qquad\qquad W^{41}_{32} = 3\sqrt{30}/5$$
$$W^{41}_{40} = -6\sqrt{5} \qquad\qquad W^{41}_{42} = -24\sqrt{5}/5 \qquad\qquad W^{41}_{50} = 10$$
$$W^{41}_{52} = 7\sqrt{70}/5$$

$$W^{42}_{21} = -\sqrt{2} \qquad\qquad W^{42}_{31} = 5\sqrt{2} \qquad\qquad W^{42}_{41} = -24\sqrt{5}/5$$
$$W^{42}_{43} = -18\sqrt{5}/5 \qquad\qquad W^{42}_{51} = 11\sqrt{10}/5 \qquad\qquad W^{42}_{53} = 18\sqrt{10}/5$$

$$W^{43}_{32} = 6\sqrt{30}/5 \qquad\qquad W^{43}_{42} = -18\sqrt{5}/5 \qquad\qquad W^{43}_{52} = 18\sqrt{70}/35$$

$$W^{43}_{54} = 20\sqrt{14}/7$$

$$W^{50}_{31} = -\sqrt{10}/2 \qquad W^{50}_{41} = 10 \qquad W^{50}_{51} = -15\sqrt{2}$$

$$W^{51}_{30} = -\sqrt{30}/2 \qquad W^{51}_{32} = -\sqrt{15}/5 \qquad W^{51}_{40} = 4\sqrt{10}$$
$$W^{51}_{42} = 11\sqrt{10}/5 \qquad W^{51}_{50} = -15\sqrt{2}$$

$$W^{52}_{31} = -\sqrt{7} \qquad W^{52}_{41} = 7\sqrt{70}/5 \qquad W^{52}_{43} = 18\sqrt{70}/35$$
$$W^{52}_{53} = -18\sqrt{35}/7$$

$$W^{53}_{32} = -3\sqrt{15}/5 \qquad W^{53}_{42} = 18\sqrt{10}/5 \qquad W^{53}_{52} = -18\sqrt{35}/7$$
$$W^{53}_{54} = -30\sqrt{7}/7$$

$$W^{54}_{43} = 20\sqrt{14}/7 \qquad W^{54}_{53} = -30\sqrt{7}/7$$

**Table D.1**: 3-dimensional matrix elements of $R^d$ in the scaled hydrogenic basis for $d = 1, 2, 3$.

| $\mu$ | $d$ | $(R^d)^{n+\mu,\ell}_{n\ell} = (R^d)^{n\ell}_{n+\mu,\ell} = \langle n+\mu, \ell, m \lvert R^d \rvert n\ell m \rangle$ |
|---|---|---|
| 0 | 1 | $n$ |
| 1 | 1 | $-(1/2)[(n - \ell)(n + \ell + 1)]^{1/2}$ |
| 0 | 2 | $(1/2)[3n^2 - \ell(\ell + 1)]$ |
| 1 | 2 | $-(1/2)(2n + 1)[(n - \ell)(n + \ell + 1)]^{1/2}$ |
| 2 | 2 | $(1/4)[(n - \ell)(n - \ell + 1)(n + \ell + 1)(n + \ell + 2)]^{1/2}$ |
| 0 | 3 | $(1/2)n[5n^2 + 1 - 3\ell(\ell + 1)]$ |
| 1 | 3 | $-(3/8)[5n^2 + 5n + 2 - \ell(\ell + 1)][(n - \ell)(n + \ell + 1)]^{1/2}$ |
| 2 | 3 | $(3/4)(n + 1)[(n - \ell)(n - \ell + 1)(n + \ell + 1)(n + \ell + 2)]^{1/2}$ |
| 3 | 3 | $-(1/8)[(n - \ell)(n - \ell + 1)(n - \ell + 2)]^{1/2}$ $\times[(n + \ell + 1)(n + \ell + 2)(n + \ell + 3)]^{1/2}$ |

**Table D.2**: Matrix elements of $Z$ in the scaled hydrogenic basis.

| $\mu$ | $\nu$ | $Z_{n\ell m}^{n+\mu,\ell+\nu} = Z_{n+\mu,\ell+\nu,m}^{n\ell} = \langle n+\mu,\ell+\nu,m|Z|n\ell m\rangle$ |
|---|---|---|
| 0 | 1 | $-[(2\ell+1)(2\ell+3)]^{-1/2}$ <br> $\times[(\ell-m+1)(\ell+m+1)(n-\ell-1)(n+\ell+1)]^{1/2}$ |
| 1 | $-1$ | $(1/2)[(2\ell-1)(2\ell+1)]^{-1/2}$ <br> $\times[(\ell-m)(\ell+m)(n-\ell+1)(n-\ell)]^{1/2}$ |
| 1 | 1 | $(1/2)[(2\ell+1)(2\ell+3)]^{-1/2}$ <br> $\times[(\ell-m+1)(\ell+m+1)(n+\ell+1)(n+\ell+2)]^{1/2}$ |

**Table D.3**: Matrix elements of $Z^2$ in the scaled hydrogenic basis.

| $\mu$ | $\nu$ | $(Z^2)_{n\ell m}^{n+\mu,\ell+\nu} = (Z^2)_{n+\mu,\ell+\nu,m}^{n\ell} = \langle n+\mu,\ell+\nu,m|Z^2|n\ell m\rangle$ |
|---|---|---|
| 0 | 0 | $(1/2)[3n^2-\ell(\ell+1)][2\ell(\ell+1)-2m^2-1]$ <br> $\times[(2\ell-1)(2\ell+3)]^{-1}$ |
| 0 | 2 | $(3/2)(2\ell+3)^{-1}[(2\ell+1)(2\ell+5)]^{-1/2}$ <br> $\times[(\ell-m+1)(\ell-m+2)(\ell+m+1)(\ell+m+2)]^{1/2}$ <br> $\times[(n-\ell-2)(n-\ell-1)(n+\ell+1)(n+\ell+2)]^{1/2}$ |
| 1 | $-2$ | $-(2\ell-1)^{-1}[(2\ell-3)(2\ell+1)]^{-1/2}$ <br> $\times[(\ell-m-1)(\ell-m)(\ell+m-1)(\ell+m)]^{1/2}$ <br> $\times[(n-\ell)(n-\ell+1)(n-\ell+2)(n+\ell)]^{1/2}$ |
| 1 | 0 | $-(1/2)(2n+1)[2\ell(\ell+1)-2m^2-1][(2\ell-1)(2\ell+3)]^{-1}$ <br> $\times[(n-\ell)(n+\ell+1)]^{1/2}$ |
| 1 | 2 | $-(2\ell+3)^{-1}[(2\ell+1)(2\ell+5)]^{-1/2}$ <br> $\times[(\ell-m+1)(\ell-m+2)(\ell+m+1)(\ell+m+2)]^{1/2}$ <br> $\times[(n-\ell-1)(n+\ell+1)(n+\ell+2)(n+\ell+3)]^{1/2}$ |
| 2 | $-2$ | $(1/4)(2\ell-1)^{-1}[(2\ell-3)(2\ell+1)]^{-1/2}$ <br> $\times[(\ell-m-1)(\ell-m)(\ell+m-1)(\ell+m)]^{1/2}$ <br> $\times[(n-\ell)(n-\ell+1)(n-\ell+2)(n-\ell+3)]^{1/2}$ |
| 2 | 0 | $(1/4)[2\ell(\ell+1)-2m^2-1][(2\ell-1)(2\ell+3)]^{-1}$ <br> $\times[(n-\ell+1)(n-\ell)(n+\ell+1)(n+\ell+2)]^{1/2}$ |
| 2 | 2 | $(1/4)(2\ell+3)^{-1}[(2\ell+1)(2\ell+5)]^{-1/2}$ <br> $\times[(\ell-m+1)(\ell-m+2)(\ell+m+1)(\ell+m+2)]^{1/2}$ <br> $\times[(n+\ell+1)(n+\ell+2)(n+\ell+3)(n+\ell+4)]^{1/2}$ |

**Table D.4**: Matrix elements of $RZ$ in the scaled hydrogenic basis.

| $\mu$ | $\nu$ | $(RZ)_{n\ell m}^{n+\mu,\ell+\nu} = (RZ)_{n+\mu,\ell+\nu,m}^{n\ell} = \langle n+\mu, \ell+\nu, m \vert RZ \vert n\ell m \rangle$ |
|---|---|---|
| 0 | 1 | $-(3/2)n[(2\ell+1)(2\ell+3)]^{-1/2}$ <br> $\times[(\ell-m+1)(\ell+m+1)(n-\ell-1)(n+\ell+1)]^{1/2}$ |
| 1 | $-1$ | $(1/2)(2n+\ell+1)[(2\ell-1)(2\ell+1)]^{-1/2}$ <br> $\times[(\ell-m)(\ell+m)(n-\ell+1)(n-\ell)]^{1/2}$ |
| 1 | 1 | $(1/2)(2n-\ell)[(2\ell+1)(2\ell+3)]^{-1/2}$ <br> $\times[(\ell-m+1)(\ell+m+1)(n+\ell+1)(n+\ell+2)]^{1/2}$ |
| 2 | $-1$ | $-(1/4)[(2\ell-1)(2\ell+1)]^{-1/2}$ <br> $\times[(\ell-m)(\ell+m)]^{1/2}$ <br> $\times[(n+\ell+1)(n-\ell)(n-\ell+1)(n-\ell+2)]^{1/2}$ |
| 2 | 1 | $-(1/4)[(2\ell+1)(2\ell+3)]^{-1/2}$ <br> $\times[(\ell-m+1)(\ell+m+1)]^{1/2}$ <br> $\times[(n-\ell)(n+\ell+1)(n+\ell+2)(n+\ell+3)]^{1/2}$ |

**Table D.5**: Matrix elements of $U = R(R^2 - Z^2)$ in the scaled hydrogenic basis. Continued in next table.

| $\mu$ | $\nu$ | $U^{n+\mu,\ell+\nu}_{n\ell m} = U^{n\ell}_{n+\mu,\ell+\nu,m} = \langle n+\mu, \ell+\nu, m|U|n\ell m\rangle$ |
|---|---|---|
| 0 | 0 | $n[(2\ell - 1)(2\ell + 3)]^{-1}[5n^2 - 3\ell(\ell + 1) + 1][\ell(\ell + 1) + m^2 - 1]$ |
| 0 | 2 | $-(5/2)n(2\ell + 3)^{-1}[(2\ell + 1)(2\ell + 5)]^{-1/2}$<br>$\times[(\ell - m + 1)(\ell - m + 2)(\ell + m + 1)(\ell + m + 2)]^{1/2}$<br>$\times[(n - \ell - 2)(n - \ell - 1)(n + \ell + 1)(n + \ell + 2)]^{1/2}$ |
| 1 | -2 | $(5/8)(3n + \ell + 1)(2\ell - 1)^{-1}[(2\ell - 3)(2\ell + 1)]^{-1/2}$<br>$\times[(\ell - m - 1)(\ell - m)(\ell + m - 1)(\ell + m)]^{1/2}$<br>$\times[(n - \ell)(n - \ell + 1)(n - \ell + 2)(n + \ell)]^{1/2}$ |
| 1 | 0 | $-(3/4)[5n^2 + 5n - \ell(\ell + 1) + 2][\ell(\ell + 1) + m^2 - 1]$<br>$\times[(2\ell - 1)(2\ell + 3)]^{-1}[(n - \ell)(n + \ell + 1)]^{1/2}$ |
| 1 | 2 | $(5/8)(3n - \ell)(2\ell + 3)^{-1}[(2\ell + 1)(2\ell + 5)]^{-1/2}$<br>$\times[(\ell - m + 1)(\ell - m + 2)(\ell + m + 1)(\ell + m + 2)]^{1/2}$<br>$\times[(n - \ell - 1)(n + \ell + 1)(n + \ell + 2)(n + \ell + 3)]^{1/2}$ |
| 2 | -2 | $-(1/4)(3n + 2\ell + 2)(2\ell - 1)^{-1}[(2\ell - 3)(2\ell + 1)]^{-1/2}$<br>$\times[(\ell - m - 1)(\ell - m)(\ell + m - 1)(\ell + m)]^{1/2}$<br>$\times[(n - \ell)(n - \ell + 1)(n - \ell + 2)(n - \ell + 3)]^{1/2}$ |
| 2 | 0 | $(3/2)(n + 1)[\ell(\ell + 1) + m^2 - 1][(2\ell - 1)(2\ell + 3)]^{-1}$<br>$\times[(n - \ell)(n - \ell + 1)(n + \ell + 1)(n + \ell + 2)]^{1/2}$ |
| 2 | 2 | $-(1/4)(3n - 2\ell)(2\ell + 3)^{-1}[(2\ell + 1)(2\ell + 5)]^{-1/2}$<br>$\times[(\ell - m + 1)(\ell - m + 2)(\ell + m + 1)(\ell + m + 2)]^{1/2}$<br>$\times[(n + \ell + 1)(n + \ell + 2)(n + \ell + 3)(n + \ell + 4)]^{1/2}$ |

**Table D.6**: Matrix elements of $U = R(R^2 - Z^2)$ in the scaled hydrogenic basis. Continuation of previous table.

| $\mu$ | $\nu$ | $U_{n\ell m}^{n+\mu,\ell+\nu} = U_{n+\mu,\ell+\nu,m}^{n\ell} = \langle n+\mu,\ell+\nu,m\vert U\vert n\ell m\rangle$ |
|---|---|---|
| 3 | $-2$ | $(1/8)(2\ell - 1)^{-1}[(2\ell - 3)(2\ell + 1)]^{-1/2}$ <br> $\times[(\ell - m - 1)(\ell - m)(\ell + m - 1)(\ell + m)]^{1/2}$ <br> $\times[(n - \ell)(n - \ell + 1)(n - \ell + 2)(n - \ell + 3)]^{1/2}$ <br> $\times[(n - \ell + 4)(n + \ell + 1)]^{1/2}$ |
| 3 | 0 | $-(1/4)[\ell(\ell + 1) + m^2 - 1][(2\ell - 1)(2\ell + 3)]^{-1}$ <br> $\times[(n - \ell)(n - \ell + 1)(n - \ell + 2)]^{1/2}$ <br> $\times[(n + \ell + 1)(n + \ell + 2)(n + \ell + 3)]^{1/2}$ |
| 3 | 2 | $(1/8)(2\ell + 3)^{-1}[(2\ell + 1)(2\ell + 5)]^{-1/2}$ <br> $\times[(\ell - m + 1)(\ell - m + 2)(\ell + m + 1)(\ell + m + 2)]^{1/2}$ <br> $\times[(n - \ell)(n + \ell + 1)(n + \ell + 2)(n + \ell + 3)]^{1/2}$ <br> $\times[(n + \ell + 4)(n + \ell + 5)]^{1/2}$ |

**Table D.7**: D-Dimensional Matrix elements of $R^d$ in the scaled hydrogenic basis for $d = 1, 2, 3$.

| $\mu$ | $d$ | $(R^d)_{qk}^{q+\mu,k} = (R^d)_{q+\mu,k}^{qk} = \langle q+\mu, k\vert R^d\vert qk\rangle$ |
|---|---|---|
| 0 | 1 | $q$ |
| 1 | 1 | $-(1/2)[(q - k)(q + k + 1)]^{1/2}$ |
| 0 | 2 | $(1/2)[3q^2 - k(k + 1)]$ |
| 1 | 2 | $-(1/2)(2q + 1)[(q - k)(q + k + 1)]^{1/2}$ |
| 2 | 2 | $(1/4)[(q - k)(q - k + 1)(q + k + 1)(q + k + 2)]^{1/2}$ |
| 0 | 3 | $(1/2)q[5q^2 + 1 - 3k(k + 1)]$ |
| 1 | 3 | $-(3/8)[5q^2 + 5q + 2 - k(k + 1)][(q - k)(q + k + 1)]^{1/2}$ |
| 2 | 3 | $(3/4)(q + 1)[(q - k)(q - k + 1)(q + k + 1)(q + k + 2)]^{1/2}$ |
| 3 | 3 | $-(1/8)[(q - k)(q - k + 1)(q - k + 2)]^{1/2}$ <br> $\times[(q + k + 1)(q + k + 2)(q + k + 3)]^{1/2}$ |

# Appendix E

# Tables of Stark Effect Energy Corrections

## E.1   Ground State Energy to Order 100

In Tables E.1 to E.4 the energy corrections for the ground state 3-D Stark effect
have been converted to the integer values $F_n = -2 \cdot 4^n E^{(n)}$ corresponding to
the series expansion (see [PR80] where the notation $e_n = F_{2n}$ is used).

$$E = \sum_{n=0}^{\infty} E^{(2n)} \lambda^{2n} = -\frac{1}{2} \sum_{n=0}^{\infty} F_{2n} \left(\frac{\lambda}{4}\right)^{2n}.$$

The floating point values of the energy corrections expressed in the Maple
internal floating point format are also given to 60 digit accuracy in Tables E.5
and E.6.

**Table E.1**: Ground state energy corrections for the 3-D Stark effect. Here we tabulate the integer values $F_n = -2 \cdot 4^n E^{(n)}$.

| $n$ | $F_n$ |
|---|---|
| 0 | 1 |
| 2 | 72 |
| 4 | 28440 |
| 6 | 40204 464 |
| 8 | 10410 22224 24 |
| 10 | 40796 30899 56240 |
| 12 | 22234 18553 14764 4656 |
| 14 | 16065 83197 06076 35384 416 |
| 16 | 14900 93419 26825 81434 33017 20 |
| 18 | 17304 59956 78806 65823 86297 85825 6 |
| 20 | 24652 33421 16532 76351 33541 30610 52496 |
| 22 | 42339 44560 43937 65310 85551 99936 02806 1088 |
| 24 | 86360 79472 08456 97060 63332 85773 60092 00480 496 |
| 26 | 20650 86548 66625 08241 13783 14631 89598 02916 49441 184 |
| 28 | 57241 05464 55691 41365 46941 33867 32846 22148 68878 81417 92 |
| 30 | 18211 11253 10624 49232 76209 20096 53334 49090 14425 46230 43964 80 |
| 32 | 65925 27588 32296 74570 62262 21917 56355 63242 67012 87268 10186 76796 0 |
| 34 | 26947 26486 73995 50582 96768 16055 51000 07195 53389 44246 68952 43008 79086 4 |
| 36 | 12352 40423 69012 45810 26672 76706 71432 88085 72311 89953 21803 92759 15016 40411 2 |
| 38 | 63110 30314 27680 83045 49968 87772 16626 97298 76694 44742 93238 40413 12540 34323 16192 |
| 40 | 35740 84638 66117 71953 59113 56622 33172 59515 24579 12136 20800 62511 96906 54022 58918 08848 |
| 42 | 22324 29795 93542 85016 26676 61045 88010 22143 26511 40042 38412 64762 25239 12709 39966 55523 60672 |
| 44 | 15309 81220 32730 54470 17106 01743 34292 68506 07339 98583 51372 02768 99492 77389 25611 82350 78639 46016 |
| 46 | 11480 05899 93885 20042 50776 07467 89277 29330 31360 26935 68975 66637 36384 10325 28126 44586 14010 95493 89120 |

**Table E.2**: Continuation of preceding table.

| $n$ | $F_n$ |
|---|---|
| 48 | 93768 01674 83758 14085 09435 06469 70029 24719 33708 35712 92571 77202 93399 62524 29022 80748 73913 63438 23611 0064 |
| 50 | 83135 93557 89076 14698 77582 53759 36931 23123 62048 04313 48354 08091 93160 80630 12546 19412 50445 88736 67928 39333 0464 |
| 52 | 79753 72759 68089 97488 10249 10852 39499 32544 94278 82030 11808 10931 64755 69006 12491 80140 00201 36156 20174 21477 72226 7936 |
| 54 | 82537 35433 13629 36480 22396 95419 18717 96243 89788 48184 19483 03232 86727 12592 16118 46114 75129 66231 56901 80266 05456 80224 6720 |
| 56 | 91894 58675 43993 33133 62818 81438 34886 77157 76432 36079 52110 53869 02494 19830 38131 88440 62489 36420 18814 05183 23281 96649 61499 6192 |
| 58 | 10978 80843 20851 30785 46995 44251 32244 72886 73186 53472 97930 93689 48683 32614 59791 85071 24614 78905 63125 73180 42124 13612 13348 13310 04736 |
| 60 | 14041 30952 00907 14418 45392 84290 54567 64561 57383 94062 25441 16479 83989 46184 09099 13153 92058 25744 64643 77641 67765 12258 60683 35227 60203 92384 |
| 62 | 19181 22027 05143 75394 42975 47892 93645 61062 89479 15172 91460 43409 13034 04905 17389 80770 07169 17869 12092 42729 86152 59857 07451 48911 04841 55529 77280 |
| 64 | 27928 75642 63507 95120 11948 86806 55924 86908 18531 04483 77423 41582 95530 79873 44765 05845 90495 67855 36225 46780 68907 35329 46342 35202 01704 19813 03718 15000 |
| 66 | 43259 35685 01308 90426 18783 92546 06623 13634 91877 87307 67288 54935 46996 38180 09340 27471 31976 26083 66669 68754 73242 12711 06250 77185 34312 50969 05626 05985 57264 |
| 68 | 71147 36360 23451 76605 24792 61509 97116 42663 80707 42998 32018 62988 58185 36422 81951 29027 83596 50693 42147 02317 38395 67655 19121 74249 25601 24202 33135 61628 05304 91152 |
| 70 | 12403 13622 25458 51300 32211 20415 53288 11393 23420 71990 50944 21219 19357 05311 48056 41543 14923 55895 41076 76436 33235 80408 97350 40086 48403 41789 52907 77467 86579 90355 691552 |

**Table E.3**: Continuation of preceding table.

| $n$ | $F_n$ |
|---|---|
| 72 | 22881 49355 96989 68529 07381 50402 80185 13383 25326 91989 76865 00077 68060 76691 60774 91423 61120 10219 25293 21596 23819 83291 05928 85808 17620 92907 95429 53570 11945 84032 12624 22286 4 |
| 74 | 44600 80042 13290 30386 45574 85209 07582 30337 77201 53960 53519 99933 23326 35359 30941 29657 96414 36553 90378 80696 62748 22154 10259 67005 68734 79423 27589 66691 75049 20003 37464 74258 02249 6 |
| 76 | 91720 74966 80787 24718 78290 82483 40739 90135 83300 73536 85071 53919 94183 29936 37675 65048 69703 61914 71737 97131 39631 19093 34909 02615 13444 36039 81590 17812 44464 45037 42349 98751 24425 43632 0 |
| 78 | 19872 54978 61803 82575 49561 64758 07351 39430 95284 27525 14985 39531 58728 54358 88148 90202 43376 65663 24182 32213 47937 09751 18081 75116 20607 21281 64422 79661 80788 32693 40867 37771 53493 41321 46688 64 |
| 80 | 45302 86795 36405 77156 25915 88915 23902 49126 53102 09554 48249 00045 11437 93720 36168 56469 41089 66958 84858 31222 09911 96936 77962 06225 10880 69350 06655 55863 54915 67001 94993 89320 43617 30342 66100 48292 64 |
| 82 | 10852 68863 87199 57224 26772 61822 39032 11455 93519 45109 14457 76442 38442 77188 39458 52427 83932 31450 92803 66209 19246 67665 68788 85648 89835 87685 13326 01021 96297 32825 46885 20905 56705 67524 08276 78581 09516 896 |
| 84 | 27287 78699 36353 92392 82653 65767 25762 91419 78517 98683 89421 45924 48412 61048 66330 88118 12420 05755 90081 72453 11938 52822 00420 94163 80759 30720 77084 32787 81829 78282 37208 47975 07315 57839 88777 16514 54266 55950 688 |

**Table E.4**: Continuation of preceding table.

| $n$ | $F_n$ |
|---|---|
| 86 | 71932 12396 53445 91476 36945 69720 44271 12346 71578 10525 01710 51402 95993 46140 83214 56622 61770 64221 65194 73990 00384 06605 01405 31472 58983 82825 50968 66537 23319 22075 86943 36888 62584 17297 92398 41331 52044 27799 11371 968 |
| 88 | 19857 66210 38107 26840 55360 90579 64725 47508 76782 86246 13493 18042 02975 45345 37114 14158 81954 45664 87337 59003 91231 14973 33908 98331 87659 12198 50500 78547 38350 52400 90848 31540 72185 89332 47283 79738 96341 40547 05298 92759 7344 |
| 90 | 57349 85843 75241 29293 20582 19341 38279 59622 62126 68568 20666 37237 28637 29644 56146 03248 96271 97107 07750 95358 21402 88301 92449 92748 72153 84615 28877 94337 47064 20849 62711 65086 88782 38371 90918 78494 50517 38286 05814 04886 90511 5328 |
| 92 | 17310 26785 10684 93693 74896 38670 63995 41984 84860 78100 44922 98171 08295 49427 51517 52884 00255 36422 49870 61497 80138 46762 61077 39460 40415 46272 02167 77314 86293 95545 95531 75254 05115 21140 50582 62642 68500 31030 93035 54986 94202 39204 96704 |
| 94 | 54554 32902 07416 67663 04974 28565 67513 74470 68513 86327 36459 39780 08948 63508 13739 64554 05780 97124 66909 91604 66813 46927 47513 71177 10717 08397 95100 77145 80880 33443 82276 71901 23184 58491 37737 70362 34630 98663 84822 98484 82996 75843 08416 47744 |
| 96 | 17935 47339 99645 71245 90017 87760 61931 26460 57370 77963 61315 41904 09838 82265 16651 31197 28262 91945 86961 45270 85378 94618 54772 99389 38735 90943 09797 86783 40398 36131 60154 28610 37848 09641 98617 53448 28555 64576 13903 64192 84388 75093 25423 38928 78872 0 |
| 98 | 61457 54442 71795 55370 87585 51753 38350 71978 40042 53190 34963 80839 66680 10760 47564 83964 39423 20400 19303 25942 08278 87516 35754 34313 81993 95033 60639 61702 49498 32048 08651 20486 75430 58910 98668 66597 82632 51043 19083 12161 94120 80651 22306 33024 25555 43299 2 |
| 100 | 21930 68433 86128 24973 77765 60580 92394 89560 24935 95981 31812 62531 42832 62245 08263 02020 07704 01657 54884 10992 67675 46707 09117 11950 13052 81537 45269 02205 67891 56924 29727 56171 44068 10604 30481 51023 47594 42558 85576 46427 10402 41858 85992 62051 44825 13746 09430 08 |

**Table E.5**: The Maple internal floating point format for the energy coefficients for the ground state 3-D Stark effect. $E^{(n)}$ is defined in terms of the integer $I_n$ and exponent $e_n$ by $E^{(n)} = I_n \times 10^{e_n}$.

| $n$ | $I_n$ | $e_n$ |
|---|---|---|
| 0 | -5000000000 0000000000 0000000000 0000000000 0000000000 0000000000 | -60 |
| 2 | -2250000000 0000000000 0000000000 0000000000 0000000000 0000000000 | -59 |
| 4 | -5554687500 0000000000 0000000000 0000000000 0000000000 0000000000 | -58 |
| 6 | -4907771484 3750000000 0000000000 0000000000 0000000000 0000000000 | -56 |
| 8 | -7942369264 5263671875 0000000000 0000000000 0000000000 0000000000 | -54 |
| 10 | -1945319604 6649932861 3281250000 0000000000 0000000000 0000000000 | -51 |
| 12 | -6626303652 3689170360 5651855468 7500000000 0000000000 0000000000 | -49 |
| 14 | -2992494398 8411939487 6360893249 5117187500 0000000000 0000000000 | -46 |
| 16 | -1734697049 5631198205 9038383886 2180709838 8671875000 0000000000 | -43 |
| 18 | -1259075329 8631655215 1860222384 5664411783 2183837890 6250000000 | -40 |
| 20 | -1121058367 5917076266 7669864918 2639928767 4576044082 6416015625 | -37 |
| 22 | -1203359420 4124757797 8013850934 0838655498 3735550194 9787139893 | -34 |
| 24 | -1534075883 5841529805 4341719651 0349001045 0030922584 1192528605 | -31 |
| 26 | -2292706634 1729705191 2966748412 4990076606 5810262972 4908183562 | -28 |
| 28 | -3971896051 3336025012 6743721995 7038162346 7191152128 3416135475 | -25 |
| 30 | -7897811107 8223603231 3595973793 9997033504 9641019478 8220182627 | -22 |
| 32 | -1786908183 3575961324 6987038926 2641298324 9878506610 7626435128 | -18 |
| 34 | -4565044236 2151408299 0205672800 4256393836 7998640605 2368392001 | -15 |
| 36 | -1307861670 8073711215 4932493260 9915675918 6325001499 1309561328 | -11 |
| 38 | -4176289536 9633736412 0944368115 3400928926 1935056418 1433451636 | -8 |
| 40 | -1478206760 3620595653 0500241913 0241240410 4469996937 7087291070 | -4 |
| 42 | -5770695769 0109322494 3006487883 5651995268 5460163732 6873063381 | -1 |
| 44 | -2473433561 7920358291 4832992736 3203001979 6840117049 4955858855 | 3 |
| 46 | -1159189725 9837761194 2801018166 8986354494 1581146822 6601608809 | 7 |
| 48 | -5917593805 8322192520 5720135650 1462619662 1720540892 8448165089 | 10 |
| 50 | -3279134469 8586391529 7775656651 7516174545 5061243789 5696612690 | 14 |

**Table E.6**: Continuation of preceding table.

| $n$ | $I_n$ | $e_n$ |
|---|---|---|
| 52 | -1966081179 5865230060 9666295202 2837006366 1478286390 6855738573 | 18 |
| 54 | -1271689297 8981712041 0017719527 7619241616 6061142592 4012710159 | 22 |
| 56 | -8849126817 8694164054 3761390392 5622508968 1966245200 7529742013 | 25 |
| 58 | -6607629972 8261974955 0921082377 1153580638 5771204449 8154521014 | 29 |
| 60 | -5281753606 0844287040 7974980454 9083735917 1454314879 6256997379 | 33 |
| 62 | -4509483213 9735195371 9437569640 0629637446 7389950541 9736742411 | 37 |
| 64 | -4103761925 5833772880 2473268564 7178010178 6605148282 8966777314 | 41 |
| 66 | -3972744499 8073661872 7166555876 5288648210 4298850986 1764455738 | 45 |
| 68 | -4083658397 9714838947 4312123108 1362453587 1248383100 7989846224 | 49 |
| 70 | -4449406909 8150435775 4601818199 8384380713 2243925606 0515269640 | 53 |
| 72 | -5130208285 8662970589 7553246546 4370538946 3259281339 6742743651 | 57 |
| 74 | -6249903313 8066024472 8597067499 7857194172 4702620255 8537106260 | 61 |
| 76 | -8033009103 5374807518 5145594001 7345080828 1057832524 8040379244 | 65 |
| 78 | -1087788027 2497282009 1044812104 8046536023 3013786500 3807210533 | 70 |
| 80 | -1549874006 1754518762 2767272417 6853642909 0753510039 3437479903 | 74 |
| 82 | -2320534656 2551369506 5129437772 7677725484 2326261437 7243015452 | 78 |
| 84 | -3646691704 6550743274 0863830921 9231477303 9968112751 9696436995 | 82 |
| 86 | -6008051326 7671911167 0791238491 5791651430 2257467568 9704127329 | 86 |
| 88 | -1036618330 0000089298 3169448933 6255164269 9004019993 1061316528 | 91 |
| 90 | -1871126437 5072954077 5236325524 0852267982 2436316534 0732386023 | 95 |
| 92 | -3529836504 7133893636 2493848255 3934504007 9273804984 2836331268 | 99 |
| 94 | -6952804822 4161557660 1878270145 9647168765 3099939156 3130469286 | 103 |
| 96 | -1428642879 7907410214 9214947541 9476999890 6142985515 4998185253 | 108 |
| 98 | -3059609903 2813459345 5945128095 7589539713 8950797223 9138231690 | 112 |
| 100 | -6823749184 6568838875 0680577243 8227015898 1304226549 6445791039 | 116 |

## E.2   Symbolic Energy Corrections for Parabolic States

The corrections $f^{(j)}$ for the perturbed separation constant to order 12 for the parabolic Stark states are given here (see Chapter 8 for details).

$$f^{(0)} = q,$$

$$4^1 f^{(1)} = 6\,q^2 - 2\,\tau,$$

$$4^2 f^{(2)} = 36\,\tau\,q - 68\,q^3 - 10\,q,$$

$$4^3 f^{(3)} = 44\,\tau^2 - 1032\,\tau\,q^2 - 120\,\tau + 1500\,q^4 + 660\,q^2,$$

$$4^4 f^{(4)} = 35640\,\tau\,q^3 - 3636\,\tau^2 q + 12960\,\tau\,q - 42756\,q^5$$
$$- 37900\,q^3 - 2210\,q,$$

$$4^5 f^{(5)} = 229200\,\tau^2 q^2 - 3680\,\tau^3 + 33712\,\tau^2 - 1364160\,\tau\,q^4$$
$$- 1029600\,\tau\,q^2 - 54240\,\tau + 1400784\,q^6 + 2093280\,q^4$$
$$+ 381264\,q^2,$$

$$4^6 f^{(6)} = 552912\,\tau^3 q - 13182480\,\tau^2 q^3 - 5881416\,\tau^2 q$$
$$+ 55719216\,\tau\,q^5 + 72439200\,\tau\,q^3 + 12051720\,\tau\,q$$
$$- 50118384\,q^7 - 113822520\,q^5 - 43002592\,q^3$$
$$- 1656500\,q,$$

$$4^7 f^{(7)} = 450696\,\tau^4 - 53877984\,\tau^3 q^2 - 8886688\,\tau^3$$
$$+ 728298480\,\tau^2 q^4 + 657768432\,\tau^2 q^2 + 46142240\,\tau^2$$
$$- 2379352416\,\tau\,q^6 - 4778464320\,\tau\,q^4 - 1663495776\,\tau\,q^2$$
$$- 60441600\,\tau + 1905807816\,q^8 + 6142374000\,q^6$$
$$+ 3997074984\,q^4 + 484419360\,q^2,$$

$$4^8 f^{(8)} = 4336206000\,\tau^3 q^3 - 98489844\,\tau^4 q + 2155524912\,\tau^3 q$$
$$- 39436837272\,\tau^2 q^5 - 60110940120\,\tau^2 q^3$$
$$- 12837436200\,\tau^2 q + 104928884208\,\tau\,q^7$$
$$+ 302909140800\,\tau\,q^5 + 182925628704\,\tau\,q^3$$
$$+ 20998605600\,\tau\,q - 75783847700\,q^9 - 330097271640\,q^7$$
$$- 331953386940\,q^5 - 83943936120\,q^3 - 2564063050\,q,$$

$$4^9 f^{(9)} = 12987661680\,\tau^4 q^2 - 66513664\,\tau^5 + 2306958208\,\tau^4$$
$$- 313746377280\,\tau^3 q^4 - 313679336640\,\tau^3 q^2$$
$$- 25354142720\,\tau^3 + 2112491924256\,\tau^2 q^6$$
$$+ 4891069777440\,\tau^2 q^4 + 2121721190976\,\tau^2 q^2$$
$$+ 104008003840\,\tau^2 - 4741154003520\,\tau\,q^8$$
$$- 18689372876160\,\tau\,q^6 - 17545111106880\,\tau\,q^4$$
$$- 4217580898560\,\tau\,q^2 - 123998092800\,\tau$$

$$+\, 3118465869552\, q^{10} + 17695955613600\, q^{8}$$
$$+\, 25598194895280\, q^{6} + 11206248868800\, q^{4}$$
$$+\, 1078065703680\, q^{2},$$

$$4^{10}\, f^{(10)} = 19093706352\, \tau^{5}q - 1343525553360\, \tau^{4}q^{3}$$
$$-\, 717606581400\, \tau^{4}q + 21237767907552\, \tau^{3}q^{5}$$
$$+\, 35612199123840\, \tau^{3}q^{3} + 8691108323328\, \tau^{3}q$$
$$-\, 112468417051680\, \tau^{2}q^{7} - 3691155567782192\, \tau^{2}q^{5}$$
$$-\, 271104353086320\, \tau^{2}q^{3} - 40501147719024\, \tau^{2}q$$
$$+\, 218322729416880\, \tau\, q^{9} + 1130767102209600\, \tau\, q^{7}$$
$$+\, 1534645588890960\, \tau\, q^{5} + 640629497858400\, \tau\, q^{3}$$
$$+\, 59428632961080\, \tau\, q - 131843747886864\, q^{11}$$
$$-\, 947164090664440\, q^{9} - 1872704078101248\, q^{7}$$
$$-\, 1272588709670760\, q^{5} - 255756162331848\, q^{3}$$
$$-\, 6733922487500\, q,$$

$$4^{11}\, f^{(11)} = 3732772967040000\, q^{2} + 45031376367043200\, q^{4}$$
$$+\, 129263525627735520\, q^{6} + 131709507058831680\, q^{8}$$
$$+\, 50643440002927200\, q^{10} + 5697798643729920\, q^{12}$$
$$-\, 15838816064578560\, \tau\, q^{2} + 3571930171113600\, \tau^{2}$$
$$-\, 405806442240000\, \tau - 815942774624256000\, \tau\, q^{4}$$
$$+\, 8983518025596288\, \tau^{2}q^{2} - 97686597736960\, \tau^{3}$$
$$-\, 125575991694955200\, \tau\, q^{6} + 29575190336065920\, \tau^{2}q^{4}$$
$$-\, 1711445923076736\, \tau^{3}q^{2} + 26427662667881664\, \tau^{2}q^{6}$$
$$+\, 11417794129664\, \tau^{4} - 3480389242659840\, \tau^{3}q^{4}$$
$$-\, 67406243137200000\, \tau\, q^{8} + 128865105585120\, \tau^{4}q^{2}$$
$$-\, 595447803776\, \tau^{5} + 10969945664\, \tau^{6}$$
$$+\, 5966527397878080\, \tau^{2}q^{8} - 3175284764352\, \tau^{5}q^{2}$$
$$-\, 1374410347836288\, \tau^{3}q^{6} + 120261151140480\, \tau^{4}q^{4}$$
$$-\, 10206901413838272\, \tau\, q^{10},$$

$$4^{12}\, f^{(12)} = -26875761111700500\, q - 250747728583221744\, q^{13}$$
$$-\, 1134082523226988520\, q^{3} - 6549523588743151632\, q^{5}$$
$$-\, 12095258403199910400\, q^{7} - 8983957949831891752\, q^{9}$$
$$-\, 2705874556442300880\, q^{11} + 249994688604883200\, \tau\, q$$
$$+\, 3096679286588838192\, \tau\, q^{3} - 180718837654280280\, \tau^{2}q$$
$$+\, 9189094783235653440\, \tau\, q^{5} - 1490737259341598928\, \tau^{2}q^{3}$$
$$+\, 43829365101036576\, \tau^{3}q - 2895797539380382560\, \tau^{2}q^{5}$$
$$+\, 254115671424258336\, \tau^{3}q^{3} + 9768565563606096192\, \tau\, q^{7}$$

$$- 4675835826310488\,\tau^4 q + 226108678695744\,\tau^5 q$$
$$- 17531168259583440\,\tau^4 q^3 + 3075554501622285248\,\tau^3 q^5$$
$$- 1820100070436354784\,\tau^2 q^7 + 3971540557994232960\,\tau\, q^9$$
$$+ 86145644160151488\,\tau^3 q^7 - 3902280325872\,\tau^6 q$$
$$- 9780498479694096\,\tau^4 q^5 + 483131045629997856\,\tau\, q^{11}$$
$$+ 401388127465248\,\tau^5 q^3 - 315868918931969040\,\tau^2 q^9.$$

The energy corrections to order 12 for the 2-D and 3-D parabolic stark states are given here (see Chapter 8 for details).

$$E^{(1)} = \frac{3}{2}\,nQ,$$

$$E^{(2)} = -\frac{1}{16}n^4(17\,n^2 - 3\,Q^2 - 9\,m^2 + 19),$$

$$E^{(3)} = \frac{3}{32}n^7 Q(23\,n^2 - Q^2 + 11\,m^2 + 39),$$

$$E^{(4)} = -\frac{1}{1024}n^{10}(5487\,n^4 + 1806\,n^2 Q^2 - 3402\,n^2 m^2 + 35182\,n^2$$
$$+ 147\,Q^4 - 1134\,Q^2 m^2 + 5754\,Q^2 - 549\,m^4 - 8622\,m^2$$
$$+ 16211),$$

$$E^{(5)} = \frac{3}{1024}n^{13} Q(10563\,n^4 + 98\,n^2 Q^2 + 772\,n^2 m^2 + 90708\,n^2$$
$$- 21\,Q^4 + 220\,Q^2 m^2 + 780\,Q^2 + 725\,m^4 + 830\,m^2$$
$$+ 59293),$$

$$E^{(6)} = -\frac{1}{8192}n^{16}(547262\,n^6 + 685152\,n^4 Q^2 - 429903\,n^4 m^2$$
$$+ 9630693\,n^4 + 390\,n^2 Q^4 - 25470\,n^2 Q^2 m^2$$
$$+ 7787370\,n^2 Q^2 - 16200\,n^2 m^4 - 4786200\,n^2 m^2$$
$$+ 22691096\,n^2 - 372\,Q^6 + 765\,Q^4 m^2 + 1185\,Q^4$$
$$- 36450\,Q^2 m^4 - 62100\,Q^2 m^2 + 7028718\,Q^2 - 6951\,m^6$$
$$- 16845\,m^4 - 4591617\,m^2 + 7335413),$$

$$E^{(7)} = \frac{3}{32768}n^{19} Q(7071885\,n^6 + 1502283\,n^4 Q^2 - 1530561\,n^4 m^2$$
$$+ 152685291\,n^4 + 1947\,n^2 Q^4 + 21410\,n^2 Q^2 m^2$$
$$+ 22015690\,n^2 Q^2 + 94915\,n^2 m^4 - 22686230\,n^2 m^2$$
$$+ 458448411\,n^2 + 957\,Q^6 - 6321\,Q^4 m^2 + 5691\,Q^4$$
$$+ 66115\,Q^2 m^4 + 64330\,Q^2 m^2 + 25667355\,Q^2 + 55937\,m^6$$
$$+ 155435\,m^4 - 261168905\,m^2 + 196899293),$$

$$E^{(8)} = -\frac{1}{4194304}n^{22}(4857120235\,n^8 + 12678910764\,n^6 Q^2$$

$$- 4746228804\,n^6 m^2 + 172582164844\,n^6$$
$$+ 645075018\,n^4 Q^4 - 1085792148\,n^4 Q^2 m^2$$
$$+ 331102439388\,n^4 Q^2 + 435500730\,n^4 m^4$$
$$- 124137275220\,n^4 m^2 + 1093789661706\,n^4$$
$$- 747012\,n^2 Q^6 - 2470860\,n^2 Q^4 m^2$$
$$+ 11959413060\,n^2 Q^4 - 79477020\,n^2 Q^2 m^4$$
$$- 19867808520\,n^2 Q^2 m^2 + 1231898249988\,n^2 Q^2$$
$$- 19492116\,n^2 m^6 + 8195443620\,n^2 m^4$$
$$- 486334179660\,n^2 m^2 + 1632167936956\,n^2$$
$$- 181341\,Q^8 + 1876644\,Q^6 m^2 - 496524\,Q^6$$
$$- 22063230\,Q^4 m^4 - 21564900\,Q^4 m^2 + 17547948306\,Q^4$$
$$- 67197564\,Q^2 m^6 - 141703380\,Q^2 m^4$$
$$- 29221748388\,Q^2 m^2 + 680368729332\,Q^2$$
$$- 6806349\,m^8 - 32695740\,m^6 + 11889823554\,m^4$$
$$- 298412555292\,m^2 + 427874233827),$$

$$E^{(9)} = \frac{3}{4194304}\,n^{25} Q\,(22112129285\,n^8 + 12598683740\,n^6 Q^2$$
$$- 10200847744\,n^6 m^2 + 917951097984\,n^6$$
$$+ 172454814\,n^4 Q^4 - 358952160\,n^4 Q^2 m^2$$
$$+ 392428448160\,n^4 Q^2 + 201412890\,n^4 m^4$$
$$- 317353315860\,n^4 m^2 + 6969252790122\,n^4$$
$$+ 24732\,n^2 Q^6 - 119616\,n^2 Q^4 m^2 + 3957836736\,n^2 Q^4$$
$$+ 38622220\,n^2 Q^2 m^4 - 8334034520\,n^2 Q^2 m^2$$
$$+ 1761026240876\,n^2 Q^2 + 40353752\,n^2 m^6$$
$$+ 4253151560\,n^2 m^4 - 1441775177880\,n^2 m^2$$
$$+ 12831312300408\,n^2 - 50331\,Q^8 + 209568\,Q^6 m^2$$
$$- 508128\,Q^6 + 2646714\,Q^4 m^4 + 6187692\,Q^4 m^2$$
$$+ 7185767946\,Q^4 + 46791608\,Q^2 m^6$$
$$+ 77010920\,Q^2 m^4 - 15084116856\,Q^2 m^2$$
$$+ 1195710089688\,Q^2 + 18392649\,m^8$$
$$+ 69937420\,m^6 + 7808504806\,m^4$$
$$- 1006368923668\,m^2 + 4299243990233),$$

$$E^{(10)} = -\frac{1}{33554432}\,n^{28}\,(800100584625\,n^{10} + 3531718242237\,n^8 Q^2$$
$$- 941836887252\,n^8 m^2 + 48684095967932\,n^8$$
$$+ 644591038770\,n^6 Q^4 - 840709724400\,n^6 Q^2 m^2$$
$$+ 169577068424400\,n^6 Q^2 + 198333399150\,n^6 m^4$$

$$- 45469881503100\, n^6 m^2 + 626373227457246\, n^6$$
$$+ 2438877042\, n^4 Q^6 - 6824616120\, n^4 Q^4 m^2$$
$$+ 23675323671720\, n^4 Q^4 + 5577327090\, n^4 Q^2 m^4$$
$$- 30829420751940\, n^4 Q^2 m^2 + 1513833667407234\, n^4 Q^2$$
$$- 2450385324\, n^4 m^6 + 7249898088540\, n^4 m^4$$
$$- 422072703577332\, n^4 m^2 + 2434676258004036\, n^4$$
$$+ 5843997\, n^2 Q^8 - 12056688\, n^2 Q^6 m^2$$
$$+ 68188173648\, n^2 Q^6 - 948879630\, n^2 Q^4 m^4$$
$$- 188824283460\, n^2 Q^4 m^2 + 125792774343042\, n^2 Q^4$$
$$- 3137863752\, n^2 Q^2 m^6 + 184465739880\, n^2 Q^2 m^4$$
$$- 164600057856888\, n^2 Q^2 m^2 + 3350330445600600\, n^2 Q^2$$
$$- 377909847\, n^2 m^8 - 60176387700\, n^2 m^6$$
$$+ 39374394170022\, n^2 m^4 - 1007444117562516\, n^2 m^2$$
$$+ 2783302471314105\, n^2 + 2495697\, Q^{10}$$
$$- 21126420\, Q^8 m^2 + 18039420\, Q^8$$
$$+ 63646830\, Q^6 m^4 - 111650940\, Q^6 m^2$$
$$+ 151641012702\, Q^6 - 1689714540\, Q^4 m^6$$
$$- 2317040100\, Q^4 m^4 - 420070516020\, Q^4 m^2$$
$$+ 102128404605060\, Q^4 - 1838749815\, Q^2 m^8$$
$$- 5617811700\, Q^2 m^6 + 404337671910\, Q^2 m^4$$
$$- 135353468181780\, Q^2 m^2 + 1392459218433945\, Q^2$$
$$- 115264584\, m^{10} - 665542440\, m^8$$
$$- 133529057424\, m^6 + 33692289582960\, m^4$$
$$- 474748699290408\, m^2 + 633573285971896),$$

$$E^{(11)} = \frac{3}{67108864} n^{31} Q (9410960157921\, n^{10} + 10154131800085\, n^8 Q^2$$
$$- 6493212136390\, n^8 m^2 + 649021962183490\, n^8$$
$$+ 695462790498\, n^6 Q^4 - 1262468040680\, n^6 Q^2 m^2$$
$$+ 558884535692280\, n^6 Q^2 + 568087787270\, n^6 m^4$$
$$- 357772658536540\, n^6 m^2 + 9630362424170230\, n^6$$
$$+ 716237298\, n^4 Q^6 - 2229000228\, n^4 Q^4 m^2$$
$$+ 29823023026188\, n^4 Q^4 + 3317640090\, n^4 Q^2 m^4$$
$$- 54076838581380\, n^4 Q^2 m^2 + 5775505963918890\, n^4 Q^2$$
$$+ 731007048\, n^4 m^6 + 24279201135960\, n^4 m^4$$
$$- 37589319599930184\, n^4 m^2 + 44061499890611496\, n^4$$
$$- 4743235\, n^2 Q^8 + 24369240\, n^2 Q^6 m^2 + 24150939960\, n^2 Q^6$$
$$+ 436268490\, n^2 Q^4 m^4 - 75851230980\, n^2 Q^4 m^2$$

$$+\, 1851113905610074\, n^2 Q^4 + 4583691280\, n^2 Q^2 m^6$$
$$+\, 71564015600\, n^2 Q^2 m^4 - 335912796015440\, n^2 Q^2 m^2$$
$$+\, 15007785638154960\, n^2 Q^2 + 2099386065\, n^2 m^8$$
$$-\, 21946605380\, n^2 m^6 + 1508667832431 90\, n^2 m^4$$
$$-\, 10094094111129700\, n^2 m^2 + 60831413232503585\, n^2$$
$$-\, 863751\, Q^{10} + 10888570\, Q^8 m^2 - 4089470\, Q^8$$
$$-\, 67596330\, Q^6 m^4 + 17289060\, Q^6 m^2$$
$$+\, 64899758022\, Q^6 + 1219104264\, Q^4 m^6$$
$$+\, 1582347480\, Q^4 m^4 - 203621159112\, Q^4 m^2$$
$$+\, 176378121030888\, Q^4 + 3433637865\, Q^2 m^8$$
$$+\, 8599490620\, Q^2 m^6 + 196331078790\, Q^2 m^4$$
$$-\, 321193872086500\, Q^2 m^2 + 74696573653 00025\, Q^2$$
$$+\, 782236742\, m^{10} + 3742954950\, m^8$$
$$-\, 58219729428\, m^6 + 144795330370780\, m^4$$
$$-\, 5299150706401266\, m^2 + 17262327593960222),$$

$$E^{(12)} = -\frac{1}{2147483648} n^{34} (1172209604185207\, n^{12}$$
$$+\, 7814178947623482\, n^{10} Q^2 - 1618417506860712\, n^{10} m^2$$
$$+\, 1102533302535625272\, n^{10} + 3057731201792421\, n^8 Q^4$$
$$-\, 3285523200438072\, n^8 Q^2 m^2$$
$$+\, 606444058246839912\, n^8 Q^2$$
$$+\, 524011843775475\, n^8 m^4 - 126356130666034830\, n^8 m^2$$
$$+\, 2458387854870199659\, n^8 + 85017525243420\, n^6 Q^6$$
$$-\, 199376946006480\, n^6 Q^4 m^2 + 191449663106346480\, n^6 Q^4$$
$$+\, 142330294848060\, n^6 Q^2 m^4$$
$$-\, 205665845342051160\, n^6 Q^2 m^2$$
$$+\, 10249548385777427100\, n^6 Q^2$$
$$-\, 28224654085320\, n^6 m^6 + 32811579865596840\, n^6 m^4$$
$$-\, 2200725198030524088\, n^6 m^2$$
$$+\, 191885583306295 29688\, n^6$$
$$+\, 24153553809\, n^4 Q^8 - 95965741872\, n^4 Q^6 m^2$$
$$+\, 4219395027159312\, n^4 Q^6 + 15338342250\, n^4 Q^4 m^4$$
$$-\, 9886399011179460\, n^4 Q^4 m^2$$
$$+\, 2262766922034353658\, n^4 Q^4$$
$$-\, 595383138504\, n^4 Q^2 m^6 + 7051691283353640\, n^4 Q^2 m^4$$
$$-\, 2450321942933003832\, n^4 Q^2 m^2$$
$$+\, 54225467781645721176\, n^4 Q^2$$

$$- 44184143619\, n^4 m^8 - 1390539507679476\, n^4 m^6$$
$$+ 398964842675433342\, n^4 m^4$$
$$- 12309651548594189460\, n^4 m^2$$
$$+ 55868270647112321805\, n^4 + 79265802\, n^2 Q^{10}$$
$$- 1724845320\, n^2 Q^8 m^2 + 940267615320\, n^2 Q^8$$
$$- 16712457300\, n^2 Q^6 m^4 - 3699329772600\, n^2 Q^6 m^2$$
$$+ 30292357570709388\, n^2 Q^6 - 712326023640\, n^2 Q^4 m^6$$
$$+ 6301875195000\, n^2 Q^4 m^4 - 70969127411868840\, n^2 Q^4 m^2$$
$$+ 6785106000564710280\, n^2 Q^4 - 866409410790\, n^2 Q^2 m^8$$
$$- 4596493948200\, n^2 Q^2 m^6 + 506403618436762200\, n^2 Q^2 m^4$$
$$- 7472149976219524200\, n^2 Q^2 m^2$$
$$+ 88307644547272913466\, n^2 Q^2$$
$$- 61913144304\, n^2 m^{10} + 851172212880\, n^2 m^8$$
$$- 9988169953482336\, n^2 m^6 + 1266807825334175520\, n^2 m^4$$
$$- 21925672124703719856\, n^2 m^2$$
$$+ 53648726696829439056\, n^2$$
$$- 74017293\, Q^{12} + 424789992\, Q^{10} m^2 - 884311032\, Q^{10}$$
$$+ 794402235\, Q^8 m^4 + 5955363810\, Q^8 m^2$$
$$+ 2994519906675\, Q^8 - 81973356696\, Q^6 m^6$$
$$- 130611345480\, Q^6 m^4 - 11813784249960\, Q^6 m^2$$
$$+ 33448395539267016\, Q^6 - 684276346215\, Q^4 m^8$$
$$- 1443370656420\, Q^4 m^6 + 19872563543910\, Q^4 m^4$$
$$- 78453204574502340\, Q^4 m^2 + 3945565377319013481\, Q^4$$
$$- 390780230544\, Q^2 m^{10} - 1568986586640\, Q^2 m^8$$
$$- 14323678538400\, Q^2 m^6 + 56158517508841440\, Q^2 m^4$$
$$- 4466784769149844944\, Q^2 m^2$$
$$+ 30397897915288281648\, Q^2$$
$$- 16641313287\, m^{12} - 111740151534\, m^{10}$$
$$+ 2808831981615\, m^8$$
$$- 11161741759419108\, m^6 + 806397825646029639\, m^4$$
$$- 8671768874861084622\, m^2 + 11023418359867957297).$$

# Appendix F

# Tables of Zeeman Effect Energy Corrections

## F.1  Ground State Energy to Order 60

In Tables F.1 to F.15 the energy corrections $E^{(n)}$ for the ground state Zeeman effect are given in rational form for the series expansion

$$E = \sum_{n=0}^{\infty} E^{(n)} \left( \frac{B^2}{8} \right)^n = \sum_{n=0}^{\infty} E^{(n)} \lambda^n,$$

where $\lambda = B^2/8$ [see Chapter 9].

The floating point values of the energy corrections expressed in the Maple internal floating point format are also given to 60 digit accuracy in Tables F.16 and F.17.

**Table F.1**: Ground state energy corrections for the Zeeman effect. The denominators are expressed in prime power factorized form.

| $n$ | $E^{(n)}$ |
|---|---|
| 0 | $-1/2^1$ |
| 1 | $2$ |
| 2 | $-53/3^1$ |
| 3 | $5581/3^2$ |
| 4 | $-21577\ 397/2^2 3^3 5^1$ |
| 5 | $31283\ 29828\ 3/2^2 3^4 5^2$ |
| 6 | $-13867\ 51316\ 0861/2^3 3^3 5^3$ |
| 7 | $53373\ 33446\ 07816\ 4463/2^3 3^5 5^4 7^2$ |
| 8 | $-99586\ 06672\ 91594\ 21112\ 3017/2^6 3^6 5^5 7^3$ |
| 9 | $86629\ 46342\ 38659\ 75592\ 74204\ 7423/2^6 3^8 5^6 7^4$ |
| 10 | $-61278\ 73544\ 61355\ 17930\ 91647\ 10303\ 3033/2^7 3^9 5^7 7^5$ |
| 11 | $28609\ 06791\ 68905\ 46438\ 86413\ 58759\ 27117\ 89049$ $/2^7 3^{10} 5^8 7^6 11^1$ |
| 12 | $-63212\ 27877\ 04554\ 40411\ 16164\ 55430\ 55820\ 82898\ 40606\ 3$ $/2^9 3^{11} 5^9 7^7 11^2$ |
| 13 | $64835\ 92275\ 41411\ 97801\ 86181\ 30859\ 58043\ 04268\ 60226\ 27637$ $303/2^9 3^{10} 5^{10} 7^8 11^4 13^1$ |
| 14 | $-33967\ 36985\ 02105\ 32149\ 38088\ 95247\ 87472\ 32823\ 19874\ 96282$ $47631\ 26549\ 93/2^{10} 3^{14} 5^{11} 7^9 11^5 13^2$ |
| 15 | $22642\ 64109\ 48230\ 76310\ 60000\ 64185\ 83018\ 78713\ 80340\ 89132$ $24993\ 14720\ 83059\ 77287/2^{11} 3^{16} 5^{12} 7^{10} 11^6 13^3$ |
| 16 | $-89127\ 61119\ 40023\ 29289\ 60050\ 53253\ 58500\ 66987\ 97584\ 11133$ $18399\ 38484\ 02093\ 66528\ 89910\ 9141/2^{14} 3^{18} 5^{13} 7^{11} 11^7 13^5$ |
| 17 | $18468\ 26604\ 84720\ 46640\ 87127\ 27486\ 80026\ 80909\ 70598\ 29685$ $38931\ 48804\ 60028\ 54928\ 79026\ 36512\ 35117\ 089$ $/2^{15} 3^{20} 5^{14} 7^{11} 11^8 13^6 17^1$ |
| 18 | $-41991\ 33085\ 85042\ 79804\ 15379\ 96924\ 73414\ 65130\ 44350\ 22887$ $57572\ 55915\ 00636\ 32969\ 28160\ 21516\ 22760\ 08867\ 58735\ 263$ $/2^{17} 3^{22} 5^{15} 7^{13} 11^9 13^7 17^2$ |
| 19 | $15180\ 92578\ 28852\ 74377\ 32059\ 80213\ 74793\ 06225\ 37970\ 33057$ $65542\ 18413\ 60394\ 98457\ 58935\ 56663\ 36217\ 49551\ 65225\ 11207$ $33809\ 553/2^{19} 3^{24} 5^{16} 7^{14} 11^{10} 13^8 17^3$ |
| 20 | $-21933\ 69107\ 02035\ 76695\ 08037\ 47959\ 30904\ 26045\ 55587\ 88464$ $32000\ 68180\ 13697\ 66912\ 67832\ 11746\ 12627\ 29358\ 70916\ 86383$ $72014\ 34323\ 04444\ 67723/2^{21} 3^{26} 5^{17} 7^{15} 11^{11} 13^9 17^4 19^2$ |

**Table F.2**: Continuation of preceding table

| $n$ | $E^{(n)}$ |
|---|---|
| 21 | 18374 84722 42289 94281 19801 48074 87931 74932 83660 03432 05332 90206 02286 00478 89070 24859 82415 14628 04311 50873 31845 20456 96140 75298 40680 73054 7 $/2^{23}3^{28}5^{18}7^{16}11^{12}13^{10}17^{5}19^{3}$ |
| 22 | $-41002$ 86553 65006 60534 29462 19548 51763 10062 41419 49085 62116 29573 74232 58085 67597 81138 01733 16647 41402 26824 42316 82149 03106 09199 20597 52967 01396 88493 73 $/2^{25}3^{30}5^{19}7^{16}11^{13}13^{11}17^{7}19^{4}$ |
| 23 | 66259 67212 89497 94112 91491 19217 17364 69058 31681 28256 22974 38880 19549 27238 06122 16182 79410 11948 77747 91364 97868 91008 83900 07192 42927 50147 20643 41512 23833 81047 32517$/2^{27}3^{32}5^{20}7^{18}11^{14}13^{12}17^{8}19^{5}23^{1}$ |
| 24 | $-23781$ 87389 44044 70514 28075 57639 55564 75813 93075 56448 02366 96805 77420 16901 51183 61756 57133 57132 12543 37224 07042 73316 03168 08339 09439 92652 52267 63000 16804 37725 43283 72736 56413 57$/2^{29}3^{34}5^{21}7^{18}11^{15}13^{13}17^{9}19^{6}23^{2}$ |
| 25 | 94716 57846 65443 49982 08998 54645 13940 25091 35681 70963 04256 58557 36727 61699 81025 29039 08940 82338 56283 60048 20799 54155 40505 98371 72428 70316 91823 84137 56047 21298 43967 11945 65862 83213 03073 4077 $/2^{31}3^{36}5^{22}7^{19}11^{16}13^{13}17^{10}19^{8}23^{3}$ |
| 26 | $-25394$ 70309 09411 40128 82757 19156 56982 85136 57649 03462 49870 13190 14444 93318 54199 37784 30740 73441 80518 39767 29011 61280 79815 21882 66878 76911 14583 57556 98275 33828 37892 84113 18987 85904 06240 87516 65049 32283 1729 $/2^{33}3^{38}5^{23}7^{21}11^{17}13^{15}17^{11}19^{9}23^{4}$ |
| 27 | 80659 91190 24121 44482 18660 67961 62153 04020 97309 53558 69263 68531 34380 83412 51408 84550 30563 68742 90553 67754 42660 09939 49045 04733 95120 78277 48047 14923 63244 06028 50566 92828 93270 51502 45871 56882 45861 37635 66798 89015 83926 3$/2^{35}3^{40}5^{24}7^{22}11^{18}13^{16}17^{12}19^{10}23^{5}$ |

**Table F.3**: Continuation of preceding table

| $n$ | $E^{(n)}$ |
|---|---|
| 28 | $-$21188 07040 40060 98218 21172 11378 65213 07920 74402 39955 87222 66057 24760 62131 58896 85576 83208 77997 87304 73529 65435 73473 58800 17896 80413 43400 60230 67473 64759 50929 41811 76826 94916 84048 05650 53696 39942 62947 71652 94120 04822 76796 76930 597 $/2^{37}3^{42}5^{25}7^{23}11^{19}13^{16}17^{13}19^{11}23^{6}$ |
| 29 | 29253 45427 46036 35759 73391 43115 58174 62663 21391 54681 65790 42748 03407 91760 49003 17085 70235 11698 73117 11879 08298 28761 15150 33983 25672 80015 81145 88737 22959 37574 61092 48573 59795 28441 68555 06456 71199 10803 39280 69512 87596 17020 42022 88232 11212 13502 643 $/2^{39}3^{44}5^{26}7^{24}11^{20}13^{18}17^{14}19^{12}23^{7}29^{1}$ |
| 30 | $-$33240 32019 38449 47630 70508 51141 91257 86259 68827 57578 82084 66450 07620 04528 41493 69876 12045 21266 92080 02797 13443 72873 60523 28431 20354 57165 57638 63995 79399 82954 03406 51567 83451 31523 47347 51801 50300 63802 06793 65345 22100 48937 00615 75411 87286 75995 78045 63021 70612 09 $/2^{41}3^{46}5^{27}7^{25}11^{21}13^{19}17^{15}19^{13}23^{8}29^{2}$ |
| 31 | 28749 67249 82691 48695 45002 14141 97974 53616 36323 03474 74469 00906 01366 03390 08440 41893 47865 65923 04039 53434 72362 30227 28274 12653 67941 62908 86454 62424 98331 24504 73628 92849 84591 00234 58133 00160 36116 49595 34001 36152 79389 33887 08716 97662 99038 75796 91152 16144 59451 84911 86968 39250 6771 $/2^{43}3^{48}5^{28}7^{26}11^{22}13^{20}17^{16}19^{14}23^{10}29^{3}31^{1}$ |
| 32 | $-$10470 67573 63227 25076 84731 51673 67917 44990 90143 51426 83835 50481 61382 60183 26848 29685 17844 91166 81391 37213 45546 72602 46217 53888 88450 00504 36008 23136 87158 80179 04292 38672 64469 60954 86471 20199 44180 25280 49834 20482 09847 99150 42249 43560 15390 33538 16924 40813 82975 40784 91612 41116 86456 14607 49611 1733 $/2^{45}3^{50}5^{29}7^{27}11^{22}13^{21}17^{17}19^{15}23^{11}29^{4}31^{2}$ |

**Table F.4**: Continuation of preceding table

| $n$ | $E^{(n)}$ |
|---|---|
| 33 | 49063 16597 51363 76528 53839 43350 05539 37923 23456 62027 91250 34189 23251 91598 79591 77394 39850 32772 09578 01387 72345 99264 06648 85258 50912 09073 69385 30275 29130 73084 30454 19844 11490 77566 83740 98778 30079 72601 12365 74333 01189 12940 73157 30715 20503 96307 46038 19645 46260 76176 64003 70718 36013 19174 71371 84220 47669 30839 04539 $/2^{47}3^{52}5^{30}7^{28}11^{24}13^{22}17^{18}19^{16}23^{12}29^{5}31^{3}$ |
| 34 | $-22182$ 28172 49909 44424 50422 28292 94067 45008 29517 99899 35744 60721 66871 18239 36348 83574 99168 63214 22173 05384 16147 27242 94051 85840 04698 15876 00026 32917 84741 54690 21411 48598 16501 51119 35595 85661 47757 96782 26969 32529 07663 49406 54709 07479 16130 30890 25674 13834 54446 86050 37131 25555 94408 24485 36183 59234 00556 17193 42983 93653 29184 35396 $7/2^{49}3^{54}5^{31}7^{29}11^{25}13^{23}17^{19}19^{17}23^{13}29^{6}31^{4}$ |
| 35 | 10626 09274 83380 25097 07735 66545 01209 41429 72417 89881 51744 09182 30530 05301 42219 33844 20680 41110 64400 55433 53371 74135 43453 86877 09603 64723 64144 10743 16931 23757 76823 32971 51218 05550 62140 19627 37040 37608 29559 34494 12010 74331 64684 48443 90922 41785 19226 44999 02358 88467 70114 89594 14306 00448 66293 08543 99947 60362 90946 54697 93289 63177 05970 89827 12328 51 $/2^{51}3^{56}5^{32}7^{30}11^{26}13^{24}17^{20}19^{18}23^{14}29^{7}31^{5}$ |
| 36 | $-10988$ 98508 82393 51315 17324 51439 83757 24881 85158 46209 55843 02717 94335 81351 26125 70983 31281 70014 06697 55773 85457 51228 85081 14670 71330 57111 85640 00972 78855 89379 94302 66095 76944 19603 44888 92646 10746 02014 05252 20736 51577 13807 31750 12406 46201 72559 39405 78111 35959 71534 47126 60971 44781 91511 93427 13104 99341 94351 94941 06996 16674 72220 96508 77186 23102 28445 42224 63124 7 $/2^{53}3^{58}5^{33}7^{29}11^{27}13^{25}17^{21}19^{19}23^{15}29^{8}31^{6}$ |

**Table F.5**: Continuation of preceding table

| $n$ | $E^{(n)}$ |
|---|---|
| 37 | 10663 07224 61502 18258 93023 93559 78882 63078 27489 27865 08577 94293 01810 41293 89920 74895 15918 72361 58150 49502 40441 99223 26273 42861 96732 10198 21970 74503 47610 02360 52077 09812 07837 04151 63505 96829 91822 02838 61398 80407 41146 25208 71321 90552 45136 68835 60263 31304 81097 10734 86179 57937 62657 51691 23444 99189 89181 93652 81251 16398 12039 61465 58139 37460 89547 94228 01274 02650 77106 81922 02393 09683 $/2^{55}3^{60}5^{34}7^{32}11^{28}13^{26}17^{22}19^{20}23^{16}29^{9}31^{7}37^{1}$ |
| 38 | $-22270$ 36950 35027 96776 21918 65661 08052 46952 91425 40016 34190 43430 13885 94668 95547 22543 43974 14521 74973 16999 54506 44267 60224 99574 80205 30960 43594 85528 44211 82019 88572 45512 82025 37506 60942 05564 56406 88553 65955 96749 73545 55603 66024 81412 43869 38320 55336 16509 84728 75918 95788 40859 45898 88938 48015 47275 59591 58341 63664 93732 13545 70190 86109 21987 17799 00623 16264 04258 70181 46146 99044 34067 51995 38294 29295 13 $/2^{57}3^{62}5^{35}7^{33}11^{29}13^{27}17^{23}19^{21}23^{17}29^{10}31^{8}37^{2}$ |
| 39 | 24494 06687 16453 16973 31796 81355 33155 59545 74433 96078 24279 95376 73967 25890 75759 81302 81111 17183 90634 31631 85642 45514 35774 63570 10906 35932 25954 70697 88443 24879 13588 99212 81329 49055 65186 05013 15609 33532 44644 27446 80831 22917 34058 82896 95053 09945 15216 58062 01270 50300 26213 67018 62895 72374 64428 74526 55030 96463 38126 86070 96918 03791 75581 84683 80893 46514 54714 67119 54415 07690 83885 61494 39236 46288 35602 13139 17025 81697 3859 $/2^{58}3^{64}5^{36}7^{34}11^{30}13^{28}17^{24}19^{22}23^{18}29^{11}31^{9}37^{3}$ |
| 40 | $-65740$ 47936 86723 67556 27355 41037 05877 24248 07763 14093 98454 92480 45879 04785 18372 75470 58287 94385 62022 23338 43183 72904 83449 51240 82589 49660 82492 82708 66172 94492 88089 59010 07316 83529 53832 48600 65847 96102 25218 92139 04588 08367 56817 25389 99404 33498 15723 15585 57112 60531 39587 69675 53897 77615 73411 55626 13402 46727 18420 87048 01288 82892 22748 32515 86954 18683 19297 94511 22224 12302 88122 58492 50944 98100 35908 11536 90951 78020 17956 84292 24002 14322 659 $/2^{62}3^{66}5^{37}7^{35}11^{31}13^{29}17^{25}19^{23}23^{19}29^{13}31^{10}37^{4}$ |

**Table F.6**: Continuation of preceding table

| $n$ | $E^{(n)}$ |
|---|---|
| 41 | 18718 59062 22923 46785 26932 22038 01287 37172 57409 23294 79370 06832 40437 96260 82291 44075 78938 67583 35815 87839 59430 33952 28856 92062 34538 10142 79685 07347 86380 19408 56845 24989 40914 56466 66360 05846 20282 98797 84822 60379 80878 91917 27874 75248 45958 39222 04836 77115 77322 59539 44602 41591 43621 64495 31096 26599 59552 30120 41743 85460 45686 33149 52333 24233 90102 07156 18895 62040 76170 99202 93706 83377 86585 88212 19032 37329 20941 97286 79878 51212 95425 88949 80944 31247 76954 44348 49 $/2^{65}3^{68}5^{38}7^{35}11^{32}13^{30}17^{26}19^{24}23^{20}29^{14}31^{11}37^{5}41^{1}$ |
| 42 | $-27403$ 48882 36457 35642 75904 49368 85181 44997 61622 38872 19940 55558 26392 15417 23725 88111 21162 15436 63958 76109 44000 96962 99372 69152 88543 91118 99435 04388 58123 07923 21872 13378 23921 64409 98240 60523 77718 38669 31392 48244 24665 39341 79557 96132 03596 35998 44434 05979 09176 03519 07084 38833 14322 14280 28322 75392 54693 38737 54401 68211 20129 53236 28153 89377 15222 60678 13729 71150 32723 82498 10044 56803 14998 99203 20131 14121 05558 15640 27872 71427 82075 30298 93127 11793 57589 27355 07732 36012 65006 15209 81$/2^{68}3^{70}5^{39}7^{37}11^{33}13^{31}17^{27}19^{25}23^{21}29^{15}31^{12}37^{6}41^{2}$ |
| 43 | 55798 82553 42843 84841 30998 13291 38982 31417 83139 21204 35883 79222 20379 86717 61131 06856 61692 23438 28911 85296 10707 09974 76737 80133 84897 15348 07552 85773 84072 28221 30724 25797 08553 81175 19223 92422 28507 68806 50267 00377 62894 66747 25571 78674 88284 02019 41059 47613 09846 68321 09860 56019 43479 28882 39736 43766 82552 13081 17010 85666 70556 64823 12702 83204 16020 08092 77894 96462 96930 71815 22774 24326 54071 59106 13235 74520 79905 52352 36230 14047 33141 14480 74748 28656 03927 65876 15755 66263 12913 01751 62909 91549 31509 92814 9 $/2^{71}3^{72}5^{39}7^{37}11^{34}13^{32}17^{28}19^{26}23^{22}29^{16}31^{14}37^{7}41^{2}43^{1}$ |

**Table F.7**: Continuation of preceding table

| $n$ | $E^{(n)}$ |
|---|---|
| 44 | $-$56441 29358 08652 79099 86046 58530 70004 93864 91815 51501 66952 59305 08767 64421 61422 11027 00557 93196 23594 44866 80541 93554 73906 57095 99455 42264 93580 34840 95076 29150 08030 06410 53687 44279 42803 71041 87953 26867 49256 68721 00734 98572 80434 11687 48387 65939 42128 43204 33071 94660 07034 00229 54406 06713 21067 55046 87923 44357 68195 04359 67271 80068 96700 88260 23943 39993 53035 01515 80788 66324 99887 87685 94557 72754 17529 38380 55789 41213 77321 19784 30603 04471 59693 91178 98873 03399 84559 61346 54882 58186 66779 92114 45844 50966 25066 75456 09001 48304 15937 $/2^{74}3^{74}5^{42}7^{38}11^{35}13^{33}17^{29}19^{27}23^{23}29^{17}31^{15}37^{8}41^{4}43^{2}$ |
| 45 | 20389 42282 93495 84444 67662 78956 56393 72884 37486 24945 10332 86323 15307 01035 69062 71880 73659 71360 49704 74629 37664 30845 49827 76584 79847 75544 82200 28430 23109 22503 53883 47662 09579 65289 47085 63128 15294 07747 38340 48984 45283 20488 49574 88569 04571 41059 46735 79720 75060 22309 31450 73962 75394 59161 82116 24305 92223 88561 35062 60658 69529 35687 49843 56801 55703 75263 08676 99342 88140 65190 44239 70633 58382 37531 39869 95312 95862 66992 23685 72312 20156 72557 83827 51690 31751 28154 84751 57323 98002 05996 97141 43348 88697 56010 63769 29665 80169 27495 81784 39679 82778 91009 73367 719 $/2^{77}3^{76}5^{44}7^{40}11^{36}13^{34}17^{30}19^{28}23^{24}29^{18}31^{16}37^{9}41^{5}43^{3}$ |
| 46 | $-$35466 46292 74461 39036 75595 92680 36974 73001 99087 83209 40609 50780 43922 54795 43053 83127 96459 75554 90748 05823 49537 39141 01107 44543 54143 70674 27845 89170 39894 75644 99849 73266 78247 52547 63553 30111 35572 18578 99845 67676 85520 70341 04522 86089 90870 20707 35410 87918 73983 70115 29936 75347 53881 43145 69098 31020 86336 63673 11745 77521 48375 87193 68868 57488 04992 46427 26278 65794 06133 35607 88500 27384 45406 57640 90403 64138 57137 18997 44391 61218 78737 61722 63425 30939 80527 69080 08159 49994 60408 90010 13185 43015 77993 86726 44290 13588 53313 98523 88274 64602 65877 23549 93825 68734 72212 01058 08502 489 $/2^{80}3^{78}5^{46}7^{41}11^{37}13^{35}17^{31}19^{29}23^{25}29^{19}31^{16}37^{10}41^{6}43^{4}$ |

**Table F.8**: Continuation of preceding table

| $n$ | $E^{(n)}$ |
|---|---|
| 47 | 29087 59965 72866 13224 08072 28259 69523 31634 44098 65659 06431 66534 81918 17397 68794 09306 84336 95857 41273 32927 20327 22882 22705 63952 30801 33422 18687 50937 58520 11645 56007 53539 30818 41843 88689 48975 69395 10234 65587 52616 78094 67208 49780 18655 23963 19680 80435 16445 85068 05228 37564 44053 67410 85684 77057 80522 64047 54921 21453 70964 46071 82149 63892 63592 33788 71338 55864 38362 86661 92219 63923 92791 75882 67618 34363 45474 00874 88585 58388 24617 31951 22407 23457 69314 71331 53435 75868 94214 13561 91369 43397 34006 76252 83635 22588 39906 96755 51768 20965 89357 19853 75712 10197 17996 46774 41801 82783 64066 20143 50086 82696 60287 753 $/2^{83}3^{80}5^{48}7^{42}11^{38}13^{36}17^{32}19^{30}23^{26}29^{20}31^{18}37^{11}41^{7}43^{5}47^{1}$ |
| 48 | $-80260$ 20475 42879 34832 53137 37967 18637 44767 37572 30228 44864 74885 84034 13367 18116 75842 95145 11905 69395 35235 96180 62899 69259 61190 50212 55169 48394 84748 29748 30660 94186 64628 29395 56089 61780 87305 28160 74144 27716 75404 34878 74677 25878 42969 80805 42811 25176 23090 47049 07441 50315 42634 30828 89327 39822 49311 16259 07164 53084 22533 82288 53082 71957 08497 13741 07885 15822 47937 25934 71447 91090 41165 35155 08450 41274 97685 13612 22019 82663 19405 95416 04273 96901 13600 30204 95179 02437 00522 68623 37436 53489 68831 86983 44755 97594 91539 61424 77195 21854 67618 43815 27573 31328 23949 77283 45768 20678 51599 99454 61530 15062 53334 54776 87193 45349 66313 470151 $/2^{86}3^{82}5^{50}7^{43}11^{39}13^{37}17^{33}19^{31}23^{27}29^{21}31^{19}37^{12}41^{8}43^{6}47^{2}$ |

**Table F.9**: Continuation of preceding table

| $n$ | $E^{(n)}$ |
|---|---|
| 49 | 12145 83712 03590 10410 55906 53361 11103 23544 99070 07083 94169 02762 77161 73075 13291 85460 34607 59613 86812 69492 01165 68911 86066 50450 50837 65558 17472 39516 47667 49794 06220 00726 56334 98304 97287 73923 60569 55384 42274 71898 11838 35482 52466 40405 10213 20838 09517 86921 12068 10252 89760 02183 35034 38554 78761 62486 44216 32324 96697 39694 40726 63756 41689 16706 50505 76967 15715 48974 94030 07083 31726 86482 31766 37875 54807 49044 61865 01607 11586 75907 58277 95996 34013 16960 43565 84779 23203 20162 20503 87443 17657 38404 72850 86978 31173 27625 43757 03729 28277 49232 76367 56005 82327 16509 73804 53468 59884 84933 26144 86551 70491 35501 09962 10819 51673 88520 48083 53198 82089 37087 76124 9843 $/2^{89}3^{84}5^{52}7^{44}11^{40}13^{38}17^{34}19^{31}23^{28}29^{22}31^{20}37^{13}41^{9}43^{7}47^{3}$ |
| 50 | $-$69085 94789 96590 45952 08551 56010 43342 96712 64419 19248 57734 89133 05931 83925 14720 09744 18726 74053 55409 45247 91103 72914 60504 02527 10062 13333 67140 14035 07112 17573 96242 91128 58059 86071 31323 71749 47945 59630 88909 16862 27545 77992 15239 08897 00000 81231 36045 32561 08902 10848 90109 26640 60524 31379 82549 57650 15190 71604 92879 03249 43209 03992 49240 04473 53282 89622 95124 43065 92359 76489 37366 34395 61089 94421 18200 02435 40030 80585 69262 34484 08989 68802 50471 37067 03144 23931 95521 14707 30210 48602 62539 73483 79410 24458 60855 98226 10248 75522 85193 32868 66491 50369 77657 23832 44489 95870 71150 15452 26571 80035 04968 14930 59549 28125 90664 84162 98873 87536 30810 45420 65032 76148 43230 33459 90943 67162 439 $/2^{92}3^{86}5^{54}7^{45}11^{41}13^{39}17^{35}19^{33}23^{29}29^{23}31^{21}37^{14}41^{10}43^{8}47^{4}$ |

**Table F.10**: Continuation of preceding table

| $n$ | $E^{(n)}$ |
|---|---|
| 51 | 11324 69087 40114 98870 41817 88539 50255 73707 03788 83869 <br> 01115 83220 57139 29387 82897 54462 98692 78171 99334 45895 <br> 68926 62020 31364 10294 15338 02779 32050 55990 28742 75900 <br> 55960 89398 38844 94016 33684 99139 48577 20471 98800 13155 <br> 91448 65689 76166 44460 50853 31766 44954 51935 62871 02180 <br> 66700 47753 87732 79000 87735 78727 23798 51045 72321 30922 <br> 86490 28735 53183 77262 73754 50492 94745 13586 95701 47054 <br> 59759 85423 61790 35224 06122 77205 97687 32590 51635 74985 <br> 29817 71032 58709 28232 57883 19654 27538 92687 46953 22766 <br> 90945 56457 13717 50424 59027 83915 80952 45952 89416 94936 <br> 94617 22215 77556 43231 22491 14595 58838 94126 60008 51006 <br> 50490 46633 69935 81246 36384 32136 66571 26607 95116 82545 <br> 82225 90697 18812 89512 98694 77572 92239 53667 09362 13336 <br> 66302 7 <br> $/2^{95}3^{88}5^{56}7^{46}11^{42}13^{40}17^{36}19^{33}23^{30}29^{24}31^{22}37^{15}41^{11}43^{9}47^{5}$ |
| 52 | $-$62869 02501 74877 43540 48546 34695 86189 51266 92243 38153 <br> 69457 26947 09243 48925 49143 36418 81516 08009 48486 70278 <br> 01775 78359 76742 81890 31457 10545 43940 67683 85919 30242 <br> 17486 25202 85857 18476 02798 70372 00401 04338 27170 65772 <br> 03047 91631 69138 05550 04127 37860 30000 76676 59992 88696 <br> 60025 36273 66676 48172 24257 64941 17264 80527 78006 17495 <br> 36863 43027 85517 05687 44502 87597 66472 45668 31261 09368 <br> 03323 59406 91997 31426 94169 36857 54601 65254 36760 32155 <br> 43526 87747 77525 28355 87723 13555 53497 12139 06315 64840 <br> 97018 60216 34527 48572 55339 18344 09271 89927 62719 90876 <br> 25202 82959 78330 60948 78049 57330 46084 36548 45725 51486 <br> 48927 64546 39084 44323 77744 19672 59889 55303 22205 43833 <br> 57234 37254 46932 11259 79553 66043 95424 15280 08225 15763 <br> 11633 69605 43447 07798 07040 68747 <br> $/2^{98}3^{90}5^{58}7^{47}11^{43}13^{41}17^{37}19^{35}23^{31}29^{25}31^{23}37^{17}41^{11}43^{10}47^{6}$ |

**Table F.11**: Continuation of preceding table

| $n$ | $E^{(n)}$ |
|---|---|
| 53 | 50996 33459 97235 50996 20590 13986 68297 92731 91117 01217<br>28275 24932 29058 59112 91950 99142 57618 63041 30870 39269<br>91670 04384 02340 46319 68381 69846 10854 93423 21804 52805<br>53999 64036 79623 61592 98570 33293 82250 62269 46880 98946<br>57484 28832 44613 26573 29864 48688 30451 58627 51160 71968<br>53831 13545 45772 38841 61076 49286 30918 27452 74161 21366<br>39903 99477 28701 39924 60974 09271 91427 46583 41540 57397<br>97901 31348 38145 06476 62553 82200 00910 09045 83498 87430<br>83690 16863 15338 87384 13272 31284 99141 73139 10759 17983<br>18189 70613 67518 95704 29916 33662 13097 84971 34791 65651<br>73960 60260 97432 14253 59131 33509 58430 87238 32283 45564<br>59526 98882 92547 62020 40665 83940 95872 42839 06563 32208<br>51748 62336 01544 60234 14239 17156 80070 91129 06336 05306<br>78024 26514 34912 41074 69597 52159 41929 35847 38049 89301<br>4077<br>$/2^{101}3^{92}5^{60}7^{48}11^{44}13^{42}17^{37}19^{36}23^{32}29^{26}31^{24}37^{18}41^{13}43^{11}47^{7}$ |
| 54 | $-$85021 15388 44975 25988 44865 55855 20641 47062 55212 53984<br>36054 34063 60755 42219 65879 99131 53213 33601 81765 77587<br>17159 82971 89869 60888 03029 58372 18360 76536 02268 31053<br>13154 98374 64007 95175 31428 40926 60675 36421 55187 51216<br>20928 12039 87524 56902 13517 72404 07891 27567 60098 79706<br>00647 27709 65299 31518 05441 17874 45683 78306 46945 65119<br>94092 39625 80502 60347 41935 68280 53124 68638 67225 51786<br>70784 77999 81222 77325 79356 06739 71141 56787 23695 40768<br>59403 20993 49797 81746 66815 89353 47643 04523 26305 78869<br>92703 04384 72395 79488 50960 00365 86361 17930 80928 00743<br>22608 68020 87791 46621 74159 23772 41777 97657 22229 14146<br>63572 11851 06221 51897 64352 11240 31083 79466 74805 92254<br>33867 36298 84936 18314 49960 16483 00925 05564 77287 27571<br>97615 30048 90777 60934 31860 16339 03222 68496 59082 18228<br>98223 11806 25015 61723 43285 67556 27<br>$/2^{104}3^{94}5^{62}7^{49}11^{45}13^{43}17^{39}19^{37}23^{33}29^{27}31^{25}37^{19}41^{14}43^{12}47^{8}53^{2}$ |

**Table F.12**: Continuation of preceding table

| $n$ | $E^{(n)}$ |
|---|---|
| 55 | 23313 91553 19077 77446 48612 14066 81409 45288 67099 26310 51311 49115 75003 64752 60902 28468 72664 52434 21289 77944 07485 93505 87424 90574 08222 88555 05157 93815 92047 84139 50717 31641 46373 94296 70856 02503 17719 97179 18960 82742 98207 14798 28444 84527 92153 34969 94910 31304 29017 22653 71288 45817 78204 47036 65214 08208 42307 40517 95759 24184 60854 96432 26987 65731 41104 24757 98573 36191 42464 89564 40900 20934 82291 75239 28512 13820 24547 51239 86002 73208 96187 26660 62761 76742 15998 39214 13167 15849 95125 81486 63339 29758 23326 96128 49225 96201 78915 71779 87244 32839 91448 10050 19394 06052 45975 19190 40630 19016 41955 44467 62778 38962 13873 40113 46075 27418 54829 93522 44221 66182 71848 14083 65587 22069 47543 43958 40422 57283 35384 02024 12838 41188 93620 07730 26115 86623 05900 56991 33505 71147 93861 87473 97805 96207 70609 09191 82217 85863 47874 91266 13929 11 <br> $/2^{107}3^{96}5^{64}7^{49}11^{46}13^{44}17^{40}19^{38}23^{34}29^{28}31^{26}37^{20}41^{15}43^{13}47^{9}53^{3}$ |
| 56 | $-$32474 02861 45657 24231 78273 79795 08188 44195 31316 30180 12429 91348 52391 11475 98226 63088 06363 05531 52951 70630 74378 83068 48392 79595 15441 53761 58978 14943 87690 00938 87937 87020 48661 45248 95141 15571 33963 87069 17827 43573 32530 28946 10626 72640 27813 84109 41295 08560 53702 31317 13215 84723 62945 19852 50079 06472 24947 35162 40508 34570 10554 15834 20774 13072 90223 10847 13449 03789 58824 29888 49209 52242 02271 50451 39547 58805 15132 69342 92493 69156 69811 52492 59668 51489 87344 92764 51086 16542 30171 73562 07516 86579 76873 44453 73956 52767 41058 73444 75324 90147 98372 46190 54352 54667 88626 42805 74510 64755 03641 09990 35363 93492 45675 66656 69315 50040 39003 74081 02770 25042 53315 70337 77976 42986 63603 48700 42434 82027 33001 68826 80260 84865 50763 24471 07482 77608 45688 07473 06815 49204 16113 40363 00648 29365 53349 67791 25789 09795 08139 55979 77542 76428 09522 16324 82874 67692 627 <br> $/2^{110}3^{98}5^{66}7^{51}11^{47}13^{45}17^{41}19^{39}23^{35}29^{29}31^{27}37^{21}41^{16}43^{14}47^{10}53^{4}$ |

**Table F.13**: Continuation of preceding table

| $n$ | $E^{(n)}$ |
|---|---|
| 57 | 66945 20244 01943 23164 10427 52279 01802 60089 21831 43219 79078 09298 10670 90883 46187 13813 93888 71648 98007 12077 01798 99148 18210 37128 50996 03512 09887 11643 49673 59290 68175 90596 93873 89786 96133 10737 88098 08972 22299 66455 68030 71722 24886 42856 84581 04472 91648 35413 32314 67853 84249 95167 12735 19993 49672 94315 61983 19516 19088 13598 50776 99515 96593 58851 18081 10951 76423 61683 96962 20845 19642 31272 91698 58869 93764 07772 16974 34946 09714 66581 81429 45592 20491 12993 12555 05364 65047 67950 45823 93723 30111 58104 28082 50334 17121 22518 22537 84015 66028 76590 56297 36473 62938 22750 01025 29999 53141 00349 85572 91544 67060 53681 40100 57417 58790 28974 24874 11110 75497 50575 22828 52260 42666 96080 77866 32973 85812 77749 87652 69930 43430 03208 97248 33334 97504 53545 64245 18824 37041 96945 57709 07161 72762 05628 66529 06728 55049 73429 80035 46707 63557 36021 58729 96889 97109 40706 57099 93379 29850 29005 30897 377 $/2^{113}3^{100}5^{68}7^{52}11^{48}13^{46}17^{42}19^{40}23^{36}29^{30}31^{28}37^{22}41^{17}43^{15}47^{11}53^{5}$ |
| 58 | $-$58584 17287 18191 01555 72570 13066 65191 92535 36065 02435 27269 16210 44382 21271 18525 61962 20360 79048 10180 00571 35009 83776 36103 13737 27396 37936 14156 34876 40289 75820 65306 56346 53206 07592 47881 87501 87244 22873 31699 98409 62038 47472 87421 64047 74127 08091 74580 67229 66930 01831 17854 60044 93385 75420 07101 43110 51617 94662 64261 21023 32517 19827 14975 12634 96416 99095 52902 35042 71753 85367 45251 62085 63840 11727 99790 88517 59110 93559 96476 50992 94250 74173 70760 57417 54483 89600 51084 78733 01392 00872 29662 43293 22165 37635 09583 89475 72812 19823 96876 40376 97085 29303 26120 42504 03162 48596 66658 78647 85935 12790 69768 14615 72294 83318 05889 76181 68566 36553 19943 92052 31311 14934 58504 43292 75418 37919 81572 97024 10736 04267 01686 86560 76888 09696 88491 49572 15425 71694 22386 58142 46141 29205 23066 28984 08931 09626 43087 41774 11546 70689 33669 92995 53383 69356 20746 75130 95184 89927 50582 80461 94110 72380 36723 14174 60990 27747 34707 $/2^{116}3^{102}5^{70}7^{53}11^{49}13^{47}17^{43}19^{41}23^{37}29^{31}31^{29}37^{23}41^{19}43^{16}47^{12}53^{6}$ |

**Table F.14**: Continuation of preceding table

| $n$ | $E^{(n)}$ |
|---|---|
| 59 | 76338 81394 98121 44858 50174 93885 40036 74246 62978 52383 08471 14801 91111 05943 10076 89998 05671 11779 01422 39125 95213 38016 07710 10926 72708 47825 30548 85145 81711 82627 86179 58648 67389 89704 29020 26272 53557 71135 01530 06179 75838 83119 52052 05871 05299 36895 57656 51233 53270 39777 12678 99195 09784 35769 78090 79375 38400 29080 62942 79936 55312 50056 53606 94125 29603 46021 24842 99627 93604 63847 36018 65074 96321 58970 22523 85720 16024 64881 63430 87609 30825 20579 25460 32384 68523 42966 00418 07971 15996 93722 52880 77186 42174 03100 25776 74283 52565 79147 01895 61330 07594 64227 64182 10295 06902 01182 05820 68154 90004 45218 36648 85974 75993 14216 33281 33576 44169 41817 26258 44358 83636 87695 07780 75097 75200 84263 40977 23034 13832 44300 74230 80781 01494 68068 11044 27166 51963 98172 33116 75961 82071 70618 35149 70110 48818 58261 68788 42861 04758 75628 12035 36238 77628 97140 54684 49862 14476 20548 75570 06971 21171 03996 28613 68084 34714 07802 30411 87142 90043 92130 99764 25879 63 $/2^{119}3^{104}5^{72}7^{54}11^{50}13^{48}17^{44}19^{42}23^{38}29^{32}31^{30}37^{24}41^{20}43^{17}47^{13}53^{7}59^{1}$ |

**Table F.15**: Continuation of preceding table

| $n$ | $E^{(n)}$ |
|---|---|
| 60 | $-$44088 04922 99867 83117 08435 27901 22925 38056 10260 29327 72194 10665 37648 47689 24461 07805 71264 80225 52745 98888 03578 33430 61216 94624 52808 08358 85201 49235 12173 38915 88968 23208 64914 37955 85979 14345 97958 52214 46835 72099 65460 09336 46227 45463 85156 83337 29546 50619 36686 75563 83554 11117 10306 92574 96771 48065 12769 56148 60189 43115 78647 93639 51939 80690 07485 27915 71194 04013 66976 05869 14494 78274 11737 49132 62062 09392 11947 23406 39411 55022 69936 03589 68923 02275 31784 10176 91039 05417 87933 67822 31525 48886 67086 56527 24266 41039 07949 84099 84486 41620 70145 84979 60369 16509 76084 61211 11447 41786 60199 50473 80765 53591 34125 42259 20894 93864 28078 77471 40206 56950 13469 48631 89502 52716 27443 01189 67288 77537 23455 40196 51459 16812 80915 75051 06393 74941 41707 17822 57034 92508 06802 70408 83375 84645 89093 54906 57343 65490 18404 18663 11605 07151 74765 15609 14543 12427 74986 80884 01167 29010 72211 08532 60678 20076 19641 29470 29914 57115 99713 94886 82917 08323 67014 95811 00156 41316 64947 3743 $/2^{122}3^{107}5^{74}7^{54}11^{51}13^{49}17^{45}19^{43}23^{39}29^{33}31^{31}37^{25}41^{21}43^{18}47^{14}53^{8}59^{2}$ |

**Table F.16**: The Maple internal floating point format for the energy coefficients for the ground state Zeeman effect. $E^{(n)}$ is defined in terms of the integer $I_n$ and exponent $e_n$ by $E^{(n)} = I_n \times 10^{e_n}$.

| $n$ | $I_n$ | $e_n$ |
|---|---|---|
| 0 | -5000000000 0000000000 0000000000 0000000000 0000000000 0000000000 | -60 |
| 1 | 2 | 0 |
| 2 | -1766666666 6666666666 6666666666 6666666666 6666666666 6666666667 | -58 |
| 3 | 6201111111 1111111111 1111111111 1111111111 1111111111 1111111111 | -57 |
| 4 | -3995814259 2592592592 5925925925 9259259259 2592592592 5925925926 | -55 |
| 5 | 3862135590 4938271604 9382716049 3827160493 8271604938 2716049383 | -53 |
| 6 | -5136115985 5040740740 7407407407 4074074074 0740740740 7407407407 | -51 |
| 7 | 8965034762 8758956294 6166120769 2953724699 7564457881 9181993785 | -49 |
| 8 | -1991346961 3544537849 3446632033 1777625806 3484064995 7807932109 | -46 |
| 9 | 5499249533 0792716107 6206863097 0393439049 3317415056 1281780613 | -44 |
| 10 | -1852372402 3975863794 1196219223 6516526578 5663436619 3426232996 | -41 |
| 11 | 7487558558 8547621547 1517816245 4941743061 8023409979 8827072277 | -39 |
| 12 | -3580931342 8750349215 7378053830 8441425966 7348265423 7529028080 | -36 |
| 13 | 2001403853 2254331208 6266785635 7932102879 0331013368 0118874766 | -33 |
| 14 | -1293188575 0382786774 2787517332 3541302533 9558292939 3381848582 | -30 |
| 15 | 9568644832 6709905344 8074764448 4241539222 7953298572 0252856100 | -28 |
| 16 | -8039993326 4373818481 8033780773 0557456163 6178344134 8191793930 | -25 |
| 17 | 7614512001 9600987415 0789313830 5200157221 5094721418 7393166173 | -22 |
| 18 | -8074619171 2170910018 1742159442 2753952005 1551528659 6940503913 | -19 |
| 19 | 9530269072 3243963162 9373697027 9655763195 5652635532 7197244133 | -16 |
| 20 | -1245248197 4957420572 3534506065 8630579730 8342093664 4348776005 | -12 |
| 21 | 1792498626 0739848589 5635515468 5659032159 4063080352 8082177897 | -9 |
| 22 | -2830018778 6212248801 1445556152 6617977510 4014761967 5525335153 | -6 |
| 23 | 4880781995 1230166177 0465611115 7109706799 3958182881 6029075699 | -3 |
| 24 | -9161093134 5707478244 7545618093 3840890456 7314253016 7126353861 | 0 |
| 25 | 1865005161 6444779042 0360264123 5340167564 2233438356 8082826690 | 4 |
| 26 | -4105053704 2402042683 5600378862 7448915393 7262453064 9452101976 | 7 |
| 27 | 9740842157 5797999581 0154097891 5716885600 7058991426 2533657830 | 10 |
| 28 | -2485058297 5605700291 3763541261 8096703970 6524470462 9504740083 | 14 |
| 29 | 6798984299 0703764440 3387479601 9571829432 0343696844 9896664535 | 17 |
| 30 | -1990201928 1150123689 7932666674 6449095899 6924783577 6421544806 | 21 |

Table F.17: Continuation of preceding table.

| $n$ | $I_n$ | $e_n$ |
|---|---|---|
| 31 | 6219283829 9816922506 9479060701 4818984573 3548179144 9875581760 | 24 |
| 32 | -2070514318 3334960441 4929070312 4585390293 8944197422 6736202011 | 28 |
| 33 | 7329422874 6512868992 8211785264 3183945793 6704888798 5765984188 | 31 |
| 34 | -2753745459 9318819840 2008423601 7401323779 4244431326 3141978775 | 35 |
| 35 | 1096211987 6999938867 3386413339 8135344394 3550857377 0836416423 | 39 |
| 36 | -4616129417 5377410308 9936190094 9658982836 7982327385 1190917889 | 42 |
| 37 | 2053091509 9108954448 5961186368 6526478792 9223902269 5283324976 | 46 |
| 38 | -9630637171 9794421713 0616512265 9963774732 1734026328 5239618314 | 49 |
| 39 | 4757952880 1743069948 8279249294 0745985070 8322076717 4763490657 | 53 |
| 40 | -2472496913 6842035019 9782443792 4357935430 5445887353 9731168262 | 57 |
| 41 | 1349776519 0915370593 7427542508 0190173594 3902076147 8364085158 | 61 |
| 42 | -7731866220 2931370692 9135801032 1316750249 1798483028 5604633095 | 64 |
| 43 | 4642084843 6490799889 1213617958 0274524487 0670195245 4725835179 | 68 |
| 44 | -2917967460 8571408595 1654472437 2219030638 0284733616 3827724407 | 72 |
| 45 | 1918402399 6649939938 4206251594 3072792969 5070462932 9166276287 | 76 |
| 46 | -1317843017 9776984298 9159136408 3433804881 4195434278 6112388621 | 80 |
| 47 | 9450253642 3348306543 1060114949 0821914242 1602935754 3187474500 | 83 |
| 48 | -7067849889 2061040008 8754037952 7855904830 9923535160 1902139551 | 87 |
| 49 | 5508325306 7068766761 7892368807 8565646047 3458588489 0176785292 | 91 |
| 50 | -4469713850 2783094169 4335584596 1073531596 4049023594 9075760835 | 95 |
| 51 | 3773300819 6609090538 0931355590 9325285524 2526733901 5110224935 | 99 |
| 52 | -3311401215 7414233983 9314473236 6946714806 2303173818 2473805936 | 103 |
| 53 | 3018773842 9514569308 7676432354 8518364241 4505978258 9555812316 | 107 |
| 54 | -2856731499 6317204852 8900076067 5547444842 9271384014 0971233146 | 111 |
| 55 | 2804344030 7020295699 2357882105 2877622192 5870778154 1738105631 | 115 |
| 56 | -2853841207 8578464607 7284231346 9356938342 0102692676 6858166169 | 119 |
| 57 | 3008767977 6126641341 1972210677 7779132374 5269727837 9575183772 | 123 |
| 58 | -3284288415 2433289891 2683406185 4439355375 4332897262 7639680663 | 127 |
| 59 | 3709624795 1582067325 2868535164 5991341371 8805831554 4389091089 | 131 |
| 60 | -4333171096 2876629344 9808003210 8866012594 9217901117 2712556102 | 135 |

# Appendix G

# Tables of Charmonium Energy Corrections

## G.1  Ground State Energy to Order 100

In Tables G.1 to G.8 the 3-dimensional ground state energy corrections $E^{(n)}$ for the charmonium potential are given to order 100 in rational form. They were obtained using program *charmgs* in Section 10.2.3 with *jmax* := *50* followed by program *symcharmgs* with *jmin* := *51* and *jmax* := *100*. They can also be obtained more efficiently using the HVHF method and procedure *hvhfpower* given in Section 10.6.

The floating point values of the energy corrections expressed in the Maple internal floating point format are also given to 60 digit accuracy in Tables G.9 to G.12.

**Table G.1**: Ground state energy corrections for charmonium. The denominators are expressed in prime power factorized form.

| $n$ | $E^{(n)}$ |
|---|---|
| 0 | $-1/2^1$ |
| 1 | $3/2^1$ |
| 2 | $-3/2^1$ |
| 3 | $27/2^2$ |
| 4 | $-795/2^4$ |
| 5 | $3843/2^3$ |
| 6 | $-5583$ |
| 7 | $95433\ 39/2^7$ |
| 8 | $-11410\ 62999/2^{10}$ |
| 9 | $18769\ 07155\ 5/2^{10}$ |
| 10 | $-13436\ 99301\ 873/2^{12}$ |
| 11 | $51910\ 28367\ 4773/2^{13}$ |
| 12 | $-43022\ 61498\ 08531\ 7/2^{15}$ |
| 13 | $23790\ 21785\ 62833\ 51/2^{13}$ |
| 14 | $-22382\ 02979\ 62938\ 9225/2^{15}$ |
| 15 | $44634\ 95181\ 25914\ 06526\ 5/2^{18}$ |
| 16 | $-18813\ 77976\ 15156\ 92676\ 7995/2^{22}$ |
| 17 | $52244\ 25597\ 30883\ 23249\ 93495/2^{22}$ |
| 18 | $-61028\ 30936\ 33587\ 59185\ 20315\ 35/2^{24}$ |
| 19 | $37405\ 45983\ 36814\ 41600\ 90979\ 6725/2^{25}$ |
| 20 | $-48022\ 88067\ 04592\ 08898\ 72593\ 70740\ 5/2^{27}$ |
| 21 | $80565\ 43787\ 20364\ 90319\ 85696\ 84534\ 95/2^{26}$ |
| 22 | $-28210\ 27134\ 05629\ 03263\ 13985\ 53259\ 5395/2^{26}$ |
| 23 | $16466\ 91999\ 91300\ 53563\ 08473\ 24961\ 13674\ 65/2^{30}$ |
| 24 | $-49998\ 52233\ 84895\ 62645\ 87487\ 53191\ 49123\ 9075/2^{33}$ |
| 25 | $19713\ 59177\ 17907\ 16091\ 88815\ 42585\ 89687\ 93788\ 7/2^{33}$ |
| 26 | $-32255\ 98687\ 64210\ 71277\ 64738\ 31833\ 67007\ 30149\ 089/2^{35}$ |
| 27 | $27343\ 77487\ 99845\ 80087\ 33774\ 35717\ 50715\ 27284\ 60273/2^{36}$ |
| 28 | $-47980\ 02704\ 37590\ 76695\ 07672\ 20980\ 11939\ 04497\ 32985\ 69/2^{38}$ |
| 29 | $67998\ 14894\ 09215\ 77085\ 24064\ 54039\ 18405\ 24736\ 22543\ 05/2^{33}$ |
| 30 | $-10191\ 23767\ 47703\ 85166\ 93063\ 18939\ 18148\ 60247\ 81496\ 14517\ 3/2^{38}$ |

**Table G.2**: Continuation of preceding table.

| | |
|---|---|
| 31 | 39397 14315 35400 47451 31647 15095 59778 32349 88125 75560 677/$2^{41}$ |
| 32 | −62795 58720 86429 50809 41925 08928 37709 54137 27233 252799 81419/$2^{46}$ |
| 33 | 32212 98492 64104 00209 70512 78358 34183 23095 53464 85239 6920 9943/$2^{46}$ |
| 34 | −68017 64040 13854 72581 89732 77163 94950 58015 15592 49022 97811 40715/$2^{48}$ |
| 35 | 73836 52219 87245 34130 17649 58242 51494 50111 74824 10759 67909 46017 57/$2^{49}$ |
| 36 | −16470 92090 81133 64195 09639 34908 78078 95671 33393 50452 08395 77530 70189/$2^{51}$ |
| 37 | 47155 86910 10064 26907 64383 93203 18660 89014 32318 68688 82176 41966 89185 7/$2^{50}$ |
| 38 | −27704 74856 40196 16686 94802 68806 05415 91219 87925 80550 48018 65468 49238 509/$2^{50}$ |
| 39 | 26704 63971 58277 91106 90959 86306 96338 27333 24004 72794 57374 75689 62512 07282 5/$2^{54}$ |
| 40 | −13189 24400 32708 61668 35303 02756 23982 80266 31766 65798 70289 70392 70665 66071 7945/$2^{57}$ |
| 41 | 83395 49529 88750 00538 03862 36818 02239 66643 55613 92165 99565 39239 17633 35659 48685/$2^{57}$ |
| 42 | −21590 61542 93637 41968 75604 99298 63165 79717 40615 80777 97334 90973 25044 77835 06097 415/$2^{59}$ |
| 43 | 28593 55351 68017 44125 25621 38260 64240 38783 18560 62841 65195 09437 00230 89744 14461 23435/$2^{60}$ |
| 44 | −77444 90137 02124 37101 30414 44641 71056 49997 22450 26809 45787 18940 22956 05832 91493 85840 75/$2^{62}$ |
| 45 | 13399 21709 30086 97604 44748 79172 87020 43134 93131 47524 19341 22772 71689 61468 65363 07789 2935/$2^{60}$ |
| 46 | −37893 69373 38660 54864 24451 22306 25054 23029 34534 53766 52728 31298 83961 32047 41169 83135 99009 5/$2^{62}$ |
| 47 | 21886 31844 63989 29818 41291 09235 82802 69016 28710 39292 82456 02708 77833 28314 32515 13874 11286 5935/$2^{65}$ |
| 48 | −25805 47292 97824 09857 12426 47428 55539 67983 78519 68446 29139 72972 59112 59448 11915 60066 35319 07040 95/$2^{69}$ |

**Table G.3**: Continuation of preceding table.

| | |
|---|---|
| 49 | 19402 45457 59184 93097 76498 59776 37221 80944 67127 69218 44486 55865 80740 46553 27619 47096 45572 85967 6075/$2^{69}$ |
| 50 | −59513 64091 27320 03474 29461 16803 57728 34209 56095 19673 37889 83844 26721 33928 68213 98707 25318 52250 24501 9 /$2^{71}$ |
| 51 | 93054 44935 85362 03894 91141 73604 60147 52029 47400 05532 77338 94026 83550 09322 71500 27826 46099 73503 57951 489 /$2^{72}$ |
| 52 | −29656 53777 91613 70314 37245 28227 17837 51175 74345 95356 95841 67164 84461 14096 82120 36766 41943 59482 98465 33204 1/$2^{74}$ |
| 53 | 12036 33226 15183 84029 33061 48529 31333 17994 59923 74153 59977 79864 87451 43943 20043 56739 36239 51125 25798 40118 351/$2^{73}$ |
| 54 | −99501 92742 43808 14956 58319 73007 22287 60779 25289 19740 03883 71651 34208 78073 69089 23091 85455 31201 43177 60375 4215/$2^{73}$ |
| 55 | 13399 26031 23811 62861 61029 25856 57197 47900 97512 50056 84224 80284 00041 09864 09995 44181 66606 99699 05386 50549 01019 797/$2^{77}$ |
| 56 | −91823 85651 86631 73442 22309 52822 24760 04455 12385 36532 08345 06186 06599 03178 09519 95484 10144 20397 85104 74502 87530 01083/$2^{80}$ |
| 57 | 80031 99519 20983 99615 22236 83241 01618 63297 61900 10670 02624 33351 96141 52024 88218 72297 34781 57445 81748 06992 76903 19168 23/$2^{80}$ |
| 58 | −28380 99547 61603 43034 90141 84130 35832 33493 31564 31590 09220 26044 60115 50251 79431 36511 48331 56119 44601 37947 55718 61366 21889/$2^{82}$ |
| 59 | 51172 21164 98130 52594 63911 74953 03853 70278 76656 70772 93168 17090 98421 35849 45752 46283 99509 66230 11903 17402 59507 17653 34509 85/$2^{83}$ |
| 60 | −18759 62060 57115 80458 73523 61411 12013 55926 32950 09795 73977 06796 77340 51933 04085 54241 36070 62173 90490 66230 59257 23451 68613 90641/$2^{85}$ |

**Table G.4**: Continuation of preceding table.

| | |
|---|---|
| 61 | 21842 47250 00237 91624 89549 66551 03597 57336 34289 44648 38403 08976 07701 07466 57733 58159 52235 83681 10157 84998 43248 83695 30458 94020 $1/2^{82}$ |
| 62 | $-16538$ 09894 18507 95318 56406 08732 84209 42529 40054 43301 84127 71321 37173 47193 83034 32569 19666 04711 08117 34014 23488 03312 97044 24286 $5333/2^{85}$ |
| 63 | 12720 00048 77765 35799 57842 34177 52768 72342 52221 99086 19749 15943 16259 62600 67730 54952 62687 91541 83351 85010 75459 02342 53720 75643 50236 $93/2^{88}$ |
| 64 | $-79486$ 14199 49385 11035 82145 75486 39356 30051 42576 26031 33564 43490 51322 57634 26922 61791 12758 43169 54689 30111 48476 35380 83057 32769 54618 $35275/2^{94}$ |
| 65 | 78800 08286 42822 39256 44887 95056 05857 45160 53773 85197 40404 73557 57205 36247 42614 44637 55258 23812 06558 82493 03619 99491 77144 43052 33336 49387 $75/2^{94}$ |
| 66 | $-31720$ 06511 77613 29037 90796 27027 69670 60938 49390 13545 82269 61827 06433 04640 05926 48280 73072 92971 80069 68285 02793 58715 06671 23906 73693 71185 $45475/2^{96}$ |
| 67 | 64792 94589 02045 11830 95650 45271 71384 10860 32235 33715 75346 25482 14415 64925 48541 97450 16833 06919 35160 34729 20086 20067 28535 50052 58886 48569 70141 $25/2^{97}$ |
| 68 | $-26858$ 05215 29342 31021 34422 74147 13148 73968 18191 63963 24083 22660 97359 13844 06437 28180 42479 18034 47522 83134 71883 79686 06301 45991 38233 54793 98464 $56525/2^{99}$ |
| 69 | 14117 70106 89713 46950 48332 16194 57851 29178 65608 22089 04699 52084 81938 00197 41814 79771 53765 12093 27093 79713 59117 46104 32589 49183 21430 97606 32279 40971 $25/2^{98}$ |
| 70 | $-15053$ 17786 79570 47450 91240 75425 88470 37251 26192 46929 69678 85974 05313 16028 75847 88383 83433 42795 69956 96552 20925 63718 12640 77886 63786 75534 75887 64964 $6325/2^{98}$ |
| 71 | 26041 83899 38118 85872 51107 55876 75560 36335 14236 81371 04870 19490 22791 78574 96907 67227 95312 48794 94650 57579 17514 21475 44282 12166 12117 19424 00971 05814 58968 25 $/2^{102}$ |

**Table G.5**: Continuation of preceding table.

| | |
|---|---|
| 72 | $-22838$ 16602 62648 75201 28872 56747 66060 75639 90837 71204 05301 03253 07069 05743 12699 85287 66535 07030 60817 87443 75301 58134 97202 11367 36514 54687 27412 35395 48358 87325 $/2^{105}$ |
| 73 | 25377 92026 09157 51790 17463 76827 51614 73345 22923 45188 65202 73387 88295 14733 40518 77875 42462 58961 97870 25724 09124 07156 10019 07700 03585 65579 66237 81405 65431 00282 $25/2^{105}$ |
| 74 | $-11432$ 13351 88435 32143 74836 96689 00117 79692 00095 99780 32655 33073 22247 96265 40063 64315 02679 02490 42160 87262 07401 03880 17920 80837 61881 85351 98180 27656 34969 75222 $60875/2^{107}$ |
| 75 | 26092 05827 69736 33676 31731 90605 26649 46599 95552 25530 92827 06973 00781 48040 53055 93661 53358 86227 48222 06720 13923 96915 48892 37036 23082 58504 05189 92564 72004 19305 14121 $35/2^{108}$ |
| 76 | $-12066$ 58807 56132 07861 59446 94061 62172 33323 03360 04036 68046 76516 48984 39880 45695 35690 32661 11979 54462 90721 24778 70562 54078 52698 90746 66562 48689 01899 27402 86430 52598 $95895/2^{110}$ |
| 77 | 35329 13280 74033 56750 47046 74379 12109 27632 37440 90165 99235 93352 28619 94398 94178 11584 80285 71062 86107 80289 08790 40722 90336 89900 53830 82178 26523 92725 32308 29706 27488 38170 $5/2^{108}$ |
| 78 | $-16761$ 88443 80752 04309 88858 82833 55130 74996 63077 37176 46355 57488 71896 58914 07062 56639 27952 64511 22850 75315 10084 46706 44539 50439 73606 71595 76010 35854 98459 82548 96607 79462 $2115/2^{110}$ |
| 79 | 16106 26846 23270 84768 69719 75304 63064 09750 56016 41055 92926 41389 39870 01507 54073 33282 35862 15202 55242 09949 55540 19038 88170 42446 56489 00460 47034 45448 05408 11357 83951 96680 84176 $75/2^{113}$ |
| 80 | $-31338$ 75923 88565 28641 49499 32767 15411 33902 32883 15615 78458 09210 58327 75157 35235 48505 59519 00809 74954 82347 50948 22167 33526 54609 82249 03181 03096 12590 34830 10084 09657 90740 41730 $99765/2^{117}$ |

**Table G.6**: Continuation of preceding table.

| | |
|---|---|
| 81 | 38580 48181 65035 38125 23068 50277 90764 07361 53210 90855 43563 17821 17115 34567 54192 62348 11181 72176 68076 92015 12899 17399 12042 58193 18590 19999 08119 18245 05921 14089 92715 76160 42313 03219 $45/2^{117}$ |
| 82 | $-19229$ 51059 06984 84096 46064 75621 79105 53606 40053 72639 31426 11252 66907 79773 61052 48549 66971 24571 01981 96196 54332 73373 58571 97017 95267 56181 12503 87913 24010 82142 60672 56279 20523 79444 $98705/2^{119}$ |
| 83 | 48498 78606 00665 31595 74077 89050 14222 42955 66434 03002 44971 13404 85332 23521 75617 90644 07062 43437 79014 47423 89872 52538 91031 80975 92235 73746 70483 32074 07929 91854 66183 83223 05087 44670 21071 $95/2^{120}$ |
| 84 | $-24754$ 50974 19675 06048 80159 73624 33664 39168 75741 91270 36497 30912 50935 44933 98313 14285 67934 69899 17111 30776 73952 68302 56236 69134 81344 47528 26447 28973 95666 79991 70433 78692 51242 31629 78306 $81475/2^{122}$ |
| 85 | 15979 33997 03185 02450 25552 20305 12241 10896 30696 67241 89692 32226 27806 37465 56368 56419 28270 21369 23302 03760 44678 14522 44616 38178 59937 80526 83571 71034 58928 64159 53504 65077 22793 75928 58924 26321 $45/2^{121}$ |
| 86 | $-20869$ 20719 66904 62215 28005 62621 61187 25221 63728 62109 42481 01183 96961 99266 95600 04178 34698 25997 36860 90464 01560 52830 74872 79556 07454 82431 10862 96449 87731 31211 93997 64711 24050 49487 92094 66414 $3085/2^{121}$ |
| 87 | 44109 13751 98339 05640 31107 72420 60538 94599 01777 20900 90097 78023 52655 30012 52324 88885 70801 98353 92429 41662 91947 33424 99232 42211 91169 28025 56826 92288 72018 44327 33489 54697 22548 57292 52550 46743 13865 $35/2^{125}$ |
| 88 | $-47143$ 39923 88112 00634 16721 23236 41101 16447 14950 08310 44305 07495 86623 71800 71843 74020 93282 88808 75413 94902 25332 09456 84122 29770 22139 86184 55369 75761 93970 34625 65888 09398 31176 93373 33293 17436 81029 $65285/2^{128}$ |

**Table G.7**: Continuation of preceding table.

| | |
|---|---|
| 89 | 63689 57838 70763 92540 75584 36408 81423 76377 83211 36926<br>79523 18993 63074 97391 52841 95705 10789 46791 17584 15995<br>73912 15161 67012 29410 71247 95213 86007 96123 52592 63718<br>05017 77650 83075 17941 22838 33151 03711 70396 $25/2^{128}$ |
| 90 | $-34799$ 06822 58607 26119 27125 06874 16728 33458 91238 66793<br>93611 86536 00596 98503 17817 07049 60598 59849 14106 19537<br>74802 20748 99667 12473 29549 73020 57819 18875 13516 28636<br>79811 95840 91313 79497 16029 69780 74010 52648 96775<br>$/2^{130}$ |
| 91 | 96111 73629 33108 40559 42528 63912 21145 83708 57612 40562<br>58045 59944 33122 54766 82908 65084 02514 04562 01732 52450<br>40157 04637 00193 21819 95893 73132 78070 19008 98201 10769<br>38982 01379 56791 59660 15835 32937 12509 77219 18647 75<br>$/2^{131}$ |
| 92 | $-53666$ 55315 68039 61014 19731 09787 84798 04776 21004 29957<br>97431 28513 05709 09927 79405 31400 16626 36805 30371 03875<br>20597 17113 19517 89198 52226 41351 26360 41400 96199 35239<br>43483 52623 42716 83223 83848 64457 17152 41005 24293 59375<br>$/2^{133}$ |
| 93 | 47324 87631 55665 37753 70447 23806 87533 06507 72201 81984<br>19172 19869 10482 57098 42602 31494 39053 80730 16122 86007<br>64779 60736 95107 93380 07941 53273 67064 22938 07891 09857<br>01301 66993 33812 23607 84728 52438 38904 80660 31515 94642<br>$5/2^{129}$ |
| 94 | $-10797$ 04155 21226 36785 44147 57074 59520 99085 47471 70898<br>06157 36308 01350 56414 08029 67814 22397 48550 32222 48923<br>76610 73448 72486 34910 42850 42809 97859 00428 03939 99672<br>77792 64497 60430 70205 75146 15620 91703 61951 06781 85422<br>$65875/2^{133}$ |
| 95 | 12446 05378 63922 41560 55298 26676 53846 83144 96345 82994<br>66840 10330 66728 78753 18502 27384 77875 81893 15858 46750<br>90205 24635 99242 20744 89715 57797 50343 29205 43475 56935<br>41407 87391 22969 86404 86123 41461 31877 65287 07983 49972<br>49782 $675/2^{136}$ |

**Table G.8**: Continuation of preceding table.

| | |
|---|---|
| 96 | $-$57984 67695 87909 92950 31176 44002 99093 55644 91384 04518 24930 01546 04716 78429 27458 07656 65378 39242 44999 52681 04872 74518 71854 29120 65900 27647 27017 80377 12086 37778 92360 19271 22504 45185 58818 75816 35187 50645 39714 15079 50778 60497 $5/2^{141}$ |
| 97 | 85289 10016 50645 14317 03912 51112 31341 75710 55426 61392 97462 06033 40671 54290 10176 45908 22370 24736 57258 89366 91850 71475 24077 67432 04935 93966 71701 49050 93131 82388 72312 64390 68568 40680 90566 34360 69489 76275 21616 67004 44700 25136 $875/2^{141}$ |
| 98 | $-$50691 77143 41739 49840 75493 14578 14096 08827 65181 10196 94879 28995 73522 73774 89595 18669 49281 97178 71448 07609 55864 89596 40623 73780 48569 08958 66431 92441 43838 66081 68728 93717 20720 06502 21650 09454 41559 16989 27300 60721 58070 89949 55657 $5/2^{143}$ |
| 99 | 15216 36498 57861 83578 21918 30174 14440 31050 10879 95863 68394 05137 40904 66194 98251 97963 88651 75283 19499 14731 52721 15996 11289 48552 23528 23637 89383 11165 44711 32035 74336 86179 75104 36328 90627 92815 15796 34023 19116 32990 49894 97714 78048 $9625/2^{144}$ |
| 100 | $-$92263 64344 17864 93050 55310 50997 65287 02660 00815 11380 49999 95933 63777 68500 17625 86815 41626 91466 89526 09257 39482 43762 44175 49506 82120 15842 43974 46631 67778 65721 35440 11737 66735 94449 90282 35603 50775 71622 66035 78680 61382 20345 83640 72826 $5/2^{146}$ |

**Table G.9**: The Maple internal floating point format for the energy coefficients for the ground state of charmonium. $E^{(n)}$ is defined in terms of the integer $I_n$ and exponent $e_n$ by $E^{(n)} = I_n \times 10^{e_n}$.

| $n$ | $I_n$ | $e_n$ |
|---|---|---|
| 0 | -5000000000 0000000000 0000000000 0000000000 0000000000 0000000000 | -60 |
| 1 | 1500000000 0000000000 0000000000 0000000000 0000000000 0000000000 | -59 |
| 2 | -1500000000 0000000000 0000000000 0000000000 0000000000 0000000000 | -59 |
| 3 | 6750000000 0000000000 0000000000 0000000000 0000000000 0000000000 | -59 |
| 4 | -4968750000 0000000000 0000000000 0000000000 0000000000 0000000000 | -58 |
| 5 | 4803750000 0000000000 0000000000 0000000000 0000000000 0000000000 | -57 |
| 6 | -5583 | 0 |
| 7 | 7455733593 7500000000 0000000000 0000000000 0000000000 0000000000 | -55 |
| 8 | -1114319334 9609375000 0000000000 0000000000 0000000000 0000000000 | -53 |
| 9 | 1832917144 0429687500 0000000000 0000000000 0000000000 0000000000 | -52 |
| 10 | -3280515873 7133789062 5000000000 0000000000 0000000000 0000000000 | -51 |
| 11 | 6336704550 1431884765 6250000000 0000000000 0000000000 0000000000 | -50 |
| 12 | -1312946013 8199819946 2890625000 0000000000 0000000000 0000000000 | -48 |
| 13 | 2904079328 1595887451 1718750000 0000000000 0000000000 0000000000 | -47 |
| 14 | -6830453429 0447669219 9707031250 0000000000 0000000000 0000000000 | -46 |
| 15 | 1702688286 3079607592 2012329101 5625000000 0000000000 0000000000 | -44 |
| 16 | -4485554638 2703048412 3218059539 7949218750 0000000000 0000000000 | -43 |
| 17 | 1245600127 5322037518 0089354515 0756835937 5000000000 0000000000 | -41 |
| 18 | -3637570700 8456444254 6386441588 4017944335 9375000000 0000000000 | -40 |
| 19 | 1114769573 0233622074 3983378186 8219375610 3515625000 0000000000 | -38 |
| 20 | -3577983429 3171174003 7247069731 7272424697 8759765625 0000000000 | -37 |
| 21 | 1200518576 3841344463 8039124687 7536177635 1928710937 5000000000 | -35 |
| 22 | -4203658005 6790863369 6136702983 8852584362 0300292968 7500000000 | -34 |
| 23 | 1533601432 9576914723 1101361067 1616592444 4794654846 1914062500 | -32 |
| 24 | -5820594068 9064518834 7842489824 9693669786 2111032009 1247558594 | -31 |
| 25 | 2294964130 4778303126 6743250407 6241508345 9345623850 8224487305 | -29 |

**Table G.10**: Continuation of preceding table.

| $n$ | $I_n$ | $e_n$ |
|---|---|---|
| 26 | -9387727732 6598621665 1723463535 5127772051 7507987096 9057083130 | -28 |
| 27 | 3979042940 7708259659 3212562900 5999547785 3266202146 1874246597 | -26 |
| 28 | -1745503215 4888277253 1493982016 4434024080 8850856046 7744246125 | -24 |
| 29 | 7916026392 5932321098 2147667136 9560673758 6500012548 6403703690 | -23 |
| 30 | -3707550667 8849288316 7441755864 9346975463 4678300198 3567373827 | -21 |
| 31 | 1791574648 1567133492 4065702168 9902572847 9870201936 4450279681 | -19 |
| 32 | -8923789665 7782230831 5422374290 0187167687 8072499448 5089766272 | -18 |
| 33 | 4577740487 3221030654 9534491236 8719751463 4798502072 0615515160 | -16 |
| 34 | -2416472014 5373576294 6086139338 3018145902 6519252483 5591264575 | -14 |
| 35 | 1311600112 0523274559 9460912132 3639557863 2717191204 8468795223 | -12 |
| 36 | -7314558251 5869397056 0026253607 9582778786 6848936691 2728655426 | -11 |
| 37 | 4188282529 7717853370 1552958848 4122748516 7564620284 2722269437 | -9 |
| 38 | -2460675979 7780256959 2747045140 5608989876 4760964189 7618595746 | -7 |
| 39 | 1482405293 8415705541 9799490396 0840977393 3983806678 8677936986 | -5 |
| 40 | -9151876481 1441150527 5783023345 9689641311 5689929142 9297554850 | -4 |
| 41 | 5786724939 4268766389 5642511414 9340037830 4185893822 6945460209 | -2 |
| 42 | -3745374744 6104353831 1658984867 4493241418 8004288175 3524830086 | 0 |
| 43 | 2480095427 3597589192 3421355351 8829420751 3279353677 4051664504 | 2 |
| 44 | -1679318606 2701988682 6310665247 1795484943 3653293554 1608686550 | 4 |
| 45 | 1162196822 5475948214 8138704353 5847237939 6456038439 8289098380 | 6 |
| 46 | -8216885013 9515845939 1949301036 5457835002 7450103096 6030066949 | 7 |
| 47 | 5932298501 8238224843 5615743726 0606766247 2780625286 2477149877 | 9 |
| 48 | -4371617158 2009318411 8077987500 7261602129 4332465621 2203173941 | 11 |
| 49 | 3286903656 6815854435 6256668662 6488340770 8015103879 6055549564 | 13 |
| 50 | -2520500733 2072748893 6720711746 0221308707 1321397593 7436880117 | 15 |

**Table G.11**: Continuation of preceding table.

| $n$ | $I_n$ | $e_n$ |
|---|---|---|
| 51 | 1970504612 3821739603 7478745061 4222804254 0746551679 3963378127 | 17 |
| 52 | -1570004037 5275973936 7340422495 4199376832 1165794606 5430970041 | 19 |
| 53 | 1274396248 7007418617 8559489608 8668287399 0209173515 0088660610 | 21 |
| 54 | -1053517635 5469596903 1135001591 1832065859 5018904713 9486164241 | 23 |
| 55 | 8866886682 3623385989 5309834303 3230446238 1696728677 6230354749 | 24 |
| 56 | -7595491388 2089127098 0423857040 4441009353 4311830663 7483496972 | 26 |
| 57 | 6620091480 6840919974 3461405188 4473450226 3831030197 0706899916 | 28 |
| 58 | -5869052305 7087549751 1415393743 0274487466 6780908199 3295297906 | 30 |
| 59 | 5291082672 2731924696 9318727053 1149125748 4072862393 2714024720 | 32 |
| 60 | -4849248269 9671578701 2196759027 4877189890 7644269256 9817134427 | 34 |
| 61 | 4516917445 5605854377 4309354475 2261926537 0951818210 6026931342 | 36 |
| 62 | -4274998379 1195999027 8175143521 4342690416 5280622187 2359277232 | 38 |
| 63 | 4110053826 2319553500 2087173355 1989516222 4431752274 2584215927 | 40 |
| 64 | -4013024635 3057967988 8543864394 9945660059 5201212592 4507949060 | 42 |
| 65 | 3978387500 7877393930 0464786959 4458143200 4384088302 3174454580 | 44 |
| 66 | -4003635085 1946627933 2814088894 2801001791 0564943885 4806443389 | 46 |
| 67 | 4089009755 7506214229 7520742461 3093695236 1944478177 4216776709 | 48 |
| 68 | -4237453466 7673684970 2780785568 5712967986 4809963608 5414617334 | 50 |
| 69 | 4454760977 9633531103 7748438766 3192350357 3982643751 0732497682 | 52 |
| 70 | -4749945407 7477987884 1430677435 7005164395 9548970574 1296217213 | 54 |
| 71 | 5135847170 5695718853 5242986100 0576801447 1633752271 9234570342 | 56 |
| 72 | -5630042601 5834586021 1196378441 0424430892 7213520256 8408291756 | 58 |
| 73 | 6256140359 2664550188 7887835794 6504401976 5269656077 6571677253 | 60 |
| 74 | -7045596247 0956092101 7302771311 1829539224 3211697257 7666030117 | 62 |
| 75 | 8040236215 4112555584 9224055356 4718919390 1687938991 8707591434 | 64 |

**Table G.12**: Continuation of preceding table.

| $n$ | $I_n$ | $e_n$ |
|---|---|---|
| 76 | -9295761320 5637714837 9046721318 5244355326 7230667074 0092122148 | 66 |
| 77 | 1088662956 5281858645 1597416096 8827662387 1160649116 2348473598 | 69 |
| 78 | -1291288606 5458980929 4144242810 3197896510 8830489613 4421579383 | 71 |
| 79 | 1550977236 1967746542 9090346997 2650273766 5361248324 8239191641 | 73 |
| 80 | -1886132963 6892477499 4135519863 0272027979 7093199984 5172582796 | 75 |
| 81 | 2321978287 4778569764 0161847694 1234290979 0752508823 0162599745 | 77 |
| 82 | -2893335176 7607573698 6089338689 9498900904 0253316979 9231324073 | 79 |
| 83 | 3648643138 2622025243 9876968030 4204379429 7845033366 9898230464 | 81 |
| 84 | -4655805817 4492936593 0677981479 3617399666 1159033371 6818858536 | 83 |
| 85 | 6010759636 7930172716 5278337564 9612750835 9012288868 7719407824 | 85 |
| 86 | -7850123253 0718289318 4000388602 5093280393 9288405785 0272135287 | 87 |
| 87 | 1037000839 4841632357 6906644008 6482240601 9348932519 8496082660 | 90 |
| 88 | -1385419987 0945573834 8593170465 8211927675 5521912257 8737705240 | 92 |
| 89 | 1871668490 0065389286 3822970985 1805620010 6153192806 3095574817 | 94 |
| 90 | -2556631757 0861654081 9861497262 4894862982 8154620101 3006613596 | 96 |
| 91 | 3530587595 6408848748 2906893321 5507142448 4717682849 9015804631 | 98 |
| 92 | -4928494536 2443012879 9916163286 0925397280 1495031309 8700737291 | 100 |
| 93 | 6953765595 2889920986 2464866021 9658879142 9234603552 1987906016 | 102 |
| 94 | -9915516679 7762987810 3003059046 5678679194 7556369004 8755985616 | 104 |
| 95 | 1428736905 8529708388 9868631694 8766195146 0570226734 5743322023 | 107 |
| 96 | -2080098272 6841452197 5090857810 2382826492 3081922527 9834782570 | 109 |
| 97 | 3059596418 1747303954 1017016070 0236006023 4491279882 8571075415 | 111 |
| 98 | -4546195293 7352054872 2217594963 0613373953 9694585613 8381032081 | 113 |
| 99 | 6823254043 9001907153 1841001648 1457042154 4311369688 6192317020 | 115 |
| 100 | -1034311214 9439031403 4559361339 4228204111 0856715671 6152813772 | 118 |

## G.2　Symbolic Energy Corrections to Order 14

The symbolic energy corrections $E^{(n)}$ for a general charmonium state have been given in (10.49) to (10.52) to order 4 as functions of $q$ and $\tau = k(k+1)$. Here we give more extensive results to order 14 obtained using program *charm* in Section 10.2.3 with *jmax* := *14* or more efficiently using procedure *hvhfpower* in Section 10.6 with *nmax* := *14*.

$$E^{(1)} = \frac{1}{2}(3\,q^2 - \tau),$$

$$E^{(2)} = -\frac{1}{8}q^2(7\,q^4 + 5\,q^2 - 3\,\tau^2),$$

$$E^{(3)} = \frac{1}{16}q^4(33\,q^6 + 75\,q^4 - 7\,q^2\tau^2 - 10\,\tau^3),$$

$$E^{(4)} = -\frac{1}{64}q^6(465\,q^8 + 2275\,q^6 - 99\,q^4\tau^2 + 440\,q^4 - 90\,q^2\tau^3 \\ - 180\,q^2\tau^2 - 84\,\tau^4),$$

$$E^{(5)} = \frac{1}{64}q^8(1995\,q^{10} + 17340\,q^8 - 465\,q^6\tau^2 + 11409\,q^6 - 364\,q^4\tau^3 \\ - 2093\,q^4\tau^2 - 264\,q^2\tau^4 - 880\,q^2\tau^3 - 198\,\tau^5),$$

$$E^{(6)} = -\frac{1}{512}q^{10}(77027\,q^{12} + 1060290\,q^{10} - 19950\,q^8\tau^2 + 1551179\,q^8 \\ - 14484\,q^6\tau^3 - 166158\,q^6\tau^2 + 170000\,q^6 - 9105\,q^4\tau^4 \\ - 84180\,q^4\tau^3 - 72000\,q^4\tau^2 - 6188\,q^2\tau^5 \\ - 27300\,q^2\tau^4 - 4004\,\tau^6),$$

$$E^{(7)} = \frac{1}{2048}q^{12}(1608201\,q^{14} + 32473350\,q^{12} - 462162\,q^{10}\tau^2 \\ + 86560293\,q^{10} - 320628\,q^8\tau^3 - 6201426\,q^8\tau^2 \\ + 32051580\,q^8 - 180975\,q^6\tau^4 - 3333740\,q^6\tau^3 \\ - 7640204\,q^6\tau^2 - 110772\,q^4\tau^5 - 1301112\,q^4\tau^4 \\ - 2040000\,q^4\tau^3 - 73440\,q^2\tau^6 - 391680\,q^2\tau^5 - 42432\,\tau^7),$$

$$E^{(8)} = -\frac{1}{16384}q^{14}(71016319\,q^{16} + 1991448850\,q^{14} - 22514814\,q^{12}\tau^2 \\ + 8591965515\,q^{12} - 15151500\,q^{10}\tau^3 - 447051150\,q^{10}\tau^2 \\ + 7050577300\,q^{10} - 7789065\,q^8\tau^4 - 246750900\,q^8\tau^3 \\ - 1088468652\,q^8\tau^2 + 552000000\,q^8 - 4284588\,q^6\tau^5 \\ - 98779296\,q^6\tau^4 - 409572240\,q^6\tau^3 - 243936000\,q^6\tau^2 \\ - 2702560\,q^4\tau^6 - 36729280\,q^4\tau^5 - 82992000\,q^4\tau^4 \\ - 1767456\,q^2\tau^7 - 10852800\,q^2\tau^6 - 930240\,\tau^8),$$

$$E^{(9)} = \frac{1}{16384}q^{16}(-10627640400\,q^8\tau^2 - 2208000000\,q^6\tau^3$$

$$- 16082366580\, q^{10}\tau^2 - 6947028400\, q^8\tau^3 - 1930127616\, q^6\tau^4$$

$$- 367234560\, q^4\tau^5 - 3936138390\, q^{12}\tau^2 - 2203128900\, q^{10}\tau^3$$

$$- 862113200\, q^8\tau^4 - 325867680\, q^6\tau^5 - 124621728\, q^4\tau^6$$

$$- 36917760\, q^2\tau^7 - 142032638\, q^{14}\tau^2 - 93518620\, q^{12}\tau^3$$

$$- 44035677\, q^{10}\tau^4 - 21521500\, q^8\tau^5 - 12592040\, q^6\tau^6$$

$$- 8330784\, q^4\tau^7 - 5383840\, q^2\tau^8 + 38680920720\, q^{10}$$

$$+ 147358364100\, q^{12} + 98578433415\, q^{14} + 15278638650\, q^{16}$$

$$+ 408787995\, q^{18} - 2615008\, \tau^9),$$

$$E^{(10)} = -\frac{1}{65536} q^{18}(-171158400000\, q^8\tau^2 - 1129090795092\, q^{10}\tau^2$$

$$- 353939772960\, q^8\tau^3 - 54381600000\, q^6\tau^4$$

$$- 848740435785\, q^{12}\tau^2 - 394268963616\, q^{10}\tau^3$$

$$- 120184396164\, q^8\tau^4 - 31260911760\, q^6\tau^5$$

$$- 6027840000\, q^4\tau^6 - 136316254536\, q^{14}\tau^2$$

$$- 76930929900\, q^{12}\tau^3 - 28779222420\, q^{10}\tau^4$$

$$- 10234167804\, q^8\tau^5 - 4068838800\, q^6\tau^6 - 1661100480\, q^4\tau^7$$

$$- 497296800\, q^2\tau^8 - 3679091955\, q^{16}\tau^2 - 2382986628\, q^{14}\tau^3$$

$$- 1029106287\, q^{12}\tau^4 - 436790808\, q^{10}\tau^5 - 229878621\, q^8\tau^6$$

$$- 149584500\, q^6\tau^7 - 103980240\, q^4\tau^8 - 66306240\, q^2\tau^9$$

$$+ 375747200000\, q^{10} + 6001318017552\, q^{12}$$

$$+ 10366146572430\, q^{14} + 4277082221727\, q^{16}$$

$$+ 469170488020\, q^{18} + 9724330239\, q^{20} - 29995680\, \tau^{10}),$$

$$E^{(11)} = \frac{1}{131072} q^{20}(-21242958605200\, q^{10}\tau^2 - 3757472000000\, q^8\tau^3$$

$$- 46800062306316\, q^{12}\tau^2 - 17527262965120\, q^{10}\tau^3$$

$$- 3924933903600\, q^8\tau^4 - 561254400000\, q^6\tau^5$$

$$- 20724382333305\, q^{14}\tau^2 - 10061341111668\, q^{12}\tau^3$$

$$- 3115304609108\, q^{10}\tau^4 - 860427312048\, q^8\tau^5$$

$$- 232920137520\, q^6\tau^6 - 47230560000\, q^4\tau^7$$

$$- 2330306306536\, q^{16}\tau^2 - 1322057550660\, q^{14}\tau^3$$

$$- 466365925410\, q^{12}\tau^4 - 147118542268\, q^{10}\tau^5$$

$$- 54452926480\, q^8\tau^6 - 24693351984\, q^6\tau^7 - 10982462400\, q^4\tau^8$$

$$- 3329664000\, q^2\tau^9 - 48621651195\, q^{18}\tau^2 - 31092267128\, q^{16}\tau^3$$

$$- 12296689965\, q^{14}\tau^4 - 4356111144\, q^{12}\tau^5 - 1936042225\, q^{10}\tau^6$$

$$- 1220651760\, q^8\tau^7 + 70944835098000\, q^{12}$$

$$+ 340059597157140\, q^{14} + 323290864183560\, q^{16}$$

$$+ 88943517401349\, q^{18} + 7207006604490\, q^{20}$$

$$+\,118718351829\,q^{22} - 174807360\,\tau^{11} - 903773052\,q^{6}\tau^{8}$$
$$-\,656432400\,q^{4}\tau^{9} - 412045920\,q^{2}\tau^{10}),$$

$$E^{(12)} = -\frac{1}{524288}q^{22}(-357086822400000\,q^{10}\tau^{2}$$
$$-\,3053365649115640\,q^{12}\tau^{2} - 8466127273 56000\,q^{10}\tau^{3}$$
$$-\,108379729920000\,q^{8}\tau^{4} - 3317253417917868\,q^{14}\tau^{2}$$
$$-\,1370579839225800\,q^{12}\tau^{3} - 348932369371344\,q^{10}\tau^{4}$$
$$-\,71596479595520\,q^{8}\tau^{5} - 10455441920000\,q^{6}\tau^{6}$$
$$-\,955928238854589\,q^{16}\tau^{2} - 477892176202584\,q^{14}\tau^{3}$$
$$-\,144713724843450\,q^{12}\tau^{4} - 37852735615464\,q^{10}\tau^{5}$$
$$-\,10845939908672\,q^{8}\tau^{6} - 3297497018112\,q^{6}\tau^{7}$$
$$-\,717704064000\,q^{4}\tau^{8} - 788637897 09936\,q^{18}\tau^{2}$$
$$-\,44900905120500\,q^{16}\tau^{3} - 14778743760924\,q^{14}\tau^{4}$$
$$-\,3851941955820\,q^{12}\tau^{5} - 1139443161780\,q^{10}\tau^{6}$$
$$-\,523326578368\,q^{8}\tau^{7} + 768283264000000\,q^{12}$$
$$+\,14396272069572120\,q^{14} + 31423206236610474\,q^{16}$$
$$+\,18444919983958930\,q^{18} + 35790412079926 35\,q^{20}$$
$$+\,221499986176800\,q^{22} + 2961221054113\,q^{24}$$
$$-\,2064391680\,\tau^{12} - 295568760960\,q^{6}\tau^{8}$$
$$-\,144816731136\,q^{4}\tau^{9} - 44429299200\,q^{2}\tau^{10}$$
$$-\,1305901870119\,q^{20}\tau^{2} - 826565878908\,q^{18}\tau^{3}$$
$$-\,298320278685\,q^{16}\tau^{4} - 82105653344\,q^{14}\tau^{5}$$
$$-\,25625460525\,q^{12}\tau^{6} - 15223046916\,q^{10}\tau^{7} - 13109573136\,q^{8}\tau^{8}$$
$$-\,11129476160\,q^{6}\tau^{9} - 8373214080\,q^{4}\tau^{10} - 5160979200\,q^{2}\tau^{11}),$$

$$E^{(13)} = \frac{1}{524288}q^{24}(-29171542767792400\,q^{12}\tau^{2}$$
$$-\,4609699584000000\,q^{10}\tau^{3} - 82703432580590172\,q^{14}\tau^{2}$$
$$-\,28034546881847920\,q^{12}\tau^{3} - 5419404704448000\,q^{10}\tau^{4}$$
$$-\,637908163200000\,q^{8}\tau^{5} - 52642756156602150\,q^{16}\tau^{2}$$
$$-\,23132372258482500\,q^{14}\tau^{3} - 6111561626233276\,q^{12}\tau^{4}$$
$$-\,1359263127247200\,q^{10}\tau^{5} - 289353866709600\,q^{8}\tau^{6}$$
$$-\,45553846272000\,q^{6}\tau^{7} - 10548280449383655\,q^{18}\tau^{2}$$
$$-\,5384007590549280\,q^{16}\tau^{3} - 1557775923476955\,q^{14}\tau^{4}$$
$$-\,350531381584448\,q^{12}\tau^{5} - 87056578286160\,q^{10}\tau^{6}$$
$$-\,30560235197184\,q^{8}\tau^{7} + 92128653733573200\,q^{14}$$
$$+\,520743951458472000\,q^{16} + 627359762166181275\,q^{18}$$
$$+\,245557751348605500\,q^{20} + 35062612485456150\,q^{22}$$

$$+\, 1702414880302500\, q^{24} + 18796729543839\, q^{26}$$
$$-\, 6162105600\, \tau^{13} - 11284182349440\, q^{6}\tau^{8}$$
$$-\, 2668289536000\, q^{4}\tau^{9} - 661712059905300\, q^{20}\tau^{2}$$
$$-\, 377686349188500\, q^{18}\tau^{3} - 114876591510912\, q^{16}\tau^{4}$$
$$-\, 21942218248860\, q^{14}\tau^{5} - 2803057502560\, q^{12}\tau^{6}$$
$$-\, 985702662000\, q^{10}\tau^{7} - 1097533982160\, q^{8}\tau^{8}$$
$$-\, 879795014912\, q^{6}\tau^{9} - 477387123840\, q^{4}\tau^{10}$$
$$-\, 147890534400\, q^{2}\tau^{11} - 8883663162339\, q^{22}\tau^{2}$$
$$-\, 5575797250200\, q^{20}\tau^{3} - 1826656925445\, q^{18}\tau^{4}$$
$$-\, 335971943076\, q^{16}\tau^{5} - 11115420225\, q^{14}\tau^{6} + 479890296\, q^{12}\tau^{7}$$
$$-\, 24513744150\, q^{10}\tau^{8} - 36023863260\, q^{8}\tau^{9} - 34901093568\, q^{6}\tau^{10}$$
$$-\, 26937204480\, q^{4}\tau^{11} - 16267958784\, q^{2}\tau^{12}),$$

$$E^{(14)} = -\frac{1}{8388608}\, q^{26}\big(-20764615173120000000\, q^{12}\tau^{2}$$
$$-\, 21417021159903130080\, q^{14}\tau^{2} - 54284569536676128000\, q^{12}\tau^{3}$$
$$-\, 610487890521600000\, q^{10}\tau^{4} - 29803747114971172200\, q^{16}\tau^{2}$$
$$-\, 11323328631874827840\, q^{14}\tau^{3} - 25445884718209856160\, q^{12}\tau^{4}$$
$$-\, 442546902748662400\, q^{10}\tau^{5} - 53118266073600000\, q^{8}\tau^{6}$$
$$-\, 12302994469639383900\, q^{18}\tau^{2} - 5637921035340508200\, q^{16}\tau^{3}$$
$$-\, 14754111152317982200\, q^{14}\tau^{4} - 3096445352867568000\, q^{12}\tau^{5}$$
$$-\, 69402693007150048\, q^{10}\tau^{6} - 17170151820307200\, q^{8}\tau^{7}$$
$$+\, 4407758978304000000\, q^{14} + 9320827615951132080\, q^{16}$$
$$+\, 24082424592334769700\, q^{18} + 1794444887250797857\,5\, q^{20}$$
$$+\, 49506187990711461100\, q^{22} + 5377662367940787210\, q^{24}$$
$$+\, 209404689389802900\, q^{26} + 1937950838975635\, q^{28}$$
$$-\, 297048701952\, \tau^{14} - 3031890301440000\, q^{6}\tau^{8}$$
$$-\, 1798032935007476700\, q^{20}\tau^{2} - 9320172009762042 00\, q^{18}\tau^{3}$$
$$-\, 253273301765925870\, q^{16}\tau^{4} - 42878329010695080\, q^{14}\tau^{5}$$
$$-\, 5199153101054640\, q^{12}\tau^{6} - 1636797762481024\, q^{10}\tau^{7}$$
$$-\, 1206616155806880\, q^{8}\tau^{8} - 603747140760192\, q^{6}\tau^{9}$$
$$-\, 156240788520960\, q^{4}\tau^{10} - 8822097896116 0500\, q^{22}\tau^{2}$$
$$-\, 50445828472486200\, q^{20}\tau^{3} - 14039528191146900\, q^{18}\tau^{4}$$
$$-\, 1464650040419280\, q^{16}\tau^{5} + 586303077017364\, q^{14}\tau^{6}$$
$$+\, 367574805160680\, q^{12}\tau^{7} + 82241476430880\, q^{10}\tau^{8}$$
$$-\, 27816432789600\, q^{8}\tau^{9} - 41924742837120\, q^{6}\tau^{10}$$
$$-\, 25216513627136\, q^{4}\tau^{11} - 7866289699840\, q^{2}\tau^{12}$$
$$-\, 977429936279628\, q^{24}\tau^{2} - 6091896291390 00\, q^{22}\tau^{3}$$

$$- 179844828679542 \, q^{20}\tau^4 - 13647733187400 \, q^{18}\tau^5$$
$$+ 15354025821420 \, q^{16}\tau^6 + 10078531669272 \, q^{14}\tau^7$$
$$+ 3248591488155 \, q^{12}\tau^8 - 331052671352 \, q^{10}\tau^9$$
$$- 1635218970840 \, q^{8}\tau^{10} - 1780907425440 \, q^{6}\tau^{11}$$
$$- 1396944550144 \, q^{4}\tau^{12} - 825135283200 \, q^{2}\tau^{13}).$$

# Appendix H

# Tables of Harmonium Energy Corrections

## H.1  Ground State Energy to Order 100

In Tables H.1 to H.13 the 3-dimensional ground state energy corrections $E^{(n)}$ for the harmonium potential are given to order 100 in rational form. They were obtained using program *harmgs* in Section 10.2.3 for $jmax := 50$ followed by program *symharmgs* with $jmin := 51$ and $jmax : 100$. They can also be obtained more efficiently using the HVHF method and procedure *hvhfpower* given in Section 10.6

The floating point values of the energy corrections expressed in the Maple internal floating point format are also given to 60 digit accuracy in Tables H.14 to H.17.

**Table H.1**: Ground state energy corrections for harmonium. The denominators are expressed in prime power factorized form.

| $n$ | $E^{(n)}$ |
|---|---|
| 0 | $-1/2^1$ |
| 1 | $3$ |
| 2 | $-129/2^2$ |
| 3 | $5451/2^2$ |
| 4 | $-66099\ 75/2^6$ |
| 5 | $73458\ 9303/2^6$ |
| 6 | $-88022\ 40553\ 89/2^9$ |
| 7 | $16996\ 02528\ 39003/2^9$ |
| 8 | $-13164\ 58151\ 74597\ 4019/2^{14}$ |
| 9 | $39139\ 88961\ 13218\ 86653\ 5/2^{14}$ |
| 10 | $-11247\ 00335\ 95272\ 72503\ 06379/2^{17}$ |
| 11 | $48115\ 56652\ 92926\ 98500\ 91825\ 29/2^{17}$ |
| 12 | $-38735\ 47640\ 30828\ 18986\ 02688\ 75632\ 3/2^{21}$ |
| 13 | $22667\ 72565\ 18268\ 48312\ 75281\ 20603\ 5947/2^{21}$ |
| 14 | $-12219\ 97346\ 24610\ 27474\ 55824\ 89330\ 01260\ 685/2^{24}$ |
| 15 | $93987\ 17678\ 40263\ 37318\ 55178\ 50896\ 06242\ 35875/2^{24}$ |
| 16 | $-52389\ 32756\ 29313\ 92374\ 68652\ 94316\ 65083\ 04163\ 71515/2^{30}$ |
| 17 | $51305\ 73520\ 91621\ 88099\ 35175\ 96730\ 19830\ 16522\ 01013\ 575$ $/2^{30}$ |
| 18 | $-44911\ 26282\ 96955\ 80863\ 65180\ 17596\ 50521\ 25187\ 16922\ 12154$ $55/2^{33}$ |
| 19 | $54594\ 70299\ 82242\ 05973\ 71342\ 94342\ 27113\ 88533\ 01344\ 65500$ $89445/2^{33}$ |
| 20 | $-11735\ 92088\ 05026\ 04326\ 55427\ 84726\ 59291\ 65740\ 91785\ 44226$ $26047\ 08745/2^{37}$ |
| 21 | $17345\ 07167\ 53224\ 06381\ 80015\ 10450\ 75743\ 10465\ 07067\ 60876$ $29762\ 09505\ 145/2^{37}$ |
| 22 | $-22463\ 49326\ 79124\ 27073\ 57217\ 28274\ 72618\ 19769\ 76120\ 75191$ $23569\ 94726\ 37375\ 95/2^{40}$ |
| 23 | $39676\ 92840\ 43936\ 82606\ 56735\ 43770\ 92833\ 99865\ 26584\ 32314$ $56603\ 75613\ 11696\ 15005/2^{40}$ |
| 24 | $-24380\ 02239\ 01287\ 37464\ 45285\ 78906\ 31378\ 97577\ 42032\ 00110$ $28494\ 35943\ 62070\ 32685\ 41975/2^{45}$ |
| 25 | $50725\ 21317\ 15512\ 88653\ 52418\ 28278\ 15195\ 84061\ 63545\ 18438$ $81011\ 86001\ 16465\ 74043\ 26701\ 499/2^{45}$ |

**Table H.2**: Continuation of preceding table.

| $n$ | $E^{(n)}$ |
|---|---|
| 26 | $-91204\ 16489\ 20577\ 78015\ 15109\ 10945\ 14035\ 64692\ 12251\ 20555$ $94183\ 53118\ 96887\ 36895\ 17323\ 50709\ 67/2^{48}$ |
| 27 | $22079\ 59120\ 32579\ 29615\ 33037\ 27765\ 81972\ 87076\ 75196\ 82297$ $27793\ 92790\ 61999\ 09738\ 37809\ 45458\ 60302\ 9/2^{48}$ |
| 28 | $-91878\ 63367\ 94021\ 70924\ 44565\ 68903\ 46301\ 40302\ 91133\ 00208$ $77052\ 91786\ 30769\ 26926\ 14359\ 69881\ 66577\ 40411/2^{52}$ |
| 29 | $25607\ 96635\ 20852\ 60850\ 20837\ 31637\ 35928\ 26282\ 28310\ 22387$ $51612\ 48041\ 94123\ 93297\ 13102\ 55121\ 25728\ 03456\ 3155/2^{52}$ |
| 30 | $-61049\ 65069\ 45130\ 07803\ 11150\ 11763\ 43252\ 16513\ 81618\ 87228$ $94148\ 96477\ 55889\ 26898\ 71795\ 28869\ 46440\ 44196\ 75585\ 549$ $/2^{55}$ |
| 31 | $19409\ 89730\ 45083\ 49772\ 43684\ 32140\ 86527\ 64423\ 61939\ 68492$ $43455\ 07420\ 34850\ 47694\ 63802\ 43044\ 06424\ 72481\ 36816\ 95363$ $39/2^{55}$ |
| 32 | $-84104\ 18904\ 74289\ 74010\ 54288\ 87388\ 00353\ 80176\ 71168\ 65067$ $55119\ 54903\ 78616\ 38926\ 52982\ 30723\ 13343\ 16250\ 78728\ 96887$ $06748\ 59/2^{62}$ |
| 33 | $30256\ 70624\ 75508\ 32422\ 74683\ 88672\ 64911\ 52339\ 03875\ 69697$ $11606\ 98034\ 74604\ 83952\ 61378\ 70852\ 89352\ 94720\ 32088\ 78858$ $29522\ 85095\ 1/2^{62}$ |
| 34 | $-92375\ 90595\ 03539\ 91011\ 77211\ 36991\ 85296\ 67830\ 84346\ 18193$ $20375\ 44906\ 47315\ 19412\ 95297\ 03227\ 50675\ 89086\ 61776\ 80549$ $94139\ 14511\ 97095/2^{65}$ |
| 35 | $37335\ 02320\ 47217\ 75057\ 39034\ 19558\ 52908\ 02334\ 97958\ 85013$ $58177\ 79490\ 20233\ 30901\ 31961\ 80140\ 72087\ 37678\ 33223\ 37507$ $89833\ 52230\ 39297\ 6461/2^{65}$ |
| 36 | $-25527\ 76806\ 57817\ 52433\ 21435\ 52919\ 87938\ 05023\ 41100\ 91468$ $73989\ 47886\ 59751\ 26173\ 43214\ 73702\ 49094\ 60537\ 61132\ 89618$ $75547\ 74357\ 12874\ 22690\ 9421/2^{69}$ |
| 37 | $11517\ 37030\ 67320\ 65584\ 02534\ 16603\ 18035\ 87184\ 64255\ 47047$ $43501\ 25681\ 37237\ 28158\ 90278\ 24873\ 41229\ 27739\ 04632\ 73681$ $15579\ 26672\ 85722\ 66545\ 26841\ 357/2^{69}$ |
| 38 | $-43825\ 56315\ 68467\ 72567\ 30557\ 11728\ 54032\ 12228\ 22984\ 64156$ $92020\ 86898\ 98897\ 50851\ 36265\ 30289\ 45091\ 23325\ 80494\ 29488$ $91438\ 54460\ 42929\ 01896\ 77760\ 53187\ 99/2^{72}$ |

**Table H.3**: Continuation of preceding table.

| $n$ | $E^{(n)}$ |
|---|---|
| 39 | 21946 57535 32152 87785 56357 80223 80402 82434 29985 09876 72001 81139 20555 08144 19621 42494 41056 61197 18985 32480 82100 94547 04491 37656 66012 14602 65842 $5/2^{72}$ |
| 40 | $-36978$ 71141 60923 03156 43861 51623 27869 76173 21325 88390 54195 44488 45119 90994 57386 95314 03522 21114 43211 01632 52431 52011 48267 88375 43844 64523 80466 32724 $5/2^{77}$ |
| 41 | 20447 98484 86785 70034 72024 31453 52846 72560 84676 86291 72123 58329 31099 33679 90159 82838 95707 56318 25276 95047 79615 11124 48900 86296 69010 18051 67942 34570 $70345/2^{77}$ |
| 42 | $-94884$ 69357 06008 13776 80893 61451 28125 50864 07801 22518 65816 98582 07605 13332 00976 87410 11452 56693 11915 67996 87962 02778 42881 35671 14772 34448 52913 38500 81892 1045 $/2^{80}$ |
| 43 | 57666 74936 65291 30625 53751 23595 08676 00160 60754 58472 89323 20819 28531 79078 17889 52263 19214 86769 38841 91098 38079 73504 23600 82352 46044 28444 86373 54756 94337 22550 $655/2^{80}$ |
| 44 | $-58693$ 10554 80086 03844 66116 34175 01320 38169 69276 62626 37781 50203 30742 68678 12469 81027 53380 43072 56042 01905 97520 33707 47653 05898 05570 88998 81424 65178 05636 37761 03655 $525/2^{84}$ |
| 45 | 39039 13569 69218 03218 31051 03566 50315 72098 75285 96450 80555 12370 56817 97779 14241 61568 43741 10751 89819 59905 29013 52846 35217 56247 82945 57444 19105 73593 81044 92232 35354 13876 $45/2^{84}$ |
| 46 | $-21699$ 66588 30321 26701 85909 17422 39514 92659 93055 16812 97303 09151 30966 90205 09130 33075 20300 19150 34432 57607 67104 03353 45310 16958 75850 98955 20621 82250 39211 47168 15261 73890 53574 $35/2^{87}$ |
| 47 | 15734 82417 57861 53171 08799 82628 74604 05974 89975 64835 50450 78790 69402 70329 29090 84938 46137 87712 83133 09022 85340 94947 99866 93644 96437 26415 40938 38410 24606 14331 76038 93294 13337 05144 $5/2^{87}$ |
| 48 | $-76139$ 44902 66330 91564 65254 72280 81678 09716 74704 23565 71787 38909 30337 32284 94616 94465 61923 44146 72392 35049 70804 09253 28500 87105 20431 28386 28552 48056 83098 01782 88103 66902 62824 75647 23209 $5/2^{93}$ |

**Table H.4**: Continuation of preceding table.

| $n$ | $E^{(n)}$ |
|---|---|
| 49 | 59974 33251 81342 18036 17029 71292 95628 19892 02361 69716 61527 37337 81170 88672 17213 96401 28067 45239 92432 88601 41732 64987 60178 88934 08121 95193 77197 94685 55005 13525 32900 28650 27368 54773 52128 91675$/2^{93}$ |
| 50 | $-$39340 74635 63968 76062 15866 59852 99996 06379 29498 85659 36968 85133 63365 20302 76196 87135 23199 69345 19763 42920 18979 98439 73093 31636 92576 58612 36453 63992 77593 39199 92489 01378 80250 17008 76221 64546 79587$/2^{96}$ |
| 51 | 33552 03765 99628 90047 33181 71707 20197 87485 32792 56077 55332 09899 53307 19581 25615 41032 49380 03355 62673 81953 26507 53423 05987 55065 80013 83824 72364 09320 91356 15014 51556 16261 12160 34192 59847 77333 21795 0897$/2^{96}$ |
| 52 | $-$47585 55136 55527 55586 98109 85977 33604 53124 32416 24466 00216 78379 41202 22394 29554 83016 02775 40557 48330 42276 92719 09259 67116 66446 03368 90367 82574 41082 74447 00565 29164 66768 54908 22922 78710 57891 02034 82461 6829$/2^{100}$ |
| 53 | 43808 10461 67047 36482 29974 53049 29321 86891 25493 79024 13124 51767 04694 19941 65020 34015 17043 92309 00846 54638 70206 29869 23697 96704 08713 27774 02433 53462 33332 33779 28603 18862 40445 06934 33860 33401 19041 44054 14190 861 $/2^{100}$ |
| 54 | $-$33485 89381 56415 79000 65110 54525 02791 20918 92669 64327 01730 03061 56506 67989 08752 56814 43802 62344 84550 12160 10030 75887 46034 41522 54622 01919 25203 78466 51864 90564 07860 38258 50008 19821 42481 24735 08667 86194 60200 28585 895$/2^{103}$ |
| 55 | 33183 59812 01538 18985 00056 41286 99714 83716 58504 27362 34363 81703 04254 69441 71603 51642 15901 29421 18098 70577 96378 85021 58873 12787 90477 68753 58027 95958 97396 02881 65940 51414 62843 24971 86404 98553 54476 91821 72604 83759 69251 69$/2^{103}$ |
| 56 | $-$10906 75895 10631 24123 40630 83856 29300 63171 50053 61581 95170 11761 74755 91805 13871 82554 55994 05336 12441 06045 84886 26940 61833 94803 06688 15504 69608 85097 20711 73740 50356 33922 75572 54157 88383 45390 21045 47182 15361 40908 34893 67510 463$/2^{108}$ |

**Table H.5**: Continuation of preceding table.

| $n$ | $E^{(n)}$ |
|---|---|
| 57 | 11603 93148 91838 89120 56830 83618 97135 68169 85213 85043 96919 29823 55488 06659 24332 48605 75250 55487 54353 06323 32718 47908 40123 94959 73403 44227 49410 26303 26441 82094 90552 00263 37483 99941 27802 22082 40203 19237 60876 85323 82062 09732 60012 51$/2^{108}$ |
| 58 | $-$10224 16138 78797 72558 16185 14575 87338 01436 28159 09547 04843 77885 30113 78953 34610 59169 71420 78143 16105 07164 85094 28355 38158 20186 03155 65062 92938 32837 68570 81633 79962 04381 78625 20527 95788 66757 61906 95040 11465 06573 53357 44017 09550 65502 87$/2^{111}$ |
| 59 | 11650 06236 06915 24088 22954 51746 46182 71058 48758 37661 92926 61117 99708 11585 67243 33635 46105 67810 93623 14005 87429 22684 54569 57516 04040 95074 71669 27884 33504 59860 50467 00081 13746 86530 23279 07700 03263 76090 11892 89423 72445 04788 13864 92753 66540 5$/2^{111}$ |
| 60 | $-$21961 91389 70941 04755 25937 62059 40910 63558 79396 34236 28610 05812 47953 37009 18309 29370 51367 09882 47114 88955 45317 59410 67796 04412 12480 35619 47982 60768 77083 82127 33451 08393 23693 55588 82292 99975 21728 61600 64899 42017 70632 73647 59329 84893 91197 46397 9$/2^{115}$ |
| 61 | 26740 83855 43231 27368 59753 80752 99896 95537 34323 00794 33304 61639 61260 20137 32107 52603 92224 90739 22659 88346 80198 55232 98724 96791 82180 90335 40710 27332 95597 86853 81700 93122 95015 51439 37159 98238 77633 06072 85896 56138 01261 06138 61639 52480 91866 34800 74019$/2^{115}$ |
| 62 | $-$26904 31375 84156 32010 50901 13176 04445 22268 45218 18308 08451 66532 31452 13113 37071 65783 84135 30664 49339 57626 46409 04098 87977 23177 23882 57043 80001 63920 91463 27835 04732 60400 52964 98276 66952 03693 82724 53996 56744 40477 56101 08959 62443 21213 43636 62211 48702 28749$/2^{118}$ |
| 63 | 34930 71068 51319 74482 30428 56024 76444 78135 95896 95515 88166 22830 59400 86767 20411 80632 77439 24041 11849 22969 30265 61653 25152 08731 48602 33953 20752 23399 39114 77649 34486 16966 36578 49317 98933 62941 89309 04316 45409 52698 28369 40738 84306 96777 27877 13345 01224 25170 3091$/2^{118}$ |

**Table H.6**: Continuation of preceding table.

| $n$ | $E^{(n)}$ |
|---|---|
| 64 | $-11979$ 67376 37750 61643 10940 76219 08222 76743 69604 46957 42565 83220 51657 57443 29906 07604 73317 89736 73576 93836 40016 11623 99431 21644 48663 75995 22731 55402 31464 06193 14069 79354 74731 42837 14297 80477 43766 02589 01345 72563 60918 24145 67884 40244 17537 06767 98663 12474 52710 17895 $5/2^{126}$ |
| 65 | 16551 78640 51763 71874 15890 22978 30354 56279 33193 62638 49620 15597 22168 22063 27673 36779 82876 50588 93039 21840 98201 00062 47448 84673 74228 42571 64128 43652 77278 71642 26613 34506 35962 01716 30470 39287 77801 09962 95942 47894 42764 47467 48131 05724 59489 22056 51082 10034 76050 52780 $96135/2^{126}$ |
| 66 | $-18859$ 64700 67515 69118 78767 57747 02959 62008 33161 08830 33590 31651 35085 34818 37390 67749 36874 78358 40868 47439 97422 77016 52188 67429 02518 75461 44289 41509 67196 66181 26744 83626 67201 48432 55254 60127 76925 45137 30674 52639 32873 49269 16814 02845 73998 71021 39920 52689 50682 65215 41126 02295/$2^{129}$ |
| 67 | 27677 93569 23551 40410 63406 72294 23403 30374 67832 77924 89798 69778 58960 93853 22947 29783 50119 05215 92335 47764 73409 90206 73269 23239 76015 33247 53465 73801 08413 23814 48199 46888 61690 14204 78696 23310 58171 00725 81211 18488 74401 10190 54063 24416 23597 78814 93941 09748 00955 54629 48035 12388 1885/$2^{129}$ |
| 68 | $-66936$ 60494 56063 37262 84493 15519 29630 54917 18298 77217 95801 25564 21730 70483 35861 33443 54063 06422 87869 23223 84045 64854 83663 36301 30633 83942 51268 24740 24536 72395 82940 57414 83048 32701 75451 74031 17538 59957 98332 91255 29981 46742 90117 73753 54102 89737 21892 31257 91092 36394 47901 05305 54633 0325/$2^{133}$ |
| 69 | 10415 91764 98649 34396 28913 21727 52953 15536 84328 27764 33623 56602 50180 72678 30653 31457 32467 15389 82149 16544 48120 53665 10900 77667 52526 82429 65993 81950 97403 40654 38842 10661 39244 37881 16816 48332 72208 29669 83398 04061 05942 68866 71495 58449 23076 00465 72699 01421 50213 38373 37943 09307 79550 38041 4165/$2^{133}$ |

**Table H.7**: Continuation of preceding table.

| $n$ | $E^{(n)}$ |
|---|---|
| 70 | $-$13343 33972 99006 83732 94244 43131 03190 33122 85908 13032 41546 36703 83439 43605 90202 03429 11511 58972 73478 47783 45883 49670 40023 71826 72006 39715 48255 24048 94880 71323 56856 88777 00533 37566 92990 36638 75557 45715 98157 36784 72356 71464 09206 47200 05489 99742 84258 37653 60938 06447 43646 24914 87210 79075 34082 $5175/2^{136}$ |
| 71 | 21979 06068 31578 57201 77402 10683 40090 35603 40350 22957 33496 20259 89769 38322 82846 59017 74886 92866 43189 81731 09493 03624 14245 33270 25094 24642 32588 03479 88016 58947 57498 84096 44832 67472 29807 73040 21957 85712 15110 70509 13946 29168 12549 33127 30275 38903 65532 63853 16585 43719 37167 73754 16191 37314 15125 56209 $425/2^{136}$ |
| 72 | $-$11912 43496 34074 43668 36046 40799 11829 38635 32446 10450 61504 75741 99754 70235 19982 44427 32540 90968 09291 63804 98558 53908 75087 27128 21565 20204 19203 47252 82135 13521 81673 43059 03357 21789 73780 80561 81540 66223 58867 76354 29058 26135 19125 43590 62834 97753 11309 23569 95239 96905 22804 45233 29512 07541 08776 69972 26256 $3025/2^{141}$ |
| 73 | 20738 28310 19480 00469 13165 29998 71186 50411 48888 14212 54143 61820 32200 19199 04383 15920 25357 42923 67631 86074 63763 28372 17081 36844 13374 33663 05188 66433 41905 30929 26145 07372 12428 46674 36339 35624 07977 30383 14028 97763 13412 54604 17407 00848 25571 70918 90387 95918 50533 36461 35182 33482 47681 51324 05792 46152 52083 17520 $925/2^{141}$ |
| 74 | $-$29675 92895 60158 16494 16017 43193 90621 67226 56860 11578 00788 89911 81125 40086 57236 41457 61206 36979 81463 49181 95956 15440 08569 13221 31288 29919 86492 51793 15343 35593 51578 63803 06119 54409 94423 60729 24145 26219 24485 32050 76309 75504 97387 70965 81602 00240 39344 62264 02231 85236 83730 98393 74313 11452 93371 38684 52173 12499 31487 625 $/2^{144}$ |

**Table H.8**: Continuation of preceding table.

| $n$ | $E^{(n)}$ |
|---|---|
| 75 | 54520 25941 26534 23784 12015 65456 21044 21353 04227 97723 27369 60643 38134 76722 49285 89494 18801 46318 35982 52838 07060 13064 09804 77569 91303 94674 20483 72515 05792 65396 70729 28543 81595 03568 63997 77624 73030 46643 56694 04176 64332 79156 98340 61274 47110 36649 12451 56029 69230 73221 52006 07404 99162 15915 49470 12085 76907 21154 55401 30156 75/$2^{144}$ |
| 76 | $-16454$ 71848 76303 33254 26477 52884 80991 65623 16951 07222 09429 06631 43871 32673 13046 17321 50581 79175 00169 45177 92560 45223 84205 80425 73426 57062 91098 40661 74776 30479 38497 34696 43688 69468 73119 34771 35955 47122 31088 99426 94035 25324 80554 47257 16308 21619 53496 59344 76182 23489 88193 15080 66825 63538 47356 33159 03425 96130 22406 48297 40162 785/$2^{148}$ |
| 77 | 31857 58738 52012 38723 61122 73154 44427 03397 25425 76851 68379 02474 47409 31260 92824 45453 42358 34778 35323 32611 02603 25506 21885 93516 90639 88240 57336 90244 48851 71669 64373 16043 32529 16702 93221 00619 04310 12095 33108 31539 15196 02329 59769 03374 02513 51986 26566 36266 63428 21227 87895 61116 28146 43160 85517 96587 11982 07439 12980 91436 93209 50808 25/$2^{148}$ |
| 78 | $-50627$ 89684 62219 77712 82196 24687 75530 01242 54349 87716 03749 97647 96155 93073 95217 36397 23738 79663 77795 96742 45329 49011 14135 55836 48114 71605 10088 10171 86510 40180 34703 99739 89589 98514 00539 40878 49983 66659 97141 21467 25542 42823 46465 91712 06468 91583 66721 25045 28639 48903 62563 01753 38028 50132 67512 80712 35027 47233 28965 34030 90234 35889 80716 95/$2^{151}$ |
| 79 | 10315 72841 00456 01956 83455 51551 82145 52644 21141 94301 59338 11021 67490 82439 59285 75086 18378 03358 70118 42771 97324 14065 20876 77412 98634 81658 38403 62424 77824 13183 79689 53872 15552 77777 52500 19220 21357 85964 93075 96790 28284 03676 60837 20557 60631 57501 01088 87280 69897 88820 55817 57569 69913 41157 71382 73589 80000 39959 75729 09756 47400 91649 11799 38903 05/$2^{151}$ |

**Table H.9**: Continuation of preceding table.

| $n$ | $E^{(n)}$ |
|---|---|
| 80 | $-$13793 51125 99659 52859 28753 91277 52501 21166 87289 02406 07398 48766 45861 33690 11988 24421 69350 99301 01608 42707 81357 39298 10503 93210 27747 75159 07546 14366 15086 45586 64598 59132 39588 25451 51607 83617 99116 61431 86312 84699 98443 86702 74352 17884 05765 40200 30893 31564 66046 22503 41665 88304 62870 94104 46347 84998 99226 04377 52267 89861 94349 90873 34188 96486 81454 $965/2^{157}$ |
| 81 | 29540 65388 98457 07683 91394 11279 04144 91646 76328 46657 10662 90256 41780 72839 28344 16606 33262 75071 62144 86812 27573 54380 96553 82763 80829 82038 63628 37830 15995 99633 61479 97132 22077 05362 05581 90114 76325 22224 19498 18290 22003 55070 56096 92196 38981 21373 37876 08784 85072 60983 51684 04271 43173 72220 28349 05994 03876 78999 13860 42905 72195 15589 31622 30588 15736 31354 $65/2^{157}$ |
| 82 | $-$51864 99298 46727 82084 84662 42626 07380 13916 76207 36478 59278 63657 58416 35873 88881 97131 38971 26927 39393 10963 73821 91083 88748 32244 71752 31702 56405 52353 70966 86036 43910 71693 69656 37236 23619 39805 94907 71030 84852 53729 46067 12985 02406 64262 12574 00223 17492 48519 23309 85579 08913 40158 44610 81632 58956 67964 29856 41358 17806 46773 03291 01327 73900 97935 14350 99616 28412 $65/2^{160}$ |
| 83 | 11660 82660 52281 29780 88982 93438 15667 07957 53932 31447 77136 00005 61819 88328 30078 96119 70525 56693 63351 76148 20905 40722 22425 63547 79753 49484 76826 07256 99589 69546 59183 65730 90075 53701 65581 17136 99545 77877 36322 91037 66159 35011 91382 37626 55101 62617 64148 82164 29266 45511 66483 69339 76861 40004 61723 67253 35583 27651 47487 73404 40875 98753 48058 77794 45662 42377 70048 26378 $35/2^{160}$ |
| 84 | $-$42960 56574 28750 05741 91290 92287 59606 99548 59917 99720 75908 20013 12110 63169 60392 76991 59595 17705 16595 47980 10654 05863 50568 07922 55671 32230 85610 44755 28652 00476 62605 50078 99390 52300 00997 67456 56077 28175 90368 50201 92713 52519 93588 72741 81348 27987 28462 19095 34779 63785 99572 29749 22386 09612 66487 64923 67901 54684 91952 70177 90601 21374 64706 65086 46882 68472 79378 60715 89635 $75/2^{164}$ |

**Table H.10**: Continuation of preceding table.

| $n$ | $E^{(n)}$ |
|---|---|
| 85 | 10128 24375 95352 78212 89486 27100 89544 56892 72948 66930 31438 34830 47482 02936 17772 52630 18081 51631 21328 89266 15019 25130 57854 90438 82628 74909 23013 42934 73046 05094 30777 32254 59286 46639 33880 65151 88736 45651 24204 60847 81141 98025 73777 24223 37259 30201 31009 58127 33043 89013 92074 23412 20298 20376 97801 51062 54490 44625 55830 87752 02281 93112 64668 90379 05338 54060 17140 35157 17230 38481 35/$2^{164}$ |
| 86 | $-$19552 95655 73904 51516 71064 11188 25257 52992 85650 91189 43852 79380 27568 73355 04448 28097 57053 51853 72663 87492 15583 88856 55514 08961 98948 35626 28044 74946 32281 83859 47092 61384 95126 71030 11165 96024 00351 70710 29862 40681 02779 70089 90280 30457 21229 33427 33043 56762 20346 79580 37348 91897 67915 02233 19818 07524 98765 66300 80211 96210 71860 37282 17097 84111 32407 76250 30048 18578 01158 46843 33777 65/$2^{167}$ |
| 87 | 48284 56288 07660 45424 34572 88766 57368 84197 88310 86848 30812 50215 31456 00210 52448 05299 93148 78553 92431 46281 43902 61437 37722 33934 00602 01121 00711 44460 19533 48935 08772 15004 51368 57211 26982 62053 15449 48533 46976 30104 94816 21127 19161 86944 33764 85620 91944 42744 77277 28226 75545 47332 37926 66552 38521 54436 98496 36528 54603 77775 98648 51713 69169 46301 27238 07322 29316 63704 46391 89877 49957 05683 5/$2^{167}$ |
| 88 | $-$39034 42488 42696 07678 24708 49312 54581 60055 90941 14759 68519 65675 39783 73068 59571 23767 42933 78516 70209 71365 65175 49315 59443 03433 50578 53207 24949 02773 05253 37528 21553 21366 56727 43710 46969 78884 29004 01048 62571 84841 31436 08651 88756 80644 22089 94958 10374 06568 87944 86606 83980 42714 21801 95642 52296 28916 26768 19400 23472 91560 61815 01913 13734 18051 37689 55006 43374 06429 68228 67810 50856 69445 78543 85/$2^{172}$ |

**Table H.11**: Continuation of preceding table.

| $n$ | $E^{(n)}$ |
|---|---|
| 89 | 10086 01369 57475 58104 76992 34930 65052 60287 61235 63724 89144 73152 97483 04959 20954 00240 17955 47908 59975 25325 01198 68597 78768 03860 11004 07006 57340 51187 08781 69261 41355 31221 17621 88203 56288 20414 45844 39974 86778 35226 84355 14072 57205 92947 70690 50028 41465 04724 08124 12469 93189 40381 21032 99112 90964 82307 80687 15915 92630 92213 56280 43092 26933 46873 75752 37290 91453 13973 49717 04175 50116 09145 49911 22656 $45/2^{172}$ |
| 90 | $-21318$ 40384 19282 60477 60698 39488 37647 14120 88751 17487 88024 42032 35881 65431 23698 31940 39022 55962 77070 56828 74834 07786 84585 92843 02029 12104 53398 66331 26020 15216 56312 10984 03162 62152 30885 09100 14202 72978 99077 82825 61568 91750 75331 99977 95188 71613 11415 10461 85398 30489 68304 81704 15281 85745 43170 14521 73898 77817 58419 31148 88795 43880 52147 52718 35001 30339 05814 83888 44022 00761 08408 81431 12512 52327 86457 $45/2^{175}$ |
| 91 | 57579 34385 24577 05669 34256 53030 86104 03231 25292 50454 12344 06403 22996 11027 36610 48579 04547 91794 05703 61424 82943 39953 85115 12674 26394 13495 73713 80056 93612 60955 03210 28653 79786 61655 70637 09877 12210 17966 32254 64077 70456 30195 17741 54379 48467 69457 64894 14428 77145 85617 33041 41987 09445 63156 73318 27683 58082 57689 59650 79189 79879 14861 77889 66342 57661 88474 59737 52718 31522 26698 22547 17080 18856 47479 73856 24035 $5/2^{175}$ |
| 92 | $-25430$ 88227 41373 19565 35379 53205 14097 01583 44167 85721 66701 07582 65078 14868 58880 34513 29885 82788 57758 50868 61598 90800 75058 51656 50856 72647 53748 32472 87044 72948 51936 74946 45386 97877 99053 66114 20470 81768 04751 73259 33433 10795 49823 18156 45623 30519 62443 27316 11727 55613 59783 11356 79277 14565 25037 78645 43913 49867 48709 63453 51552 72389 43342 20056 97513 85025 18096 33695 44598 41898 91971 43561 13219 23547 26982 12233 89471 $65/2^{179}$ |

**Table H.12**: Continuation of preceding table.

| $n$ | $E^{(n)}$ |
|---|---|
| 93 | 71729 35653 33397 32384 72971 08909 31703 16734 39483 30136 72019 88919 42581 46190 55371 49617 54336 91521 58186 45023 68104 13148 22613 96091 49072 30975 02695 33711 44612 95363 37110 88166 56217 58817 96080 63114 46744 87335 23681 58865 61119 89262 62845 28433 99867 42614 23933 02636 58183 95965 48368 91020 67241 95827 67780 03106 38707 99184 25957 99261 93643 63352 38249 13416 45660 14212 69124 48237 91500 62555 42545 17805 19590 33654 92184 55110 74683 82592 $5/2^{179}$ |
| 94 | $-16534$ 20975 68606 10228 63701 97922 70556 45393 51895 29625 31368 15223 10402 98160 49652 63218 41410 05375 02319 88268 49433 17225 91674 89635 44697 52914 92302 62907 32068 88296 87655 88914 63832 34595 26908 10759 37698 98291 30972 11711 39117 16213 23307 01408 22277 47134 22488 53660 93142 35949 66594 12646 86233 94789 63196 07615 98638 15387 93297 18075 00093 79759 21940 88252 46786 25882 87493 94090 34066 36987 91667 80909 07007 87484 53446 05238 09963 09456 40665 55 $/2^{182}$ |
| 95 | 48656 77492 98365 44187 98348 64961 22797 57966 44234 00487 87856 69079 05407 13789 63716 26539 71779 55884 22036 03529 51465 17752 49860 64074 21676 27058 29116 74505 88790 29854 16645 07503 44768 32324 52040 09728 14701 64969 50619 00494 33688 80249 97094 98208 07227 70220 15891 92012 56305 98675 86066 25308 25365 85363 62215 90617 23923 93360 75132 69781 22469 96891 53512 66574 22267 72457 66406 21179 11422 01424 55954 38532 72607 67558 97778 51697 58909 98697 44991 65872 $5/2^{182}$ |
| 96 | $-18714$ 61517 29717 27843 40201 42077 87628 16946 12428 46936 53768 81401 04207 96207 62939 79148 97708 43898 07807 85811 17501 50725 54194 41875 56851 65988 51141 26584 08762 97159 57964 19576 62868 68505 76525 46284 28800 81677 65437 02854 71222 90629 09919 18279 97420 23671 12652 55084 81174 56459 03543 02700 46531 89614 40222 57585 90650 59298 14483 84782 45626 16843 41426 82806 73608 66523 97026 37590 73376 98957 18725 03741 07103 40089 90229 92926 04461 16623 42821 01897 69361 $775/2^{189}$ |

**Table H.13**: Continuation of preceding table.

| $n$ | $E^{(n)}$ |
|---|---|
| 97 | 57409 34631 22590 31485 58698 31898 65475 65046 84042 60610 34059 19037 84514 64897 23572 33178 45651 79476 84994 61695 63099 72341 65404 41117 35757 34511 32023 09318 53860 85597 65713 18320 35962 76683 18978 84391 70648 51702 05580 94271 44340 36214 80815 29343 63602 30978 97523 44022 42493 75346 94381 29329 99773 86972 54775 74225 56339 99135 79654 11016 01230 71947 92695 43726 96190 47207 26990 91448 16948 00832 94987 23907 68980 52765 40487 36509 20552 40601 08193 90743 27232 11716 $75/2^{189}$ |
| 98 | $-14379$ 92581 09655 73613 82646 99638 85304 74046 10664 91247 08272 67702 66839 37982 16986 03315 85690 62007 83452 11515 36319 98430 86916 51806 41719 77812 57388 13351 16865 55625 70461 40064 94905 69315 63132 95108 57330 53953 95149 68057 31753 12087 37680 25816 21680 29613 72845 60676 63625 25738 94781 47453 01309 95066 85210 61743 44691 72657 42546 56795 98003 68793 11262 92800 08582 75553 83056 85531 82619 12193 83361 81675 03395 46833 64958 23803 17600 55718 86946 22005 23864 33822 60168 $475/2^{192}$ |
| 99 | 45944 46310 74763 08307 88744 86696 82778 96752 83226 23480 35229 64878 22444 61698 32307 22773 52090 87414 43657 83051 17682 27239 50097 95498 50655 19888 52651 62140 62772 71781 62500 55602 48732 90655 72159 35024 64856 43946 93394 80581 80325 79594 98079 55607 77803 32272 04533 08806 61982 47756 26501 62311 10972 52940 51985 84834 30595 24027 38451 09459 47234 60595 23678 07074 05492 72356 53579 31538 87787 40102 71611 67774 89495 82428 10731 55177 37801 99295 54099 92251 32445 08920 54358 90716 $25/2^{192}$ |
| 100 | $-23962$ 60895 12166 25275 88439 93741 16644 60762 51638 64151 94708 70569 96813 83740 22530 07780 63539 64010 36899 86722 35147 03139 47656 19288 73972 32076 18355 45943 90618 52286 11392 14761 78620 78247 02743 25530 50247 28823 94894 41946 53948 98111 60824 96730 88541 48138 48258 56438 93919 73270 29566 52326 10047 62987 72454 54423 30780 00875 35201 70604 83425 00492 28017 17284 31558 16321 42321 94371 20824 92806 78472 50630 88120 95419 74086 08312 91435 66475 47034 84956 18597 06570 13421 58125 68860 $185/2^{196}$ |

**Table H.14**: The Maple internal floating point format for the energy coefficients for the ground state of harmonium. $E^{(n)}$ is defined in terms of the integer $I_n$ and exponent $e_n$ by $E^{(n)} = I_n \times 10^{e_n}$.

| $n$ | $I_n$ | $e_n$ |
|---|---|---|
| 0 | -5000000000 0000000000 0000000000 0000000000 0000000000 0000000000 | -60 |
| 1 | 3 | 0 |
| 2 | -3225000000 0000000000 0000000000 0000000000 0000000000 0000000000 | -58 |
| 3 | 1362750000 0000000000 0000000000 0000000000 0000000000 0000000000 | -56 |
| 4 | -1032808593 7500000000 0000000000 0000000000 0000000000 0000000000 | -54 |
| 5 | 1147795785 9375000000 0000000000 0000000000 0000000000 0000000000 | -52 |
| 6 | -1719187608 1816406250 0000000000 0000000000 0000000000 0000000000 | -50 |
| 7 | 3319536188 2617773437 5000000000 0000000000 0000000000 0000000000 | -48 |
| 8 | -8035022898 8401734558 1054687500 0000000000 0000000000 0000000000 | -46 |
| 9 | 2388909278 0347831209 4116210937 5000000000 0000000000 0000000000 | -43 |
| 10 | -8580782592 4127750420 1033782958 9843750000 0000000000 0000000000 | -41 |
| 11 | 3670926401 4658125687 3460791778 5644531250 0000000000 0000000000 | -38 |
| 12 | -1847051448 9690217488 3017003837 7285003662 1093750000 0000000000 | -35 |
| 13 | 1080881388 2745193630 5774746228 9784908294 6777343750 0000000000 | -32 |
| 14 | -7283671773 9468976703 6333616554 8122337460 5178833007 8125000000 | -30 |
| 15 | 5602072285 6537304710 4786545572 3180911466 4793014526 3671875000 | -27 |
| 16 | -4879136342 8282916894 8219431859 1159120354 5361291617 1550750732 | -24 |
| 17 | 4778218940 7536935153 7448909813 7228765079 7610504087 0606899261 | -21 |
| 18 | -5228359115 9730661734 5332654188 9752407077 6640684523 7943343818 | -18 |
| 19 | 6355659919 5841939623 5996734283 7144840197 8861616809 0310529806 | -15 |
| 20 | -8539006288 9947179985 4238522869 0757981828 3400849222 8201857833 | -12 |
| 21 | 1262020063 2461933405 8662578500 4852740141 4108143127 1578337695 | -8 |
| 22 | -2043042811 0478194024 4467230763 5539091070 0964082199 0939907415 | -5 |
| 23 | 3608595616 6419857806 0380481266 2277436052 3779643985 6229313239 | -2 |
| 24 | -6929219122 7898278681 4732943714 4713601746 3330972845 7247262962 | 1 |
| 25 | 1441697269 5571328317 0792302677 8394041702 3232645999 2038163023 | 5 |

**Table H.15**: Continuation of preceding table.

| $n$ | $I_n$ | $e_n$ |
|---|---|---|
| 26 | -3240222841 7559008027 7910555731 0189867583 0492322999 9665449384 | 8 |
| 27 | 7844246569 0137657984 3456680996 7084875682 8148688304 0469822925 | 11 |
| 28 | -2040115491 6394530697 1536324015 1846411297 2564739092 1784494597 | 15 |
| 29 | 5686110771 5822669317 8030800884 4910419648 5952616740 8801384888 | 18 |
| 30 | -1694468196 1592880038 1897404371 1100057195 0892270364 4808431944 | 22 |
| 31 | 5387328723 2687332916 1723935989 5954365974 3781553163 1404040781 | 25 |
| 32 | -1823718889 6070812267 8179931650 4988806232 0837548112 0247983626 | 29 |
| 33 | 6560877329 1700694469 5941457886 0101432314 0171138409 5152544248 | 32 |
| 34 | -2503853947 9172608289 5460272449 3190315719 0872433687 5952502776 | 36 |
| 35 | 1011967831 7089017120 6453988551 2287772693 9882593304 0475408354 | 40 |
| 36 | -4324572124 3166655394 2315418809 6438044240 2529805892 4618857256 | 43 |
| 37 | 1951118423 1060853250 4233101400 3665190879 1763811668 9267911379 | 47 |
| 38 | -9280423981 4558501414 4247757588 7964911467 7756519297 9800817789 | 50 |
| 39 | 4647368100 8931331730 4253972037 4636697397 9444516118 6050624285 | 54 |
| 40 | -2447045852 8467895014 0918148510 7563995027 6263420824 1205475365 | 58 |
| 41 | 1353134130 6083974775 4202042129 3009442391 1953826219 2120489658 | 62 |
| 42 | -7848677894 9635907813 8343167549 0827348470 6579069375 6759840911 | 65 |
| 43 | 4770081706 4947486906 2239475069 9263473408 7212668818 9535966099 | 69 |
| 44 | -3034362437 4900790001 0560937147 7227003553 4512663074 2720560394 | 73 |
| 45 | 2018276011 2075341341 0773334585 9421780201 2425256589 3215039152 | 77 |
| 46 | -1402308040 4158242773 1914340685 9412051513 8910815759 7635703681 | 81 |
| 47 | 1016839179 6985139835 6498102136 9274578145 1730365117 9509605915 | 85 |
| 48 | -7688119639 3188949513 0497704518 2894713528 4665078983 8966347345 | 88 |
| 49 | 6055859998 7560094284 3863142009 0464943216 0708421204 2589020794 | 92 |
| 50 | -4965500285 2443938508 8631638913 4197798943 0790948049 5934061464 | 96 |

**Table H.16**: Continuation of preceding table.

| $n$ | $I_n$ | $e_n$ |
|---|---|---|
| 51 | 4234862528 0716486604 1990032304 2679687091 4795969795 9030422518 | 100 |
| 52 | -3753838112 5670901216 4030757846 3425446860 5541267886 6761366444 | 104 |
| 53 | 3455850106 3950484832 2172885861 1250798567 7092701095 7357517379 | 108 |
| 54 | -3301964063 4427121267 6287603280 2701864405 6136704170 4724981633 | 112 |
| 55 | 3272155406 4443509142 8827051256 5791274142 9993300683 9456974483 | 116 |
| 56 | -3360904585 6105583183 9165002424 0662741023 0225048978 9698033896 | 120 |
| 57 | 3575737460 4219555981 9129266555 5989902959 0530967475 0149308840 | 124 |
| 58 | -3938203714 6330828996 9820128556 0612330899 2524618373 3081845270 | 128 |
| 59 | 4487440790 8868913249 8326030389 3004205172 0245190471 5431955621 | 132 |
| 60 | -5287138451 3084639245 8874959414 2947804923 2594368395 9622261454 | 136 |
| 61 | 6437622713 7244491349 9042800667 1677782611 9172380042 1411734852 | 140 |
| 62 | -8096222421 9565879024 7656076552 9424645004 7668534847 6081265330 | 144 |
| 63 | 1051157838 8628570124 5816196304 2965997898 6550384727 3423836664 | 149 |
| 64 | -1408203883 4011555717 8698751176 5545058211 3232462935 5473987872 | 153 |
| 65 | 1945653141 5301963087 1159916240 0144089172 4852195156 6547523254 | 157 |
| 66 | -2771176064 3673666863 9014759954 6964057084 7073852378 9087831113 | 161 |
| 67 | 4066907131 0982533140 4283511158 4340680849 2442156126 1209071308 | 165 |
| 68 | -6147156326 3112063163 4993144727 4695504987 4780828014 1970737566 | 169 |
| 69 | 9565509653 1612403230 2621220549 9756772240 4203216858 8173566721 | 173 |
| 70 | -1531740280 6250821644 2295112055 8899492779 4131291751 7980160285 | 178 |
| 71 | 2523072428 6554988608 8278146153 5492568651 5060501657 9828886944 | 182 |
| 72 | -4273376466 0715799682 6200969389 3402285668 7326955860 7387402661 | 186 |
| 73 | 7439494211 4542174306 3738654083 9216789784 3131143122 8939600724 | 190 |
| 74 | -1330714677 5512747370 6900054781 7815984803 0665937039 3007347467 | 195 |
| 75 | 2444772985 2653406682 0635601619 0496375056 8505428888 9570733071 | 199 |

**Table H.17**: Continuation of preceding table.

| $n$ | $I_n$ | $e_n$ |
|---|---|---|
| 76 | -4611594349 5227124902 0719548071 1813306967 9427245694 5569440731 | 203 |
| 77 | 8928397655 9952320750 9031140256 3017371931 5389243136 6904802255 | 207 |
| 78 | -1773619852 5651526683 1595496349 6260400864 2118170647 6947449466 | 212 |
| 79 | 3613853594 8471972196 6584273515 7165456309 1354985121 2129506874 | 216 |
| 80 | -7550322713 3739696715 7486727728 3006295790 5658233599 2181773438 | 220 |
| 81 | 1617002848 8668642204 7394393126 8075542360 6574700965 5920976280 | 225 |
| 82 | -3548746827 2284853604 9138127916 9503848541 5153679267 4443067950 | 229 |
| 83 | 7978661335 2852286005 9241831571 4231551411 8774432221 3684024412 | 233 |
| 84 | -1837175744 6904324224 8415278188 8383928581 7053785949 5747190989 | 238 |
| 85 | 4331265999 3115836136 2183475565 8239042741 4507948572 7186025037 | 242 |
| 86 | -1045209045 2911797193 7193258105 2558754541 5565899467 0692886763 | 247 |
| 87 | 2581065514 1987810089 9707654797 4741907677 3377383408 1278700748 | 251 |
| 88 | -6520614581 8387778469 9333219707 6107151822 5185651052 9013026341 | 255 |
| 89 | 1684846341 9687898075 4783003040 6114009586 4928197587 5019890709 | 260 |
| 90 | -4451490426 8906044510 2555960700 2510100001 9694107811 2124089848 | 264 |
| 91 | 1202312799 0555700569 6415001461 5208559076 4114140947 9603346892 | 269 |
| 92 | -3318884994 5714545923 5635587981 5525006393 7656494149 2572721481 | 273 |
| 93 | 9361117813 4732182388 4696687894 0762835488 5562350765 0668956828 | 277 |
| 94 | -2697268820 0316834448 3172511100 5030769842 1874254433 6000257290 | 282 |
| 95 | 7937506771 2015249131 9389678265 6327433067 0895042368 1200670639 | 286 |
| 96 | -2385128164 1614335081 9172506362 2521442090 4381092722 9162655048 | 291 |
| 97 | 7316669218 6769233470 5449619982 6457180265 2589135970 5647122787 | 295 |
| 98 | -2290854349 2134024340 8307468282 0344727128 9737921142 1809219641 | 300 |
| 99 | 7319375253 7843879144 8922191028 0093192637 1122200192 4798860832 | 304 |
| 100 | -2385914905 6133313409 2942835188 1787421020 5248459913 8364816660 | 309 |

## H.2  Symbolic Energy Corrections to Order 10

The symbolic energy corrections for a general harmonium state have been given in (10.53) to (10.56) to order 4 as functions of $q$ and $\tau = k(k+1)$. Here we give more extensive results to order 10 obtained using program *harm* in Section 10.2.3 with *jmax := 10* or more efficiently with procedure *hvhfpower* in Section 10.6 with *nmax := 10* and *d := 2*.

$$E^{(1)} = \frac{1}{2}q^2(5\,q^2 - 3\,\tau + 1),$$

$$E^{(2)} = -\frac{1}{16}q^6(143\,q^4 - 90\,q^2\tau + 345\,q^2 - 21\,\tau^2 - 126\,\tau + 28),$$

$$E^{(3)} = \frac{1}{16}q^{10}(1530\,q^6 - 1305\,q^4\tau + 11145\,q^4 - 6825\,q^2\tau + 8645\,q^2$$
$$- 33\,\tau^3 + 33\,\tau^2 - 2706\,\tau + 484),$$

$$E^{(4)} = -\frac{1}{1024}q^{14}(1502291\,q^8 - 1640100\,q^6\tau + 22937530\,q^6$$
$$+ 251370\,q^4\tau^2 - 19742520\,q^4\tau + 54811295\,q^4 - 3060\,q^2\tau^3$$
$$+ 2184330\,q^2\tau^2 - 31859700\,q^2\tau + 25371140\,q^2 - 4005\,\tau^4$$
$$- 7260\,\tau^3 + 1425540\,\tau^2 - 7286640\,\tau + 1137344),$$

$$E^{(5)} = \frac{1}{1024}q^{18}(27669770\,q^{10} - 36942507\,q^8\tau + 738100629\,q^8$$
$$+ 10748400\,q^6\tau^2 - 818853750\,q^6\tau + 3796517920\,q^6$$
$$- 346242\,q^4\tau^3 + 187966626\,q^4\tau^2 - 3192023007\,q^4\tau$$
$$+ 5300835461\,q^4 - 6930\,q^2\tau^4 - 4502190\,q^2\tau^3$$
$$+ 458172330\,q^2\tau^2 - 2944640160\,q^2\tau + 1820247660\,q^2$$
$$- 8379\,\tau^5 - 17955\,\tau^4 - 4433688\,\tau^3$$
$$+ 165916284\,\tau^2 - 492400656\,\tau + 70057408),$$

$$E^{(6)} = -\frac{3}{8192}q^{22}(-640381106480\,q^6\tau + 164689108353\,q^4\tau^2$$
$$- 1430376269490\,q^4\tau - 7886005400\,q^2\tau^3 + 221330307900\,q^2\tau^2$$
$$- 949563212520\,q^2\tau - 86105853394\,q^8\tau + 30280493840\,q^6\tau^2$$
$$- 2363069520\,q^4\tau^3 + 11579750\,q^2\tau^4 - 2399789700\,q^{10}\tau$$
$$+ 1015072091\,q^8\tau^2 - 99282080\,q^6\tau^3 + 623889\,q^4\tau^4$$
$$- 45500\,q^2\tau^5 + 17262526720 + 583854083483\,q^8$$
$$- 129722804160\,\tau + 1777016373300\,q^4 + 502265307440\,q^2$$
$$+ 16259896\,\tau^4 - 3937115160\,\tau^3 + 56786744976\,\tau^2$$
$$- 129214\,\tau^5 + 1518336155\,q^{12} - 49703\,\tau^6$$
$$+ 1749218303660\,q^6 + 63393364650\,q^{10}),$$

$$E^{(7)} = \frac{1}{8192}q^{26}(-522962253144870\,q^6\tau + 139762652943507\,q^4\tau^2$$

$$- 8363355777316358\, q^4\tau - 8508808592940\, q^2\tau^3$$
$$+ 131887711729980\, q^2\tau^2 - 447473931577560\, q^2\tau$$
$$- 121991813347635\, q^8\tau + 45877477646220\, q^6\tau^2$$
$$- 4801867864563\, q^4\tau^3 + 83251027320\, q^2\tau^4$$
$$- 9852064808820\, q^{10}\tau + 4679187532845\, q^8\tau^2$$
$$- 678390832020\, q^6\tau^3 + 19085833767\, q^4\tau^4 - 18670200\, q^2\tau^5$$
$$- 184491014565\, q^{12}\tau + 101903505600\, q^{10}\tau^2$$
$$- 17676727005\, q^8\tau^3 + 614328660\, q^6\tau^4 - 852159\, q^4\tau^5$$
$$- 348000\, q^2\tau^6 + 6706247297280 + 485651234257405\, q^8$$
$$- 53025299553600\, \tau + 866076004563108\, q^4$$
$$+ 213223231554000\, q^2 + 54367159608\, \tau^4 - 2913414281352\, \tau^3$$
$$+ 26974389458160\, \tau^2 - 35741790\, \tau^5 + 6129773123305\, q^{12}$$
$$- 1052541\, \tau^6 - 345951\, \tau^7 + 1051180670572670\, q^6$$
$$+ 90295906154220\, q^{10} + 100977902060\, q^{14}),$$

$$E^{(8)} = -\frac{1}{4194304}\, q^{30}(594204250476429312$$
$$- 78125322929429248200\, q^6\tau + 21059948889078261960\, q^4\tau^2$$
$$- 100073398486962479040\, q^4\tau - 1421633697707002560\, q^2\tau^3$$
$$+ 15727724851189248480\, q^2\tau^2 - 459481187777707484800\, q^2\tau$$
$$- 25471579777432757280\, q^8\tau + 9816883396020120420\, q^6\tau^2$$
$$- 1168004616414294120\, q^4\tau^3 + 31571147636285400\, q^2\tau^4$$
$$- 3543802955282041200\, q^{10}\tau + 1762322608743286980\, q^8\tau^2$$
$$- 301494175483704840\, q^6\tau^3 + 14174187079413690\, q^4\tau^4$$
$$- 89606370351960\, q^2\tau^5 - 187881788567547360\, q^{12}\tau$$
$$+ 112493961213303180\, q^{10}\tau^2 - 24515511450738360\, q^8\tau^3$$
$$+ 1599545316614700\, q^6\tau^4 - 16175781583920\, q^4\tau^5$$
$$+ 1490296500\, q^2\tau^6 - 2507933204497800\, q^{14}\tau$$
$$+ 1706018271083940\, q^{12}\tau^2 - 431074490161560\, q^{10}\tau^3$$
$$+ 33403509754290\, q^8\tau^4 - 411328477560\, q^6\tau^5$$
$$- 102913020\, q^4\tau^6 - 465517800\, q^2\tau^7$$
$$+ 74039990765605923747\, q^8 - 4892215995275360256\, \tau$$
$$+ 915971846344442058480\, q^4 + 20311060777074397440\, q^2$$
$$+ 13631589661850160\, \tau^4 - 3801658553322250368\, \tau^3$$
$$+ 2737908587619986688\, \tau^2 - 74473619966592\, \tau^5$$
$$+ 2209203410900611994\, q^{12} + 4424409528\, \tau^6$$
$$- 1474764984\, \tau^7 + 1209106249546731\, q^{16}$$
$$- 423795141\, \tau^8 + 129064684435547846440\, q^6$$

$$+ 19094535217676629420\, q^{10} + 101214248995905300\, q^{14}),$$

$$\begin{aligned}
E^{(9)} = \frac{1}{4194304}\, q^{34}(& -25814146312845876753600\, q^6\tau \\
& + 69239933968170012156 72\, q^4\tau^2 \\
& - 282088061532688378911 84\, q^4\tau \\
& - 489436751303964915360\, q^2\tau^3 \\
& + 4375791106917473077920\, q^2\tau^2 \\
& - 11538709576991932483200\, q^2\tau \\
& - 10536073661405744149863\, q^8\tau \\
& + 4081401630633420208500\, q^6\tau^2 \\
& - 5160029643156604544 64\, q^4\tau^3 \\
& + 17129844727027436040\, q^2\tau^4 \\
& - 2046049483298294493900\, q^{10}\tau \\
& + 1035730559479051124436\, q^8\tau^2 \\
& - 1921000848223429757 40\, q^6\tau^3 \\
& + 11441535593246037630\, q^4\tau^4 - 1439898981161068960\, q^2\tau^5 \\
& - 185540570384991723762\, q^{12}\tau \\
& + 1149844831007366223 00\, q^{10}\tau^2 \\
& - 27920308338823244652\, q^8\tau^3 + 2414349602975836680\, q^6\tau^4 \\
& - 52866280306087794\, q^4\tau^5 + 118810876946580\, q^2\tau^6 \\
& - 6875210871080888700\, q^{14}\tau + 4969583971210418532\, q^{12}\tau^2 \\
& - 1461572357237327700\, q^{10}\tau^3 + 1614264903996660 30\, q^8\tau^4 \\
& - 4885774343654820\, q^6\tau^5 + 17263487650500\, q^4\tau^6 \\
& - 4087874700\, q^2\tau^7 + 24967232274976587192509\, q^8 \\
& - 1128669070553841223680\,\tau + 23611717163095060880400\, q^4 \\
& + 4816847647217148143360\, q^2 + 5660516676902782800\,\tau^4 \\
& - 1097611578727226707 20\,\tau^3 + 674395434813222931200\,\tau^2 \\
& - 76084458466518768\,\tau^5 + 12870187171919924696 78\, q^{12} \\
& + 13259189908250524057 6 + 123226696670280\,\tau^6 \\
& - 8765961120\,\tau^7 + 3265591352230432797\, q^{16} \\
& - 4063522995\,\tau^8 + 29439272409650730\, q^{18} \\
& - 1038064335\,\tau^9 - 68351957131283631\, q^{16}\tau \\
& + 551776690881684000\, q^{14}\tau^2 - 18400407216980964\, q^{12}\tau^3 \\
& + 2346113399519700\, q^{10}\tau^4 - 83645792066298\, q^8\tau^5 \\
& + 355099612560\, q^6\tau^6 - 579423780\, q^4\tau^7 \\
& - 1237877550\, q^2\tau^8 + 372575985113063521332 40\, q^6 \\
& + 8021872386582406004430\, q^{10} + 99943774907337685240\, q^{14}),
\end{aligned}$$

$$\begin{aligned}
E^{(10)} = -\frac{1}{33554432}q^{38}(&-7881999418808955899855280\,q^6\tau \\
&+ 209009075561910457960193376\,q^4\tau^2 \\
&- 76330207211168409554762784\,q^4\tau \\
&- 1504989428729927565063360\,q^2\tau^3 \\
&+ 1163867225513397062452800\,q^2\tau^2 \\
&- 284869046322388446931200000\,q^2\tau \\
&- 378339578081424111180439306\,q^8\tau \\
&+ 1458841358561421912331068080\,q^6\tau^2 \\
&- 1896844322866615243106832\,q^4\tau^3 \\
&+ 70029642241102314072960\,q^2\tau^4 \\
&- 920525436312856021359348080\,q^{10}\tau \\
&+ 46769215283771723584219530\,q^8\tau^2 \\
&- 90331331095104309347976080\,q^6\tau^3 \\
&+ 6089006984336082756734440\,q^4\tau^4 \\
&- 10606380529882672735200\,q^2\tau^5 \\
&- 116245943036133341582314880\,q^{12}\tau \\
&+ 730518003379744477989000\,q^{10}\tau^2 \\
&- 1876773174747532628466240\,q^8\tau^3 \\
&+ 18810058549127374745100\,q^6\tau^4 \\
&- 592132893691543799364\,q^4\tau^5 \\
&+ 3515847328666328040\,q^2\tau^6 \\
&- 7316804004054901814700080\,q^{14}\tau \\
&+ 54352720848717170041266\,q^{12}\tau^2 \\
&- 17270475838742681973600\,q^{10}\tau^3 \\
&+ 2283233578581999430926\,q^8\tau^4 \\
&- 10466943776941852116080\,q^6\tau^5 \\
&+ 1074970943571787878\,q^4\tau^6 \\
&- 82806177562416080\,q^2\tau^7 \\
&+ 7774307478151774117933520480\,q^8 \\
&- 25984401283807182600683520\,\tau \\
&+ 596697085408450547953505920\,q^4 \\
&+ 1136057224943169806063104080\,q^2 \\
&+ 1914370575115439990764880\,\tau^4 \\
&- 29543793188671469935718480\,\tau^3 \\
&+ 16285428992291247481804800\,\tau^2 \\
&- 41998749924174456067280\,\tau^5
\end{aligned}$$

$$+\, 586104178165940043794390 7\, q^{12}$$

$$+\, 29674979167976207790899 2 + 2232074007469920432\, \tau^{6}$$

$$-\, 1053610521588144\, \tau^{7} + 34705378442365672338990\, q^{16}$$

$$-\, 177867803412\, \tau^{8} + 843187971747245957840\, q^{18}$$

$$-\, 90087252450\, \tau^{9} + 59173624613545168 99\, q^{20}$$

$$-\, 20721643839\, \tau^{10} - 1985244095915674215930\, q^{16}\tau$$

$$+\, 1681626116022930233640\, q^{14}\tau^{2}$$

$$-\, 625896996064290484560\, q^{12}\tau^{3}$$

$$+\, 100464361507158503400\, q^{10}\tau^{4}$$

$$-\, 5877020173201344468\, q^{8}\tau^{5}$$

$$+\, 82825025031214200\, q^{6}\tau^{6}$$

$$-\, 98704083050496\, q^{4}\tau^{7} - 87460896600\, q^{2}\tau^{8}$$

$$-\, 15205758562221894240\, q^{18}\tau + 14197453680723157845\, q^{16}\tau^{2}$$

$$-\, 58931616246006160 80\, q^{14}\tau^{3} + 10695290575906598 22\, q^{12}\tau^{4}$$

$$-\, 71867091085629840\, q^{10}\tau^{5} + 1183390343326170\, q^{8}\tau^{6}$$

$$-\, 1679177790000\, q^{6}\tau^{7} - 13676757633\, q^{4}\tau^{8}$$

$$-\, 26681701200\, q^{2}\tau^{9} + 1030436771159479373196590 40\, q^{6}$$

$$+\, 292826932774317352822811 40\, q^{10}$$

$$+\, 63021398160827265250938 0\, q^{14}).$$

# Appendix I

# Tables of Screened Coulomb Energy Corrections

## I.1  Ground State Energy to Order 100

In Tables I.1 to I.8 the 3-dimensional ground state energy corrections $E^{(n)}$ for the Yukawa potential $e^{-\lambda r}/r$ are given to order 100 in rational form. The floating point values of the energy corrections expressed in the Maple internal floating point format are also given to 60 digit accuracy in Tables I.9 to I.12.

**Table I.1**: Ground state energy corrections for the Yukawa potential. The denominators are expressed in prime power factorized form.

| $n$ | $E^{(n)}$ |
|---|---|
| 0 | $-1/2^1$ |
| 1 | $1$ |
| 2 | $-3/2^2$ |
| 3 | $1/2^1$ |
| 4 | $-11/2^4$ |
| 5 | $21/2^4$ |
| 6 | $-145/2^4 3^1$ |
| 7 | $757/2^5 3^1$ |
| 8 | $-69433/2^{10} 3^1$ |
| 9 | $32144\,9/2^9 3^2$ |
| 10 | $-23439\,67/2^{11} 5^1$ |
| 11 | $24316\,577/2^{11} 3^1 5^1$ |
| 12 | $-25360\,41607/2^{15} 3^3$ |
| 13 | $47860\,81153\,7/2^{15} 3^3 5^1$ |
| 14 | $-14592\,37850\,51/2^{15} 3^1 5^1 7^1$ |
| 15 | $15995\,72488\,09633/2^{16} 3^4 5^2 7^1$ |
| 16 | $-42949\,29463\,45844\,21/2^{22} 3^4 5^2 7^1$ |
| 17 | $46466\,86497\,54309\,73/2^{20} 3^4 5^2 7^1$ |
| 18 | $-23670\,72185\,02703\,471/2^{23} 3^4 5^2$ |
| 19 | $25325\,03612\,32201\,4333/2^{23} 3^3 5^2 7^1$ |
| 20 | $-57276\,15455\,10214\,99199\,3/2^{27} 3^4 5^2 7^1$ |
| 21 | $21549\,03296\,64682\,80269\,87/2^{27} 3^2 5^2 7^2$ |
| 22 | $-31757\,01402\,26589\,04122\,57977/2^{27} 3^5 5^2 7^2 11^1$ |
| 23 | $64608\,37466\,35701\,41425\,16729/2^{28} 3^5 5^1 7^2 11^1$ |
| 24 | $-10771\,24814\,77467\,74201\,20851\,827/2^{33} 3^5 5^1 7^2 11^1$ |
| 25 | $10771\,06412\,85444\,85198\,92862\,39432\,7/2^{32} 3^6 5^4 7^2 11^1$ |
| 26 | $-55564\,55763\,96743\,64737\,31350\,81890\,3/2^{34} 3^6 5^3 7^2 13^1$ |
| 27 | $15344\,97204\,84076\,80571\,90421\,49898\,89581/2^{34} 3^8 5^4 7^2 11^1 13^1$ |
| 28 | $-32663\,22590\,08090\,86036\,07725\,23196\,42800\,71/2^{38} 3^7 5^4 7^3 11^1 13^1$ |
| 29 | $19022\,47435\,94919\,42077\,28480\,29801\,67651\,521/2^{38} 3^7 5^4 7^3 11^1 13^1$ |
| 30 | $-72709\,02876\,86525\,51676\,77699\,90699\,38769\,3327/2^{38} 3^9 5^5 7^2 11^1 13^1$ |

**Table I.2**: Continuation of preceding table.

| $n$ | $E^{(n)}$ |
|---|---|
| 31 | 61761 40432 17338 30903 08267 40905 45487 89063 $3/2^{39}3^{9}5^{5}7^{3}$ $11^{1}13^{1}$ |
| 32 | $-$32616 15550 42142 04234 65124 14426 22644 82034 $147/2^{46}3^{8}5^{4}$ $7^{3}11^{1}13^{1}$ |
| 33 | 12733 56381 37314 72017 98783 24735 96174 71109 96009 $3/2^{43}$ $3^{10}5^{5}7^{3}11^{2}13^{1}$ |
| 34 | $-$22271 08736 32478 73387 25394 13678 31183 97179 36187 5729 $/2^{47}3^{10}5^{5}7^{3}11^{2}13^{1}17^{1}$ |
| 35 | 72941 76451 30383 00457 06795 32680 73448 28799 10322 85439 $/2^{47}3^{10}5^{6}7^{3}11^{2}13^{1}17^{1}$ |
| 36 | $-$23353 52475 05474 18436 37762 90140 61854 69314 29706 91684 $859/2^{51}3^{11}5^{6}7^{3}11^{2}13^{1}17^{1}$ |
| 37 | 21572 66725 30984 16848 03381 35940 07753 96852 76491 88157 $3/2^{51}3^{10}5^{5}7^{1}11^{2}13^{1}17^{1}$ |
| 38 | $-$69375 40749 47614 63278 69149 12945 23802 16276 68130 65734 $26233/2^{51}3^{10}5^{6}7^{3}11^{2}13^{1}17^{1}19^{1}$ |
| 39 | 11407 53312 08055 26005 87771 52339 09978 02115 00930 85089 07636 $9037/2^{52}3^{12}5^{6}7^{3}11^{2}13^{2}17^{1}19^{1}$ |
| 40 | $-$18630 42643 01465 94048 18894 44827 19210 51651 73503 86541 17987 75813 $9/2^{57}3^{12}5^{7}7^{2}11^{2}13^{2}17^{1}19^{1}$ |
| 41 | 84417 32429 71757 53914 24493 68930 52369 60417 72721 97585 89605 $6537/2^{56}3^{11}5^{7}7^{3}11^{1}13^{2}19^{1}$ |
| 42 | $-$29358 41457 25863 57336 61611 47474 18931 80637 10603 66214 80079 97654 $65281/2^{58}3^{13}5^{7}7^{4}11^{2}13^{2}17^{1}19^{1}$ |
| 43 | 22009 61232 40931 26460 39823 45341 07739 80475 78027 83933 75128 96878 46183 $3/2^{58}3^{13}5^{7}7^{4}11^{2}13^{2}17^{1}19^{1}$ |
| 44 | $-$29491 45094 22211 14196 56910 01506 09492 19814 79352 93584 31407 44305 14187 $0021/2^{62}3^{13}5^{7}7^{4}11^{3}13^{2}17^{1}19^{1}$ |
| 45 | 72364 04324 56742 09057 25225 45976 42553 36187 65592 02500 48019 38428 08053 $97/2^{62}3^{11}5^{6}7^{3}11^{3}13^{2}17^{1}19^{1}$ |

**Table I.3**: Continuation of preceding table.

| $n$ | $E^{(n)}$ |
|---|---|
| 46 | $-31637$ 37667 82061 22470 78566 73652 64163 88829 15056 69322 95677 89923 49035 78668 $1/2^{62}3^{13}5^{7}7^{4}11^{3}13^{1}17^{1}19^{1}23^{1}$ |
| 47 | 65476 45177 62503 60195 77052 23819 97783 20542 92545 76712 36167 35641 16144 39436 $633/2^{63}3^{13}5^{7}7^{4}11^{3}13^{2}17^{1}19^{1}23^{1}$ |
| 48 | $-23884$ 97585 41272 05499 96253 59645 19931 10508 42578 62562 0471 84420 08374 58214 $22981/2^{69}3^{14}5^{5}7^{4}11^{3}13^{2}19^{1}23^{1}$ |
| 49 | 20033 84289 42922 17061 88083 07669 57774 72262 42728 84985 20104 89813 45743 47156 $25941/2^{67}3^{14}5^{7}7^{6}11^{3}13^{1}19^{1}$ |
| 50 | $-12918$ 47385 28698 80570 93510 76026 30917 52588 29617 64244 19488 46690 37222 19401 14503 $4253/2^{70}3^{14}5^{9}7^{5}11^{2}13^{2}19^{1}23^{1}$ |
| 51 | 29034 72319 24983 27246 31295 94783 15265 14810 75804 57520 97522 18688 47497 43099 92227 96571 $237/2^{70}3^{15}5^{7}$ $7^{6}11^{3}13^{2}17^{2}19^{1}23^{1}$ |
| 52 | $-42927$ 75533 44261 21617 24679 10002 56594 22515 09050 07074 90340 74492 62822 15449 61159 74989 83469 $57/2^{74}3^{14}5^{9}$ $7^{6}11^{3}13^{3}17^{2}19^{1}23^{1}$ |
| 53 | 37100 78076 23842 85220 73339 36326 93370 29800 79592 46163 82292 99676 78158 47127 01123 96556 09493 $233/2^{74}3^{14}$ $5^{9}7^{6}11^{3}13^{3}17^{2}19^{1}23^{1}$ |
| 54 | $-97445$ 86310 15810 42050 55235 21486 40709 60061 49903 77643 38793 54511 65420 95493 39093 77097 98802 $6009/2^{74}3^{15}$ $5^{9}7^{6}11^{3}13^{3}17^{2}19^{1}23^{1}$ |
| 55 | 12397 48275 02479 17347 07740 35007 37471 58600 16436 77295 19118 72465 67722 12211 20827 28122 54488 79362 $23/2^{75}$ $3^{16}5^{10}7^{6}11^{4}13^{3}17^{2}19^{1}$ |
| 56 | $-63721$ 97063 68629 36548 47304 55059 67556 82878 08117 87998 75936 99626 59189 42956 80935 71631 51381 59580 05569 $/2^{80}3^{14}5^{10}7^{7}11^{4}13^{3}17^{2}19^{1}23^{1}$ |
| 57 | 43065 97087 69478 31141 77424 82844 97157 97742 82297 00485 42896 31983 27344 34413 53155 65795 38645 27627 64687 1 $/2^{79}3^{16}5^{9}7^{7}11^{4}13^{3}17^{2}19^{2}$ |
| 58 | $-22654$ 95033 31601 84217 05366 02810 44239 62017 62169 21580 26225 75842 33386 36355 91081 97974 61385 20802 26142 14185 $1/2^{81}3^{17}5^{10}7^{6}11^{4}13^{3}17^{2}19^{2}23^{1}29^{1}$ |

**Table I.4**: Continuation of preceding table.

| $n$ | $E^{(n)}$ |
|---|---|
| 59 | 14767 43599 52954 03511 92981 42885 40990 08415 28673 08842 25946 99621 43318 54824 61725 33147 43454 88662 18914 71300 233/$2^{81}3^{17}5^{10}7^{7}11^{4}13^{3}17^{2}19^{2}23^{1}29^{1}$ |
| 60 | $-$48398 14190 56742 59506 49302 26118 74911 67203 60567 65196 10779 01167 55207 24729 13046 42696 71724 29309 83027 72303 6581/$2^{85}3^{17}5^{11}7^{7}11^{4}13^{3}17^{2}19^{2}29^{1}$ |
| 61 | 45476 21250 14868 12095 85172 34021 95956 16147 05830 48360 86364 80499 10297 36036 02920 33638 75012 74055 40735 07648 30044 9/$2^{85}3^{18}5^{11}7^{6}11^{4}13^{3}17^{2}19^{2}23^{1}29^{1}$ |
| 62 | $-$14638 81977 47104 94821 87613 67400 97522 44734 54338 18551 39424 20053 96578 85797 34058 73831 94502 07478 30801 63504 52018 213/$2^{85}3^{18}5^{10}7^{7}11^{4}13^{2}17^{2}19^{2}23^{1}29^{1}31^{1}$ |
| 63 | 13484 87453 72094 34980 79188 64237 43012 25913 81444 21049 15937 60233 14877 35616 70323 47689 86913 95065 54079 70797 41098 93761/$2^{86}3^{19}5^{11}7^{8}11^{4}13^{3}19^{2}23^{1}29^{1}31^{1}$ |
| 64 | $-$89433 55000 32365 34503 77600 49723 69129 33025 58244 15590 94357 60950 89262 95989 98460 65530 16425 15054 20560 44993 99557 80514 9999/$2^{94}3^{19}5^{11}7^{8}11^{3}13^{3}17^{2}19^{2}23^{1}29^{1}31^{1}$ |
| 65 | 13281 27419 49967 87817 02114 26193 88303 91447 53286 90854 57952 14446 66104 04987 31061 36299 39808 02935 63328 61603 18123 81782 18761 03/$2^{90}3^{18}5^{12}7^{8}11^{4}13^{4}17^{2}19^{2}23^{1}29^{1}31^{1}$ |
| 66 | $-$42403 41743 70780 50656 70325 00015 17723 71879 19427 34716 24483 65361 75570 59758 79355 72709 38289 42173 46743 60037 47362 37454 67505 06325 7/$2^{95}3^{20}5^{12}7^{8}11^{5}13^{4}17^{2}19^{2}23^{1}29^{1}31^{1}$ |
| 67 | 20568 19414 77304 23896 88803 68520 43002 64425 50227 44414 69605 94662 45645 29796 53763 29840 90825 08590 72935 59081 03676 24082 19866 27427 9/$2^{95}3^{19}5^{12}7^{7}11^{5}13^{4}17^{2}19^{2}23^{1}29^{1}31^{1}$ |
| 68 | $-$12094 18098 57853 35108 38351 35842 77242 78048 18353 68908 96505 08773 90119 90981 12377 86462 50543 74512 81256 69102 84624 86862 32889 11624 07070 9/$2^{99}3^{20}5^{12}7^{8}11^{5}13^{4}17^{3}19^{2}23^{1}29^{1}31^{1}$ |
| 69 | 15753 75225 71112 44277 99073 90302 36355 70696 62109 99946 37125 58202 56560 78502 90392 08206 16959 10329 84553 32940 09346 02969 00749 94495 89648 7/$2^{99}3^{21}5^{12}7^{8}11^{5}13^{4}17^{3}19^{1}23^{2}31^{1}$ |

**Table I.5**: Continuation of preceding table.

| $n$ | $E^{(n)}$ |
|---|---|
| 70 | $-12771\ 25717\ 64800\ 76593\ 52912\ 39389\ 85932\ 75111\ 32803\ 83854$ $91928\ 90328\ 50462\ 83191\ 98643\ 86442\ 93777\ 71511\ 05679\ 75425$ $85852\ 15719\ 42372\ 37385\ 99472\ 37881/2^{99}3^{21}5^{11}7^{9}11^{5}13^{4}17^{3}19^{2}$ $23^{2}29^{1}31^{1}$ |
| 71 | $79755\ 04398\ 67884\ 72363\ 48629\ 79504\ 44621\ 09429\ 65275\ 65327$ $03213\ 28358\ 63467\ 04165\ 23571\ 59370\ 65778\ 19437\ 82435\ 67575$ $90716\ 77906\ 66137\ 17363\ 70686\ 89749/2^{100}3^{21}5^{12}7^{9}11^{5}13^{4}17^{2}$ $19^{2}23^{2}29^{1}31^{1}$ |
| 72 | $-13957\ 25694\ 98456\ 33514\ 37907\ 72837\ 36613\ 21077\ 20224\ 20538$ $31758\ 14834\ 25658\ 00181\ 35216\ 28732\ 84075\ 45507\ 38083\ 43904$ $21531\ 03190\ 75476\ 30826\ 85683\ 73428\ 31399/2^{105}3^{22}5^{12}7^{9}11^{5}$ $13^{4}17^{3}19^{2}23^{2}29^{1}31^{1}$ |
| 73 | $75576\ 44329\ 46196\ 09456\ 70685\ 15776\ 49406\ 61660\ 50655\ 68493$ $31679\ 85082\ 66853\ 83290\ 96805\ 87383\ 01462\ 75230\ 16152\ 77560$ $26151\ 20052\ 84631\ 63675\ 05017\ 35461\ 68391/2^{104}3^{22}5^{12}7^{9}11^{5}$ $13^{4}17^{3}19^{2}23^{2}29^{1}31^{1}$ |
| 74 | $-40774\ 20461\ 82443\ 84725\ 78975\ 88746\ 25995\ 80769\ 35772\ 47477$ $19256\ 90537\ 32217\ 14522\ 14511\ 07077\ 32716\ 15563\ 44232\ 83867$ $25575\ 17680\ 60713\ 02131\ 27077\ 32047\ 23537\ 953/2^{106}3^{21}5^{12}7^{9}$ $11^{5}13^{4}17^{3}19^{2}23^{2}29^{1}31^{1}37^{1}$ |
| 75 | $11988\ 39358\ 52482\ 21197\ 40613\ 64500\ 81670\ 70352\ 75688\ 66232$ $04823\ 20376\ 94700\ 00511\ 43671\ 64801\ 12818\ 03272\ 35675\ 85255$ $21806\ 35651\ 36820\ 27605\ 50077\ 37086\ 42167\ 9701/2^{106}3^{23}5^{13}7^{9}$ $11^{5}13^{2}17^{3}19^{2}23^{2}29^{1}31^{1}37^{1}$ |
| 76 | $-68661\ 32448\ 16837\ 53556\ 06606\ 02136\ 86919\ 54735\ 22514\ 26612$ $78744\ 80151\ 70246\ 38682\ 44289\ 70588\ 36796\ 81940\ 37322\ 44195$ $15478\ 45973\ 53997\ 23707\ 69696\ 73044\ 01484\ 66154\ 2577/2^{110}$ $3^{23}5^{13}7^{9}11^{5}13^{4}17^{3}19^{3}23^{2}29^{1}31^{1}37^{1}$ |
| 77 | $17447\ 44866\ 38853\ 91929\ 43744\ 48102\ 05207\ 20912\ 99333\ 23263$ $79434\ 65428\ 01978\ 78778\ 10473\ 58388\ 96785\ 44323\ 24147\ 01565$ $18760\ 27227\ 58304\ 19217\ 54396\ 99879\ 97862\ 73840\ 23979/2^{110}$ $3^{23}5^{13}7^{10}11^{5}13^{4}17^{3}19^{3}23^{2}29^{1}37^{1}$ |
| 78 | $-23959\ 98777\ 40289\ 94097\ 14971\ 94309\ 27464\ 38260\ 89495\ 70049$ $68220\ 02552\ 69818\ 54936\ 68099\ 13917\ 51141\ 45023\ 16074\ 64037$ $61537\ 26485\ 01702\ 59544\ 47565\ 24881\ 69292\ 23063\ 47729\ 4927$ $/2^{110}3^{24}5^{13}7^{10}11^{5}13^{5}17^{3}19^{3}23^{2}29^{1}31^{1}37^{1}$ |

**Table I.6**: Continuation of preceding table.

| $n$ | $E^{(n)}$ |
|---|---|
| 79 | 26159 18217 39975 24383 35301 76860 53558 38341 98076 83447 32274 75116 67574 76823 84944 87801 14477 80543 96034 24948 12816 39269 00462 03799 80671 44156 17204 90615 29725 3831 $/2^{111}3^{23}5^{13}7^{9}11^{5}13^{5}17^{3}19^{3}23^{2}29^{1}31^{1}37^{1}$ |
| 80 | $-13557$ 57988 32385 51112 62877 49469 19510 59385 57223 91463 85921 45733 06800 86731 47374 20447 08330 54180 62488 97824 66221 97378 08403 02878 51637 15045 08850 59568 70570 62754 $561/2^{117}3^{23}5^{13}7^{10}11^{5}13^{5}17^{3}19^{3}23^{2}29^{1}31^{1}37^{1}$ |
| 81 | 53412 72339 27908 98135 06892 00708 81753 07046 84805 47054 99501 25939 86686 77757 75895 40530 89878 83172 64546 99674 85435 50924 95596 61916 59632 69037 42265 72203 55785 20563 $15011/2^{115}3^{26}5^{14}7^{10}11^{5}13^{5}17^{3}19^{3}23^{2}29^{1}31^{1}37^{1}$ |
| 82 | $-68778$ 31143 39282 03550 54808 96747 42096 51377 25819 85845 95779 12524 71039 10632 89392 22698 83696 33331 22649 25086 64445 35327 40714 10760 51312 10232 99624 21035 45159 35698 42507 769/$2^{118}3^{25}5^{14}7^{10}11^{5}13^{5}17^{3}19^{3}23^{2}29^{1}31^{1}37^{1}41^{1}$ |
| 83 | 81720 07266 54729 33684 96436 33081 74553 47810 78876 70563 00739 65419 58818 31426 23038 25311 57979 26926 81824 43226 06389 85974 39621 33038 08534 84062 07813 70531 71027 98112 22474 9527/$2^{118}3^{25}5^{14}7^{10}11^{5}13^{5}17^{3}19^{3}23^{2}29^{1}31^{1}37^{1}41^{1}$ |
| 84 | $-33186$ 47472 68412 51773 36995 18436 32889 15335 95754 63654 81675 29407 48911 76958 76738 47608 35715 83091 62945 54886 88798 44501 62701 02323 24936 84512 94149 76216 91632 36859 71285 66201/$2^{122}3^{27}5^{12}7^{10}11^{5}13^{5}17^{2}19^{3}23^{2}29^{1}31^{1}37^{1}41^{1}$ |
| 85 | 14493 35127 23799 95294 91244 16374 72868 97589 36133 36806 39313 06129 38753 92590 68409 40644 46063 56527 69015 51748 21501 94697 34536 33544 78094 34696 88543 87109 31162 56013 92511 39956 07225 7/$2^{122}3^{27}5^{15}7^{10}11^{5}13^{5}17^{4}19^{3}23^{2}29^{1}31^{1}37^{1}41^{1}$ |
| 86 | $-50657$ 90203 45337 85411 16088 54932 02423 39088 43098 41054 84035 70803 30983 61104 21380 84622 15620 92446 59881 74269 89372 04376 29569 81534 03802 71336 32639 88391 43379 30919 54391 49186 97614 23/$2^{122}3^{26}5^{14}7^{10}11^{5}13^{5}17^{4}19^{3}23^{2}29^{1}31^{1}37^{1}41^{1}43^{1}$ |

**Table I.7**: Continuation of preceding table.

| $n$ | $E^{(n)}$ |
|---|---|
| 87 | 32514 82304 02792 92942 94389 02879 38835 40497 19213 36741 36173 04188 64074 06372 13570 93162 46144 50547 12704 05868 27320 75045 09857 72882 32392 63419 37190 86903 03558 25610 66309 01819 06417 52866 $3/2^{123}3^{28}5^{14}7^{10}11^{5}13^{5}17^{4}19^{3}23^{2}29^{2}31^{1}37^{1}41^{1}43^{1}$ |
| 88 | $-$54580 67348 08900 17401 19895 98244 91073 92484 49618 10461 25799 15531 98516 07042 84341 61701 95056 97744 94386 03694 60095 22590 85950 70343 11782 54832 91337 15231 32504 79162 44548 91264 84677 19938 $1503/2^{128}3^{28}5^{15}7^{10}11^{6}13^{4}17^{4}19^{3}23^{2}29^{2}31^{1}37^{1}41^{1}43^{1}$ |
| 89 | 44353 47877 93085 79177 28954 37746 42041 07730 36547 69775 35509 30683 61810 88859 89160 54007 24151 69428 60099 64495 84275 15793 48034 05919 06531 38701 44260 23868 99871 72657 09707 98380 44919 07171 71573 $7/2^{127}3^{28}5^{15}7^{10}11^{6}13^{5}17^{4}19^{3}23^{2}29^{2}31^{1}37^{1}41^{1}43^{1}$ |
| 90 | $-$43562 66686 30856 85882 54329 72153 46064 20366 65362 13192 54922 53621 19775 95108 16871 57964 61083 73222 99633 79022 85782 32651 46389 15515 12344 91326 55206 13373 26905 55314 50297 74909 60771 68253 30693 $61/2^{129}3^{29}5^{16}7^{9}11^{5}13^{5}17^{4}19^{3}23^{2}29^{2}31^{1}37^{1}41^{1}43^{1}$ |
| 91 | 12929 05124 38965 48827 69624 64640 41227 18743 89216 31861 58831 90245 39517 20972 87212 84844 71862 02257 02573 52735 90616 27361 11128 74932 31141 50022 27122 93751 48846 69251 52884 48296 10933 82120 06243 05801 $53/2^{129}3^{28}5^{16}7^{11}11^{6}13^{6}17^{4}19^{3}23^{2}29^{2}31^{1}37^{1}41^{1}43^{1}$ |
| 92 | $-$24377 87803 28410 80371 57474 73579 14456 07373 29084 37984 54874 43813 77457 15622 43100 41101 81332 44915 44336 29953 24025 24131 75533 92515 35046 67415 70664 04729 41310 84702 97991 05981 43660 11273 85734 72779 $7401/2^{133}3^{28}5^{14}7^{11}11^{6}13^{6}17^{4}19^{3}23^{3}29^{2}31^{1}37^{1}41^{1}43^{1}$ |
| 93 | 53940 24841 77265 50720 41626 10224 82552 47504 76997 31970 07846 37881 37188 58463 50049 96886 04665 78920 81048 55661 01903 83050 66380 46399 57243 45204 05754 63121 13904 69129 24823 07780 44548 91783 43610 89887 64527 $1/2^{133}3^{30}5^{16}7^{11}11^{5}13^{6}17^{4}19^{3}23^{3}29^{2}31^{2}41^{1}43^{1}$ |

**Table I.8**: Continuation of preceding table.

| $n$ | $E^{(n)}$ |
|---|---|
| 94 | $-19181$ 67832 03063 67326 62231 52140 41404 54677 01873 87110 61489 68481 91508 07129 51930 92961 70345 45965 81077 96219 15345 23598 89308 87995 62154 34098 81277 93800 80035 04649 44396 42283 58751 25090 60546 80466 69507 05035 $7/2^{133}3^{30}5^{16}7^{10}11^{6}13^{6}17^{4}19^{3}23^{3}29^{2}31^{2}37^{1}41^{1}43^{1}47^{1}$ |
| 95 | 11152 58818 02347 49375 91148 40546 21166 65836 57339 07129 29524 10831 74827 28144 23994 52083 84638 31756 43795 51076 78938 33669 47274 87951 70870 13767 65317 58664 65334 36749 97803 12740 80791 02701 61237 17171 54998 15719 $08223/2^{134}3^{29}5^{17}7^{11}11^{6}13^{6}17^{4}19^{4}23^{3}29^{2}31^{2}37^{1}41^{1}43^{1}47^{1}$ |
| 96 | $-33959$ 69224 34050 91169 87083 19971 36243 59210 65147 48316 28407 61126 13851 59537 11016 01422 71070 43870 11943 82002 89369 51720 02836 67462 32740 17477 18556 34504 20273 41037 43517 74869 08071 18498 15723 60217 80076 87520 72730 461 $/2^{141}3^{31}5^{16}7^{11}11^{6}13^{6}17^{4}19^{4}23^{3}29^{2}31^{2}37^{1}41^{1}43^{1}47^{1}$ |
| 97 | 56532 03284 96196 73431 78419 25549 59590 11433 87689 86811 65052 96007 98027 71352 03011 27963 39758 69769 64765 79311 61506 42718 75707 78723 92577 73444 28418 52694 87047 94334 98612 19090 25456 21476 06953 13864 12489 82963 12555 879 $/2^{138}3^{31}5^{16}7^{11}11^{6}13^{6}17^{4}19^{4}23^{3}29^{2}31^{2}37^{1}41^{1}43^{1}47^{1}$ |
| 98 | $-29736$ 21341 16465 22099 52877 45495 74981 77049 85037 42176 35832 88552 87603 35743 25645 80671 61707 72505 33930 35879 81112 83737 61461 87507 59881 08768 27293 38163 14296 89761 83647 59987 33262 47249 93379 08125 45598 34673 80089 07244 $901/2^{142}3^{31}5^{17}7^{13}11^{6}13^{6}17^{4}19^{4}23^{3}29^{2}31^{2}37^{1}41^{1}43^{1}47^{1}$ |
| 99 | 63174 08444 59326 55041 86762 79626 64069 68262 28137 22341 84447 21775 08709 04565 39318 51388 83800 97105 56018 29065 35758 65173 99192 81571 52668 51836 85350 43044 01322 06306 16898 60602 80458 13287 50960 39038 89045 48420 16776 55445 $8723/2^{142}3^{31}5^{17}7^{12}11^{7}13^{6}17^{4}19^{4}23^{3}29^{2}31^{2}37^{1}41^{1}43^{1}47^{1}$ |
| 100 | $-48184$ 72128 88432 24371 60025 52804 31301 30741 32116 72296 83998 96134 15127 81306 36539 56253 23247 89603 06144 88993 71544 68059 94335 44135 63583 62336 45042 45401 17970 17682 35184 61618 87006 70318 44844 12369 51603 93886 72130 54729 90868 $719/2^{146}3^{31}5^{18}7^{13}11^{7}13^{6}17^{4}19^{4}23^{3}29^{2}31^{2}37^{1}41^{1}43^{1}47^{1}$ |

**Table I.9**: The Maple internal floating point format for the energy coefficients for the ground state of Yukawa potential. $E^{(n)}$ is defined in terms of the integer $I_n$ and exponent $e_n$ by $E^{(n)} = I_n \times 10^{e_n}$.

| $n$ | $I_n$ | $e_n$ |
|---|---|---|
| 0 | -5000000000 0000000000 0000000000 0000000000 0000000000 0000000000 | -60 |
| 1 | 1 | 0 |
| 2 | -7500000000 0000000000 0000000000 0000000000 0000000000 0000000000 | -60 |
| 3 | 5000000000 0000000000 0000000000 0000000000 0000000000 0000000000 | -60 |
| 4 | -6875000000 0000000000 0000000000 0000000000 0000000000 0000000000 | -60 |
| 5 | 1312500000 0000000000 0000000000 0000000000 0000000000 0000000000 | -59 |
| 6 | -3020833333 3333333333 3333333333 3333333333 3333333333 3333333333 | -59 |
| 7 | 7885416666 6666666666 6666666666 6666666666 6666666666 6666666667 | -59 |
| 8 | -2260188802 0833333333 3333333333 3333333333 3333333333 3333333333 | -58 |
| 9 | 6975889756 9444444444 4444444444 4444444444 4444444444 4444444444 | -58 |
| 10 | -2289030273 4375000000 0000000000 0000000000 0000000000 0000000000 | -57 |
| 11 | 7915552408 8541666666 6666666666 6666666666 6666666666 6666666667 | -57 |
| 12 | -2866438809 9952980324 0740740740 7407407407 4074074074 0740740741 | -56 |
| 13 | 1081923003 8565176504 6296296296 2962962962 9629629629 6296296296 | -55 |
| 14 | -4241181438 6567615327 3809523809 5238095238 0952380952 3809523810 | -55 |
| 15 | 1721872252 9834350370 6459435626 1022927689 5943562610 2292768959 | -54 |
| 16 | -7223922257 6310262057 7265556106 7019400352 7336860670 1940035273 | -54 |
| 17 | 3126226151 0395523017 4661733906 5255731922 3985890652 5573192240 | -53 |
| 18 | -1393466634 5308060763 7946988329 4753086419 7530864197 5308641975 | -52 |
| 19 | 6389374738 8295126541 5817341476 5211640211 6402116402 1164021164 | -52 |
| 20 | -3010515737 3139414206 6465273525 7695381393 2980599647 2663139330 | -51 |
| 21 | 1456261287 0969658988 5858181112 2072526927 4376417233 5600907029 | -50 |
| 22 | -7225946310 9846876353 0906717161 8240120999 2975865991 7390076120 | -50 |
| 23 | 3675224678 1280740695 8527609035 7100600349 9183062675 1261671897 | -49 |
| 24 | -1914745824 9886869203 5052459967 8625042168 7191202940 2108767188 | -48 |
| 25 | 1021180326 8812565453 4860848700 0579611372 7503107256 7448228824 | -47 |

**Table I.10**: Continuation of preceding table.

| $n$ | $I_n$ | $e_n$ |
|---|---|---|
| 26 | -5571870767 5028140467 2495837489 0880739584 9313276820 3039367061 | -47 |
| 27 | 3108594515 9692778078 4040582603 7160060389 3695297089 0957736931 | -46 |
| 28 | -1772394041 7376650021 1835193941 4349681125 2495341603 3403981998 | -45 |
| 29 | 1032210361 4706309986 4326396972 1497587747 0290553783 4373825512 | -44 |
| 30 | -6137267834 3397979676 1233562845 9684983823 0467029658 4442091298 | -44 |
| 31 | 3723710074 9058589698 4207546159 4106757832 0618758386 0698716229 | -43 |
| 32 | -2304479005 2884781454 2352067178 2251137786 2611772213 5189654773 | -42 |
| 33 | 1454034609 8515224646 5806224571 6546766070 5158670455 1630274804 | -41 |
| 34 | -9349691740 4813316185 9806972108 6881554388 4816552550 3915478950 | -41 |
| 35 | 6124380027 6147131339 5224820049 7509124421 7879858772 5102924824 | -40 |
| 36 | -4085047089 4916419629 9972306116 6910238468 6322851731 8096051017 | -39 |
| 37 | 2773548808 9861079728 1336009406 0895488757 1513647023 6830072213 | -38 |
| 38 | -1916098496 0916910847 2033503873 2129127885 3369927481 4143405814 | -37 |
| 39 | 1346443592 4178555794 4666227437 9508672021 4382840960 6806984962 | -36 |
| 40 | -9620491903 0613519812 6503253941 5372298482 9987106943 5355093024 | -36 |
| 41 | 6987163121 5042759585 0877053405 8811613681 7031291981 9019935093 | -35 |
| 42 | -5156556152 2808541008 5550221668 7977903817 0617067583 6385973304 | -34 |
| 43 | 3865801457 3135232822 8662100588 1003102578 5962145745 6166667928 | -33 |
| 44 | -2943137988 9828276519 6334010239 4941260038 3653951840 0413562853 | -32 |
| 45 | 2274824321 6635866232 0578345463 2575879962 2698903073 8053274655 | -31 |
| 46 | -1784557172 8058689884 8272079439 4242188298 0064952494 4547206962 | -30 |
| 47 | 1420501803 8737486242 2442486291 2242479099 4681190247 6657684756 | -29 |
| 48 | -1147015167 7217242471 4794814165 6476377333 1318098402 9739400674 | -28 |
| 49 | 9392986734 0759103655 0363089054 0472562753 3654398239 9157236895 | -28 |
| 50 | -7799022925 7372222303 0908638097 8520412968 4641236887 5604282240 | -27 |

Appendix I. Tables of Screened Coulomb Energy Corrections

**Table I.11**: Continuation of preceding table.

| $n$ | $I_n$ | $e_n$ |
|---|---|---|
| 51 | 6564125298 5699542825 0603795541 5012840327 6000059358 9226840736 | -26 |
| 52 | -5599061833 1233900599 7063473914 8148179190 0676546785 5498568262 | -25 |
| 53 | 4839050258 4502625580 3885553565 6793685884 0820488447 4178464780 | -24 |
| 54 | -4236617301 4585613020 7346336206 4894816234 3036852482 5866195729 | -23 |
| 55 | 3756671652 0520242037 0410681029 3156798723 6983772669 1770929010 | -22 |
| 56 | -3373071741 0918361095 1464766947 2932177138 4212766001 6940340296 | -21 |
| 57 | 3066213193 8171300723 4692408052 0554384413 3188711937 1537696714 | -20 |
| 58 | -2821318884 4186777982 0289250556 4572475443 7412613452 6625551277 | -19 |
| 59 | 2627218046 3413035792 5652702962 0450511873 6707296271 1575010648 | -18 |
| 60 | -2475469381 4849654195 8930533709 3436891137 4046253615 5952512472 | -17 |
| 61 | 2359728875 2104765426 8897404743 6098817406 9015497435 5790659570 | -16 |
| 62 | -2275293921 0428896449 2880718714 3155724207 2250990528 8074718743 | -15 |
| 63 | 2218776453 5934823509 0966887074 0175122482 9045864521 6489339648 | -14 |
| 64 | -2187872378 4103640010 8431433713 9369801236 7530980726 2715638264 | -13 |
| 65 | 2181204845 3903021239 8087675326 1405082654 9208031311 7955187666 | -12 |
| 66 | -2198226288 3272065177 6511466306 1333368848 3915816784 7663675187 | -11 |
| 67 | 2239169633 1246518359 5841728180 9876761979 3918610115 2161359171 | -10 |
| 68 | -2305043349 5850634531 5926875012 0898000805 8373689783 9642834478 | -9 |
| 69 | 2397668587 5801039582 6879690229 3384735525 9610947343 6914253441 | -8 |
| 70 | -2519759886 8515199882 0537442553 9094737106 0008890651 2403521034 | -7 |
| 71 | 2675054212 1221277027 9934658411 1202463997 5975219086 7071406920 | -6 |
| 72 | -2868496658 3814451886 9919972421 8423526649 2965206512 4924051079 | -5 |
| 73 | 3106495435 6288299714 4416178738 3610315141 3475264348 2226876912 | -4 |
| 74 | -3397264081 6758019074 9451268946 7502658752 7130824216 8292485478 | -3 |
| 75 | 3751275778 7405616710 7781893495 8719169337 5971081426 5381965714 | -2 |

**Table I.12**: Continuation of preceding table.

| $n$ | $I_n$ | $e_n$ |
|---|---|---|
| 76 | -4181863845 4809952457 7368095793 4032538222 3945885412 6962249198 | -1 |
| 77 | 4706014869 2346222240 7518317438 1153981343 1827504203 2956172282 | 0 |
| 78 | -5345417820 1178153561 2556072392 5182482659 3554986664 7186570805 | 1 |
| 79 | 6127855651 0521666111 0549780360 2667067637 1327365520 6393675969 | 2 |
| 80 | -7089057878 2016598141 8816784929 1181617148 9291025849 8862498067 | 3 |
| 81 | 8275177063 9907935061 0138060477 1988598773 1914519542 0221872655 | 4 |
| 82 | -9746114140 2910042620 0634921232 5161865761 5443835015 4571316671 | 5 |
| 83 | 1158000449 7721434151 0640730662 1596494559 6551304934 5770419276 | 7 |
| 84 | -1387929936 6956095935 9355417593 5546016943 1681677776 7407001394 | 8 |
| 85 | 1677905062 4410226920 1769414858 9237251074 0803064072 8350386725 | 9 |
| 86 | -2045825414 5808957225 8144175759 3098422426 3430061813 8712306069 | 10 |
| 87 | 2515545988 2175696648 5894746367 6269423894 2687134853 8957510579 | 11 |
| 88 | -3119036386 9278459626 6314958626 4846789908 7460964003 8610943946 | 12 |
| 89 | 3899382901 4736862348 5581106053 7210173118 1382582809 3557469887 | 13 |
| 90 | -4914984224 3630285416 5446546205 6667419690 5630299330 0034555989 | 14 |
| 91 | 6245446080 0726969063 0354372098 6058747625 8495996429 2383764607 | 15 |
| 92 | -7999906226 9668703322 6211123464 8060245418 4666359520 0924884233 | 16 |
| 93 | 1032885393 6525895553 1472200045 3431822950 6088489656 0690520453 | 18 |
| 94 | -1344099966 0689421842 1010354428 1374765447 1434771776 8382569595 | 19 |
| 95 | 1762748062 9727605538 9560451982 4873164686 3872738597 6300735558 | 20 |
| 96 | -2329677747 5375611047 1598832621 0854458580 5058449982 7571777035 | 21 |
| 97 | 3102535040 8679876610 8190895730 7771331096 7954990403 0801516905 | 22 |
| 98 | -4163146468 0340393408 4795341042 2040796314 9105395821 2198248227 | 23 |
| 99 | 5628340160 1971048335 6288811460 8245564598 7891926675 8340492729 | 24 |
| 100 | -7665892145 3895500898 7291899588 2285274325 2675111975 8646396432 | 25 |

## I.2  Ground State Yukawa Energy for D Dimensions

Here we give the ground state energy corrections, $E^{(n)}$, to order 20 for the Yukawa potential as a function of the dimensionality, $D$, of space, where $q = (D - 1)/2$. The results were obtained using program *yukawa* given in Chapter 10. For $D = 3$, corresponding to $q = 1$, these results agree with those in Section I.1. These results can also be obtained as a special case of the general results obtained in Section 10.7 using the HVHF method.

$$E^{(1)} = 1,$$

$$E^{(2)} = -\frac{1}{4}(2q+1)q,$$

$$E^{(3)} = \frac{1}{12}(2q+1)(1+q)q^2,$$

$$E^{(4)} = -\frac{1}{96}(2q+1)(1+q)q^3(8q+3),$$

$$E^{(5)} = \frac{1}{160}(2q+1)(1+q)q^4(14q^2+19q+2),$$

$$E^{(6)} = -\frac{1}{2880}(2q+1)(1+q)q^5(352q^3+717q^2+366q+15),$$

$$E^{(7)} = \frac{1}{8064}(2q+1)(1+q)q^6(1412q^4+4266q^3+3927q^2 + 975q+18),$$

$$E^{(8)} = -\frac{1}{645120}(2q+1)(1+q)q^7(179712q^5+700380q^4 + 963200q^3+516270q^2+69963q+630),$$

$$E^{(9)} = \frac{1}{829440}(2q+1)(1+q)q^8(378488q^6+1861548q^5 + 3453110q^4+2887365q^3+984317q^2+78282q+360),$$

$$E^{(10)} = -\frac{1}{58060800}(2q+1)(1+q)q^9(45541376q^7+269644956q^6 + 636292328q^5+741330030q^4+421778021q^3 + 95748786q^2+4701978q+11340),$$

$$E^{(11)} = \frac{1}{141926400}(2q+1)(1+q)q^{10}(196064784q^8 + 1372223304q^7+3962644672q^6+5980091880q^5 + 4890737523q^4+2000150418q^3+312021459q^2 + 9817650q+12600),$$

$$E^{(12)} = -\frac{1}{6131220480}(2q+1)(1+q)q^{11}(15326904320q^9 + 124019751600q^8+426139362240q^7+797478701040q^6 + 863182841892q^5+525757069656q^4+159351215149q^3$$

$$+ 17509257756\, q^2 + 362702952\, q + 249480),$$

$$E^{(13)} = \frac{1}{36229939200}\, (2\, q + 1)\, (1 + q)\, q^{12} (166949954272\, q^{10}$$
$$+ 1541173562160\, q^9 + 6171449660400\, q^8$$
$$+ 13873578404040\, q^7 + 18879951230070\, q^6$$
$$+ 15570332081199\, q^5 + 7316565979727\, q^4$$
$$+ 1675816502205\, q^3 + 132348556512\, q^2$$
$$+ 1841137020\, q + 680400),$$

$$E^{(14)} = -\frac{1}{1859803545600}\, (2\, q + 1)\, (1 + q)\, q^{13} (16078947876864\, q^{11}$$
$$+ 167158696194864\, q^{10} + 767369086212960\, q^9$$
$$+ 2025166513843440\, q^8 + 3345250122027748\, q^7$$
$$+ 3521457388273440\, q^6 + 2295020331986063\, q^5$$
$$+ 850716320483700\, q^4 + 149354063253549\, q^3$$
$$+ 8620608484422\, q^2 + 81700391340\, q + 16216200),$$

$$E^{(15)} = \frac{1}{12875563008000}\, (2\, q + 1)\, (1 + q)\, q^{14} (211784322060352\, q^{12}$$
$$+ 2455815303942432\, q^{11} + 12756642751520976\, q^{10}$$
$$+ 38819998729394520\, q^9 + 75865193413291788\, q^8$$
$$+ 98006675279010498\, q^7 + 82867906409813129\, q^6$$
$$+ 43705981252775820\, q^5 + 12969971722783614\, q^4$$
$$+ 1765290925315929\, q^3 + 75492178431552\, q^2$$
$$+ 492409520820\, q + 52390800),$$

$$E^{(16)} = -\frac{1}{5356234211328000}\, (2\, q + 1)\, (1 + q)\, q^{15} ($$
$$169749761163788288\, q^{13} + 2176599231105743808\, q^{12}$$
$$+ 12655851427488482304\, q^{11} + 43784354951710332960\, q^{10}$$
$$+ 99277565322019100016\, q^9 + 152994975117927238296\, q^8$$
$$+ 160604807816113191388\, q^7 + 111801333041251309170\, q^6$$
$$+ 48496447489477110123\, q^5 + 11641854579250570818\, q^4$$
$$+ 1240206570337639554\, q^3 + 39733997269820490\, q^2$$
$$+ 179622954548100\, q + 10216206000),$$

$$E^{(17)} = \frac{1}{30351993864192000}\, (2\, q + 1)\, (1 + q)\, q^{16} ($$
$$18733216904661600000\, q^{14} + 263731616980740720000\, q^{13}$$
$$+ 1701133135527276360000\, q^{12} + 6613620934713827580000\, q^{11}$$
$$+ 17133071715978640422000\, q^{10} + 30833510319892599719400\, q^9$$
$$+ 38950302980842869630300\, q^8 + 34076588586528178655250\, q^7$$

$$+ 1986391646728795857551\, q^6 + 716721321129445657311\, q^5$$
$$+ 140271827719069738263\, q^4 + 11792303844034846860\, q^3$$
$$+ 2855766996518872940\, q^2 + 898853056693200\, q$$
$$+ 27243216000),$$

$$E^{(18)} = -\frac{1}{1639007668666368000}\,(2\,q+1)\,(1+q)\,q^{17}($$
$$198904642891512020992\, q^{15} + 30552121461215222191040\, q^{14}$$
$$+ 21694517471759885737600\, q^{13}$$
$$+ 9387431932595521577136 0\, q^{12}$$
$$+ 2744133662118114535665 76\, q^{11}$$
$$+ 567187444402158264316632\, q^{10}$$
$$+ 842422484344642603672556\, q^9$$
$$+ 895065769176713338505730\, q^8$$
$$+ 664168540663544166865043\, q^7$$
$$+ 328115053289611522858638\, q^6$$
$$+ 99282918398368947665774\, q^5$$
$$+ 15940912965847874801550\, q^4$$
$$+ 1065045416083112633004\, q^3 + 19635105716799489720\, q^2$$
$$+ 43149256270772400\, q + 694702008000),$$

$$E^{(19)} = \frac{1}{3663664200548352000}\,(2\,q+1)\,(1+q)\,q^{18}($$
$$881605439885227737344\, q^{16}$$
$$+ 146948036772640624016 64\, q^{15}$$
$$+ 114115414032009083181440\, q^{14}$$
$$+ 545133889614053232615360\, q^{13}$$
$$+ 1779757013450346294168896\, q^{12}$$
$$+ 4168839186435149280355392\, q^{11}$$
$$+ 7150315912864453182077224\, q^{10}$$
$$+ 8996034435941621650733580\, q^9$$
$$+ 8184565357878197887999543\, q^8$$
$$+ 5215973653026445067791890\, q^7$$
$$+ 2202972656085763695241576\, q^6$$
$$+ 562416901251700328587260\, q^5$$
$$+ 74470381307307898518357\, q^4$$
$$+ 3978377582931916886484\, q^3 + 56149366111315320060\, q^2$$
$$+ 862754606950068 00\, q + 735566832000),$$

$$E^{(20)} = -\frac{1}{276810184041431040000}(2\,q+1)(1+q)\,q^{19}($$

$$133081708841429781970944\,q^{17}$$
$$+\,23955750516813413592 06144\,q^{16}$$
$$+\,2022883329732243628545 3312\,q^{15}$$
$$+\,10594159867364238663803 1360\,q^{14}$$
$$+\,38297666745062123939125 6896\,q^{13}$$
$$+\,1005534024110422885923770016\,q^{12}$$
$$+\,1963311318369062370992653968\,q^{11}$$
$$+\,2868734324193790566252864840\,q^{10}$$
$$+\,3113455457733881465175745060\,q^{9}$$
$$+\,2456976315499961667233189940\,q^{8}$$
$$+\,1357163757766366551421919785\,q^{7}$$
$$+\,493215269077034604861187440\,q^{6}$$
$$+\,106740735683529448099891990\,q^{5}$$
$$+\,11708421745993697427421770\,q^{4}$$
$$+\,502746416539161286926960\,q^{3}$$
$$+\,5455782033056001995400\,q^{2}$$
$$+\,5864192432401665600\,q + 26398676304000),$$

## I.3 Energy Series for Screened Coulomb Potential

Here we give the energy corrections $E^{(n)}$ for a general state of the screened Coulomb potential to order 10 as functions of $q$ and $\tau$ (see (10.6) to (10.8)) and the coefficients $V_j$ defining the screened Coulomb potential (see (10.62) to (10.67)) These results were obtained using the HVHF method of Section 10.7 and procedure *hvhfscr*. They extend the results to order 5 reported in (10.137).

$$E^{(1)} = -V_1,$$

$$E^{(2)} = -\frac{1}{2}V_2(3\,q^2 - \tau),$$

$$E^{(3)} = -\frac{1}{2}V_3 q^2(5\,q^2 - 3\,\tau + 1),$$

$$E^{(4)} = \frac{1}{8}q^2(-7\,V_2^2 q^4 - 35\,V_4 q^4 + 30\,V_4 q^2\tau - 25\,V_4 q^2 - 5\,V_2^2 q^2$$
$$+\,3\,V_2^2\tau^2 - 3\,V_4\tau^2 + 6\,V_4\tau),$$

$$E^{(5)} = \frac{1}{8}q^4(-45\,V_2 V_3 q^4 - 63\,V_5 q^4 + 70\,V_5 q^2\tau + 14\,V_2 V_3 q^2\tau$$
$$-\,105\,V_5 q^2 - 63\,V_2 V_3 q^2 - 15\,V_5\tau^2 + 15\,V_2 V_3\tau^2$$

$$+ 10\,V_2 V_3 \tau + 50\,V_5 \tau - 12\,V_5),$$

$$
\begin{aligned}
E^{(6)} = \frac{1}{16} q^4 (&-143\,V_3^2 q^6 - 231\,V_6 q^6 - 33\,V_2^3 q^6 - 231\,V_2 V_4 q^6 \\
&+ 90\,V_3^2 q^4 \tau + 315\,V_6 q^4 \tau + 135\,V_2 V_4 q^4 \tau - 735\,V_6 q^4 \\
&- 585\,V_4 V_2 q^4 - 345\,V_3^2 q^4 - 75\,V_2^3 q^4 + 7\,V_2^3 q^2 \tau^2 \\
&+ 63\,V_2 V_4 q^2 \tau^2 + 21\,V_3^2 q^2 \tau^2 - 105\,V_6 q^2 \tau^2 + 189\,V_2 V_4 q^2 \tau \\
&+ 126\,V_3^2 q^2 \tau + 525\,V_6 q^2 \tau - 84\,V_2 V_4 q^2 - 28\,V_3^2 q^2 \\
&- 294\,V_6 q^2 - 15\,V_2 V_4 \tau^3 + 10\,V_2^3 \tau^3 + 5\,V_6 \tau^3 \\
&+ 30\,V_2 V_4 \tau^2 - 40\,V_6 \tau^2 + 60\,V_6 \tau),
\end{aligned}
$$

$$
\begin{aligned}
E^{(7)} = \frac{1}{16} q^6 (&-364\,V_2^2 V_3 q^6 - 546\,V_5 V_2 q^6 - 429\,V_7 q^6 - 728\,V_4 V_3 q^6 \\
&+ 99\,V_2^2 V_3 q^4 \tau + 462\,V_2 V_5 q^4 \tau + 660\,V_3 V_4 q^4 \tau + 693\,V_7 q^4 \tau \\
&- 2310\,V_2 V_5 q^4 - 2860\,V_3 V_4 q^4 - 2310\,V_7 q^4 - 1298\,V_2^2 V_3 q^4 \\
&+ 90\,V_2 V_5 q^2 \tau^2 + 90\,V_2^2 V_3 q^2 \tau^2 - 315\,V_7 q^2 \tau^2 + 2205\,V_7 q^2 \tau \\
&+ 1170\,V_2 V_5 q^2 \tau + 225\,V_2^2 V_3 q^2 \tau + 1620\,V_3 V_4 q^2 \tau - 972\,V_3 V_4 q^2 \\
&- 1104\,V_2 V_5 q^2 - 186\,V_2^2 V_3 q^2 - 2121\,V_7 q^2 + 35\,V_7 \tau^3 \\
&- 70\,V_2 V_5 \tau^3 - 28\,V_3 V_4 \tau^3 + 63\,V_2^2 V_3 \tau^3 + 140\,V_2 V_5 \tau^2 \\
&- 28\,V_3 V_4 \tau^2 + 84\,V_2^2 V_3 \tau^2 - 385\,V_7 \tau^2 + 168\,V_3 V_4 \tau \\
&+ 882\,V_7 \tau + 168\,V_2 V_5 \tau - 180\,V_7),
\end{aligned}
$$

$$
\begin{aligned}
E^{(8)} = \frac{1}{128} q^6 (&-7365\,V_4^2 q^8 - 13680\,V_3 V_5 q^8 - 9900\,V_2 V_6 q^8 \\
&- 10440\,V_2 V_3^2 q^8 - 930\,V_2^4 q^8 - 8400\,V_2^2 V_4 q^8 - 6435\,V_8 q^8 \\
&+ 16016\,V_3 V_5 q^6 \tau + 12012\,V_8 q^6 \tau + 5824\,V_2 V_3^2 q^6 \tau \\
&+ 8736\,V_4^2 q^6 \tau + 4368\,V_2^2 V_4 q^6 \tau + 10920\,V_2 V_6 q^6 \tau \\
&- 54054\,V_8 q^6 - 45500\,V_2^2 V_4 q^6 - 65520\,V_6 V_2 q^6 - 4550\,V_2^4 q^6 \\
&- 43316\,V_4^2 q^6 - 54600\,V_2 V_3^2 q^6 - 82992\,V_3 V_5 q^6 \\
&- 2640\,V_3 V_5 q^4 \tau^2 + 1848\,V_2 V_3^2 q^4 \tau^2 - 1650\,V_4^2 q^4 \tau^2 \\
&+ 2112\,V_2^2 V_4 q^4 \tau^2 + 198\,V_2^4 q^4 \tau^2 - 6930\,V_8 q^4 \tau^2 \\
&+ 34320\,V_4^2 q^4 \tau + 46200\,V_2 V_6 q^4 \tau + 20768\,V_2 V_3^2 q^4 \tau \\
&+ 64240\,V_3 V_5 q^4 \tau + 64680\,V_8 q^4 \tau + 15576\,V_2^2 V_4 q^4 \tau \\
&- 21648\,V_2 V_3^2 q^4 - 22660\,V_2^2 V_4 q^4 - 69300\,V_2 V_6 q^4 \\
&- 32395\,V_4^2 q^4 - 880\,V_2^4 q^4 - 93555\,V_8 q^4 - 68640\,V_3 V_5 q^4 \\
&+ 720\,V_2^2 V_4 q^2 \tau^3 - 720\,V_3 V_5 q^2 \tau^3 + 720\,V_2 V_3^2 q^2 \tau^3 \\
&+ 1260\,V_8 q^2 \tau^3 - 1800\,V_2 V_6 q^2 \tau^3 + 180\,V_2^4 q^2 \tau^3 \\
&- 360\,V_4^2 q^2 \tau^3 - 18270\,V_8 q^2 \tau^2 - 5760\,V_3 V_5 q^2 \tau^2 \\
&+ 4500\,V_2^2 V_4 q^2 \tau^2 + 360\,V_2^4 q^2 \tau^2 + 3600\,V_2 V_3^2 q^2 \tau^2 \\
&- 3600\,V_4^2 q^2 \tau^2 + 1800\,V_2 V_6 q^2 \tau^2 + 59388\,V_8 q^2 \tau
\end{aligned}
$$

$$+\ 22080\,V_2 V_6 q^2\tau + 2232\,V_2^2 V_4 q^2\tau + 24384\,V_3 V_5 q^2\tau$$
$$+\ 2976\,V_2 V_3^2 q^2\tau + 11664\,V_4^2 q^2\tau - 6480\,V_2 V_6 q^2 - 2484\,V_4^2 q^2$$
$$-\ 27396\,V_8 q^2 - 4608\,V_3 V_5 q^2 + 63\,V_4^2\tau^4 + 140\,V_2 V_6\tau^4$$
$$-\ 336\,V_2^2 V_4\tau^4 - 35\,V_8\tau^4 + 168\,V_2^4\tau^4 - 252\,V_4^2\tau^3$$
$$+\ 700\,V_8\tau^3 - 1120\,V_2 V_6\tau^3 + 672\,V_2^2 V_4\tau^3 - 3780\,V_8\tau^2$$
$$+\ 252\,V_4^2\tau^2 + 1680\,V_2 V_6\tau^2 + 5040\,V_8\tau),$$

$$E^{(9)} = \frac{1}{128}q^8\big(-12155\,V_9 q^8 - 21879\,V_7 V_2 q^8 - 12240\,V_3^3 q^8$$
$$-\ 14484\,V_2^3 V_3 q^8 - 59058\,V_2 V_3 V_4 q^8 - 30855\,V_6 V_3 q^8$$
$$-\ 34425\,V_4 V_5 q^8 - 22032\,V_2^2 V_5 q^8 + 49980\,V_4 V_5 q^6\tau$$
$$+\ 48120\,V_2 V_3 V_4 q^6\tau + 44100\,V_3 V_6 q^6\tau + 29700\,V_2 V_7 q^6\tau$$
$$+\ 25740\,V_9 q^6\tau + 3720\,V_2^3 V_3 q^6\tau + 10440\,V_3^3 q^6\tau$$
$$+\ 16800\,V_2^2 V_5 q^6\tau - 214830\,V_2 V_7 q^6 - 99420\,V_2^3 V_3 q^6$$
$$-\ 89160\,V_3^3 q^6 - 175140\,V_2^2 V_5 q^6 - 292950\,V_4 V_5 q^6$$
$$-\ 276150\,V_3 V_6 q^6 - 150150\,V_9 q^6 - 443160\,V_2 V_4 V_3 q^6$$
$$+\ 4368\,V_2^2 V_5 q^4\tau^2 + 3640\,V_2^3 V_3 q^4\tau^2 - 18018\,V_9 q^4\tau^2$$
$$+\ 5460\,V_2 V_3 V_4 q^4\tau^2 - 16926\,V_4 V_5 q^4\tau^2 - 13650\,V_3 V_6 q^4\tau^2$$
$$-\ 4914\,V_2 V_7 q^4\tau^2 + 216216\,V_9 q^4\tau + 273000\,V_3 V_6 q^4\tau$$
$$+\ 54600\,V_3^3 q^4\tau + 297752\,V_4 V_5 q^4\tau + 91000\,V_2^2 V_5 q^4\tau$$
$$+\ 196560\,V_2 V_7 q^4\tau + 254800\,V_2 V_3 V_4 q^4\tau + 18200\,V_2^3 V_3 q^4\tau$$
$$-\ 416325\,V_4 V_5 q^4 - 52416\,V_2^3 V_3 q^4 - 69160\,V_3^3 q^4$$
$$-\ 435435\,V_9 q^4 - 416871\,V_2 V_7 q^4 - 195468\,V_2^2 V_5 q^4$$
$$-\ 396942\,V_2 V_3 V_4 q^4 - 440895\,V_3 V_6 q^4 - 1100\,V_3 V_6 q^2\tau^3$$
$$+\ 4620\,V_9 q^2\tau^3 - 4620\,V_2 V_7 q^2\tau^3 - 660\,V_4 V_5 q^2\tau^3$$
$$+\ 2024\,V_2^3 V_3 q^2\tau^3 + 264\,V_3^3 q^2\tau^3 - 264\,V_2 V_3 V_4 q^2\tau^3$$
$$+\ 17820\,V_2^2 V_5 q^2\tau^2 + 11572\,V_2^3 V_3 q^2\tau^2 - 264\,V_3^3 q^2\tau^2$$
$$-\ 85470\,V_9 q^2\tau^2 - 48950\,V_3 V_6 q^2\tau^2 + 20856\,V_2 V_3 V_4 q^2\tau^2$$
$$-\ 60390\,V_4 V_5 q^2\tau^2 - 11550\,V_2 V_7 q^2\tau^2 + 21648\,V_3^3 q^2\tau$$
$$+\ 232540\,V_4 V_5 q^2\tau + 240900\,V_3 V_6 q^2\tau + 45320\,V_2^2 V_5 q^2\tau$$
$$+\ 3520\,V_2^3 V_3 q^2\tau + 374220\,V_9 q^2\tau + 207900\,V_2 V_7 q^2\tau$$
$$+\ 110264\,V_2 V_3 V_4 q^2\tau - 102300\,V_4 V_5 q^2 - 132660\,V_2 V_7 q^2$$
$$-\ 108900\,V_3 V_6 q^2 - 30360\,V_2 V_3 V_4 q^2 - 18480\,V_2^2 V_5 q^2$$
$$-\ 3872\,V_3^3 q^2 - 289300\,V_9 q^2 + 225\,V_3 V_6\tau^4$$
$$-\ 1170\,V_2 V_3 V_4\tau^4 - 1440\,V_2^2 V_5\tau^4 - 315\,V_9\tau^4$$
$$+\ 945\,V_2 V_7\tau^4 + 1260\,V_2^3 V_3\tau^4 + 495\,V_4 V_5\tau^4$$
$$-\ 1440\,V_2 V_3 V_4\tau^3 - 8820\,V_2 V_7\tau^3 + 2280\,V_2^3 V_3\tau^3$$
$$+\ 7980\,V_9\tau^3 + 1440\,V_2^2 V_5\tau^3 - 540\,V_4 V_5\tau^3$$

$$- 300\, V_3 V_6 \tau^3 + 7560\, V_2 V_3 V_4 \tau^2 + 9324\, V_2 V_7 \tau^2$$

$$+ 6912\, V_2^2 V_5 \tau^2 - 9324\, V_4 V_5 \tau^2 - 56532\, V_9 \tau^2$$

$$- 9300\, V_3 V_6 \tau^2 + 16848\, V_4 V_5 \tau + 18000\, V_3 V_6 \tau$$

$$+ 109584\, V_9 \tau + 19440\, V_2 V_7 \tau - 20160\, V_9),$$

$$E^{(10)} = \frac{1}{256} q^8 \left(-154660\, V_4 V_6 q^{10} - 95095\, V_2 V_8 q^{10} - 135850\, V_3 V_7 q^{10}\right.$$

$$- 109725\, V_2^2 V_6 q^{10} - 304950\, V_2 V_3 V_5 q^{10} - 46189\, V_{10} q^{10}$$

$$- 7980\, V_2^5 q^{10} - 204535\, V_3^2 V_4 q^{10} - 80123\, V_5^2 q^{10}$$

$$- 164445\, V_2 V_4^2 q^{10} - 165110\, V_2^2 V_3^2 q^{10} - 88350\, V_2^3 V_4 q^{10}$$

$$+ 109395\, V_{10} q^8 \tau + 324360\, V_2 V_3 V_5 q^8 \tau + 86904\, V_2^2 V_3^2 q^8 \tau$$

$$+ 110160\, V_2^2 V_6 q^8 \tau + 228888\, V_3 V_7 q^8 \tau + 264690\, V_4 V_6 q^8 \tau$$

$$+ 137700\, V_5^2 q^8 \tau + 177174\, V_2 V_4^2 q^8 \tau + 153153\, V_2 V_8 q^8 \tau$$

$$+ 228276\, V_3^2 V_4 q^8 \tau + 43452\, V_2^3 V_4 q^8 \tau - 3205860\, V_2 V_3 V_5 q^8$$

$$- 1238790\, V_2^2 V_6 q^8 - 1725636\, V_3 V_7 q^8 - 1327326\, V_2 V_8 q^8$$

$$- 1525716\, V_2^2 V_3^2 q^8 - 1685754\, V_2 V_4^2 q^8 - 802230\, V_{10} q^8$$

$$- 1846200\, V_4 V_6 q^8 - 839460\, V_2^3 V_4 q^8 - 2045202\, V_3^2 V_4 q^8$$

$$- 69360\, V_2^5 q^8 - 935850\, V_5^2 q^8 - 14100\, V_2 V_3 V_5 q^6 \tau^2$$

$$- 66570\, V_5^2 q^6 \tau^2 - 35730\, V_3^2 V_4 q^6 \tau^2 + 23340\, V_2^3 V_4 q^6 \tau^2$$

$$+ 1860\, V_2^5 q^6 \tau^2 - 126000\, V_4 V_6 q^6 \tau^2 - 90090\, V_{10} q^6 \tau^2$$

$$- 11070\, V_2 V_4^2 q^6 \tau^2 - 48510\, V_2 V_8 q^6 \tau^2 + 9450\, V_2^2 V_6 q^6 \tau^2$$

$$+ 33780\, V_2^2 V_3^2 q^6 \tau^2 - 102060\, V_3 V_7 q^6 \tau^2 + 875700\, V_2^2 V_6 q^6 \tau$$

$$+ 1503810\, V_2 V_8 q^6 \tau + 1329480\, V_2 V_4^2 q^6 \tau + 596520\, V_2^2 V_3^2 q^6 \tau$$

$$+ 2086560\, V_3 V_7 q^6 \tau + 1688760\, V_3^2 V_4 q^6 \tau + 1171800\, V_5^2 q^6 \tau$$

$$+ 298260\, V_2^3 V_4 q^6 \tau + 2293200\, V_4 V_6 q^6 \tau + 2473200\, V_2 V_3 V_5 q^6 \tau$$

$$+ 1351350\, V_{10} q^6 \tau - 3600597\, V_{10} q^6 - 960150\, V_2^3 V_4 q^6$$

$$- 4666410\, V_3 V_7 q^6 - 4455360\, V_4 V_6 q^6 - 2932515\, V_3^2 V_4 q^6$$

$$- 1520430\, V_2^2 V_3^2 q^6 - 2539425\, V_2^2 V_6 q^6 - 45636\, V_2^5 q^6$$

$$- 4228455\, V_2 V_8 q^6 - 2649645\, V_2 V_4^2 q^6 - 5430750\, V_2 V_3 V_5 q^6$$

$$- 2167431\, V_5^2 q^6 - 9100\, V_2^2 V_6 q^4 \tau^3 + 10920\, V_2^2 V_3^2 q^4 \tau^3$$

$$- 19110\, V_2 V_8 q^4 \tau^3 - 1820\, V_3^2 V_4 q^4 \tau^3 - 10920\, V_2 V_4^2 q^4 \tau^3$$

$$+ 5460\, V_5^2 q^4 \tau^3 + 3640\, V_3 V_7 q^4 \tau^3 + 1456\, V_2^5 q^4 \tau^3$$

$$+ 7280\, V_2^3 V_4 q^4 \tau^3 + 30030\, V_{10} q^4 \tau^3 - 18200\, V_2 V_3 V_5 q^4 \tau^3$$

$$+ 9100\, V_4 V_6 q^4 \tau^3 - 173810\, V_3^2 V_4 q^4 \tau^2 + 86450\, V_2^2 V_6 q^4 \tau^2$$

$$- 580580\, V_3 V_7 q^4 \tau^2 - 35490\, V_2 V_4^2 q^4 \tau^2 + 165620\, V_2^2 V_3^2 q^4 \tau^2$$

$$- 248430\, V_2 V_8 q^4 \tau^2 - 690690\, V_{10} q^4 \tau^2 - 700700\, V_4 V_6 q^4 \tau^2$$

$$- 27300\, V_2 V_3 V_5 q^4 \tau^2 + 8372\, V_2^5 q^4 \tau^2 - 366730\, V_5^2 q^4 \tau^2$$

$$+ 118300\, V_2^3 V_4 q^4 \tau^2 + 1416324\, V_3^2 V_4 q^4 \tau + 2369640\, V_2 V_3 V_5 q^4 \tau$$

$$+ 3918915\, V_{10}q^4\tau + 1665300\, V_5^2 q^4\tau + 2918097\, V_2 V_8 q^4\tau$$
$$+ 3404310\, V_4 V_6 q^4\tau + 977340\, V_2^2 V_6 q^4\tau + 3479112\, V_3 V_7 q^4\tau$$
$$+ 314496\, V_2^2 V_3^2 q^4\tau + 1190826\, V_2 V_4^2 q^4\tau + 157248\, V_2^3 V_4 q^4\tau$$
$$- 116584\, V_2^2 V_3^2 q^4 - 2708264\, V_3 V_7 q^4 - 4467320\, V_{10} q^4$$
$$- 692796\, V_2 V_4^2 q^4 - 1190020\, V_5^2 q^4 - 1433640\, V_2 V_3 V_5 q^4$$
$$- 631748\, V_3^2 V_4 q^4 - 3017924\, V_2 V_8 q^4 - 92040\, V_2^3 V_4 q^4$$
$$- 2440100\, V_4 V_6 q^4 - 824460\, V_2^2 V_6 q^4 - 1551\, V_3^2 V_4 q^2\tau^4$$
$$- 7590\, V_2 V_3 V_5 q^2\tau^4 - 8085\, V_2^2 V_6 q^2\tau^4 + 2310\, V_3 V_7 q^2\tau^4$$
$$+ 8085\, V_2 V_8 q^2\tau^4 + 1485\, V_5^2 q^2\tau^4 + 1056\, V_2^5 q^2\tau^4$$
$$+ 5346\, V_2^2 V_3^2 q^2\tau^4 - 3465\, V_{10} q^2\tau^4 + 2178\, V_2^3 V_4 q^2\tau^4$$
$$+ 3300\, V_4 V_6 q^2\tau^4 - 3069\, V_2 V_4^2 q^2\tau^4 + 108570\, V_{10} q^2\tau^3$$
$$- 6644\, V_3^2 V_4 q^2\tau^3 - 80850\, V_2 V_8 q^2\tau^3 - 16720\, V_2^2 V_6 q^2\tau^3$$
$$- 50600\, V_2 V_3 V_5 q^2\tau^3 + 3520\, V_2^5 q^2\tau^3 + 17820\, V_5^2 q^2\tau^3$$
$$+ 15400\, V_3 V_7 q^2\tau^3 + 24816\, V_2^2 V_3^2 q^2\tau^3 + 28600\, V_4 V_6 q^2\tau^3$$
$$- 33924\, V_2 V_4^2 q^2\tau^3 + 21164\, V_2^3 V_4 q^2\tau^3 - 64680\, V_2 V_8 q^2\tau^2$$
$$- 372680\, V_3 V_7 q^2\tau^2 - 28292\, V_3^2 V_4 q^2\tau^2 + 52404\, V_2 V_4^2 q^2\tau^2$$
$$- 400400\, V_4 V_6 q^2\tau^2 + 121880\, V_2 V_3 V_5 q^2\tau^2 - 200420\, V_5^2 q^2\tau^2$$
$$- 979440\, V_{10} q^2\tau^2 + 138380\, V_2^2 V_6 q^2\tau^2 + 38280\, V_2^3 V_4 q^2\tau^2$$
$$+ 40744\, V_2^2 V_3^2 q^2\tau^2 + 92400\, V_2^2 V_6 q^2\tau + 928620\, V_2 V_8 q^2\tau$$
$$+ 409200\, V_5^2 q^2\tau + 2603700\, V_{10} q^2\tau + 91080\, V_2 V_4^2 q^2\tau$$
$$+ 195360\, V_2 V_3 V_5 q^2\tau + 95568\, V_3^2 V_4 q^2\tau + 838200\, V_4 V_6 q^2\tau$$
$$+ 918720\, V_3 V_7 q^2\tau - 221760\, V_2 V_8 q^2 - 1062864\, V_{10} q^2$$
$$- 69696\, V_5^2 q^2 - 158400\, V_3 V_7 q^2 - 150480\, V_4 V_6 q^2$$
$$- 315\, V_2 V_8 \tau^5 - 270\, V_4 V_6 \tau^5 + 810\, V_2 V_4^2 \tau^5$$
$$+ 900\, V_2^2 V_6 \tau^5 + 792\, V_2^5 \tau^5 - 1980\, V_2^3 V_4 \tau^5$$
$$+ 63\, V_{10} \tau^5 + 3960\, V_2^3 V_4 \tau^4 - 7200\, V_2^2 V_6 \tau^4$$
$$+ 6300\, V_2 V_8 \tau^4 + 2700\, V_4 V_6 \tau^4 - 3240\, V_2 V_4^2 \tau^4$$
$$- 2520\, V_{10} \tau^4 - 34020\, V_2 V_8 \tau^3 - 7560\, V_4 V_6 \tau^3$$
$$+ 3240\, V_2 V_4^2 \tau^3 + 10800\, V_2^2 V_6 \tau^3 + 32004\, V_{10} \tau^3$$
$$+ 6480\, V_4 V_6 \tau^2 + 45360\, V_2 V_8 \tau^2 - 145152\, V_{10} \tau^2 + 181440\, V_{10}\tau).$$

## I.4   Energy Series for Yukawa Potential

Here we give the energy corrections $E^{(n)}$ for a general state of the Yukawa potential to order 14 as functions of $q$ and $\tau$ (see (10.6) to (10.8)). These results are special cases of the screened Coulomb results of the preceding section with $V_j = (-1)^j / j!$ and were obtained using the HVHF method of Section 10.7 and procedure *hvhfscr* . They extend the results to order 6 reported in (10.138).

$$E^{(1)} = 1,$$

$$E^{(2)} = -\frac{1}{4}(3\,q^2 - \tau),$$

$$E^{(3)} = \frac{1}{12}q^2(5\,q^2 - 3\,\tau + 1),$$

$$E^{(4)} = -\frac{1}{192}q^2(77\,q^4 - 30\,q^2\tau + 55\,q^2 - 15\,\tau^2 - 6\,\tau),$$

$$E^{(5)} = \frac{1}{320}q^4(171\,q^4 - 70\,q^2\tau + 245\,q^2 - 45\,\tau^2 - 50\,\tau + 4),$$

$$
\begin{aligned}
E^{(6)} = -\frac{1}{5760}q^4(&4763\,q^6 - 2070\,q^4\tau + 11580\,q^4 - 945\,q^2\tau^2 \\
&- 2940\,q^2\tau + 1057\,q^2 - 340\,\tau^3 - 205\,\tau^2 - 30\,\tau),
\end{aligned}
$$

$$
\begin{aligned}
E^{(7)} = \frac{1}{16128}q^6(&22763\,q^6 - 10857\,q^4\tau + 84700\,q^4 - 4095\,q^2\tau^2 \\
&- 26145\,q^2\tau + 19677\,q^2 - 2163\,\tau^3 - 3843\,\tau^2 \\
&- 2058\,\tau + 36),
\end{aligned}
$$

$$
\begin{aligned}
E^{(8)} = -\frac{1}{5160960}q^6(&13283265\,q^8 - 6922188\,q^6\tau + 70720286\,q^6 \\
&- 2150610\,q^4\tau^2 - 25530120\,q^4\tau + 32003125\,q^4 \\
&- 1044540\,q^2\tau^3 - 4270770\,q^2\tau^2 - 5463612\,q^2\tau \\
&+ 640764\,q^2 - 290535\,\tau^4 - 233940\,\tau^3 - 60900\,\tau^2 - 5040\,\tau),
\end{aligned}
$$

$$
\begin{aligned}
E^{(9)} = \frac{1}{6635520}q^8(&32694383\,q^8 - 18584460\,q^6\tau + 237169110\,q^6 \\
&- 4889430\,q^4\tau^2 - 98129304\,q^4\tau + 180758487\,q^4 \\
&- 2153580\,q^2\tau^3 - 16367010\,q^2\tau^2 - 41938380\,q^2\tau \\
&+ 12261700\,q^2 - 1011105\,\tau^4 - 2301180\,\tau^3 \\
&- 1867740\,\tau^2 - 634896\,\tau + 2880),
\end{aligned}
$$

$$
\begin{aligned}
E^{(10)} = -\frac{1}{464486400}q^8(&4546296155\,q^{10} - 2804511420\,q^8\tau \\
&+ 43305010860\,q^8 - 618494310\,q^6\tau^2 - 20191096800\,q^6\tau \\
&+ 50710286067\,q^6 - 240704100\,q^4\tau^3 - 3118055850\,q^4\tau^2 \\
&- 14743588860\,q^4\tau + 7721142910\,q^4 - 125921565\,q^2\tau^4 \\
&- 575000580\,q^2\tau^3 - 1047390960\,q^2\tau^2 - 865398600\,q^2\tau \\
&+ 39607128\,q^2 - 28002744\,\tau^5 - 28206990\,\tau^4 \\
&- 10363752\,\tau^3 - 1628424\,\tau^2 - 90720\,\tau),
\end{aligned}
$$

$$E^{(11)} = \frac{1}{1135411200}q^{10}(22729725165\,q^{10} - 15145344885\,q^8\tau$$

$$+\, 276158552065\, q^8 - 2770818270\, q^6\tau^2 - 143264808390\, q^6\tau$$
$$+\, 466085135747\, q^6 - 946155210\, q^4\tau^3 - 19655194020\, q^4\tau^2$$
$$-\, 162150984765\, q^4\tau + 130650413355\, q^4 - 516170655\, q^2\tau^4$$
$$-\, 3909515610\, q^2\tau^3 - 13446563130\, q^2\tau^2 - 22452675960\, q^2\tau$$
$$+\, 3116758788\, q^2 - 214199601\, \tau^5 - 576537885\, \tau^4$$
$$-\, 582288168\, \tau^3 - 305918316\, \tau^2 - 78843600\, \tau + 100800),$$

$$E^{(12)} = -\frac{1}{49049763840}\, q^{10}(2056261008581\, q^{12} - 1473038695128\, q^{10}\tau$$
$$+\, 31128322638028\, q^{10} - 217800867735\, q^8\tau^2$$
$$-\, 17782661848050\, q^8\tau + 72227185413009\, q^8$$
$$-\, 62705236440\, q^6\tau^3 - 2079419934900\, q^6\tau^2$$
$$-\, 29184556564548\, q^6\tau + 33095679279058\, q^6$$
$$-\, 35419732425\, q^4\tau^4 - 396832359000\, q^4\tau^3$$
$$-\, 2445642316425\, q^4\tau^2 - 7633362971850\, q^4\tau$$
$$+\, 2087740280780\, q^4 - 18669386736\, q^2\tau^5$$
$$-\, 91903691880\, q^2\tau^4 - 193387842648\, q^2\tau^3$$
$$-\, 235346209956\, q^2\tau^2 - 148243033224\, q^2\tau$$
$$+\, 2958072624\, q^2 - 3396620997\, \tau^6 - 4107581478\, \tau^5$$
$$-\, 1938537216\, \tau^4 - 442960056\, \tau^3 - 48465648\, \tau^2$$
$$-\, 1995840\, \tau),$$

$$E^{(13)} = \frac{1}{5796790027200}\, q^{12}(51890307003725\, q^{12} - 39793115247834\, q^{10}\tau$$
$$+\, 960050625454063\, q^{10} - 4565198302665\, q^8\tau^2$$
$$-\, 598840800828270\, q^8\tau + 2953046826532803\, q^8$$
$$-\, 1025996884380\, q^6\tau^3 - 56887506883350\, q^6\tau^2$$
$$-\, 1357045598966382\, q^6\tau + 2042956234227649\, q^6$$
$$-\, 630689275125\, q^4\tau^4 - 9390288638640\, q^4\tau^3$$
$$-\, 104670051798585\, q^4\tau^2 - 589260132419010\, q^4\tau$$
$$+\, 261364587881072\, q^4 - 390560354730\, q^2\tau^5$$
$$-\, 2894914119825\, q^2\tau^4 - 10488331389900\, q^2\tau^3$$
$$-\, 25352426004360\, q^2\tau^2 - 32615196434184\, q^2\tau$$
$$+\, 2372151822768\, q^2 - 142348402287\, \tau^6$$
$$-\, 435041410986\, \tau^5 - 512458344216\, \tau^4$$
$$-\, 336139263720\, \tau^3 - 136687717296\, \tau^2$$
$$-\, 29490851520\, \tau + 10886400),$$

$$E^{(14)} = -\frac{1}{14878428364800}\, q^{12}(2890766140116003\, q^{14}$$

$$- 2363788786249800\, q^{12}\tau + 64316389459335650\, q^{12}$$
$$- 195943893650505\, q^{10}\tau^2 - 43493627918184444\, q^{10}\tau$$
$$+ 254777440062167022\, q^{10} - 26618046054600\, q^{8}\tau^3$$
$$- 3129130174170000\, q^{8}\tau^2 - 131087803469434812\, q^{8}\tau$$
$$+ 251344167374562834\, q^{8} - 21524804316015\, q^{6}\tau^4$$
$$- 361309864375460\, q^{6}\tau^3 - 85819908550225585\, q^{6}\tau^2$$
$$- 86825991925620948\, q^{6}\tau + 56082658483686835\, q^{6}$$
$$- 16201168807824\, q^{4}\tau^5 - 153539945649090\, q^{4}\tau^4$$
$$- 882899979393432\, q^{4}\tau^3 - 4136914428600954\, q^{4}\tau^2$$
$$- 10116187814480748\, q^{4}\tau + 1609062509131896\, q^{4}$$
$$- 8185809907275\, q^{2}\tau^6 - 42889756269360\, q^{2}\tau^5$$
$$- 98596648955805\, q^{2}\tau^4 - 140198008810860\, q^{2}\tau^3$$
$$- 134708283489780\, q^{2}\tau^2 - 70869179017728\, q^{2}\tau$$
$$+ 658142740080\, q^{2} - 1240465100160\, \tau^7$$
$$- 1750668087168\, \tau^6 - 1007034308280\, \tau^5$$
$$- 300561036204\, \tau^4 - 48716327088\, \tau^3$$
$$- 4020690960\, \tau^2 - 129729600\, \tau).$$

# Appendix J

# Solutions to Exercises

## J.1   Solutions to Chapter 1 Exercises

■ **Solution 1.1**

Use bilinearity to expand the commutator

$$
\begin{aligned}
[A+B, A+B] &= [A+B, A] + [A+B, B] \\
&= [A, A] + [B, A] + [A, B] + [B, B] \\
&= [B, A] + [A, B] = 0.
\end{aligned}
$$

Therefore $[A, B] = -[B, A]$.

■ **Solution 1.2**

Expand each of the three terms in rule (5) to obtain

$$
\begin{aligned}
[A, [B, C]] &= ABC - BCA - ACB + CBA, \\
[B, [C, A]] &= BCA - CAB - BAC + ACB, \\
[C, [A, B]] &= CAB - ABC - CBA + BAC.
\end{aligned}
$$

Add these results to obtain the Jacobi identity.

■ **Solution 1.3**

$$
\begin{aligned}
[AB, C] &= ABC - CAB = ABC - ACB + ACB - CAB \\
&= A[B, C] + [A, C]B.
\end{aligned}
$$

The second identity can be obtained in a similar manner. To prove the third
identity write

$$
\begin{aligned}
[AB, C] &= ABC - CAB = ABC + ACB - ACB - CAB \\
&= A\{B, C\} - \{A, C\}B.
\end{aligned}
$$

### ■ Solution 1.4

Use the definition

$$e_j \times e_k = \sum_\ell \epsilon_{jk\ell} e_\ell$$

of the vector cross product in terms of the Levi-Civita symbol $\epsilon_{jk\ell}$ (see Appendix A). The defining commutation relations follow from this definition:

$$[e_j, e_k] = \sum_\ell \epsilon_{jk\ell} e_\ell.$$

The Jacobi identity is a consequence of the vector identity (A.6).

### ■ Solution 1.5

From (1.4)

$$\begin{aligned}
(E_{ij} E_{k\ell})_{rs} &= \sum_t (E_{ij})_{rt}(E_{k\ell})_{ts} = \delta_{ir}\left(\sum_t \delta_{jt}\delta_{kt}\right)\delta_{\ell s} \\
&= \delta_{ir}\delta_{jk}\delta_{\ell s} = \delta_{jk}(E_{i\ell})_{rs}.
\end{aligned}$$

Similarly $(E_{k\ell} E_{ij})_{rs} = \delta_{\ell i}(E_{kj})_{rs}$. Therefore

$$[E_{ij}, E_{k\ell}]_{rs} = \delta_{jk}(E_{i\ell})_{rs} - \delta_{\ell i}(E_{kj})_{rs} = (\delta_{jk}E_{i\ell} - \delta_{\ell i}E_{kj})_{rs}.$$

### ■ Solution 1.6

Identities (a) and (b) can be obtained by direct calculation of the matrix products. Thus, $\sigma_1^2 = \sigma_2^2 = \sigma_3^2 = I_2$, the $2 \times 2$ identity matrix. These results are special cases of the first identity for $k = j$ ($\epsilon_{jj\ell} = 0$). Similarly, $\sigma_1\sigma_2 = i\sigma_3$, $\sigma_2\sigma_1 = -i\sigma_3$, $\sigma_2\sigma_3 = i\sigma_1$, $\sigma_3\sigma_2 = -i\sigma_1$, $\sigma_3\sigma_1 = i\sigma_2$, $\sigma_1\sigma_3 = -i\sigma_2$, which give the remaining cases of identity (a). Subtracting we obtain the commutators $[\sigma_1, \sigma_2] = 2i\sigma_3$, $[\sigma_2, \sigma_3] = 2i\sigma_1$ and $[\sigma_3, \sigma_1] = 2i\sigma_2$, which gives identity (b). Identity (c) follows in a similar manner since

$$\{\sigma_1, \sigma_2\} = \{\sigma_2, \sigma_3\} = \{\sigma_3, \sigma_1\} = 0,$$
$$\{\sigma_1, \sigma_1\} = \{\sigma_2, \sigma_2\} = \{\sigma_3, \sigma_3\} = 2.$$

Identity (d) follows from (a) and the formula for the components of the vector cross product $(u \times v)_\ell = \sum_{jk} \epsilon_{jk\ell} u_j v_k$ in terms of the Levi-Civita symbol (Appendix A):

$$\begin{aligned}
(u \cdot \sigma)(v \cdot \sigma) &= \sum_{jk} u_j \sigma_j v_k \sigma_k \\
&= \sum_{jk} u_j v_k \left(i\sum_\ell \epsilon_{jk\ell}\sigma_\ell + \delta_{jk}\right)
\end{aligned}$$

$$
\begin{aligned}
&= i \sum_\ell \left( \sum_{jk} \epsilon_{jk\ell} u_j v_k \right) \sigma_\ell + \sum_j u_j v_j \\
&= i \sum_\ell (u \times v)_\ell \sigma_\ell + u \cdot v \\
&= i(u \times v) \cdot \sigma + u \cdot v.
\end{aligned}
$$

Identities (e) and (f) now follow from (d):

$$
\begin{aligned}
[u \cdot \sigma, v \cdot \sigma] &= u \cdot v + i(u \times v) \cdot \sigma - v \cdot u - i(v \times u) \cdot \sigma \\
&= 2i(u \times v) \cdot \sigma, \\
\{u \cdot \sigma, v \cdot \sigma\} &= u \cdot v + i(u \times v) \cdot \sigma + v \cdot u + i(v \times u) \cdot \sigma \\
&= 2u \cdot v,
\end{aligned}
$$

since $u \cdot v = v \cdot u$ and $v \times u = -u \times v$ if the components of $u$ and $v$ commute.

## ■ Solution 1.7

Here we use the assumption that the components of the orbital angular momentum $L$ commute with the components of the spin $S$. Therefore

$$
\begin{aligned}
[J_j, J_k] &= [L_j + S_j, L_k + S_k] = [L_j, L_k] + [S_j, S_k] \\
&= i \sum_\ell \epsilon_{jk\ell} L_\ell + i \sum_\ell \epsilon_{jk\ell} S_\ell \\
&= i \sum_\ell \epsilon_{jk\ell} (L_\ell + S_\ell) = i \sum_\ell \epsilon_{jk\ell} J_\ell.
\end{aligned}
$$

## ■ Solution 1.8

$$
\begin{aligned}
[L^2, L_k] &= \sum_j [L_j L_j, L_k] \\
&= \sum_j L_j [L_j, L_k] + \sum_j [L_j, L_k] L_j \\
&= i \sum_{j\ell} \epsilon_{jk\ell} L_j L_\ell + i \sum_{j\ell} \epsilon_{jk\ell} L_\ell L_j \\
&= i \sum_{j\ell} \epsilon_{jk\ell} L_j L_\ell + i \sum_{j\ell} \epsilon_{\ell k j} L_j L_\ell \\
&= 0 \text{ since } \epsilon_{\ell k j} = -\epsilon_{jk\ell}.
\end{aligned}
$$

## ■ Solution 1.9

$[E_{12}, E_{21}] = E_{11} - E_{22}$ is derived in the text. The others can be obtained as follows:

$$
[E_{12}, E_{11}]
$$

$$= [a_p^+ a_n, a_p^+ a_p] = a_p^+ [a_n, a_p^+ a_p] + [a_p^+, a_p^+ a_p] a_n$$
$$= a_p^+(\{a_n, a_p^+\}a_p - a_p^+\{a_n, a_p\}) + (\{a_p^+, a_p^+\}a_p - a_p^+\{a_p^+, a_p\})a_n$$
$$= -a_p^+ a_n = -E_{12},$$

$[E_{12}, E_{22}]$

$$= [a_p^+ a_n, a_n^+ a_n] = a_p^+ [a_n, a_n^+ a_n] + [a_p^+, a_n^+ a_n] a_n$$
$$= a_p^+(\{a_n, a_n^+\}a_n - a_n^+\{a_n, a_n\}) + (\{a_p^+, a_n^+\}a_n - a_n^+\{a_p^+, a_n\})a_n$$
$$= a_p^+ a_n = E_{12},$$

$[E_{21}, E_{11}]$

$$= [a_n^+ a_p, a_p^+ a_p] = a_n^+ [a_p, a_p^+ a_p] + [a_n^+, a_p^+ a_p] a_p$$
$$= a_n^+(\{a_p, a_p^+\}a_p - a_p^+\{a_p, a_p\}) + (\{a_n^+, a_p^+\}a_p - a_p^+\{a_n^+, a_p\})a_p$$
$$= a_n^+ a_p = E_{21},$$

$[E_{21}, E_{22}]$

$$= [a_n^+ a_p, a_n^+ a_n] = a_n^+ [a_p, a_n^+ a_n] + [a_n^+, a_n^+ a_n] a_p$$
$$= a_n^+(\{a_p, a_n^+\}a_n - a_n^+\{a_p, a_n\}) + (\{a_n^+, a_n^+\}a_n - a_n^+\{a_n^+, a_n\})a_p$$
$$= -a_n^+ a_p = -E_{21},$$

$[E_{11}, E_{22}]$

$$= [a_p^+ a_p, a_n^+ a_n] = a_p^+ [a_p, a_n^+ a_n] + [a_p^+, a_n^+ a_n] a_p$$
$$= a_p^+(\{a_p, a_n^+\}a_n - a_n^+\{a_p, a_n\}) + (\{a_p^+, a_n^+\}a_n - a_n^+\{a_p^+, a_n\})a_p$$
$$= 0.$$

## ■ Solution 1.10

The set of all $n \times n$ skew hermitian matrices is a real Lie algebra if it is closed under commutation and if any real linear combination of hermitian matrices is also hermitian. Thus, if $A$ and $B$ are skew hermitian then

$$[A, B]^\dagger = (AB - BA)^\dagger = (AB)^\dagger - (BA)^\dagger = B^\dagger A^\dagger - A^\dagger B^\dagger$$
$$= BA - AB = -[A, B],$$

and the commutator is also skew hermitian. If $A = \sum_j \alpha_j A_j$ is a real linear combination of skew hermitian matrices then A is also skew hermitian since $\alpha_j^* = \alpha_j$. This is not true for complex linear combinations so the set of $n \times n$ skew hermitian matrices do not form a complex Lie algebra.

## ■ Solution 1.11

An $n \times n$ complex matrix $A$ depends on $2n^2$ real parameters. The skew hermitian condition $A^\dagger = -A$ gives $n^2$ conditions, one for each matrix element, and the trace condition, $\mathrm{Tr}(A) = \sum_j (A)_{jj} = 0$, gives another condition so there are $2n^2 - n^2 - 1 = n^2 - 1$ independent real parameters.

The matrices $G_{jk}$, $H_{jk}$ and $D_j$ are clearly traceless and they are skew hermitian since

$$
\begin{aligned}
G_{jk}^\dagger &= -i(E_{jk}^\dagger + E_{kj}^\dagger) = -i(E_{kj} + E_{jk}) = -G_{jk}, \\
H_{jk}^\dagger &= E_{jk}^\dagger - E_{kj}^\dagger = E_{kj} - E_{jk} = -H_{jk}, \\
D_j^\dagger &= -i(E_{jj}^\dagger - E_{j+1,j+1}^\dagger) = -i(E_{jj} - E_{j+1,j+1}) = -D_j.
\end{aligned}
$$

They are also closed under commutation (see Exercise 1.10).

Finally we must show that the $n^2 - 1$ matrices $G_{jk}$, $H_{jk}$ and $D_j$ are linearly independent. Let

$$
A = \sum_{j<k} \alpha_{jk} G_{jk} + \sum_{j<k} \beta_{jk} H_{jk} + \sum_{j=1}^{n-1} \gamma_j D_j = 0,
$$

where the coefficients $\alpha_{jk}$, $\beta_{jk}$ and $\gamma_j$ are real. Expressing this linear combination in terms of the linearly independent matrices $E_{jk}$:

$$
\begin{aligned}
\sum_{j<k} (i\alpha_{jk} + \beta_{jk}) E_{jk} + \sum_{j<k} (i\alpha_{jk} - \beta_{jk}) E_{kj} + i\gamma_1 E_{11} \\
+ i \sum_{j=2}^{n-1} (\gamma_j - \gamma_{j-1}) E_{jj} - i\gamma_{n-1} E_{nn} = 0.
\end{aligned}
$$

Therefore $i\alpha_{jk} + \beta_{jk} = 0$ and $i\alpha_{jk} - \beta_{jk} = 0$ for $j < k$ which implies that $\alpha_{jk} = 0$ and $\beta_{jk} = 0$. Also $\gamma_1 = 0$ and $\gamma_j - \gamma_{j-1} = 0$, $j = 2, \ldots, n-1$ implies that $\gamma_j = 0$ for $j = 1, \ldots, n-1$. Therefore the $n^2 - 1$ matrices are linearly independent and form a basis for the Lie algebra su$(n)$.

## ■ Solution 1.12

Because of the antisymmetry property of the commutator there are 28 commutators $[E_j, E_k]$ to evaluate for $j < k$. For example

$$
\begin{aligned}
[E_1, E_2] &= [G_{12}, H_{12}] = i[E_{12} + E_{21}, E_{12} - E_{21}] \\
&= -i[E_{12}, E_{21}] + i[E_{21}, E_{12}] = -2i[E_{12}, E_{21}] \\
&= -2i(E_{11} - E_{22}) = -2D_1 = -2E_3, \\
[E_4, E_6] &= [G_{13}, G_{23}] = -[E_{13} + E_{31}, E_{23} + E_{32}] \\
&= -[E_{13}, E_{32}] - [E_{31}, E_{23}] = -E_{12} + E_{21} \\
&= -H_{12} = -E_2, \\
[E_4, E_5] &= [G_{13}, H_{13}] = i[E_{13} + E_{31}, E_{13} - E_{31}]
\end{aligned}
$$

$$= -i[E_{13}, E_{31}] + i[E_{31}, E_{13}]$$
$$= -i(E_{11} - E_{33}) + i(E_{33} - E_{11}) = -2i(E_{11} - E_{33})$$
$$= -2i(E_{11} - E_{22} + E_{22} - E_{33})$$
$$= -2D_1 - 2D_2 = -E_3 - \sqrt{3}E_8.$$

The complete set of 28 commutators is

$$
\begin{array}{lll}
[E_1, E_2] = -2E_3 & [E_1, E_3] = 2E_2 & [E_1, E_4] = -E_7 \\
[E_1, E_5] = E_6 & [E_1, E_6] = -E_5 & [E_1, E_7] = E_4 \\
[E_1, E_8] = 0 & [E_2, E_3] = -2E_1 & [E_2, E_4] = -E_6 \\
[E_2, E_5] = -E_7 & [E_2, E_6] = E_4 & [E_2, E_7] = E_5 \\
[E_2, E_8] = 0 & [E_3, E_4] = -E_5 & [E_3, E_5] = E_4 \\
[E_3, E_6] = -E_7 & [E_3, E_7] = -E_6 & [E_3, E_8] = 0 \\
[E_4, E_5] = -E_3 - \sqrt{3}E_8 & [E_4, E_6] = -E_2 & [E_4, E_7] = -E_1 \\
[E_4, E_8] = \sqrt{3}E_5 & [E_5, E_6] = E_1 & [E_5, E_7] = -E_2 \\
[E_5, E_8] = -\sqrt{3}E_4 & [E_6, E_7] = E_3 - \sqrt{3}E_8 & [E_6, E_8] = \sqrt{3}E_7 \\
[E_7, E_8] = -\sqrt{3}E_6.
\end{array}
$$

These results can be used the show that $e_{jk\ell} = -e_{kj\ell}$ (from antisymmetry of the commutator) and $e_{jk\ell} = -e_{j\ell k}$. It follows that $e_{jk\ell}$ is antisymmetric in all three indices. All non-zero values can be obtained from the following ones using antisymmetry:

$$e_{123} = -2,$$
$$e_{147} = e_{165} = e_{246} = e_{257} = e_{345} = e_{376} = -1,$$
$$e_{458} = e_{678} = -\sqrt{3}.$$

$E_1$, $E_2$ and $E_3$ form a subalgebra since they are closed under commutation. If we define $F_j = -\frac{1}{2}E_j$, $j = 1, 2, 3$ then

$$[F_1, F_2] = F_3, \quad [F_2, F_3] = F_1, \quad [F_3, F_1] = F_2,$$

which are just the defining commutation relations (1.6) of su(2).

Finally if $C = \sum_j E_j^2$ then using Exercise 1.3

$$\sum_j [E_j^2, E_k] = \sum_j E_j [E_j, E_k] + \sum_j [E_j, E_k] E_j$$
$$= \sum_{j\ell} e_{jk\ell} E_j E_\ell + \sum_{j\ell} e_{jk\ell} E_\ell E_j$$
$$= \sum_{j\ell} e_{jk\ell} E_j E_\ell + \sum_{j\ell} e_{\ell k j} E_j E_\ell$$
$$= 0, \text{ since } e_{\ell k j} = -e_{jkl}.$$

## ■ Solution 1.13

Verify that properties (1.7) and (1.8) hold for $T'$ and $T''$. From (1.7), letting $C = \alpha A + \beta B$,

$$\begin{pmatrix} T'(C) & X(C) \\ Z & T''(C) \end{pmatrix} = \begin{pmatrix} \alpha T'(A) + \beta T'(B) & \alpha X(A) + \beta X(B) \\ Z & \alpha T''(A) + \beta T''(B) \end{pmatrix}.$$

Therefore property (1.7) holds for $T'$ and $T''$. From (1.8), letting $C = [A, B]$,

$$\begin{pmatrix} T'(C) & X(C) \\ Z & T''(C) \end{pmatrix} = \begin{pmatrix} [T'(A), T'(B)] & X(A, B) \\ Z & [T''(A), T''(B)] \end{pmatrix},$$

where

$$X(A, B) = T'(A)X(B) + X(A)T''(B) - T'(B)X(A) - X(B)T''(A),$$

and property (1.8) also holds for $T'$ and $T''$. Therefore they are representations.

## ■ Solution 1.14

Using the commutator identities of Exercise 1.3 and the boson commutation relations

$$\begin{aligned}
[J_1, J_2] &= -\frac{i}{4}[a_1^+ a_2 + a_1 a_2^+, a_1^+ a_2 - a_1 a_2^+] \\
&= -\frac{i}{4}(-[a_1^+ a_2, a_1 a_2^+] + [a_1 a_2^+, a_1^+ a_2]) \\
&= -\frac{i}{4}(-a_1^+ a_1 [a_2, a_2^+] - [a_1^+, a_1]a_2^+ a_2 + a_1 a_1^+ [a_2^+, a_2] \\
&\qquad + [a_1, a_1^+]a_2 a_2^+) \\
&= -\frac{i}{4}(-a_1^+ a_1 + a_2^+ a_2 - a_1 a_1^+ + a_2 a_2^+) \\
&= -\frac{i}{4}(-2a_1^+ a_1 + 2a_2^+ a_2) = iJ_3.
\end{aligned}$$

The other commutators are obtained in a similar manner.

## J.2 Solutions to Chapter 2 Exercises

## ■ Solution 2.1

$$\begin{aligned}
A = \sum_{jk} a_{jk} s_{jk} &= \frac{1}{2}\sum_{jk} a_{jk} s_{jk} + \frac{1}{2}\sum_{jk} a_{jk} s_{jk} \\
&= \frac{1}{2}\sum_{jk} a_{jk} s_{jk} + \frac{1}{2}\sum_{jk} a_{kj} s_{kj} \\
&= \frac{1}{2}\sum_{jk} a_{jk} s_{jk} - \frac{1}{2}\sum_{jk} a_{jk} s_{jk} = 0,
\end{aligned}$$

since $a_{kj} = -a_{jk}$ and $s_{kj} = s_{jk}$.

### ■ Solution 2.2

**Derivation of (2.11):** Let $g$ be an arbitrary function of $\boldsymbol{r}$:

$$
\begin{aligned}
[p_j, f(r)]g &= -i\left[\frac{\partial}{\partial x_j}, f(r)\right]g \\
&= -i\frac{\partial}{\partial x_j}(f(r)g) + if(r)\frac{\partial g}{\partial x_j} \\
&= -if(r)\frac{\partial g}{\partial x_j} - i\frac{\partial f(r)}{\partial x_j}g + if(r)\frac{\partial g}{\partial x_j} \\
&= -i\frac{\partial f(r)}{\partial x_j}g.
\end{aligned}
$$

Therefore,

$$
[p_j, f(r)] = -i\frac{\partial f(r)}{\partial x_j} = -if'(r)\frac{x_j}{r} = -ix_j r^{-1}f'(r).
$$

**Derivation of (2.12):** This is a special case of (2.11) with $f(r) = r^{-n}$ so

$$
[p_j, r^{-n}] = -ix_j r^{-1}(-nr^{-n-1}) = inx_j r^{-n-2}.
$$

**Derivation of (2.13):** Use (2.12) and (2.9) to obtain

$$
\begin{aligned}
[p_j, x_k r^{-n}] &= x_k[p_j, r^{-n}] + [p_j, x_k]r^{-n} \\
&= x_k(inx_j r^{-n-2}) + (-i\delta_{jk})r^{-n} \\
&= ir^{-n}(nx_j x_k r^{-2} - \delta_{jk}).
\end{aligned}
$$

**Derivation of (2.14):** Let $k = j$ in the preceding result and sum over $j$ to obtain

$$
[p_j, x_j r^{-n}] = ir^{-n}(nx_j x_j r^{-2} - 3) = ir^{-n}(n - 3).
$$

### ■ Solution 2.3

**Derivation of (2.17):** Evaluate the components $[x_j, p^2]$:

$$
\begin{aligned}
[x_j, p^2] &= [x_j, p_k p_k], \quad \text{(summed over } k) \\
&= p_k[x_j, p_k] + [x_j, p_k]p_k, \quad \text{from (2.1)} \\
&= ip_k\delta_{jk} + i\delta_{jk}p_k = 2ip_k, \quad \text{from (2.9).}
\end{aligned}
$$

**Derivation of (2.19):**

$$
\begin{aligned}
[\boldsymbol{r}\cdot\boldsymbol{p}, p^2] &= [x_j p_j, p_k p_k], \quad \text{(summed over } j, k) \\
&= [x_j, p_k p_k]p_j, \quad \text{from (2.4)} \\
&= (p_k[x_j, p_k] + [x_j, p_k]p_k)p_j, \quad \text{from (2.1)} \\
&= p_k(i\delta_{jk})p_j + i\delta_{jk}p_k p_j = 2ip^2, \quad \text{from (2.9).}
\end{aligned}
$$

**Derivation of (2.20):** Evaluate the components $[x_j, \boldsymbol{r} \cdot \boldsymbol{p}]$:

$$[x_j, \boldsymbol{r} \cdot \boldsymbol{p}] = [x_j, x_k p_k], \quad \text{(summed over } k)$$
$$= x_k [x_j, p_k] = x_k(i\delta_{jk}) = ix_j, \quad \text{from (2.3)}.$$

**Derivation of (2.21):** Evaluate the components $[p_j, \boldsymbol{r} \cdot \boldsymbol{p}]$:

$$[p_j, \boldsymbol{r} \cdot \boldsymbol{p}] = [p_j, x_k p_k], \quad \text{(summed over } k)$$
$$= [p_j, x_k] p_k = -i\delta_{jk} p_k = -ip_j.$$

**Derivation of (2.22):** Use (2.11) to obtain

$$[f(r), \boldsymbol{r} \cdot \boldsymbol{p}] = [f(r), x_j p_j], \quad \text{(summed over } j)$$
$$= x_j [f(r), p_j] = ix_j x_j r^{-1} f'(r)$$
$$= ir f'(r).$$

### ■ Solution 2.4

**Derivation of (2.23):**

$$[L_j, x_k] = [(\boldsymbol{r} \times \boldsymbol{p})_j, x_k]$$
$$= [\epsilon_{jlm} x_l p_m, x_k], \quad \text{(summed over } l \text{ and } m)$$
$$= \epsilon_{jlm} x_l [p_m, x_k] = \epsilon_{jlm} x_l(-i\delta_{km}), \quad \text{from (2.9)}$$
$$= -i\epsilon_{jlk} x_l = i\epsilon_{jkl} x_l.$$

**Derivation of (2.24):**

$$[L_j, p_k] = [(\boldsymbol{r} \times \boldsymbol{p})_j, p_k]$$
$$= [\epsilon_{jlm} x_l p_m, p_k], \quad \text{(summed over } l \text{ and } m)$$
$$= \epsilon_{jlm} [x_l, p_k] p_m = \epsilon_{jlm}(i\delta_{lk}) p_m, \quad \text{from (2.9)}$$
$$= i\epsilon_{jkm} p_m.$$

**Derivation of (2.26):** Here we use (2.11) and the rule derived in Exercise 2.1:

$$[L_j, f(r)] = \epsilon_{jkl} [x_k p_l, f(r)]$$
$$= \epsilon_{jkl} x_k [p_l, f(r)]$$
$$= \epsilon_{jkl} x_k(-ix_l r^{-1} f'(r)), \quad \text{from (2.26)}$$
$$= -i\epsilon_{jkl} x_k x_l r^{-1} f'(r) = 0.$$

### ■ Solution 2.5

**Derivation of (2.29):**

$$[L^2, p_j] = [L_k L_k, p_j], \quad \text{(summed over } k)$$
$$= L_k [L_k, p_j] + [L_k, p_j] L_k, \quad \text{from (2.2)}$$

$$
\begin{aligned}
&= L_k(i\epsilon_{kj\ell}p_\ell) + (i\epsilon_{kj\ell}p_\ell)L_k, \quad \text{from (2.24)} \\
&= -i\epsilon_{jk\ell}L_k p_\ell - i\epsilon_{jk\ell}p_l L_k \\
&= -i\epsilon_{jk\ell}(p_\ell L_k + i\epsilon_{k\ell m}p_m) - i\epsilon_{jk\ell}p_\ell L_k, \quad \text{from (2.24)} \\
&= -2i\epsilon_{jk\ell}p_\ell L_k + \epsilon_{jk\ell}\epsilon_{mk\ell}p_m \\
&= 2i\epsilon_{j\ell k}p_\ell L_k + 2\delta_{jm}p_m, \quad \text{from (A.10)} \\
&= 2i(\boldsymbol{p}\times\boldsymbol{L})_j + 2p_j.
\end{aligned}
$$

**Derivation of (2.30):** Evaluate the components $[L_j, p^2]$ and use the rule derived in Exercise 2.1:

$$
\begin{aligned}
[L_j, p^2] &= [L_j, p_k p_k], \quad \text{(summed over } k) \\
&= p_k[L_j, p_k] + [L_j, p_k]p_k, \quad \text{from (2.1)} \\
&= p_k(i\epsilon_{jk\ell}p_\ell) + (i\epsilon_{jk\ell}p_\ell)p_k, \quad \text{from (2.24)} \\
&= i\epsilon_{jk\ell}(p_k p_\ell + p_\ell p_k) = 0, \quad \text{from Exercise 2.1.}
\end{aligned}
$$

■ **Solution 2.6**

**Derivation of (2.32):** Use the rule derived in Exercise 2.1:

$$
\begin{aligned}
\boldsymbol{p}\cdot\boldsymbol{L} &= p_j L_j = \epsilon_{jk\ell}p_j x_k p_l = 0, \\
\boldsymbol{L}\cdot\boldsymbol{p} &= L_j p_j = \epsilon_{jk\ell}x_k p_l p_j = 0,
\end{aligned}
$$

since $p_j x_k p_l$ and $x_k p_\ell p_j$ are symmetric in $j$ and $\ell$.
**Derivation of (2.33):** Evaluate the components $(\boldsymbol{p}\times\boldsymbol{L})_j$:

$$
\begin{aligned}
(\boldsymbol{p}\times\boldsymbol{L})_j &= \epsilon_{jk\ell}p_k L_\ell = \epsilon_{jk\ell}p_k(\epsilon_{\ell mn}x_m p_n) \\
&= \epsilon_{jk\ell}\epsilon_{mn\ell}p_k x_m p_n \\
&= (\delta_{jm}\delta_{kn} - \delta_{jn}\delta_{mk})p_k x_m p_n, \quad \text{from (A.9)} \\
&= p_k x_j p_k - p_k x_k p_j \\
&= (x_j p_k - i\delta_{jk})p_k - (\boldsymbol{p}\cdot\boldsymbol{r})p_j, \quad \text{from (2.9)} \\
&= x_j p^2 - ip_j - (\boldsymbol{r}\cdot\boldsymbol{p} - 3i)p_j, \quad \text{from (2.10)} \\
&= x_j p^2 - (\boldsymbol{r}\cdot\boldsymbol{p})p_j + 2ip_j \\
&= x_j p^2 - (p_j(\boldsymbol{r}\cdot\boldsymbol{p}) + ip_j) + 2ip_j, \quad \text{from (2.21)} \\
&= x_j p^2 - p_j(\boldsymbol{r}\cdot\boldsymbol{p} - i).
\end{aligned}
$$

**Derivation of (2.34):** Evaluate the components of $(\boldsymbol{r}\times\boldsymbol{L})_j$:

$$
\begin{aligned}
(\boldsymbol{r}\times\boldsymbol{L})_j &= \epsilon_{jk\ell}x_k L_\ell \\
&= \epsilon_{jk\ell}\epsilon_{\ell mn}x_k x_m p_n \\
&= \epsilon_{jk\ell}\epsilon_{mn\ell}x_k x_m p_n \\
&= (\delta_{jm}\delta_{kn} - \delta_{jn}\delta_{km})x_k x_m p_n, \quad \text{from (A.9)} \\
&= x_k x_j p_k - x_k x_k p_j
\end{aligned}
$$

$$\begin{aligned}
&= x_k(p_k x_j + i\delta_{kj}) - r^2 p_j, \quad \text{from (2.9)}\\
&= (\boldsymbol{r}\cdot\boldsymbol{p})x_j + ix_j - r^2 p_j\\
&= (\boldsymbol{r}\cdot\boldsymbol{p})x_j + ix_j - (p_j r^2 + 2ix_j), \quad \text{from (2.12)}\\
&= (\boldsymbol{r}\cdot\boldsymbol{p})x_j - ix_j - p_j r^2\\
&= (\boldsymbol{r}\cdot\boldsymbol{p} - i)x_j - p_j r^2.
\end{aligned}$$

**Derivation of (2.35):** Evaluate the components using the rule in Exercise 2.1.

$$\begin{aligned}
(\boldsymbol{r}\times\boldsymbol{L})_j + (\boldsymbol{L}\times\boldsymbol{r})_j &= \epsilon_{jk\ell}x_k L_\ell + \epsilon_{jk\ell}L_k x_\ell\\
&= \epsilon_{jk\ell}x_k L_\ell + \epsilon_{jk\ell}(x_\ell L_k + i\epsilon_{k\ell m}x_m), \quad \text{from (2.23)}\\
&= \epsilon_{jk\ell}(x_k L_\ell + x_\ell L_k) + i\epsilon_{jk\ell}\epsilon_{mk\ell}x_m\\
&= 2i\delta_{jm}x_m = 2ix_j, \quad \text{from (A.9).}
\end{aligned}$$

**Derivation of (2.36):** Evaluate the components using the rule in Exercise 2.1.

$$\begin{aligned}
(\boldsymbol{p}\times\boldsymbol{L})_j + (\boldsymbol{L}\times\boldsymbol{p})_j &= \epsilon_{jk\ell}p_k L_\ell + \epsilon_{jk\ell}L_k p_\ell\\
&= \epsilon_{jk\ell}p_k L_\ell + \epsilon_{jk\ell}(p_\ell L_k + i\epsilon_{k\ell m}p_m), \quad \text{from (2.24)}\\
&= \epsilon_{jk\ell}(p_k L_\ell + p_\ell L_k) + i\epsilon_{jk\ell}\epsilon_{k\ell m}p_m\\
&= 2i\delta_{jm}p_m = 2ip_j, \quad \text{from (A.9).}
\end{aligned}$$

## J.3   Solutions to Chapter 3 Exercises

■ **Solution 3.1**

$$\begin{aligned}
J_j V_k - V_k J_j &= i\epsilon_{jk\ell}V_\ell\\
\epsilon_{mjk}J_j V_k - \epsilon_{mjk}V_k J_j &= i\epsilon_{mjk}\epsilon_{jk\ell}V_\ell\\
(\boldsymbol{J}\times\boldsymbol{V})_m + (\boldsymbol{V}\times\boldsymbol{J})_m &= 2\delta_{m\ell}V_\ell = 2V_m.
\end{aligned}$$

■ **Solution 3.2**

$$\begin{aligned}
[\boldsymbol{J}\cdot\boldsymbol{V}, J_j] &= J_k V_k J_j - J_j J_k V_k\\
&= J_k V_k J_j - (J_k J_j + i\epsilon_{jk\ell}J_\ell)V_k\\
&= J_k V_k J_j - i\epsilon_{jk\ell}J_\ell V_k - J_k(V_k J_j + i\epsilon_{jk\ell}V_\ell)\\
&= -i\epsilon_{jk\ell}J_\ell V_k - i\epsilon_{jk\ell}J_k V_\ell\\
&= -i\epsilon_{jk\ell}J_\ell V_k + i\epsilon_{j\ell k}J_\ell V_k = 0.
\end{aligned}$$

■ **Solution 3.3**

$$\begin{aligned}
[J^2, V_j] &= [J_k J_k, V_j]\\
&= J_k[J_k, V_j] + [J_k, V_j]J_k\\
&= i\epsilon_{kj\ell}J_k V_\ell + i\epsilon_{kj\ell}V_\ell J_k\\
&= -i\epsilon_{jk\ell}J_k V_\ell + i\epsilon_{j\ell k}V_\ell J_k\\
&= i(\boldsymbol{V}\times\boldsymbol{J} - \boldsymbol{J}\times\boldsymbol{V})_j.
\end{aligned}$$

■ **Solution 3.4**

$$
\begin{aligned}
[J^2, (\boldsymbol{J} \times \boldsymbol{V})_j] &= \epsilon_{jk\ell}[J^2, J_k V_\ell] \\
&= \epsilon_{jk\ell} J_k [J^2, V_\ell] \quad \text{since} \quad [J^2, J_k] = 0 \\
&= i\epsilon_{jk\ell} J_k (\boldsymbol{V} \times \boldsymbol{J} - \boldsymbol{J} \times \boldsymbol{V})_\ell \\
&= 2i\epsilon_{jk\ell} J_k (i\boldsymbol{V} - \boldsymbol{J} \times \boldsymbol{V})_\ell \\
&= -2\epsilon_{jk\ell} J_k V_\ell - 2i\epsilon_{jk\ell}\epsilon_{\ell mn} J_k J_m V_n \\
&= -2\epsilon_{jk\ell} J_k V_\ell - 2i(\delta_{jm}\delta_{kn} - \delta_{jn}\delta_{km}) J_k J_m V_n \\
&= -2\epsilon_{jk\ell} J_k V_\ell - 2i(J_k J_j V_k - J_k J_k V_j) \\
&= -2\epsilon_{jk\ell} J_k V_\ell - 2i J_k (V_k J_j + i\epsilon_{jk\ell} V_\ell) + 2i J^2 V_j \\
&= 2i(J^2 V_j - (\boldsymbol{J} \cdot \boldsymbol{V}) J_j).
\end{aligned}
$$

■ **Solution 3.5**

$$
\begin{aligned}
[J^2, [J^2, \boldsymbol{V}]] &= [J^2, i(\boldsymbol{V} \times \boldsymbol{J} - \boldsymbol{J} \times \boldsymbol{V})] \\
&= [J^2, -2\boldsymbol{V} - 2i\boldsymbol{J} \times \boldsymbol{V}] \\
&= -2[J^2, \boldsymbol{V}] - 2i(2i)(J^2\boldsymbol{V} - (\boldsymbol{J} \cdot \boldsymbol{V})\boldsymbol{J}) \\
&= -2[J^2, \boldsymbol{V}] + 4J^2\boldsymbol{V} - 4(\boldsymbol{J} \cdot \boldsymbol{V})\boldsymbol{J} \\
&= 2(J^2\boldsymbol{V} - 2(\boldsymbol{J} \cdot \boldsymbol{V})\boldsymbol{J} + \boldsymbol{V}J^2).
\end{aligned}
$$

■ **Solution 3.6**

Let $c_j = \omega_j e^{i\alpha_j}$ and define $|jm\rangle = e^{i\gamma_j}|jm\rangle'$. Then from (3.57) and (3.60)

$$
\begin{aligned}
V_3 e^{i\gamma_j}|jm\rangle' &= \sqrt{(j-m)(j+m)}\,\omega_j e^{i\alpha_j} e^{i\gamma_j - 1}|j-1, m\rangle' \\
&\quad - m a_j e^{i\gamma_j}|jm\rangle' \\
&\quad + \sqrt{(j-m+1)(j+m+1)}\,\omega_{j+1} e^{-i\alpha_{j+1}} e^{i\gamma_{j+1}}|j+1, m\rangle', \\
V_3|jm\rangle' &= \sqrt{(j-m)(j+m)}\,\omega_j e^{i(\alpha_j + \gamma_{j-1} - \gamma_j)}|j-1, m\rangle' \\
&\quad - m a_j|jm\rangle' \\
&\quad + \sqrt{(j-m+1)(j+m+1)}\,\omega_{j+1} e^{-i(\alpha_{j+1} + \gamma_j - \gamma_{j+1})}|j+1, m\rangle'.
\end{aligned}
$$

We can now choose the $\gamma_j$ such that $\gamma_{j+1} - \gamma_j = \alpha_{j+1}$. In fact if $j_0$ is the smallest value of $j$ then $\gamma_{j+1} = \gamma_{j_0} + \alpha_{j_0+1} + \cdots + \alpha_{j+1}$. Therefore

$$
\begin{aligned}
V_3|jm\rangle' &= \sqrt{(j-m)(j+m)}\,\omega_j|j-1, m\rangle' \\
&\quad - m a_j|jm\rangle' \\
&\quad + \sqrt{(j-m+1)(j+m+1)}\,\omega_{j+1}|j+1, m\rangle'.
\end{aligned}
$$

Similar results are obtained for (3.58) and (3.59).

## ■ Solution 3.7

To prove (a) write $|n\rangle = N_n (a^+)^n |0\rangle$ where $N_n$ is a normalization factor. Then

$$\langle n|n \rangle = N_n^2 \langle 0|a^n (a^+)^n|0\rangle.$$

Now use $[a, a^+] = 1$ to move one annihilation operator to the right of all creation operators. The result is

$$a^n (a^+)^n = a^{n-1}(a^+)^n a + n a^{n-1}(a^+)^{n-1}.$$

For $n = 1$ this is just the commutation relation $[a, a^+] = 1$ and assuming this formula the general result is proved by induction since

$$\begin{aligned}
a^{n+1}(a^+)^{n+1} &= a(a^n(a^+)^n)a^+ \\
&= a(a^{n-1}(a^+)^n a + n a^{n-1}(a^+)^{n-1})a^+ \\
&= a^n(a^+)^n a a^+ + n a^n(a^+)^n \\
&= a^n(a^+)^n(a^+ a + 1) + n a^n(a^+)^n \\
&= a^n(a^+)^{n+1} a + (n+1)a^n(a^+)^n,
\end{aligned}$$

which is the same identity with $n$ replaced by $n + 1$. Therefore

$$\langle n|n \rangle = N_n^2 \langle 0|a^{n-1}(a^+)^n a|0\rangle + n N_n^2 \langle 0|a^{n-1}(a^+)^{n-1}|0\rangle.$$

The first term is zero since $a|0\rangle = 0$ so

$$\langle n|n \rangle = n \frac{N_n^2}{N_{n-1}^2} = 1,$$

and

$$N_n^2 = \frac{1}{n}N_{n-1}^2, \quad N_0 = 1.$$

Therefore $N_n^2 = 1/n!$ and we can choose $N_n = 1/\sqrt{n!}$.

To prove (b) let $a^+|n\rangle = \alpha_n |n + 1\rangle$ and use (a) to obtain

$$a^+ N_n (a^+)^n |0\rangle = \alpha_n N_{n+1}(a^+)^{n+1}|0\rangle.$$

Therefore $\alpha_n = N_n/N_{n+1} = \sqrt{n + 1}$.

To prove (c) let $a|n\rangle = \beta_n |n - 1\rangle$ and apply $[a, a^+] = 1$ to the general state $|n\rangle$ to obtain

$$(aa^+ - a^+ a)|n\rangle = (\alpha_n \beta_{n+1} - \beta_n \alpha_{n-1})|n\rangle = |n\rangle.$$

Therefore $\sqrt{n + 1}\beta_{n+1} - \sqrt{n}\beta_n = 1$ and we can choose $\beta_n = \sqrt{n}$.

## ■ Solution 3.8

Since the annihilation and creation operators for the two states commute
($[a_j, a_k^+] = \delta_{jk}$) we can extend the results of the preceding exercise to the
two state case by writing

$$|n_1, n_2\rangle = \frac{1}{\sqrt{n_1!}} \frac{1}{\sqrt{n_2!}} (a_1^+)^{n_1} (a_2^+)^{n_2} |0, 0\rangle,$$

$$a_1^+ |n_1, n_2\rangle = \sqrt{n_1 + 1} \, |n_1 + 1, n_2\rangle,$$

$$a_2^+ |n_1, n_2\rangle = \sqrt{n_2 + 1} \, |n_1, n_2 + 1\rangle,$$

$$a_1 |n_1, n_2\rangle = \sqrt{n_1} \, |n_1 - 1, n_2\rangle,$$

$$a_2 |n_1, n_3\rangle = \sqrt{n_2} \, |n_1, n_2 - 1\rangle.$$

Using the definitions from Exercise 1.15 for $J_+$, $J_-$ and $J_3$ it follows that

$$J_+ |n_1, n_2\rangle = \sqrt{(n_1 + 1)n_2} \, |n_1 + 1, n_2 - 1\rangle,$$

$$J_- |n_1, n_2\rangle = \sqrt{n_1(n_2 + 1)} \, |n_1 - 1, n_2 + 1\rangle,$$

$$J_3 |n_1, n_2\rangle = \frac{1}{2}(n_1 - n_2)|n_1, n_2\rangle.$$

None of these operators change the total number $n = n_1 + n_2$ of bosons in the
two states so the subspaces

$$V_n = \{|n_1, n_2\rangle \,|\, n_1 + n_2 = n\}, \quad n = 0, 1, 2, \ldots$$

are irreducible invariant subspaces and we obtain all unirreps of su(2).  If we
define

$$m = \frac{1}{2}(n_1 - n_2), \quad j = \frac{1}{2}(n_1 + n_2),$$

then $2j$ is a non-negative integer, $m = -j, -j + 1, \ldots, j$ and we obtain the
representation (3.19) to (3.21) if we denote the state $|n_1, n_2\rangle$ by $|jm\rangle$.
The number operator $N = a_1^+ a_1 + a_2^+ a_2$ gives the total number of bosons since

$$N|n_1, n_2\rangle = (n_1 + n_2)|n_1, n_2\rangle.$$

It can be chosen as the Casimir operator since

$$\begin{aligned}
[N, J_+] &= [a_1^+ a_1 + a_2^+ a_2, a_1^+ a_2] \\
&= a_1^+ [a_1, a_1^+ a_2] + [a_2^+, a_1^+ a_2] a_2 \\
&= a_1^+ a_2 - a_1^+ a_2 = 0. \\
[N, J_-] &= [a_1^+ a_1 + a_2^+ a_2, a_2^+ a_1] \\
&= [a_1^+, a_2^+ a_1] a_1 + a_2^+ [a_2, a_2^+ a_1] \\
&= -a_2^+ a_1 + a_2^+ a_1 = 0, \\
[N, J_3] &= \frac{1}{2}[a_1^+ a_1 + a_2^+ a_2, a_1^+ a_1 - a_2^+ a_2] = 0.
\end{aligned}$$

The connection between $N$ and the usual Casimir operator $J^2$ defined in (3.7) is

$$
\begin{aligned}
J^2 &= \frac{1}{2}(J_+ J_- + J_- J_+) + J_3^2 \\
&= \frac{1}{2}(a_1^+ a_2 a_2^+ a_1 + a_2^+ a_1 a_1^+ a_2) + \frac{1}{4}(a_1^+ a_1 - a_2^+ a_2)^2 \\
&= \frac{1}{2}a_1^+ a_1(a_2^+ a_2 + 1) + \frac{1}{2}(a_1^+ a_1 + 1)a_2^+ a_2 + \frac{1}{4}(a_1^+ a_1 - a_2^+ a_2)^2 \\
&= a_1^+ a_1 a_2^+ a_2 + \frac{1}{2}(a_1^+ a_1 + a_2^+ a_2) \\
&\qquad + \frac{1}{4}(a_1^+ a_1 a_1^+ a_1 + a_2^+ a_2 a_2^+ a_2 - 2a_1^+ a_1 a_2^+ a_2) \\
&= a_1^+ a_1 a_2^+ a_2 + \frac{1}{2}N + \frac{1}{4}(N^2 - 4a_1^+ a_1 a_2^+ a_2) \\
&= \frac{1}{2}N + \frac{1}{4}N^2 = \frac{1}{2}N\left(\frac{1}{2}N + 1\right).
\end{aligned}
$$

so the eigenvalues of $J^2$ are $j(j+1)$. One advantage of the boson formalism here is that it is obvious that $N$ is a Casimir operator since it conserves the total number of bosons.

## J.4  Solutions to Chapter 4 Exercises

### ■ Solution 4.1

By direct calculation using $[a, a^+] = 1$ and the general commutation rules (2.1) and (2.2)

$$
\begin{aligned}
[T_1, T_2] &= 2(\alpha_2 \beta_1 - \alpha_1 \beta_2)(a^+ a + a a^+), \\
[T_2, T_3] &= 2(\alpha_3 + \beta_3)(-\alpha_2 a^+ a^+ + \beta_2 a a), \\
[T_3, T_1] &= 2(\alpha_3 + \beta_3)(\alpha_1 a^+ a^+ - \beta_1 a a).
\end{aligned}
$$

Substituting into (4.1) we obtain

$$
\begin{aligned}
2(\alpha_2 \beta_1 - \alpha_1 \beta_2)(a^+ a + a a^+) &= -i(\alpha_3 a^+ a + \beta_3 a a^+), \\
2(\alpha_3 + \beta_3)(-\alpha_2 a^+ a^+ + \beta_2 a a) &= i(\alpha_1 a^+ a^+ + \beta_1 a a), \\
2(\alpha_3 + \beta_3)(\alpha_1 a^+ a^+ - \beta_1 a a) &= i(\alpha_2 a^+ a^+ + \beta_2 a a).
\end{aligned}
$$

Compare coefficients of the bilinear products to obtain the system of equations

$$
\beta_3 = \alpha_3, \ 2(\alpha_2 \beta_1 - \alpha_1 \beta_2) = -i\alpha_3, \ -2\alpha_2(\alpha_3 + \beta_3) = i\alpha_1,
$$
$$
2\beta_2(\alpha_3 + \beta_3) = i\beta_1, \ 2\alpha_1(\alpha_3 + \beta_3) = i\alpha_2, \ -2\beta_1(\alpha_3 + \beta_3) = i\beta_2,
$$

whose solution has the form

$$
\alpha_1 = \alpha, \quad \alpha_2 = \pm i\alpha, \quad \alpha_3 = \pm\frac{1}{4},
$$
$$
\beta_1 = \frac{1}{16\alpha}, \quad \beta_2 = \pm\frac{i}{16\alpha}, \quad \beta_3 = \pm\frac{1}{4},
$$

for any real constant $\alpha$. A standard choice of the upper sign and $\alpha = 1/4$ gives the realization

$$T_1 = \frac{1}{4}(a^+a^+ + aa),\ T_2 = -\frac{i}{4}(a^+a^+ - aa),\ T_3 = \frac{1}{4}(a^+a + aa^+).$$

The Casimir operator is evaluated as follows

$$16T^2 = (a^+a + aa^+)^2 - (a^+a^+ + aa)^2 + (a^+a^+ - aa)^2$$
$$= a^+aa^+a + a^+aaa^+ + aa^+a^+a + aa^+aa^+$$
$$- 2a^+a^+aa - 2aaa^+a^+.$$

Now use $aa^+ - a^+a = 1$ to put each term in the canonical form with all creation operators to the left of all annihilation operators and obtain the simple result $T^2 = -3/16$.

Therefore $k(k+1) = -3/16$ and we obtain the two $T_3$ eigenvalue spectra from (4.23): one for $k = -1/4$ and the other for $k = -3/4$, corresponding to $q = k+1+\mu = 3/4+\mu$ and $q = 1/4+\mu$ respectively. This is analogous to the 1-dimensional harmonic oscillator considered in Section 4.9. The two eigenvalue spectra correspond to states with an odd number of bosons ($k = -1/4$) or an even number ($k = -3/4$). This also follows directly from the realization of $T_1$ and $T_2$ which implies that $T_+ = T_1 + iT_2 = \frac{1}{2}a^+a^+$ and $T_- = T_1 - iT_2 = \frac{1}{2}aa$ so the raising and lowering operators can only change the total number of bosons by 2 at at time. Since $T_3$ conserves the number of bosons it follows that the subspace of states with even numbers of bosons and the subspace with odd numbers of bosons are separately invariant under the action of the so(2,1) generators. They are also irreducible subspaces.

## ■ Solution 4.2

Using the definition (4.106) of the radial momentum (see also Section 2.2)

$$[R, P_R] = -i\left[R, \frac{\partial}{\partial R} + \frac{M}{R}\right] = -i\left[R, \frac{\partial}{\partial R}\right] = i.$$

## ■ Solution 4.3

Use (4.106) and apply (4.45) to an arbitrary function $f$ to obtain

$$[R^{-n}, P_R]f = -i\left[R^{-n}, \frac{\partial}{\partial R} + \frac{M}{R}\right]f = -i\left[R^{-n}, \frac{\partial}{\partial R}\right]f$$
$$= -iR^{-n}\frac{\partial f}{\partial R} + i\frac{\partial}{\partial R}(R^{-n}f) = -inR^{-n-1}f.$$

Note that this result also follows from the derivation of (2.11) and (2.12) given in Exercise 2.2.

For part (b) expand the identity $[R^{-n}R^n, P_R] = 0$ and use the identity (4.31) for positive powers of $R$ to obtain

$$R^{-n}[R^n, P_R] + [R^{-n}, P_R]R^n = 0.$$

Now postmultiply by $R^{-n}$ to obtain

$$\begin{aligned}
[R^{-n}, P_R] &= -R^{-n}[R^n, P_R]R^{-n} \\
&= -R^{-n}(inR^{n-1})R^{-n} = -inR^{-n-1}.
\end{aligned}$$

## ■ Solution 4.4

Using the definition (4.98) and the canonical commutation relations (2.7) to (2.9) for the coordinate and conjugate momentum vector components

$$\begin{aligned}
[L_{jk}, L_{\ell m}] &= [x_j p_k - x_k p_j, x_\ell p_m - x_m p_\ell] \\
&= x_j[p_k, x_\ell p_m] + [x_j, x_\ell p_m]p_k - x_j[p_k, x_m p_\ell] - [x_j, x_m p_\ell]p_k \\
&\quad - x_k[p_j, x_\ell p_m] - [x_k, x_\ell p_m]p_j + x_k[p_j, x_m p_\ell] + [x_k, x_m p_\ell]p_j \\
&= -i\delta_{k\ell}x_j p_m + i\delta_{jm}x_\ell p_k + i\delta_{km}x_j p_\ell - i\delta_{j\ell}x_m p_k \\
&\quad + i\delta_{j\ell}x_k p_m - i\delta_{km}x_\ell p_j - i\delta_{jm}x_k p_\ell + i\delta_{k\ell}x_m p_j \\
&= -i\delta_{k\ell}(x_j p_m - x_m p_j) + i\delta_{jm}(x_\ell p_k - x_k p_\ell) \\
&\quad + i\delta_{km}(x_j p_\ell - x_\ell p_j) + i\delta_{j\ell}(x_k p_m - x_m p_k) \\
&= i(-\delta_{k\ell}L_{jm} + \delta_{jm}L_{\ell k} + \delta_{km}L_{j\ell} + \delta_{j\ell}L_{km}) \\
&= i(-\delta_{k\ell}L_{jm} - \delta_{jm}L_{k\ell} + \delta_{km}L_{j\ell} + \delta_{j\ell}L_{km}).
\end{aligned}$$

## ■ Solution 4.5

With an implied summation over $j, k = 1, \ldots, D$

$$\begin{aligned}
2[L^2, L_{\ell m}] &= [L_{jk}L_{jk}, L_{\ell m}] \\
&= L_{jk}[L_{jk}, L_{\ell m}] + [L_{jk}, L_{\ell m}]L_{jk} \\
&= i(L_{\ell k}L_{km} + L_{jm}L_{j\ell} - L_{mk}L_{k\ell} - L_{j\ell}L_{jm} \\
&\quad + L_{km}L_{\ell k} + L_{j\ell}L_{jm} - L_{k\ell}L_{mk} - L_{jm}L_{j\ell}) = 0,
\end{aligned}$$

using the antisymmetry property (4.99).

## ■ Solution 4.6

We need to use the canonical commutation relation given in (2.9), which is also valid in $D$-dimensions for $j, k = 1, \ldots, D$ and the generalization of (2.10) to $D$-dimensions given by $[x_j, p_j] = \boldsymbol{r} \cdot \boldsymbol{p} - \boldsymbol{p} \cdot \boldsymbol{r} = Di$, since $\delta_{jj} = D$ (summed over $j$). Substituting (4.98) into the definition (4.101)

$$L^2 = \frac{1}{2}L_{jk}L_{jk} = \frac{1}{2}(x_j p_k - x_k p_j)^2$$

$$
\begin{aligned}
&= \frac{1}{2}\left( x_j p_k x_j p_k - x_j p_k x_k p_j - x_k p_j x_j p_k + x_k p_j x_k p_j \right) \\
&= \frac{1}{2}\left( x_j(x_j p_k - i\delta_{jk})p_k - x_j(x_k p_k - Di)p_j \right. \\
&\qquad \left. - x_k(x_j p_j - Di)p_k + x_k(x_k p_j - i\delta_{jk})p_j \right) \\
&= r^2 p^2 - i\mathbf{r}\cdot\mathbf{p} - x_j x_k p_j p_k + iD\mathbf{r}\cdot\mathbf{p} \\
&= r^2 p^2 - x_j(p_j x_k + i\delta_{jk})p_k + i(D-1)\mathbf{r}\cdot\mathbf{p} \\
&= r^2 p^2 - (\mathbf{r}\cdot\mathbf{p})^2 + i(D-2)\mathbf{r}\cdot\mathbf{p}.
\end{aligned}
$$

## ■ Solution 4.7

Let $u$ be a homogeneous polynomial of degree $\ell$. Then

$$
u(\lambda\mathbf{r}) = u(\lambda x_1, \ldots, \lambda x_D) = \lambda^\ell u(x_1, \ldots, x_D),
$$

and we first show that $\mathbf{r}\cdot\nabla u(\mathbf{r}) = \ell u(\mathbf{r})$. Let $\mathbf{y} = \lambda\mathbf{r}$ and differentiate the homogeneity condition with respect to $\lambda$ to obtain

$$
\frac{du(\lambda\mathbf{r})}{d\lambda} = \sum_j \frac{\partial u(\lambda\mathbf{r})}{\partial y_j}\frac{dy_j}{d\lambda} = \ell\lambda^{\ell-1}u(\mathbf{r}),
$$

or

$$
\sum_j x_j \frac{\partial u(\lambda\mathbf{r})}{\partial y_j} = \ell\lambda^{\ell-1}u(\mathbf{r}).
$$

Let $\lambda = 1$ to obtain $\mathbf{r}\cdot\nabla u(\mathbf{r}) = \ell u(\mathbf{r})$ and substitute into (4.103):

$$
\begin{aligned}
L^2 u(\mathbf{r}) &= (r^2 p^2 - (\mathbf{r}\cdot\mathbf{p})^2 + i(D-2)\mathbf{r}\cdot\mathbf{p})u(\mathbf{r}) \\
&= (-r^2\nabla^2 + (\mathbf{r}\cdot\nabla)^2 + (D-2)\mathbf{r}\cdot\nabla)u(\mathbf{r}) \\
&= (\ell^2 + (D-2)\ell)u(\mathbf{r}) \\
&= \ell(\ell + D - 2)u(\mathbf{r}).
\end{aligned}
$$

## ■ Solution 4.8

The radial equation for the $D$-dimensional harmonic oscillator is

$$
\left[ \frac{1}{2}p_r^2 + \frac{(D-1)(D-3)}{8r^2} + \frac{\ell(\ell+D-2)}{2r^2} + \frac{1}{2}\omega^2 r^2 - E_{\text{osc}} \right] R(r) = 0.
$$

Substitute the definition (4.106) of $p_r$ to obtain

$$
\left[ -\frac{1}{2}\left( \frac{d^2}{dr^2} + \frac{D-1}{r}\frac{d}{dr} \right) + \frac{\ell(\ell+D-2)}{2r^2} + \frac{1}{2}\omega^2 r^2 - E_{\text{osc}} \right] R(r) = 0.
$$

Replace the derivatives by

$$\frac{d}{dr} = \frac{d\rho}{dr}\frac{d}{d\rho} = \frac{c}{a}\rho^{1-a}\frac{d}{d\rho}$$

$$\frac{d^2}{dr^2} = \frac{c}{a}\frac{d\rho}{dr}\frac{d}{d\rho}\left(\rho^{1-a}\frac{d}{d\rho}\right) = \frac{c^2}{a^2}\rho^{1-a}\left(\frac{d}{d\rho}\rho^{1-a}\right)\frac{d}{d\rho}$$

$$= \frac{c^2}{a^2}\rho^{1-a}\left(\rho^{1-a}\frac{d}{d\rho} + (1-a)\rho^{-a}\right)\frac{d}{d\rho}$$

$$= \frac{c^2}{a^2}\rho^{2-2a}\frac{d^2}{d\rho^2} + \frac{c^2}{a^2}(1-a)\rho^{1-2a}\frac{d}{d\rho},$$

to obtain the differential equation

$$\left[-\frac{c^2}{2a^2}\rho^{2-2a}\frac{d^2}{d\rho^2} - \frac{c^2}{2a^2}(1+(D-2)a)\rho^{1-2a}\frac{d}{d\rho} + \frac{c^2\ell(\ell+D-2)}{2\rho^a}\right.$$

$$\left. + \frac{\omega^2}{2c^2}\rho^{2a} - E_{\text{osc}}\right]\rho^b P(\rho) = 0.$$

Now use the operator identities

$$\frac{d}{d\rho}\rho^b = \rho^b\frac{d}{d\rho} + b\rho^{b-1},$$

$$\frac{d^2}{d\rho^2}\rho^b = \rho^b\frac{d^2}{d\rho^2} + 2b\rho^{b-1}\frac{d}{d\rho} + b(b-1)\rho^{b-2},$$

to obtain the transformed radial differential equation

$$\left[-\frac{d^2}{d\rho^2} - \frac{1+(D-2)a+2b}{\rho}\frac{d}{d\rho} + \frac{a^2\ell(\ell+D-2) - b(b+(D-2)a)}{\rho^2}\right.$$

$$\left. + \frac{\omega^2 a^2}{c^4}\rho^{4a-2} - \frac{2a^2}{c^2}E_{\text{osc}}\rho^{2a-2}\right]P(\rho) = 0.$$

We can now compare this result with the radial equation for the $D$-dimensional hydrogen atom using $\rho$ as independent variable:

$$\left[-\frac{d^2}{d\rho^2} - \frac{D-1}{\rho}\frac{d}{d\rho} + \frac{s(s+D-2)}{\rho^2} - \frac{2\mathcal{Z}}{\rho} - 2E_H\right]P(\rho) = 0.$$

These equations have the same form if we let $4a-2 = 0$ and $1+(D-2)a+2b = D-1$ to obtain $a = 1/2$ and $b = (D-2)/4$. Now equate coefficients of $\rho^{-2}$ to obtain

$$\frac{1}{4}\ell(\ell+D-2) - \frac{3}{16}(D-2)^2 = s(s+D-2),$$

which can be solved for

$$s = \frac{1}{2}\left(\ell - \frac{D-2}{2}\right).$$

Finally a comparison of the constant term and the $1/\rho$ term gives the connection between the energy eigenvalues

$$\frac{\omega^2 a^2}{c^4} = -2E_H, \quad \frac{2a^2}{c^2}E_{\mathrm{osc}} = 2Z.$$

Therefore

$$E_H = -\frac{2\omega^2 Z^2}{E_{\mathrm{osc}}^2},$$

and using (4.121) for the harmonic oscillator energy levels

$$E_H = -\frac{2\omega^2 Z^2}{\omega^2(\ell + 2\mu + D/2)} = -\frac{2\omega^2 Z^2}{\omega^2(2s + 2\mu + D - 1)}$$
$$= -\frac{Z^2}{2\left(s + \frac{D-1}{2} + \mu\right)^2},$$

which is just (4.115).

## ■ Solution 4.9

(4.236) follows immediately from the orthogonality relation

$$\int_0^\infty e^{-x} x^\alpha L_p^{(\alpha)} L_q^{(\alpha)}\, dx = \frac{(p+\alpha)!}{p!}\delta_{pq}.$$

To obtain (4.237) write (4.238) in the factored form

$$\frac{2\pi}{4\alpha^3} N^2_{n_1,n_2,m}\left[\int_0^\infty \xi f_{n_1,|m|}(\xi)^2\, d\xi + \int_0^\infty \eta f_{n_2,|m|}(\eta)^2\, d\eta\right] = 1.$$

Since the Laguerre polynomials satisfy the recurrence relation

$$(n+1)L_{n+1}^{(\alpha)}(x) = (2n + \alpha + 1 - x)L_n^{(\alpha)}(x) - (n+\alpha)L_{n-1}^{(\alpha)}(x),$$

and are orthogonal then

$$\int_0^\infty (2n_1 + |m| + 1 - \xi)f_{n_1,|m|}(\xi)^2\, d\xi = 0,$$

so

$$\int_0^\infty \xi f_{n_1,|m|}(\xi)^2\, d\xi = 2n_1 + |m| + 1.$$

Therefore

$$\frac{2\pi}{4\alpha^3} N^2_{n_1,n_2,m}(2n_1 + |m| + 1 + 2n_2 + |m| + 1) = 1,$$

and the result (4.237) follows since $n = n_1 + n_2 + |m| + 1$.

## J.5   Solutions to Chapter 5 Exercises

### ■ Solution 5.1

From the Jacobi identity

$$[[J_+, V_-], V_+] + [[V_-, V_+], J_+] + [[V_+, J_+], V_-] = 0,$$

but $[J_+, V_-] = 2V_3$ and $[V_+, J_+] = 0$ so $2[V_3, V_+] = -[[V_-, V_+], J_+]$, which is (5.10). Similarly

$$[[J_-, V_+], V_-] + [[V_+, V_-], J_-] + [[V_-, J_-], V_+] = 0,$$

but $[J_-, V_+] = -2V_3$ and $[V_-, J_-] = 0$ so $2[V_3, V_-] = [[V_+, V_-], J_-]$, which is the first part of (5.11). Again from the Jacobi identity

$$[[J_-, V_3], V_+] + [[V_3, V_+], J_-] + [[V_+, J_-], V_3] = 0,$$

but $[J_-, V_3] = V_-$ and $[V_+, J_-] = 0$ so $[V_+, V_-] = [[V_3, V_+], J_-]$, which is the second part of (5.11).

### ■ Solution 5.2

According to the Wigner-Eckart theorem the reduced matrix elements of $\boldsymbol{J}$ do not depend on the particular component so apply it to $J_3$. We know that $\langle \gamma' j' m | J_3 | \gamma j m \rangle = \delta_{\gamma' \gamma} \delta_{j' j} m$ but from the Wigner-Eckart theorem

$$\langle \gamma' j' m | J_3 | \gamma j m \rangle = (-1)^{j-m} \begin{pmatrix} j & 1 & j \\ -m & 0 & m \end{pmatrix} \langle \gamma' j' \| \boldsymbol{J} \| \gamma j \rangle$$

From tables of 3-j symbols [RO59]

$$\begin{pmatrix} j & 1 & j \\ -m & 0 & m \end{pmatrix} = (-1)^{j-m} \frac{m}{[j(j+1)(2j+1)]^{1/2}}.$$

Therefore $\langle \gamma' j' \| \boldsymbol{J} | \gamma j \rangle = \delta_{\gamma' \gamma} \delta_{j' j} [j(j+1)(2j+1)]^{1/2}$.

The formula (5.83) for reduced matrix elements of $\boldsymbol{M}$ is obtained in exactly the same way. The reduced matrix elements of $\boldsymbol{I}$ are obtained from

$$\langle j_2' m_2 | \boldsymbol{I} | j_2 m_2 \rangle = \delta_{j_2' j_2} = (-1)^{j_2 - m_2} \begin{pmatrix} j_2 & 0 & j_2 \\ -m_2 & 0 & m_2 \end{pmatrix} \langle j_2 \| \boldsymbol{I} \| j_2 \rangle.$$

Since

$$\begin{pmatrix} j & 0 & j \\ -m & 0 & m \end{pmatrix} = (-1)^{j-m}(2j+1)^{-1/2},$$

the result (5.84) follows.

### ■ Solution 5.3

To show that $L$ is a constant of motion calculate

$$\frac{dL}{dt} = \frac{d}{dt}(r \times p) = m\frac{d}{dt}\left(r \times \frac{dr}{dt}\right) = r \times \frac{dp}{dt}$$
$$= \frac{f(r)}{r} r \times r = 0.$$

To show that $U$ is a constant of motion first calculate

$$\frac{d}{dt}(p \times L) = \frac{dp}{dt} \times L = \frac{f(r)}{r} r \times (r \times p).$$

Next use the vector identity $A \times (B \times C) = (A \cdot C)B - (A \cdot B)C$ to obtain

$$r \times (r \times p) = (r \cdot p)r - r^2 p = m\left(r \cdot \frac{dr}{dt}\right)r - mr^2\frac{dr}{dt}.$$

Next note that

$$r \cdot \frac{dr}{dt} = \frac{1}{2}\frac{d}{dt}(r \cdot r) = \frac{1}{2}\frac{d}{dt}(r^2) = r\frac{dr}{dt},$$

so that

$$r \times (r \times p) = mrr\frac{dr}{dt} - mr^2\frac{dr}{dt} = -mr^3\frac{d}{dt}\left(\frac{r}{r}\right).$$

Therefore

$$\frac{d}{dt}(p \times L) = -mr^2 f(r)\frac{d}{dt}\left(\frac{r}{r}\right),$$

and substituting $f(r) = -k/r^2$

$$\frac{d}{dt}\left(p \times L - mk\frac{r}{r}\right) = 0,$$

so $U$ is also a constant of motion.

### ■ Solution 5.4

$$mU \cdot L = (p \times L) \cdot L - \frac{mk}{r} r \cdot (r \times p)$$

$$= (L \times L) \cdot p - \frac{mk}{r} p \cdot (r \times r) = 0,$$

$$m^2 U^2 = \left(p \times L - mk\frac{r}{r}\right) \cdot \left(p \times L - mk\frac{r}{r}\right)$$

$$= (p \times L) \cdot (p \times L) - \frac{2mk}{r}(p \times L) \cdot r + m^2 k^2$$

$$= (\boldsymbol{L} \times (\boldsymbol{p} \times \boldsymbol{L})) \cdot \boldsymbol{p} - \frac{2mk}{r}(\boldsymbol{r} \times \boldsymbol{p}) \cdot \boldsymbol{L} + m^2 k^2$$

$$= ((\boldsymbol{L} \cdot \boldsymbol{L})\boldsymbol{p} - (\boldsymbol{L} \cdot \boldsymbol{p})\boldsymbol{L}) \cdot \boldsymbol{p} - \frac{2mk}{r}L^2 + m^2 k^2$$

$$= L^2 p^2 - \frac{2mk}{r}L^2 + m^2 k^2$$

$$= 2m \left( \frac{1}{2m}p^2 - \frac{k}{r} \right) L^2 + m^2 k^2$$

$$= 2mEL^2 + m^2 k^2.$$

The orbit equation is obtained by calculating $\boldsymbol{U} \cdot \boldsymbol{r}$:

$$mU \cdot r = mUr \cos\theta = (\boldsymbol{p} \times \boldsymbol{L}) \cdot \boldsymbol{r} - mkr$$
$$= (\boldsymbol{r} \times \boldsymbol{p}) \cdot \boldsymbol{L} - mkr = L^2 - mkr.$$

Therefore the orbit equation is given by

$$\frac{1}{r} = \frac{mk}{L^2} \left( 1 + \frac{U}{k} \cos\theta \right),$$

so the eccentricity is $e = U/k$.


### ■ Solution 5.5

From the orbit equation derived in the preceding exercise $\theta = 0$ corresponds to $r = a(1 - e)$, where $a$ is the semi-major axis of the ellipse in the case of a closed orbit. Therefore $mkea(1 - e) = L^2 - mka(1 - e)$ so $L^2 = mka(1 - e^2)$. To obtain the energy substitute this result and $U = ke$ into (5.100) and solve for $E$.


### ■ Solution 5.6

$$\boldsymbol{L} \cdot \boldsymbol{U} = \frac{1}{2}\boldsymbol{L} \cdot (\boldsymbol{p} \times \boldsymbol{L} - \boldsymbol{L} \times \boldsymbol{p}) - \mathcal{Z}\boldsymbol{r} \cdot \frac{\boldsymbol{r}}{r}$$
$$= \frac{1}{2}\boldsymbol{L} \cdot (\boldsymbol{p} \times \boldsymbol{L} - \boldsymbol{L} \times \boldsymbol{p}),$$

since $\boldsymbol{L}$ commutes with functions of $r$ and $\boldsymbol{L} \cdot \boldsymbol{r} = 0$ (see (2.26) and (2.31)). Using (2.33)

$$\boldsymbol{L} \cdot (\boldsymbol{p} \times \boldsymbol{L}) = (\boldsymbol{L} \cdot \boldsymbol{r})p^2 - (\boldsymbol{L} \cdot \boldsymbol{p})(\boldsymbol{r} \cdot \boldsymbol{p} - i) = 0,$$

since $\boldsymbol{L} \cdot \boldsymbol{r} = 0$ and $\boldsymbol{L} \cdot \boldsymbol{p} = 0$ (see (2.32)).

Therefore, using (2.36), $\boldsymbol{L} \cdot (\boldsymbol{L} \times \boldsymbol{p}) = \boldsymbol{L} \cdot (2i\boldsymbol{p} - \boldsymbol{p} \times \boldsymbol{L}) = 0$ and $\boldsymbol{L} \cdot \boldsymbol{U} = 0$. Similarly

$$\boldsymbol{U} \cdot \boldsymbol{L} = \frac{1}{2}(\boldsymbol{p} \times \boldsymbol{L} - \boldsymbol{L} \times \boldsymbol{p}) \cdot \boldsymbol{L} - \mathcal{Z}\frac{1}{r}\boldsymbol{r} \cdot \boldsymbol{L}$$
$$= \frac{1}{2}(\boldsymbol{p} \times \boldsymbol{L} - \boldsymbol{L} \times \boldsymbol{p}) \cdot \boldsymbol{L},$$

since $\boldsymbol{r} \cdot \boldsymbol{L} = 0$ (see (2.31)). Using the second equality in (2.33)

$$(\boldsymbol{p} \times \boldsymbol{L}) \cdot \boldsymbol{L} = p^2(\boldsymbol{r} \cdot \boldsymbol{L}) - (\boldsymbol{p} \cdot \boldsymbol{r} - i)\boldsymbol{p} \cdot \boldsymbol{L} = 0,$$

since $\boldsymbol{r} \cdot \boldsymbol{L} = 0$ and $\boldsymbol{p} \cdot \boldsymbol{L} = 0$ (see (2.32)). Therefore, using (2.36) again

$$(\boldsymbol{L} \times \boldsymbol{p}) \cdot \boldsymbol{L} = (2i\boldsymbol{p} - \boldsymbol{p} \times \boldsymbol{L}) \cdot \boldsymbol{L} = 0,$$

and $\boldsymbol{U} \cdot \boldsymbol{L} = 0$.

### ■ Solution 5.7

Using (A.9)

$$\begin{aligned}
(\boldsymbol{p} \times \boldsymbol{L}) \cdot (\boldsymbol{p} \times \boldsymbol{L}) &= \epsilon_{jk\ell}\epsilon_{jmn}p_k L_\ell p_m L_n \\
&= (\delta_{km}\delta_{\ell n} - \delta_{kn}\delta_{m\ell})p_k L_\ell p_m L_n \\
&= p_k L_\ell p_k L_\ell - p_k L_\ell p_\ell L_k.
\end{aligned}$$

The last term is zero since from (2.32) $L_\ell p_\ell = \boldsymbol{L} \cdot \boldsymbol{p} = 0$. Therefore, using (2.24)

$$\begin{aligned}
(\boldsymbol{p} \times \boldsymbol{L}) \cdot (\boldsymbol{p} \times \boldsymbol{L}) &= p_k(p_k L_\ell + i\epsilon_{\ell kj}p_j)L_\ell \\
&= p^2 L^2 + i\epsilon_{\ell kj}p_k p_j L_\ell \\
&= p^2 L^2 + i(\boldsymbol{p} \times \boldsymbol{p}) \cdot \boldsymbol{L} \\
&= p^2 L^2.
\end{aligned}$$

To derive the identity of part (b) first use (2.33) to obtain

$$\begin{aligned}
(\boldsymbol{p} \times \boldsymbol{L}) \cdot \boldsymbol{p} &= (\boldsymbol{r}p^2 - \boldsymbol{p}(\boldsymbol{r} \cdot \boldsymbol{p}) + i\boldsymbol{p}) \cdot \boldsymbol{p} \\
&= \boldsymbol{r}p^2 \cdot \boldsymbol{p} - \boldsymbol{p} \cdot (\boldsymbol{r} \cdot \boldsymbol{p})\boldsymbol{p} + ip^2.
\end{aligned}$$

Now use (2.17) and (2.21) to obtain

$$\begin{aligned}
(\boldsymbol{p} \times \boldsymbol{L}) \cdot \boldsymbol{p} &= (p^2\boldsymbol{r} + 2i\boldsymbol{p}) \cdot \boldsymbol{p} - \boldsymbol{p} \cdot (\boldsymbol{p}(\boldsymbol{r} \cdot \boldsymbol{p}) + i\boldsymbol{p}) + ip^2 \\
&= p^2(\boldsymbol{r} \cdot \boldsymbol{p}) + 2ip^2 - p^2(\boldsymbol{r} \cdot \boldsymbol{p}) - ip^2 + ip^2 = 2ip^2.
\end{aligned}$$

An alternate derivation using the Levi-Civita symbol is

$$\begin{aligned}
(\boldsymbol{p} \times \boldsymbol{L}) \cdot \boldsymbol{p} &= \epsilon_{jk\ell}p_k L_\ell p_j \\
&= \epsilon_{jk\ell}p_k(p_j L_\ell + i\epsilon_{\ell jm}p_m) \\
&= (\boldsymbol{p} \times \boldsymbol{p}) \cdot \boldsymbol{L} + 2i\delta_{km}p_k p_m = 2ip^2,
\end{aligned}$$

where we have used (2.24) and (A.10).

The identity of part (c) is obtained in a similar manner using (2.33), (2.10) and (2.19):

$$\begin{aligned}
\boldsymbol{p} \cdot (\boldsymbol{p} \times \boldsymbol{L}) &= \boldsymbol{p} \cdot (\boldsymbol{r}p^2 - \boldsymbol{p}(\boldsymbol{r} \cdot \boldsymbol{p}) + i\boldsymbol{p}) \\
&= (\boldsymbol{p} \cdot \boldsymbol{r})p^2 - p^2(\boldsymbol{r} \cdot \boldsymbol{p}) + ip^2 \\
&= (\boldsymbol{r} \cdot \boldsymbol{p} - 3i)p^2 - p^2(\boldsymbol{r} \cdot \boldsymbol{p}) + ip^2 \\
&= [\boldsymbol{r} \cdot \boldsymbol{p}, p^2] - 2ip^2 = 0.
\end{aligned}$$

An alternate and simpler derivation using the Levi-Civita symbol is

$$\boldsymbol{p}\cdot(\boldsymbol{p}\times\boldsymbol{L}) = \epsilon_{jk\ell}p_j p_k L_\ell = (\boldsymbol{p}\times\boldsymbol{p})\cdot\boldsymbol{L} = 0.$$

The identity of part (d) can be obtained as follows using the basic commutation relations (2.9) and (2.33) and the identity (A.10):

$$\begin{aligned}
(\boldsymbol{p}\times\boldsymbol{L})\cdot\boldsymbol{r} &= \epsilon_{jk\ell}p_k L_\ell x_j \\
&= \epsilon_{jk\ell}p_k(x_j L_\ell + i\epsilon_{\ell jm}x_m) \\
&= \epsilon_{jk\ell}p_k x_j L_\ell + i\epsilon_{k\ell j}\epsilon_{m\ell j}p_k x_m \\
&= \epsilon_{jk\ell}(x_j p_k - i\delta_{jk})L_\ell + 2i\delta_{km}p_k x_m \\
&= L_\ell L_\ell + 2i p_k x_k = L^2 + 2i\boldsymbol{p}\cdot\boldsymbol{r}.
\end{aligned}$$

Finally, the identity of part (e) is easily derived since

$$\boldsymbol{r}\cdot(\boldsymbol{p}\times\boldsymbol{L}) = \epsilon_{jk\ell}x_j p_k L_\ell = L_\ell L_\ell = L^2.$$

### ■ Solution 5.8

$$U^2 = \left(\frac{1}{2}(\boldsymbol{p}\times\boldsymbol{L} - \boldsymbol{L}\times\boldsymbol{p}) - \mathcal{Z}\frac{\boldsymbol{r}}{r}\right)\cdot\left(\frac{1}{2}(\boldsymbol{p}\times\boldsymbol{L} - \boldsymbol{L}\times\boldsymbol{p}) - \mathcal{Z}\frac{\boldsymbol{r}}{r}\right).$$

Substitute $\boldsymbol{L}\times\boldsymbol{p} = 2i\boldsymbol{p} - \boldsymbol{p}\times\boldsymbol{L}$ from (2.36) to obtain

$$U^2 - \mathcal{Z}^2 = (\boldsymbol{p}\times\boldsymbol{L} - i\boldsymbol{p})^2 - \mathcal{Z}(\boldsymbol{p}\times\boldsymbol{L} - i\boldsymbol{p})\cdot\frac{\boldsymbol{r}}{r} - \mathcal{Z}\frac{\boldsymbol{r}}{r}\cdot(\boldsymbol{p}\times\boldsymbol{L} - i\boldsymbol{p}).$$

From the preceding exercise

$$\begin{aligned}
(\boldsymbol{p}\times\boldsymbol{L} - i\boldsymbol{p})^2 &= (\boldsymbol{p}\times\boldsymbol{L})^2 - i(\boldsymbol{p}\times\boldsymbol{L})\cdot\boldsymbol{p} - i\boldsymbol{p}\cdot(\boldsymbol{p}\times\boldsymbol{L}) - p^2 \\
&= p^2 L^2 - i(2ip^2) - p^2 = p^2(L^2 + 1), \\
(\boldsymbol{p}\times\boldsymbol{L} - i\boldsymbol{p})\cdot\frac{\boldsymbol{r}}{r} &= (\boldsymbol{p}\times\boldsymbol{L})\cdot\boldsymbol{r}\frac{1}{r} - i\boldsymbol{p}\cdot\boldsymbol{r}\frac{1}{r} \\
&= (L^2 + 2i\boldsymbol{p}\cdot\boldsymbol{r})\frac{1}{r} - i\boldsymbol{p}\cdot\boldsymbol{r}\frac{1}{r} \\
&= \frac{1}{r}L^2 + i\boldsymbol{p}\cdot\boldsymbol{r}\frac{1}{r}, \\
\frac{\boldsymbol{r}}{r}\cdot(\boldsymbol{p}\times\boldsymbol{L} - i\boldsymbol{p}) &= \frac{1}{r}\boldsymbol{r}\cdot(\boldsymbol{p}\times\boldsymbol{L}) - i\frac{1}{r}\boldsymbol{r}\cdot\boldsymbol{p} \\
&= \frac{1}{r}L^2 - i\frac{1}{r}\boldsymbol{r}\cdot\boldsymbol{p}.
\end{aligned}$$

Therefore

$$\begin{aligned}
U^2 - \mathcal{Z}^2 &= p^2(L^2 + 1) - \frac{2\mathcal{Z}}{r}L^2 - \mathcal{Z}i(\boldsymbol{p}\cdot\boldsymbol{r})\frac{1}{r} + \mathcal{Z}i\frac{1}{r}(\boldsymbol{r}\cdot\boldsymbol{p}) \\
&= p^2(L^2 + 1) - \frac{2\mathcal{Z}}{r}L^2 - \mathcal{Z}i(\boldsymbol{r}\cdot\boldsymbol{p} - 3i)\frac{1}{r} + \mathcal{Z}i\frac{1}{r}(\boldsymbol{r}\cdot\boldsymbol{p}) \\
&= p^2(L^2 + 1) - \frac{2\mathcal{Z}}{r}L^2 - \frac{3\mathcal{Z}}{r} - \mathcal{Z}i[\boldsymbol{r}\cdot\boldsymbol{p}, r^{-1}].
\end{aligned}$$

Finally use (2.22) to obtain

$$U^2 - \mathcal{Z}^2 = p^2(L^2 + 1) - \frac{2\mathcal{Z}}{r}L^2 - \frac{2\mathcal{Z}}{r}$$
$$= 2\left(\frac{1}{2}p^2 - \frac{\mathcal{Z}}{r}\right)(L^2 + 1) = 2H(L^2 + 1).$$

## ■ Solution 5.9

By direct calculation using (2.36), (2.24) and (2.23)

$$[L_\jmath, U_k] = \tfrac{1}{2}[L_\jmath, (\boldsymbol{p} \times \boldsymbol{L})_k - (\boldsymbol{L} \times \boldsymbol{p})_k] - \mathcal{Z}[L_\jmath, x_k r^{-1}]$$
$$= [L_\jmath, (\boldsymbol{p} \times \boldsymbol{L})_k] - i[L_\jmath, p_k] - \mathcal{Z}r^{-1}[L_\jmath, x_k]$$
$$= [L_\jmath, (\boldsymbol{p} \times \boldsymbol{L})_k] - i(i\epsilon_{\jmath km}p_m) - \mathcal{Z}r^{-1}(i\epsilon_{\jmath km}x_m).$$

Now using (2.25), (2.24) and (A.8)

$$[L_\jmath, (\boldsymbol{p} \times \boldsymbol{L})_k] = \epsilon_{k\ell m}[L_\jmath, p_\ell L_m]$$
$$= \epsilon_{k\ell m}(p_\ell[L_\jmath, L_m] + [L_\jmath, p_\ell]L_m)$$
$$= i\epsilon_{k\ell m}\epsilon_{\jmath mn}p_\ell L_n + i\epsilon_{k\ell m}\epsilon_{\jmath \ell n}p_n L_m$$
$$= i\epsilon_{k\ell m}\epsilon_{\jmath mn}p_\ell L_n + i\epsilon_{kmn}\epsilon_{\jmath m\ell}p_\ell L_n$$
$$= i(\epsilon_{k\ell m}\epsilon_{n\jmath m} + \epsilon_{nkm}\epsilon_{\ell \jmath m})p_\ell L_n$$
$$= -i\epsilon_{\ell nm}\epsilon_{k\jmath m}p_\ell L_n = i\epsilon_{\jmath km}(\boldsymbol{p} \times \boldsymbol{L})_m.$$

Therefore

$$[L_\jmath, U_k] = i\epsilon_{\jmath km}\left((\boldsymbol{p} \times \boldsymbol{L})_m - ip_m - \frac{\mathcal{Z}}{r}x_m\right)$$
$$= i\epsilon_{\jmath km}U_m.$$

An even simpler derivation is obtained by observing that $\boldsymbol{p} \times \boldsymbol{L}$, $\boldsymbol{p}$, and $\boldsymbol{r}$ are so(3) vector operators and $r^{-1}$ is a scalar operator so $U$ must also be an so(3) vector operator.

## ■ Solution 5.10

Using the expression (5.107) for $U$

$$[U_\jmath, U_k] = \tfrac{1}{4}[x_\jmath p^2, x_k p^2] + [p_\jmath(\boldsymbol{r} \cdot \boldsymbol{p}), p_k(\boldsymbol{r} \cdot \boldsymbol{p})] + [x_\jmath H, x_k H]$$
$$- \tfrac{1}{2}(A_{\jmath k} - A_{k\jmath}) + \tfrac{1}{2}(B_{\jmath k} - B_{k\jmath}) - (C_{\jmath k} - C_{k\jmath}),$$

where we have defined

$$A_{\jmath k} = [x_\jmath p^2, p_k(\boldsymbol{r} \cdot \boldsymbol{p})],$$
$$B_{\jmath k} = [x_\jmath p^2, x_k H],$$
$$C_{\jmath k} = [p_\jmath(\boldsymbol{r} \cdot \boldsymbol{p}), x_k H].$$

Now evaluate each commutator. Using (2.17)

$$[x_jp^2, x_kp^2] = x_j[p^2, x_kp^2] + [x_j, x_kp^2]p^2$$
$$= x_j[p^2, x_k]p^2 + x_k[x_j, p^2]p^2$$
$$= -2i(x_jp_k - x_kp_j)p^2.$$

Using (2.21)

$$[p_j(\boldsymbol{r}\cdot\boldsymbol{p}), p_k(\boldsymbol{r}\cdot\boldsymbol{p})] = p_j[\boldsymbol{r}\cdot\boldsymbol{p}, p_k](\boldsymbol{r}\cdot\boldsymbol{p}) + p_k[p_j, \boldsymbol{r}\cdot\boldsymbol{p}](\boldsymbol{r}\cdot\boldsymbol{p})$$
$$= p_j(ip_k)(\boldsymbol{r}\cdot\boldsymbol{p}) + p_k(-ip_j)(\boldsymbol{r}\cdot\boldsymbol{p}) = 0.$$

Using (2.17)

$$[x_jH, x_kH] = x_j[H, x_k]H + x_k[x_j, H]H$$
$$= \tfrac{1}{2}x_j[p^2, x_k]H + \tfrac{1}{2}x_k[x_j, p^2]H$$
$$= -i(x_jp_k - x_kp_j)H.$$

Using (2.9), (2.19) and (2.20)

$$A_{jk} = x_j[p^2, p_k(\boldsymbol{r}\cdot\boldsymbol{p})] + [x_j, p_k(\boldsymbol{r}\cdot\boldsymbol{p})]p^2$$
$$= x_jp_k[p^2, \boldsymbol{r}\cdot\boldsymbol{p}] + p_k[x_j, \boldsymbol{r}\cdot\boldsymbol{p}]p^2 + [x_j, p_k](\boldsymbol{r}\cdot\boldsymbol{p})p^2$$
$$= -2ix_jp_kp^2 + ip_kx_jp^2 + i\delta_{jk}(\boldsymbol{r}\cdot\boldsymbol{p})p^2$$
$$= -2ix_jp_kp^2 + i(x_jp_k - i\delta_{jk})p^2 + i\delta_{jk}(\boldsymbol{r}\cdot\boldsymbol{p})p^2$$
$$= -ix_jp_kp^2 + \delta_{jk}(1 + i\boldsymbol{r}\cdot\boldsymbol{p})p^2.$$

Therefore

$$A_{jk} - A_{kj} = -i(x_jp_k - x_kp_j)p^2.$$

Using (2.17)

$$B_{jk} = x_j[p^2, x_kH] + [x_j, x_kH]p^2$$
$$= x_jx_k[p^2, H] + x_j[p^2, x_k]H + x_k[x_j, H]p^2$$
$$= x_jx_k[p^2, H] - 2ix_jp_kH + \tfrac{1}{2}x_k[x_j, p^2]p^2$$
$$= x_jx_k[p^2, H] - 2ix_jp_kH + ix_kp_jp^2.$$

Therefore

$$B_{jk} - B_{kj} = i(x_jp_k - x_kp_j)(-2H - p^2).$$

Finally, using (2.9) and (2.20)

$$C_{jk} = p_j[\boldsymbol{r}\cdot\boldsymbol{p}, x_kH] + [p_j, x_kH](\boldsymbol{r}\cdot\boldsymbol{p})$$
$$= p_jx_k[\boldsymbol{r}\cdot\boldsymbol{p}, H] + p_j[\boldsymbol{r}\cdot\boldsymbol{p}, x_k] + x_k[p_j, H](\boldsymbol{r}\cdot\boldsymbol{p})$$
$$+ [p_j, x_k]H(\boldsymbol{r}\cdot\boldsymbol{p})$$
$$= p_jx_k[\boldsymbol{r}\cdot\boldsymbol{p}, H] + p_j(-ix_k)H + x_k[p_j, H](\boldsymbol{r}\cdot\boldsymbol{p}) - i\delta_{jk}H(\boldsymbol{r}\cdot\boldsymbol{p}).$$

Using (2.19), (2.22) and (2.12)

$$[\mathbf{r}\cdot\mathbf{p}, H] = \tfrac{1}{2}[\mathbf{r}\cdot\mathbf{p}, p^2] - \mathcal{Z}[\mathbf{r}\cdot\mathbf{p}, r^{-1}]$$
$$= \tfrac{1}{2}(2ip^2) + \mathcal{Z}ir^{-1} = \tfrac{1}{2}ip^2 + iH$$
$$[p_j, H] = -\mathcal{Z}[p_j, r^{-1}] = -\mathcal{Z}(ix_j r^{-3}).$$

Therefore

$$C_{jk} = \tfrac{1}{2}ip_j x_k p^2 - i\mathcal{Z}x_j x_k r^{-3}(\mathbf{r}\cdot\mathbf{p}) - i\delta_{jk}H(\mathbf{r}\cdot\mathbf{p})$$
$$= \tfrac{1}{2}i(x_k p_j - i\delta_{jk})p^2 - i\mathcal{Z}x_j x_k r^{-3}(\mathbf{r}\cdot\mathbf{p}) - i\delta_{jk}H(\mathbf{r}\cdot\mathbf{p}),$$

and

$$C_{jk} - C_{kj} = -\tfrac{1}{2}i(x_j p_k - x_k p_j)p^2.$$

Collect these results to obtain

$$[U_j, U_k] = -2i(x_j p_k - x_k p_j)H.$$

and, since $x_j p_k - x_k p_j = \epsilon_{jk\ell}L_\ell$ and $[L_k, H] = 0$,

$$[U_j, U_k] = (-2H)i\epsilon_{jk\ell}L_\ell.$$

### ■ Solution 5.11

This identity was proved more generally in Section 4.5 (let $a = 1$ and $\xi = L^2$ in (4.57)). An alternate derivation can be obtained as follows, using (4.26) and (4.27). Since

$$P^2 = P_R^2 + \frac{L^2}{R^2}, \quad [L^2, f(R)] = 0$$

then

$$L^2 = R^2 P^2 - R^2 P_R^2 = R^2 P^2 - R(P_R R + i)P_R$$
$$= R^2 P^2 - RP_R RP_R - iRP_R.$$

From (5.134) to (5.136), $R = T_3 - T_1$, $RP_R = T_2$ and $RP^2 = T_3 + T_1$ so

$$L^2 = (T_3 - T_1)(T_3 + T_1) - T_2^2 - iT_2$$
$$= T_3^2 - T_1^2 - T_2^2 + [T_3, T_1] - iT_2^z$$
$$= T_3^2 - T_1^2 - T_2^2, \quad \text{from (4.1).}$$

### ■ Solution 5.12

$$C_\alpha \cdot C_\beta$$
$$= \tfrac{1}{4}(RP^2)\cdot(RP^2) - \tfrac{1}{2}RP^2\cdot P(R\cdot P) - \tfrac{1}{2}P(R\cdot P)\cdot RP^2$$
$$+ P(R\cdot P)\cdot P(R\cdot P) - \tfrac{1}{4}\beta RP^2\cdot R + \tfrac{1}{2}\beta P(R\cdot P)\cdot R$$
$$- \tfrac{1}{4}\alpha R^2 P^2 + \tfrac{1}{2}\alpha(R\cdot P)^2 + \tfrac{1}{4}\alpha\beta R^2.$$

Now evaluate each term using (2.10) and (2.17) to (2.22):

$$(RP^2)\cdot(RP^2) = R\cdot P^2 RP^2 = R\cdot(RP^2 - 2iP)P^2$$
$$= R^2 P^4 - 2i(R\cdot P)P^2,$$

$$(RP^2)\cdot P(R\cdot P) = R\cdot P((R\cdot P)P^2 - 2iP^2)$$
$$= (R\cdot P)^2 - 2i(R\cdot P)P^2,$$

$$P(R\cdot P)\cdot RP^2 = ((R\cdot P)P - iP)\cdot RP^2$$
$$= (R\cdot P)(R\cdot P - 3i)P^2 - i(R\cdot P - 3i)P^2$$
$$= (R\cdot P)^2 P^2 - 4i(R\cdot P)P^2 - 3P^2,$$

$$P(R\cdot P)\cdot P(R\cdot P) = ((R\cdot P)P - iP)\cdot P(R\cdot P)$$
$$= (R\cdot P)P\cdot P(R\cdot P) - iP^2(R\cdot P)$$
$$= (R\cdot P)P\cdot((R\cdot P)P - iP) - i((R\cdot P)P^2 - 2iP^2)$$
$$= (R\cdot P)((R\cdot P)P - iP)\cdot P - 2i(R\cdot P)P^2 - 2P^2$$
$$= (R\cdot P)^2 P^2 - 3i(R\cdot P)P^2 - 2P^2,$$

$$RP^2\cdot R = R\cdot P^2 R = R\cdot(RP^2 - 2iP)$$
$$= R^2 P^2 - 2iR\cdot P,$$

$$P(R\cdot P)\cdot R = P\cdot(R\cdot P)R = P\cdot(R(R\cdot P) - iR)$$
$$= (R\cdot P - 3i)(R\cdot P) - i(R\cdot P - 3i)$$
$$= (R\cdot P)^2 - 4i(R\cdot P) - 3.$$

Therefore

$$C_\alpha \cdot C_\beta = \tfrac{1}{4}R^2 P^4 - \tfrac{1}{4}(\alpha + \beta)R^2 P^2 + \tfrac{1}{4}\alpha\beta R^2 - \tfrac{3}{2}\beta$$
$$= -\tfrac{1}{2}i(R\cdot P)P^2 + \tfrac{1}{2}(\alpha + \beta)(R\cdot P)^2 - \tfrac{3}{2}\beta i(R\cdot P) - \tfrac{1}{2}P^2.$$

Now note that

$$R^2 P^4 = R(RP^2)P^2 = R(P^2 R + 2iR^{-1}(R\cdot P - i))P^2$$
$$= (RP^2)^2 + 2i(R\cdot P - i)P^2,$$

and

$$\beta R^2 P^2 = \beta R(P^2 R + 2iR^{-1}(R\cdot P - i))$$
$$= \beta(RP^2)R + 2i\beta(R\cdot P - i).$$

Therefore

$$C_\alpha \cdot C_\beta = \tfrac{1}{4}(RP^2)^2 - \tfrac{1}{4}\beta(RP^2)R - \tfrac{1}{4}\alpha R(RP^2) + \tfrac{1}{4}\alpha\beta R^2$$
$$+ \tfrac{1}{2}(\alpha + \beta)(\boldsymbol{R}\cdot\boldsymbol{P})^2 - 2\beta i(\boldsymbol{R}\cdot\boldsymbol{P}) - 2\beta$$
$$= \tfrac{1}{4}(RP^2 - \alpha R)(RP^2 - \beta R) + \tfrac{1}{2}(\alpha + \beta)(\boldsymbol{R}\cdot\boldsymbol{P})^2$$
$$- 2\beta i(\boldsymbol{R}\cdot\boldsymbol{P}) - 2\beta.$$

Finally, substituting $\boldsymbol{R}\cdot\boldsymbol{P} = T_2 + i$ gives the required identity. Substituting $\alpha = \beta = 1$ and using the realization (5.114) of $T_1$ we obtain $A^2 = T_1^2 + T_2^2 - 1$, and from the preceding exercise $T_1^2 + T_2^2 = T_3^2 - L^2$. Therefore $A^2 + L^2 + 1 = T_3^2$.

■ **Solution 5.13**

$$[L_j, T_3] = \epsilon_{jk\ell}[X_k P_\ell, T_3]$$
$$= \epsilon_{jk\ell}(X_k[P_\ell, T_3] + [X_k, T_3]P_\ell),$$

$$[P_\ell, T_3] = \tfrac{1}{2}[P_\ell, RP^2 + R] = \tfrac{1}{2}[P_\ell, R] + \tfrac{1}{2}[P_\ell, R]P^2$$
$$= -\tfrac{1}{2}i X_\ell R^{-1}(1 + P^2), \quad \text{from (2.11)},$$

$$[X_k, T_3] = \tfrac{1}{2}[X_k, RP^2 + R] = \tfrac{1}{2}R[X_k, P^2]$$
$$= \tfrac{1}{2}R(2iP_k), \quad \text{from (2.17)}.$$

Therefore

$$[L_j, T_3] = -\tfrac{1}{2}i(\epsilon_{jk\ell}X_k X_\ell)R^{-1}(1 + P^2) + iR(\epsilon_{jk\ell}P_k P_\ell).$$

Both terms on the right are zero, since $X_k X_\ell$ and $P_k P_\ell$ are symmetric in $k$ and $\ell$ and $\epsilon_{jk\ell}$ is antisymmetric in $k$ and $\ell$ (see Exercise 2.1).

To derive $[A_j, T_3] = 0$ use commutation relations (2.11) and (2.17) to (2.22):

$$[A, T_3] = \tfrac{1}{4}[\boldsymbol{R}P^2, RP^2] + \tfrac{1}{4}[\boldsymbol{R}P^2, R] - \tfrac{1}{2}[P(\boldsymbol{R}\cdot\boldsymbol{P}), RP^2]$$
$$- \tfrac{1}{2}[P(\boldsymbol{R}\cdot\boldsymbol{P}), R] - \tfrac{1}{4}[\boldsymbol{R}, RP^2].$$

Evaluate each commutator and put the terms in a canonical order with position operators to the left of all momentum operators:

$$[\boldsymbol{R}P^2, RP^2] = \boldsymbol{R}[P^2, R]P^2 + R[\boldsymbol{R}, P^2]P^2$$
$$= \boldsymbol{R}(-2iR^{-1}\boldsymbol{R}\cdot\boldsymbol{P} - 2R^{-1})P^2 + 2iR\boldsymbol{P}P^2, \quad \text{from (2.17),(2.18)}$$
$$= -2iR^{-1}\boldsymbol{R}(\boldsymbol{R}\cdot\boldsymbol{P})P^2 - 2R^{-1}\boldsymbol{R}P^2 + 2iR\boldsymbol{P}P^2,$$

$$[\boldsymbol{R}P^2, R] = \boldsymbol{R}[P^2, R]$$
$$= \boldsymbol{R}(-2iR^{-1}\boldsymbol{R}\cdot\boldsymbol{P} - 2R^{-1}), \quad \text{from (2.18)}$$
$$= -2iR^{-1}\boldsymbol{R}(\boldsymbol{R}\cdot\boldsymbol{P}) - 2R^{-1}\boldsymbol{R},$$

$$[P(R \cdot P), RP^2]$$
$$= PR[R \cdot P, P^2] + P[R \cdot P, R]P^2 + [P, R]P^2 R \cdot P$$
$$= PR(2iP^2) - iPRP^2$$
$$\quad - iR^{-1}RP^2(R \cdot P), \quad \text{from } (2.19),(2.20),(2.11)$$
$$= i(RP - iR^{-1}R)P^2 - iR^{-1}RP^2(R \cdot P), \quad \text{from } (2.11)$$
$$= i(RP - iR^{-1}R)P^2 - iR^{-1}R((R \cdot P)P^2 - 2iP^2), \quad \text{from } (2.19)$$
$$= iRPP^2 - R^{-1}RP^2 - iR^{-1}R(R \cdot P)P^2,$$

$$[P(R \cdot P), R] = P[R \cdot P, R] + [P, R]R \cdot P$$
$$= P(-iR) + (-iR^{-1}R)(R \cdot P), \quad \text{from } (2.22),(2.11)$$
$$= -i(RP - iR^{-1}R) - iR^{-1}R(R \cdot P), \quad \text{from } (2.11),$$

$$[R, RP^2] = R[R, P^2] = 2iRP, \quad \text{from } (2.17)$$

Collect these results together to get $[A, T_3] = 0$.


## J.6  Solutions to Chapter 6 Exercises

### ■ Solution 6.1

From (2.23) and (2.24)

$$[T_2, L_j] = [R \cdot P - i, L_j] = [X_k P_k, L_j]$$
$$= X_k[P_k, L_j] + [X_k, L_j]P_k$$
$$= X_k(-i\epsilon_{jk\ell}P_\ell) + (-i\epsilon_{jk\ell}X_\ell)P_k$$
$$= -i\epsilon_{jk\ell}(X_k P_\ell + X_\ell P_k) = 0, \text{ from Exercise 2.1.}$$

That $[T_3, L_j] = 0$ was shown in Exercise 5.13.  A similar derivation using $RP^2 - R$ in place of $RP^2 + R$ shows that $[T_1, L_j] = 0$.


### ■ Solution 6.2

Using the definition of $A$ given in (5.128) and the commutation relations (2.19) to (2.21)

$$[T_2, A] = \tfrac{1}{2}[R \cdot P, RP^2] - [R \cdot P, P(R \cdot P)] - \tfrac{1}{2}[R \cdot P, R],$$

where

$$[R \cdot P, RP^2] = R[R \cdot P, P^2] + [R \cdot P, R]P^2$$
$$= R(2iP^2) + (-iR)P^2 = iRP^2$$
$$[R \cdot P, P(R \cdot P)] = [R \cdot P, P](R \cdot P) = iP(R \cdot P)$$
$$[R \cdot P, R] = -iR.$$

Therefore

$$[T_2, A] = i(\tfrac{1}{2}RP^2 - P(R \cdot P) + \tfrac{1}{2}R) = iB.$$

### ■ Solution 6.3

First evaluate $[L_j, B_k]$ by expressing $\boldsymbol{B} = \boldsymbol{A} + \boldsymbol{R}$ from (6.12) and using the commutation relations (6.4) and (2.23) to obtain

$$
\begin{aligned}
[L_j, B_k] &= [L_j, X_k + A_k] = [L_j, X_k] + [L_j, A_k] \\
&= i\epsilon_{jk\ell}(A_\ell + X_\ell) = i\epsilon_{jk\ell}B_\ell.
\end{aligned}
$$

Next evaluate $[B_j, B_k]$ in a similar manner to obtain

$$
\begin{aligned}
[B_j, B_k] &= [X_j + A_j, X_k + A_k] \\
&= [X_j, A_k] + [A_j, X_k] + [A_j, A_k] \\
&= [X_j, A_k] - [X_k, A_j] + i\epsilon_{jk\ell}L_\ell.
\end{aligned}
$$

Using (2.17), (2.20) and (2.9)

$$
\begin{aligned}
[X_j, A_k] &= \tfrac{1}{2}[X_j, X_k P^2] - [X_j, P_k(\boldsymbol{R}\cdot\boldsymbol{P})] \\
&= \tfrac{1}{2}X_k[X_j, P^2] - P_k[X_j, \boldsymbol{R}\cdot\boldsymbol{P}] - [X_j, P_k](\boldsymbol{R}\cdot\boldsymbol{P}) \\
&= \tfrac{1}{2}X_k(2iP_j) - P_k(iX_j) - i\delta_{jk}\boldsymbol{R}\cdot\boldsymbol{P} \\
&= iX_k P_j - i(X_j P_k - i\delta_{jk}) - i\delta_{jk}\boldsymbol{R}\cdot\boldsymbol{P} \\
&= -i(X_j P_k - X_k P_j) - i\delta_{jk}(\boldsymbol{R}\cdot\boldsymbol{P} - i) \\
&= -i\epsilon_{jk\ell}L_\ell - i\delta_{jk}T_2.
\end{aligned}
$$

Therefore

$$
\begin{aligned}
[B_j, B_k] &= -i\epsilon_{jk\ell}L_\ell + i\epsilon_{kj\ell}L_\ell + i\epsilon_{jk\ell}L_\ell \\
&= -i\epsilon_{jk\ell}L_\ell.
\end{aligned}
$$

Next evaluate $[A_j, B_k]$ using (6.12) and (6.5):

$$
\begin{aligned}
[A_j, B_k] &= [A_j, X_k + A_k] = [A_j, X_k] + [A_j, A_k] \\
&= -i\epsilon_{jk\ell}L_\ell + i\delta_{jk}T_2 + i\epsilon_{jk\ell}L_\ell \\
&= i\delta_{jk}T_2.
\end{aligned}
$$

Finally evaluate $[T_2, B_j]$ using (6.12), (6.10) and (2.20):

$$
\begin{aligned}
[T_2, B_j] &= [T_2, X_j + A_j] = [T_2, X_j] + [T_2, A_j] \\
&= [\boldsymbol{R}\cdot\boldsymbol{P}, X_j] + iB_j = -iX_j + iB_j \\
&= iA_j.
\end{aligned}
$$

### ■ Solution 6.4

$$
\begin{aligned}
\boldsymbol{C}_\alpha &\times \boldsymbol{C}_\beta \\
&= (\tfrac{1}{2}\boldsymbol{R}P^2 - \boldsymbol{P}(\boldsymbol{R}\cdot\boldsymbol{P}) - \tfrac{1}{2}\alpha\boldsymbol{R}) \times (\tfrac{1}{2}\boldsymbol{R}P^2 - \boldsymbol{P}(\boldsymbol{R}\cdot\boldsymbol{P}) - \tfrac{1}{2}\beta\boldsymbol{R}) \\
&= \tfrac{1}{4}\boldsymbol{R}P^2 \times \boldsymbol{R}P^2 - \tfrac{1}{2}\boldsymbol{R}P^2 \times \boldsymbol{P}(\boldsymbol{R}\cdot\boldsymbol{P}) - \tfrac{1}{4}\beta\boldsymbol{R}P^2 \times \boldsymbol{R} \\
&\quad - \tfrac{1}{2}\boldsymbol{P}(\boldsymbol{R}\cdot\boldsymbol{P}) \times \boldsymbol{R}P^2 + \boldsymbol{P}(\boldsymbol{R}\cdot\boldsymbol{P}) \times \boldsymbol{P}(\boldsymbol{R}\cdot\boldsymbol{P}) \\
&\quad + \tfrac{1}{2}\beta\boldsymbol{P}(\boldsymbol{R}\cdot\boldsymbol{P}) \times \boldsymbol{R} - \tfrac{1}{4}\alpha\boldsymbol{R} \times \boldsymbol{R}P^2 \\
&\quad + \tfrac{1}{2}\alpha\boldsymbol{R} \times \boldsymbol{P}(\boldsymbol{R}\cdot\boldsymbol{P}) + \tfrac{1}{4}\alpha\beta\boldsymbol{R} \times \boldsymbol{R}.
\end{aligned}
$$

Now use commutation relations (2.17) to (2.21) to obtain

$$\boldsymbol{R}P^2 \times \boldsymbol{R}P^2$$
$$= \boldsymbol{R} \times P^2 \boldsymbol{R}P^2 = \boldsymbol{R} \times (\boldsymbol{R}P^2 - 2i\boldsymbol{P})P^2$$
$$= -2i(\boldsymbol{R} \times \boldsymbol{P})P^2 = -2i\boldsymbol{L}P^2,\ \text{since } \boldsymbol{R} \times \boldsymbol{R} = 0,$$

$$\boldsymbol{R}P^2 \times \boldsymbol{P}(\boldsymbol{R}\cdot\boldsymbol{P})$$
$$= (\boldsymbol{R} \times \boldsymbol{P})P^2(\boldsymbol{R}\cdot\boldsymbol{P})$$
$$= (\boldsymbol{R} \times \boldsymbol{P})((\boldsymbol{R}\cdot\boldsymbol{P})P^2 - 2i P^2)$$
$$= \boldsymbol{L}(\boldsymbol{R}\cdot\boldsymbol{P})P^2 - 2i\boldsymbol{L}P^2,$$

$$\boldsymbol{R}P^2 \times \boldsymbol{R}$$
$$= \boldsymbol{R} \times P^2 \boldsymbol{R}$$
$$= \boldsymbol{R} \times (\boldsymbol{R}P^2 - 2i\boldsymbol{P})$$
$$= (\boldsymbol{R} \times \boldsymbol{R})P^2 - 2i\boldsymbol{R} \times \boldsymbol{P} = -2i\boldsymbol{L},$$

$$\boldsymbol{P}(\boldsymbol{R}\cdot\boldsymbol{P}) \times \boldsymbol{R}P^2$$
$$= \boldsymbol{P} \times (\boldsymbol{R}\cdot\boldsymbol{P})\boldsymbol{R}P^2$$
$$= \boldsymbol{P} \times (\boldsymbol{R}(\boldsymbol{R}\cdot\boldsymbol{P}) - i\boldsymbol{R})P^2$$
$$= (\boldsymbol{P} \times \boldsymbol{R})(\boldsymbol{R}\cdot\boldsymbol{P})P^2 - i(\boldsymbol{P} \times \boldsymbol{R})P^2$$
$$= -\boldsymbol{L}(\boldsymbol{R}\cdot\boldsymbol{P})P^2 + i\boldsymbol{L}P^2,\ \text{since } \boldsymbol{P} \times \boldsymbol{R} = -\boldsymbol{R} \times \boldsymbol{P},$$

$$\boldsymbol{P}(\boldsymbol{R}\cdot\boldsymbol{P}) \times \boldsymbol{P}(\boldsymbol{R}\cdot\boldsymbol{P})$$
$$= \boldsymbol{P} \times (\boldsymbol{R}\cdot\boldsymbol{P})\boldsymbol{P}(\boldsymbol{R}\cdot\boldsymbol{P})$$
$$= \boldsymbol{P} \times (\boldsymbol{P}(\boldsymbol{R}\cdot\boldsymbol{P}) + i\boldsymbol{P})(\boldsymbol{R}\cdot\boldsymbol{P})$$
$$= 0,\ \text{since } \boldsymbol{P} \times \boldsymbol{P} = 0,$$

$$\boldsymbol{P}(\boldsymbol{R}\cdot\boldsymbol{P}) \times \boldsymbol{R}$$
$$= \boldsymbol{P} \times (\boldsymbol{R}\cdot\boldsymbol{P})\boldsymbol{R}$$
$$= \boldsymbol{P} \times (\boldsymbol{R}(\boldsymbol{R}\cdot\boldsymbol{P}) - i\boldsymbol{R})$$
$$= (\boldsymbol{P} \times \boldsymbol{R})(\boldsymbol{R}\cdot\boldsymbol{P}) - i\boldsymbol{P} \times \boldsymbol{R}$$
$$= -\boldsymbol{L}(\boldsymbol{R}\cdot\boldsymbol{P}) + i\boldsymbol{L}.$$

Collect these results to obtain the required identity.  Now the four identities follow by substituting $\alpha = \pm 1$ and $\beta = \pm 1$.  Note that $\boldsymbol{A} \times \boldsymbol{B} = \boldsymbol{L}T_2$ gives $\boldsymbol{D} = 0$ in (6.14) and (6.15).  Also since $\boldsymbol{L}\cdot\boldsymbol{A} = 0$, $\boldsymbol{L}\cdot\boldsymbol{R} = 0$, from (2.31), and $\boldsymbol{R} = \boldsymbol{B} - \boldsymbol{A}$, then $\boldsymbol{L}\cdot\boldsymbol{B} = \boldsymbol{L}\cdot(\boldsymbol{R} + \boldsymbol{A}) = 0$, so $W = 0$.

### ■ Solution 6.5

Letting $k = j$ in (6.4) and (6.6) and summing over $j$ directly gives (a) and (b). Identities (c) and (d) are easily obtained as follows

$$[B_j, \boldsymbol{L}\cdot\boldsymbol{A}] = [B_j, L_k L_k] = L_k[B_j, L_k] + [B_j, L_k]L_k$$

$$\begin{aligned}
&= L_k(-i\delta_{jk}T_2) + i\epsilon_{jk\ell}B_\ell A_k \\
&= -iT_2 L_j + i\epsilon_{jk\ell}(A_k B_\ell - i\delta_{k\ell}T_2) \\
&= -iT_2 L_j + i\epsilon_{jk\ell}A_k B_\ell = -iD_j.
\end{aligned}$$

Similarly $[A_j, \boldsymbol{L}\cdot\boldsymbol{B}] = iD_j$.

To derive (e) calculate

$$\begin{aligned}
[A_j, D_k] &= [A_j, T_2 L_k - \epsilon_{kmn}A_m B_n] \\
&= T_2[A_j, L_k] + [A_j, T_2]L_k \\
&\quad - \epsilon_{kmn}(A_m[A_j, B_n] - [A_j, A_m]B_n) \\
&= i\epsilon_{jk\ell}T_2 A_\ell - iB_j L_k - \epsilon_{kmn}A_m(i\delta_{jn}T_2) - \epsilon_{nkm}\epsilon_{\ell jm}L_\ell B_n \\
&= i\epsilon_{jk\ell}T_2 A_\ell - iB_j L_k - \epsilon_{kmj}A_m T_2 - i(\delta_{n\ell}\delta_{kj} - \delta_{nj}\delta_{\ell k})L_\ell B_n \\
&= i\epsilon_{jk\ell}[T_2, A_\ell] - iB_j L_k - i\delta_{jk}\boldsymbol{L}\cdot\boldsymbol{B} + iL_k B_j \\
&= -\epsilon_{jk\ell}B_\ell + i[L_k, B_j] - i\delta_{jk}\boldsymbol{L}\cdot\boldsymbol{B} \\
&= -\epsilon_{jk\ell}B_\ell - \epsilon_{kj\ell}B_\ell - i\delta_{jk}\boldsymbol{L}\cdot\boldsymbol{B} = -i\delta_{jk}\boldsymbol{L}\cdot\boldsymbol{B}.
\end{aligned}$$

As a special case let $k = j$ and sum over $j$ to obtain (6.20). Similarly (f) and (6.21) can be derived. Finally (g) and (h) are easily derived:

$$\begin{aligned}
[T_2, \boldsymbol{L}\cdot\boldsymbol{A}] &= L_j[T_2, A_j] + [T_2, L_j]A_j = L_j(iB_j) \\
&= i\boldsymbol{L}\cdot\boldsymbol{B}, \\
[T_2, \boldsymbol{L}\cdot\boldsymbol{B}] &= L_j[T_2, B_j] + [T_2, L_j]B_j = L_j(iA_j) \\
&= i\boldsymbol{L}\cdot\boldsymbol{A}.
\end{aligned}$$

### ■ Solution 6.6

From Exercise 5.11 and Exercise 5.12 with $\alpha = \beta = 1$ and $\alpha = \beta = -1$ it follows that

$$\begin{aligned}
L^2 &= T_3^2 - T_1^2 - T_2^2, \\
A^2 &= T_1^2 + T_2^2 - 1, \\
B^2 &= T_3^2 - T_2^2 + 1.
\end{aligned}$$

Substitution into (6.13) gives $Q = 2$.

### ■ Solution 6.7

Define

$$D_\alpha = \tfrac{1}{2}(RP^2 + \alpha R),$$

and the vector operators

$$C_\beta = \tfrac{1}{2}\boldsymbol{R}P^2 - \boldsymbol{P}(\boldsymbol{R}\cdot\boldsymbol{P}) - \tfrac{1}{2}\beta\boldsymbol{R},$$

so that $T_1 = D_{-1}$, $T_3 = D_1$, $\boldsymbol{A} = \boldsymbol{C}_1$ and $\boldsymbol{B} = \boldsymbol{C}_{-1}$. Then

$$[D_\alpha, C_\beta] = \tfrac{1}{4}[RP^2, RP^2] - \tfrac{1}{2}[RP^2, \boldsymbol{P}(\boldsymbol{R} \cdot \boldsymbol{P})] - \tfrac{1}{4}\beta[RP^2, \boldsymbol{R}]$$
$$+ \tfrac{1}{4}\alpha[R, RP^2] - \tfrac{1}{2}\alpha[R, \boldsymbol{P}(\boldsymbol{R} \cdot \boldsymbol{P})].$$

These commutators have already been evaluated in Exercise 5.13:

$$[RP^2, RP^2] = -2iR\boldsymbol{P}P^2 + 2iR^{-1}\boldsymbol{R}(\boldsymbol{R} \cdot \boldsymbol{P} - i)P^2,$$
$$[RP^2, \boldsymbol{P}(\boldsymbol{R} \cdot \boldsymbol{P})] = -iR\boldsymbol{P}P^2 + R^{-1}\boldsymbol{R}P^2 + iR^{-1}\boldsymbol{R}(\boldsymbol{R} \cdot \boldsymbol{P})P^2,$$
$$[RP^2, \boldsymbol{R}] = -2iR\boldsymbol{P},$$
$$[R, RP^2] = 2iR^{-1}\boldsymbol{R}(\boldsymbol{R} \cdot \boldsymbol{P}) + 2R^{-1}\boldsymbol{R},$$
$$[R, \boldsymbol{P}(\boldsymbol{R} \cdot \boldsymbol{P})] = i(R\boldsymbol{P} - iR^{-1}\boldsymbol{R}) + iR^{-1}\boldsymbol{R}(\boldsymbol{R} \cdot \boldsymbol{P}).$$

Therefore

$$[D_\alpha, C_\beta] = \tfrac{1}{2}(\beta - \alpha)iR\boldsymbol{P} = \tfrac{1}{2}(\beta - \alpha)i\boldsymbol{\Gamma},$$

and the four cases $\alpha = -1$, $\beta = 1$, $\alpha = -1$, $\beta = -1$, $\alpha = 1$, $\beta = 1$, $\alpha = 1$, $\beta = -1$ give the commutators (6.26) to (6.29).


### ■ Solution 6.8

From (2.9), (2.11) and Exercise 2.1

$$[L_j, \Gamma_k] = \epsilon_{k\ell m}[RP_j, X_\ell P_m]$$
$$= \epsilon_{k\ell m}R[P_j, X_\ell]P_m + \epsilon_{k\ell m}X_\ell[R, P_m]P_j$$
$$= -i\epsilon_{kjm}RP_m + iR^{-1}(\epsilon_{k\ell m}X_\ell X_m)P_j$$
$$= i\epsilon_{jkm}RP_m = i\epsilon_{jkm}\Gamma_m.$$

To evaluate (6.32) and (6.33) use the definition of $\boldsymbol{C}_\alpha$ in the preceding exercise and (2.11), (2.18), (2.21), (2.22) to obtain

$$[\Gamma_j, C_\alpha] = \tfrac{1}{2}[RP_j, RP^2] - [RP_j, \boldsymbol{P}(\boldsymbol{R} \cdot \boldsymbol{P})] - \tfrac{1}{2}\alpha[RP_j, \boldsymbol{R}]$$
$$= \tfrac{1}{2}R[P_j, \boldsymbol{R}]P^2 + \tfrac{1}{2}R[R, P^2]P_j - RP[P_j, \boldsymbol{R} \cdot \boldsymbol{P}]$$
$$- \boldsymbol{P}[R, \boldsymbol{R} \cdot \boldsymbol{P}]P_j - [R, \boldsymbol{P}](\boldsymbol{R} \cdot \boldsymbol{P})P_j - \tfrac{1}{2}\alpha R[P_j, \boldsymbol{R}]$$
$$= \tfrac{1}{2}R[P_j, \boldsymbol{R}]P^2 + \tfrac{1}{2}R(2iR^{-1}(\boldsymbol{R} \cdot \boldsymbol{P} - i))P_j - RP(-iP_j)$$
$$- \boldsymbol{P}(iR)P_j - iR^{-1}\boldsymbol{R}(\boldsymbol{R} \cdot \boldsymbol{P})P_j - \tfrac{1}{2}\alpha R[P_j, \boldsymbol{R}]$$
$$= -\tfrac{1}{2}(RP^2 - \alpha R)[\boldsymbol{R}, P_j] + R^{-1}\boldsymbol{R}P_j + iR\boldsymbol{P}P_j - i\boldsymbol{P}RP_j$$
$$= -\tfrac{1}{2}(RP^2 - \alpha R)[\boldsymbol{R}, P_j] + R^{-1}\boldsymbol{R}P_j + i[R, \boldsymbol{P}]P_j$$
$$= -\tfrac{1}{2}(RP^2 - \alpha R)[\boldsymbol{R}, P_j].$$

Therefore from (2.9) we obtain (6.32) for $\alpha = 1$ and (6.33) for $\alpha = -1$.

To derive (6.34) and (6.36) use the definition of $D_\alpha$ in the preceding exercise and (2.11), (2.18) and (2.21) to obtain

$$
\begin{aligned}
[\Gamma_j, D_\alpha] &= \tfrac{1}{2}[RP, RP^2] + \tfrac{1}{2}\alpha[RP, R] \\
&= \tfrac{1}{2}R[P, R]P^2 + \tfrac{1}{2}R[R, P^2] + \tfrac{1}{2}\alpha R[P, R] \\
&= \tfrac{1}{2}R(-iR^{-1}\boldsymbol{R})P^2 + \tfrac{1}{2}R(2iR^{-1}(\boldsymbol{R}\cdot\boldsymbol{P} - i))\boldsymbol{P} + \tfrac{1}{2}\alpha R(-iR^{-1}\boldsymbol{R}) \\
&= -\tfrac{1}{2}iRP^2 + i(\boldsymbol{R}\cdot\boldsymbol{P})\boldsymbol{P} + \boldsymbol{P} - \tfrac{1}{2}\alpha i\boldsymbol{R} \\
&= -\tfrac{1}{2}iRP^2 + i(\boldsymbol{P}(\boldsymbol{R}\cdot\boldsymbol{P}) + i\boldsymbol{P}) + \boldsymbol{P} - \tfrac{1}{2}\alpha i\boldsymbol{R} \\
&= -i(\tfrac{1}{2}RP^2 - \boldsymbol{P}(\boldsymbol{R}\cdot\boldsymbol{P}) + \tfrac{1}{2}\alpha\boldsymbol{R}).
\end{aligned}
$$

Therefore $\alpha = -1$ gives (6.34) and $\alpha = 1$ gives (6.36). To derive (6.35) use (2.21) and (2.22) to obtain

$$
\begin{aligned}
[\boldsymbol{\Gamma}, T_2] &= [RP, \boldsymbol{R}\cdot\boldsymbol{P} - i] = R[\boldsymbol{P}, \boldsymbol{R}\cdot\boldsymbol{P}] + [R, \boldsymbol{R}\cdot\boldsymbol{P}]\boldsymbol{P} \\
&= R(-i\boldsymbol{P}) + iR\boldsymbol{P} = 0.
\end{aligned}
$$

Finally to derive (6.37) use (2.11) to obtain

$$
\begin{aligned}
[\Gamma_j, \Gamma_k] &= [RP_j, RP_k] = R[P_j, R]P_k + R[R, P_k]P_j \\
&= R(-iR^{-1}X_j)P_k + R(iR^{-1}X_k)P_j \\
&= -i(X_j P_k - X_k P_j) = -i\epsilon_{jk\ell}L_\ell.
\end{aligned}
$$

## ■ Solution 6.9

From (B.78) and (B.84)

$$
\begin{aligned}
Q_2 &= (L_{12}^2 + L_{13}^2 + L_{14}^2 - L_{15}^2 - L_{16}^2) \\
&\quad + (L_{23}^2 + L_{24}^2 - L_{25}^2 - L_{26}^2) \\
&\quad + (L_{34}^2 - L_{35}^2 - L_{36})^2 \\
&\quad + (-L_{45}^2 - L_{46}^2) + (-L_{56}^2) \\
&= L^2 + A^2 - B^2 - \Gamma^2 + T_3^2 - T_1^2 - T_2^2.
\end{aligned}
$$

From the so(4,1) results (6.13) and (6.24) for our particular realization, $T_2^2 + B^2 - A^2 - L^2 = 2$ so $Q_2 = -2 - \Gamma^2 + T_3^2 - T_1^2$. Since $\boldsymbol{\Gamma} = R\boldsymbol{P}$ then from (2.11) and the so(2,1) realization (6.42) to (6.44)

$$
\begin{aligned}
\Gamma^2 &= R\boldsymbol{P}\cdot R\boldsymbol{P} = RPR\cdot\boldsymbol{P} \\
&= R(R\boldsymbol{P} - iR^{-1}\boldsymbol{R})\cdot\boldsymbol{P} \\
&= R^2P^2 - i\boldsymbol{R}\cdot\boldsymbol{P} = (T_3 - T_1)(T_3 + T_1) - i(T_2 + i) \\
&= T_3^2 - T_1^2 + [T_3, T_1] - iT_2 + 1 = T_3^2 - T_1^2 + 1.
\end{aligned}
$$

Therefore $Q_2 = -3$.

For $Q_3$ we obtain from (B.85) after some simplification

$$48Q_3 = 8\epsilon_{abcdef}L^{ab}L^{cd}L^{ef}, \ a < b, c < d, e < f$$
$$= 48\epsilon_{jk\ell}(-L_{j4}L_{k5}L_{\ell6} + L_{45}L_{j6}L_{k\ell}$$
$$- L_{46}L_{j5}L_{k\ell} + L_{56}L_{j4}L_{k\ell}),$$

where a summation over $j, k, \ell = 1, 2, 3$ is implied. Therefore

$$48Q_3 = -48\boldsymbol{A}\cdot(\boldsymbol{B}\times\boldsymbol{G})$$
$$+ 48L_{45}(L_{16}L_{23} - L_{26}L_{13} + L_{36}L_{12})$$
$$- 48L_{46}(L_{15}L_{23} - L_{25}L_{13} + L_{35}L_{12})$$
$$+ 48L_{56}(L_{14}L_{23} - L_{24}L_{13} + L_{34}L_{12}),$$
$$Q_3 = -\boldsymbol{A}\cdot(\boldsymbol{B}\times\boldsymbol{\Gamma}) + T_2(\boldsymbol{\Gamma}\cdot\boldsymbol{L}) - T_1(\boldsymbol{B}\cdot\boldsymbol{L}) + T_3(\boldsymbol{A}\cdot\boldsymbol{L}).$$

From Exercise 6.4 for the scaled hydrogenic realization

$$\boldsymbol{A}\cdot(\boldsymbol{B}\times\boldsymbol{\Gamma}) = (\boldsymbol{A}\times\boldsymbol{B})\cdot\boldsymbol{\Gamma} = T_2(\boldsymbol{L}\cdot\boldsymbol{\Gamma}).$$

But from (2.26) and (2.31)

$$\boldsymbol{L}\cdot\boldsymbol{\Gamma} = \boldsymbol{L}\cdot\boldsymbol{R}\boldsymbol{P} = R(\boldsymbol{L}\cdot\boldsymbol{P}) = 0,$$
$$\boldsymbol{\Gamma}\cdot\boldsymbol{L} = R(\boldsymbol{P}\cdot\boldsymbol{L}) = 0,$$

and since $\boldsymbol{A}\cdot\boldsymbol{L} = \boldsymbol{B}\cdot\boldsymbol{L} = 0$ then $Q_3 = 0$.

## J.7  Solutions to Chapter 7 Exercises

### ■ Solution 7.1

From (7.9) and (7.10) we have the two identities

$$\langle\psi^{(2)}|E_0 - H_0|\psi^{(1)}\rangle = \langle\psi^{(2)}|V|\phi_0\rangle,$$
$$\langle\psi^{(1)}|E_0 - H_0|\psi^{(2)}\rangle = \langle\psi^{(1)}|V|\psi^{(1)}\rangle - E^{(1)}\langle\psi^{(1)}|\psi^{(1)}\rangle.$$

Since $H_0$ and $V$ are hermitian the left hand sides of these two identities are equal so the required formula for $E^{(3)}$ is

$$E^{(3)} = \langle\phi_0|V|\psi^{(2)}\rangle = \langle\psi^{(1)}|V|\psi^{(1)}\rangle - E^{(1)}\langle\psi^{(1)}|\psi^{(1)}\rangle.$$

Similarly, from (7.9) and (7.11)

$$\langle\psi^{(3)}|E_0 - H_0|\psi^{(1)}\rangle = \langle\psi^{(3)}|V|\phi_0\rangle,$$
$$\langle\psi^{(1)}|E_0 - H_0|\psi^{(3)}\rangle = \langle\psi^{(1)}|V|\psi^{(2)}\rangle - E^{(2)}\langle\psi^{(1)}|\psi^{(1)}\rangle - E^{(1)}\langle\psi^{(1)}|\psi^{(2)}\rangle.$$

Since $E^{(4)} = \langle\psi^{(3)}|V|\phi_0\rangle$ we obtain

$$E^{(4)} = \langle\psi^{(1)}|V|\psi^{(2)}\rangle - E^{(1)}\langle\psi^{(1)}|\psi^{(2)}\rangle - E^{(2)}\langle\psi^{(1)}|\psi^{(1)}\rangle.$$

From (7.9) and (7.8) for $n = 4$

$$\langle\psi^{(4)}|E_0 - H_0|\psi^{(1)}\rangle = \langle\psi^{(4)}|V|\psi^{(0)}\rangle,$$
$$\langle\psi^{(1)}|E_0 - H_0|\psi^{(4)}\rangle = \langle\psi^{(1)}|V - E^{(1)}|\psi^{(3)}\rangle - E^{(3)}\langle\psi^{(1)}|\psi^{(1)}\rangle$$
$$- E^{(2)}\langle\psi^{(1)}|\psi^{(2)}\rangle.$$

Therefore

$$E^{(5)} = \langle\psi^{(1)}|V - E^{(1)}|\psi^{(3)}\rangle - E^{(3)}\langle\psi^{(1)}|\psi^{(1)}\rangle - E^{(2)}\langle\psi^{(1)}|\psi^{(2)}\rangle.$$

From (7.10) and (7.11)

$$\langle\psi^{(3)}|E_0 - H_0|\psi^{(2)}\rangle = \langle\psi^{(3)}|V - E^{(1)}|\psi^{(1)}\rangle,$$
$$\langle\psi^{(2)}|E_0 - H_0|\psi^{(3)}\rangle = \langle\psi^{(2)}|V - E^{(1)}|\psi^{(2)}\rangle - E^{(2)}\langle\psi^{(2)}|\psi^{(1)}\rangle.$$

Therefore

$$\langle\psi^{(1)}|V - E^{(1)}|\psi^{(3)}\rangle = \langle\psi^{(2)}|V - E^{(1)}|\psi^{(2)}\rangle - E^{(2)}\langle\psi^{(2)}|\psi^{(1)}\rangle,$$

and $\psi^{(3)}$ can be eliminated from $E^{(5)}$. Therefore

$$E^{(5)} = \langle\psi^{(2)}|V - E^{(1)}|\psi^{(2)}\rangle - E^{(3)}\langle\psi^{(1)}|\psi^{(1)}\rangle$$
$$- E^{(2)}(\langle\psi^{(1)}|\psi^{(2)}\rangle + \langle\psi^{(2)}|\psi^{(1)}\rangle)$$
$$= \langle\psi^{(2)}|V - E^{(1)}|\psi^{(2)}\rangle - E^{(3)}\langle\psi^{(1)}|\psi^{(1)}\rangle - 2\,\mathrm{Re}\langle\psi^{(1)}|\psi^{(2)}\rangle.$$

In our applications the wavefunction corrections will be real and we can replace $\mathrm{Re}\langle\psi^{(1)}|\psi^{(2)}\rangle$ by $\langle\psi^{(1)}|\psi^{(2)}\rangle$.

### ■ Solution 7.2

Let $\langle A\rangle$ denote a ground state expectation value of the form $\langle\psi_0|A|\psi_0\rangle$ for some operator $A$. Then

$$E^{(n)} = \langle VGVG\cdots GV\rangle + \text{all bracketed terms,}$$

where the leading term has $n$ factors of $V$. The bracketed terms are obtained by inserting one or more nested bracket pairs $\langle\rangle$ having a $V$ factor at either end in all possible ways and attaching a minus sign if an odd number of bracket pairs were inserted. Brackets can be inserted beginning with $E^{(3)}$ so we have

$$E^{(1)} = \langle V\rangle,$$
$$E^{(2)} = \langle VGV\rangle,$$
$$E^{(3)} = \langle VGVGV\rangle - \langle VG\langle V\rangle GV\rangle$$
$$= \langle VGVGV\rangle - \langle V\rangle\langle VG^2V\rangle,$$
$$E^{(4)} = \langle VGVGVGV\rangle - \langle VG\langle V\rangle GVGV\rangle - \langle VGVG\langle V\rangle GV\rangle$$
$$+ \langle VG\langle V\rangle G\langle V\rangle GV\rangle - \langle VG\langle VGV\rangle GV\rangle$$
$$= \langle VGVGVGV\rangle - \langle V\rangle\langle VG^2VGV\rangle - \langle V\rangle\langle VGVG^2V\rangle$$
$$+ \langle V\rangle^2\langle VG^3V\rangle - \langle VGV\rangle\langle VG^2V\rangle,$$

which give the results in (7.17) to (7.23). It is clear that higher order corrections have many terms giving complicated formulas: there are $(2n-2)!/[n!(n-1)!]$ terms in $E^{(n)}$. Nevertheless the bracketing technique is a convenient way to generate the formulas to order 4 (there are 14 terms in order 5 and 42 terms in order 6). Other explicit formulas can also be developed [SI70].

### ■ Solution 7.3

The unperturbed ground state wavefunction is

$$\phi_0 = \phi_{100} = \frac{1}{\sqrt{\pi}}e^{-r} = 2Y_{00}e^{-r},$$

and the perturbation is

$$V = z = r\cos\vartheta = \sqrt{\frac{4\pi}{3}}\,rY_{10},$$

where we have used the spherical harmonic functions

$$Y_{00} = \frac{1}{\sqrt{4\pi}}, \quad Y_{10} = \sqrt{\frac{3}{4\pi}}\cos\vartheta.$$

The unperturbed energy is $E_0 = -1/2$. Since $E^{(1)} = 0$, (7.38) can be expressed as

$$\left(-\frac{1}{2}\nabla^2 - \frac{1}{r} + \frac{1}{2}\right)\psi^{(1)} = -\frac{2}{\sqrt{3}}re^{-r}Y_{10}.$$

Now use

$$\nabla^2 = \frac{1}{r^2}\frac{\partial}{\partial r}\left(r^2\frac{\partial}{\partial r}\right) - \frac{L^2}{r^2}$$

to obtain

$$\left[\frac{\partial^2}{\partial r^2} + \frac{2}{r}\frac{\partial}{\partial r} - \frac{L^2}{r^2} + \frac{2}{r} - 1\right]\psi^{(1)} = \frac{4}{\sqrt{3}}re^{-r}Y_{10}.$$

If we assume that $\psi^{(1)} = f(r)Y_{10}$ then $\langle\phi_0|\psi^{(1)}\rangle = 0$ since $Y_{00}$ and $Y_{10}$ are orthogonal. Since the spherical harmonic functions $Y_{\ell m}(\vartheta,\varphi)$ are eigenfunctions of $L^2$ with eigenvalues $\ell(\ell+1)$ it follows that $L^2\psi^{(1)} = 2\psi^{(1)}$. Therefore we have the inhomogeneous radial differential equation for $f_1$

$$\frac{d^2 f_1}{dr^2} + \frac{2}{r}\frac{df_1}{dr} + \left(-\frac{2}{r^2} + \frac{2}{r} - 1\right)f_1 = \frac{4}{\sqrt{3}}re^{-r}.$$

Now assume that $f_1(r) = g_1(r)e^{-r}$ to obtain

$$\frac{d^2 g_1}{dr^2} + 2\left(\frac{1}{r} - 1\right)\frac{dg_1}{dr} - \frac{2g_1}{r^2} = \frac{4}{\sqrt{3}}r.$$

Since $f_1$ should be finite at the origin we look for a polynomial solution. Substituting $g_1(r) = \sum a_k r^k$ gives the solution

$$g_1(r) = -\tfrac{1}{\sqrt{3}}(2r + r^2).$$

Therefore the first order correction to the wavefunction is

$$\psi^{(1)} = g_1(r)e^{-r}Y_{10},$$

and the second order correction to the energy is

$$\begin{aligned}
E^{(2)} &= \langle \phi_0 | V | \psi^{(1)} \rangle \\
&= \frac{2}{\sqrt{3}} \int_0^\infty r^3 g_1(r) e^{-2r}\, dr \int_\Omega Y_{10}Y_{10}\, d\Omega \\
&= \frac{2}{\sqrt{3}} \int_0^\infty r^3 g_1(r) e^{-2r}\, dr = -\frac{9}{4}.
\end{aligned}$$

## ■ Solution 7.4

Since $E^{(1)} = 0$ we now have to solve (7.10) which is

$$(H_0 - E_0)\psi^{(2)} = -V\psi^{(1)} - \frac{9}{4}\phi_0 = -\left[\sqrt{\frac{4\pi}{3}}rg_1(r)Y_{10}Y_{10} + \frac{9}{2}Y_{00}\right]e^{-r}.$$

Since

$$Y_{20} = \sqrt{\frac{5}{16\pi}}(3\cos^2 \vartheta - 1) = \sqrt{\frac{5}{16\pi}}(4\pi Y_{10}Y_{10} - 1),$$

then

$$Y_{10}Y_{10} = \frac{1}{\sqrt{5\pi}}Y_{20} + \frac{1}{2\sqrt{\pi}}Y_{00},$$

and

$$(H_0 - E_0)\psi^{(2)} = [h_0(r)Y_{00} + h_2(r)Y_{20}]\, e^{-r},$$

where we have defined

$$h_0 = \frac{1}{6}(4r^2 + 2r^3 - 27), \qquad h_2 = \frac{2}{3\sqrt{5}}(2r^2 + r^3).$$

As in the previous exercise we can substitute for $\nabla^2$ to obtain

$$\left[\frac{\partial^2}{\partial r^2} + \frac{2}{r}\frac{\partial}{\partial r} - \frac{L^2}{r^2} + \frac{2}{r} - 1\right]\psi^{(2)} = -2(h_0 Y_{00} + h_2 Y_{20})e^{-r}.$$

Now let $\psi^{(2)} = g e^{-r}$ to eliminate $e^{-r}$ and obtain

$$\frac{\partial^2 g}{\partial r^2} + 2\left(\frac{1}{r} - 1\right)\frac{\partial g}{\partial r} - \frac{L^2 g}{r^2} = -2(h_0 Y_{00} + h_2 Y_{20}).$$

Now substitute $g = g_0 Y_{00} + g_2 Y_{20}$ and use $L^2 Y_{00} = 0$, $L^2 Y_{20} = 6 Y_{20}$ to obtain the pair of differential equations (the spherical harmonics are linearly independent angular functions)

$$\frac{d^2 g_0}{dr^2} + 2\left(\frac{1}{r} - 1\right)\frac{dg_0}{dr} = -2h_0,$$

$$\frac{d^2 g_2}{dr^2} + 2\left(\frac{1}{r} - 1\right)\frac{dg_2}{dr} - \frac{6g_2}{r^2} = -2h_2.$$

Again we look for polynomial solutions $g_0(r) = \sum b_k r^k$, $g_2(r) = \sum c_k r^k$ and the results are

$$g_0(r) = b_0 + \tfrac{3}{2}r^2 + \tfrac{1}{2}r^3 + \tfrac{1}{12}r^4, \quad g_2(r) = \tfrac{1}{\sqrt{5}}\left(\tfrac{5}{4}r^2 + \tfrac{5}{6}r^3 + \tfrac{1}{6}r^4\right).$$

The arbitrary constant $b_0$ is determined by requiring that $\langle \phi_0 | \psi^{(2)} \rangle = 0$:

$$\begin{aligned}
\langle \phi_0 | \psi^{(2)} \rangle &= 2\int_0^\infty r^2 g_0(r) e^{-2r}\, dr \int_\Omega Y_{00} Y_{00}\, d\Omega \\
&\quad + 2\int_0^\infty r^2 g_2(r) e^{-r}\, dr \int_\Omega Y_{00} Y_{20},\, d\Omega \\
&= 2\int_0^\infty r^2 g_0(r) e^{-2r}\, dr = 2\left(\frac{81}{32} + \frac{b_0}{4}\right),
\end{aligned}$$

since the spherical harmonic functions are orthonormal. Therefore $\langle \phi_0 | \psi^{(2)} \rangle = 0$ for $b_0 = -81/8$.

We can now evaluate $E^{(4)}$ using the formula derived in Exercise 7.1 as follows:

$$\begin{aligned}
\langle \psi^{(1)} | V | \psi^{(2)} \rangle &= \sqrt{\frac{4\pi}{3}} \int_0^\infty r^3 g_0(r) g_1(r) e^{-2r}\, dr \int_\Omega Y_{10} Y_{10} Y_{00}\, d\Omega \\
&\quad + \sqrt{\frac{4\pi}{3}} \int_0^\infty r^3 g_1(r) g_2(r) e^{-2r}\, dr \int_\Omega Y_{10} Y_{10} Y_{20}\, d\Omega.
\end{aligned}$$

The angular integrals are easily evaluated:

$$\int_\Omega Y_{10} Y_{10} Y_{00}\, d\Omega = \frac{1}{\sqrt{4\pi}} \int_\Omega Y_{10} Y_{10}\, d\Omega = \frac{1}{\sqrt{4\pi}},$$

$$\int_\Omega Y_{10} Y_{10} Y_{20}\, d\Omega = \int_\Omega \left(\tfrac{1}{\sqrt{5\pi}} Y_{20} + \tfrac{1}{2\sqrt{\pi}} Y_{00}\right) Y_{20}\, d\Omega = \frac{1}{\sqrt{5\pi}}.$$

Therefore

$$\begin{aligned}
\langle \psi^{(1)} | V | \psi^{(2)} \rangle &= \frac{1}{\sqrt{3}} \int_0^\infty r^3 g_0(r) g_1(r) e^{-2r}\, dr \\
&\quad + \frac{2}{\sqrt{15}} \int_0^\infty r^3 g_1(r) g_2(r) e^{-2r}\, dr = -\frac{4329}{64}, \\
\langle \psi^{(1)} | \psi^{(1)} \rangle &= \int_0^\infty r^2 g_1(r) g_1(r) e^{-2r}\, dr \int_\Omega Y_{10} Y_{10}\, d\Omega = \frac{43}{8}.
\end{aligned}$$

Finally $E^{(4)} = -\frac{4329}{64} - (-\frac{9}{4})(\frac{43}{8}) = -\frac{3555}{64}$ and to fourth order the ground state energy is

$$E = -\frac{1}{2} - \frac{9}{4}\lambda^3 - \frac{3555}{64}\lambda^2.$$

It is clear that the generalization of the Dalgarno and Lewis method to higher orders is possible but not very systematic and not easily automated.

## ■ Solution 7.5

Here we show only the Maple statements and not their output. The line numbers are not part of the Maple program but are included for reference. First define the differential equation for $f_1(r)$.

```
01   # dalgarno
02   de1 := diff(f1(r), r, r) + (2/r)*diff(f1(r), r)
03          + (-2/r^2 + 2/r - 1)*f1(r) - (4/3^(1/2))*r*exp(-r):
```

Now factor out the $e^{-r}$ part and multiply by $r^2$ to get a differential equation with polynomial coefficients.

```
04   de1 := simplify(subs(f1(r) = exp(-r)*g1(r), de1)):
05   de1 := simplify(de1*exp(r)):
06   de1 := simplify(de1*r^2);
```

Maple will not directly find the polynomial solution using its differential equation solver *dsolve* but it can easily be found directly by substituting a 4th degree trial polynomial into the differential equation, using *collect* to collect coefficients of the various powers of $r$.

```
07   g1_trial := a0 + a1*r + a2*r^2 + a3*r^3 + a4*r^4:
08   result := collect(simplify( subs(g1(r) = g1_trial, de1)), r);
```

Now set the coefficients of the resulting polynomial to zero, solve the resulting system of equations, assign the results to $a_0$ to $a_4$ using *assign*, and make a function $g_1(r)$ from the result.

```
09   equations := { seq(coeff(result, r, i) = 0, i = 0..5) };
10   solve(equations, {a0, a1, a2, a3, a4});
11   assign(");
12   g1 := r -> g1_trial:
```

Now repeat this procedure to find the solutions $g_0(r)$ and $g_2(r)$ of Exercise 7.4. Here we also make use of the operator " which represents the result of the last Maple statement.

```
13   h0 := (4*r^2 + 2*r^3 - 27)/6:
14   de0 := diff(g0(r), r, r) + 2*(1/r-1)*diff(g0(r), r) + 2*h0:
15   g0_trial := b0 + b1*r + b2*r^2 + b3*r^3 + b4*r^4:
16   collect(simplify(subs(g0(r) = g0_trial, r*de0)), r);
```

```
17    equations := { seq(coeff(", r, i) = 0, i = 0..4) };
18    solve(equations, {b0, b1, b2, b3, b4});
19    assign(");
20    g0 := r -> g0_trial:
21
22    h2 := (2/3/sqrt(5))*(2*r^2 + r^3):
23    de2 := diff(g2(r), r, r) + 2*(1/r-1)*diff(g2(r), r)
24           - 6*g2(r)/r^2 + 2*h2:
25    g2_trial := c0 + c1*r + c2*r^2 + c3*r^3 + c4*r^4:
26    collect(simplify(subs(g2(r) = g2_trial, r^2*de2)), r);
27    equations := { seq(coeff(", r, i) = 0, i = 0..5) };
28    solve(equations, {c0, c1, c2, c3, c4});
29    assign(");
30    g2 := r -> g2_trial:
```

Define the spherical harmonics needed to evaluate angular parts of matrix elements.

```
31    Y00 := sqrt(1/4/Pi):
32    Y10 := sqrt(3/4/Pi)*cos(theta):
33    Y20 := sqrt(5/16/Pi)*(3*cos(theta)^2 - 1):
```

Define the radial parts of $\phi_0$, $\psi^{(1)}$ and $\psi^{(2)}$.

```
34    R0 := 2*exp(-r):
35    R1 := g1(r)*exp(-r):
36    R2_00 := g0(r)*exp(-r):
37    R2_20 := g2(r)*exp(-r):
```

Now evaluate $\langle \phi_0 | \psi^{(2)} \rangle$. Setting it to zero will determine $b_0$.

```
38    psi0_psi2 := int(R0*R2_00*r^2, r = 0..infinity)
39           * 2*Pi*int(Y00*Y00*sin(theta), theta = 0..Pi)
40           + int(R0*R2_20*r^2, r = 0..infinity)
41           * 2*Pi*int(Y00*Y20*sin(theta), theta = 0..Pi):
42    simplify(");
43    solve({"});
44    assign("):
```

Define the radial and angular parts of the perturbation.

```
45    VR := r:
46    VA := cos(theta):
```

Calculate $E^{(2)} = \langle \phi_0 | V | \psi^{(2)} \rangle$.

```
47    deltaE2 := int(R0*VR*R1*r^2, r = 0..infinity)
48           * 2*Pi*int(Y00*VA*Y10*sin(theta), theta = 0..Pi);
```

Finally calculate $E^{(4)} = \langle \psi^{(1)} | V | \psi^{(2)} \rangle - E^{(2)} \langle \psi^{(1)} | \psi^{(1)} \rangle$.

```
49   psi1_psi1 := int(R1*R1*r^2, r = 0..infinity)
50        * 2*Pi*int(Y10*Y10*sin(theta), theta = 0..Pi);
51   psi1_V_psi2 := int(R1*VR*R2_00*r^2, r = 0..infinity)
52        * 2*Pi*int(Y10*VA*Y00*sin(theta), theta = 0..Pi)
53        + int(R1*VR*R2_20*r^2, r = 0..infinity)
54        * 2*Pi*int(Y10*VA*Y20*sin(theta), theta = 0..Pi);
55   deltaE4 := psi1_V_psi2 - deltaE2*psi1_psi1;
```

## ■ Solution 7.6

First project (7.66) onto $\Psi^{(m+1)}$ to obtain

$$\langle \Psi^{(m+1)}|\kappa_0 - K_0|\Psi^{(n)}\rangle = \langle \Psi^{(m+1)}|W|\Psi^{(n-1)}\rangle$$
$$- \sum_{s=0}^{n-1} E^{(n-s)}\langle \Psi^{(m+1)}|S|\Psi^{(s)}\rangle.$$

Now substitute $n-1$ for $m$ and $m+1$ for $n$ to obtain

$$\langle \Psi^{(n)}|\kappa_0 - K_0|\Psi^{(m+1)}\rangle = \langle \Psi^{(n)}|W|\Psi^{(m)}\rangle - \sum_{r=0}^{m} E^{(m+1-r)}\langle \Psi^{(n)}|S|\Psi^{(r)}\rangle.$$

The left sides of these results are identical since $K_0$ is hermitian.  We also
assume that the matrix elements of $W$ and $S$ are real so

$$\langle \Psi^{(m)}|W|\Psi^{(n)}\rangle = \langle \Psi^{(m+1)}|W|\Psi^{(n-1)}\rangle + \sum_{r=0}^{m} E^{(m+1-r)}\langle \Psi^{(n)}|S|\Psi^{(r)}\rangle$$
$$- \sum_{s=0}^{n-1} E^{(n-s)}\langle \Psi^{(m+1)}|S|\Psi^{(s)}\rangle. \tag{A}$$

Substitute $m = 0$ into (A) to obtain

$$\langle \Phi_0|W|\Psi^{(n)}\rangle = \langle \Psi^{(1)}|W|\Psi^{(n-1)}\rangle + E^{(1)}\langle \Psi^{(n)}|S|\Phi_0\rangle$$
$$- \sum_{s=0}^{n-1} E^{(n-s)}\langle \Psi^{(1)}|S|\Psi^{(s)}\rangle.$$

Substitute for $\langle \Psi^{(1)}|W|\Psi^{(n-1)}\rangle$ using (A) with 1 for $m$ and $n-1$ for $n$ to obtain

$$\langle \Phi_0|W|\Psi^{(n)}\rangle = \langle \Psi^{(2)}|W|\Psi^{(n-2)}\rangle + \sum_{r=0}^{1} E^{(2-r)}\langle \Psi^{(n-1)}|S|\Psi^{(r)}\rangle$$
$$- \sum_{s=0}^{n-2} E^{(n-1-s)}\langle \Psi^{(2)}|S|\Psi^{(s)}\rangle + E^{(1)}\langle \Psi^{(n)}|S|\Phi_0\rangle$$
$$- \sum_{s=0}^{n-1} E^{(n-s)}\langle \Psi^{(1)}|S|\Psi^{(s)}\rangle.$$

Now make both sums over $s$ go from 1 to $n-2$ to obtain

$$\langle \Phi_0|W|\Psi^{(n)}\rangle = \langle \Psi^{(2)}|W|\Psi^{(n-2)}\rangle + \sum_{r=1}^{2} E^{(r)}\langle \Psi^{(n+1-r)}|S|\Phi_0\rangle$$
$$- \sum_{r=1}^{2} E^{(n+1-r)}\langle \Psi^{(r)}|S|\Phi_0\rangle - \sum_{r=1}^{2}\sum_{s=1}^{n-2} E^{(n+1-r-s)}\langle \Psi^{(r)}|S|\Psi^{(s)}\rangle.$$

This suggests the following general result for $p \geq 1$.

$$\langle \Phi_0 | W | \Psi^{(n)} \rangle = \langle \Psi^{(p)} | W | \Psi^{(n-p)} \rangle + \sum_{r=1}^{p} E^{(r)} \langle \Psi^{(n+1-r)} | S | \Phi_0 \rangle$$

$$- \sum_{r=1}^{p} E^{(n+1-r)} \langle \Psi^{(r)} | S | \Phi_0 \rangle - \sum_{r=1}^{p} \sum_{s=1}^{n-p} E^{(n+1-r-s)} \langle \Psi^{(r)} | S | \Psi^{(s)} \rangle, \qquad (B)$$

which we can prove by induction as follows.

Assume (B) and substitute for $\langle \Psi^{(p)} | W | \Psi^{(n-p)} \rangle$ using (A) with $p$ for $m$ and $n - p$ for $n$ to obtain

$$\langle \Phi_0 | W | \Psi^{(n)} \rangle = \langle \Psi^{(p+1)} | W | \Psi^{(n-1-p)} \rangle + \sum_{r=0}^{p} E^{(p+1-r)} \langle \Psi^{(n-p)} | S | \Psi^{(r)} \rangle$$

$$- \sum_{s=0}^{n+1-p} E^{(n-p-s)} \langle \Psi^{(p+1)} | S | \Psi^{(s)} \rangle + \sum_{r=1}^{p} E^{(r)} \langle \Psi^{(n+1-r)} | S | \Phi_0 \rangle$$

$$- \sum_{r=1}^{p} E^{(n+1-r)} \langle \Psi^{(r)} | S | \Phi_0 \rangle - \sum_{r=1}^{p} \sum_{s=1}^{n-p} E^{(n+1-r-s)} \langle \Psi^{(r)} | S | \Psi^{(s)} \rangle.$$

Now rewrite the double sum for $s = 1$ to $n - p - 1$ to obtain

$$\langle \Phi_0 | W | \Psi^{(n)} \rangle = \langle \Psi^{(p+1)} | W | \Psi^{(n-1-p)} \rangle + \sum_{r=0}^{p} E^{(p+1-r)} \langle \Psi^{(n-p)} | S | \Psi^{(r)} \rangle$$

$$- \sum_{s=0}^{n+1-p} E^{(n-p-s)} \langle \Psi^{(p+1)} | S | \Psi^{(s)} \rangle + \sum_{r=1}^{p} E^{(r)} \langle \Psi^{(n+1-r)} | S | \Phi_0 \rangle$$

$$- \sum_{r=1}^{p} E^{(n+1-r)} \langle \Psi^{(r)} | S | \Phi_0 \rangle - \sum_{r=1}^{p} E^{(p+1-r)} \langle \Psi^{(r)} | S | \Psi^{(n-p)} \rangle$$

$$- \sum_{r=1}^{p} \sum_{s=1}^{n-p-1} E^{(n+1-r-s)} \langle \Psi^{(r)} | S | \Psi^{(s)} \rangle$$

$$= \langle \Psi_{p+1} | W | \Psi_{n-1-p} \rangle + E^{(p+1)} \langle \Psi^{(n-p)} | S | \Phi_0 \rangle - E^{(n-p)} \langle \Psi^{(p+1)} | S | \Phi_0 \rangle$$

$$+ \sum_{r=1}^{p} E^{(r)} \langle \Psi^{(n+1-r)} | S | \Phi_0 \rangle - \sum_{r=1}^{p} E^{(n+1-r)} \langle \Psi^{(r)} | S | \Phi_0 \rangle$$

$$- \sum_{r=1}^{p+1} \sum_{s=1}^{n-p-1} E^{(n+1-r-s)} \langle \Psi^{(r)} | S | \Psi^{(s)} \rangle$$

$$= \langle \Psi^{(p+1)} | W | \Psi^{(n-p-1)} \rangle + \sum_{r=1}^{p+1} E^{(r)} \langle \Psi^{(n+1-r)} | S | \Phi_0 \rangle$$

$$- \sum_{r=1}^{p+1} E^{(n+1-r)} \langle \Psi^{(r)} | S | \Phi_0 \rangle$$

$$- \sum_{r=1}^{p+1} \sum_{s=1}^{n-p-1} E^{(n+1-r-s)} \langle \Psi^{(r)} | S | \Psi^{(s)} \rangle,$$

which is just (B) with $p$ replaced by $p+1$ so (B) is true by induction.  Now replace the left side of (B) using the energy formula (7.74) to obtain

$$\langle\Phi_0|S|\Phi_0\rangle E^{(n+1)} = \langle\Psi^{(p)}|W|\Psi^{(n-p)}\rangle - \sum_{s=1}^{n} E^{(s)}\langle\Phi_0|S|\Psi^{(n+1-s)}\rangle$$
$$+ \sum_{r=1}^{p} E^{(r)}\langle\Psi^{(n+1-r)}|S|\Phi_0\rangle - \sum_{r=1}^{p} E^{(n+1-r)}\langle\Psi^{(r)}|S|\Phi_0\rangle$$
$$- \sum_{r=1}^{p}\sum_{s=1}^{n-p} E^{(n+1-r-s)}\langle\Psi^{(r)}|S|\Psi^{(s)}\rangle.$$

Combine the first two sums ($n \geq p$) to get a sum over $r$ from $p+1$ to $n$ and then change the summation index to go from 1 to $n-p$ to get

$$\langle\Psi_0|S|\Psi_0\rangle E^{(n+1)} = \langle\Psi^{(p)}|W|\Psi^{(n-p)}\rangle - \sum_{r=1}^{p}\langle\Phi_0|S|\Psi^{(r)}\rangle E^{(n+1-r)}$$
$$- \sum_{s=1}^{n-p}\langle\Phi_0|S|\Psi^{(s)}\rangle E^{(n+1-s)} - \sum_{r=1}^{p}\sum_{s=1}^{n-p}\langle\Psi^{(r)}|S|\Psi^{(s)}\rangle E^{(n+1-r-s)}.$$

Finally let $n = p + q$ to obtain the symmetric formula (7.75).

## J.8    Solutions to Chapter 8 Exercises

### ■ Solution 8.1

Denoting by $\Delta E(\mathcal{Z}, \lambda)$ the energy correction for nuclear charge $\mathcal{Z}$ and perturbation parameter $\lambda$ we can write (8.2) in the form

$$\left[T_3 - n + \lambda\left(\frac{n}{\mathcal{Z}}\right)^3 RZ - \left(\frac{n}{\mathcal{Z}}\right)^2 R\Delta E(\mathcal{Z}, \lambda)\right]\Psi = 0,$$

which can also be expressed as

$$\left[T_3 - n + \left(\frac{\lambda}{\mathcal{Z}^3}\right) n^3 RZ - n^2 R\Delta E(1, \lambda\mathcal{Z}^{-3})\right]\Psi = 0.$$

Therefore

$$\Delta E(\mathcal{Z}, \lambda) = \mathcal{Z}^2 \Delta E(1, \lambda Z^{-3}).$$

## J.9    Solutions to Chapter 9 Exercises

### ■ Solution 9.1

Denoting by $\Delta E(\mathcal{Z}, \lambda)$ the energy correction for nuclear charge $\mathcal{Z}$ and perturbation parameter $\lambda$ we can write (9.3) in the form

$$\left[T_3 - n + \lambda\left(\frac{n}{\mathcal{Z}}\right)^4 R(R^2 - Z^2) - \left(\frac{n}{\mathcal{Z}}\right)^2 R\Delta E(\mathcal{Z}, \lambda)\right]\Psi = 0,$$

which can also be expressed as

$$\left[ T_3 - n + \left(\frac{\lambda}{\mathcal{Z}^4}\right) n^4 R(R^2 - Z^2) - n^2 R \Delta E(1, \lambda \mathcal{Z}^{-4}) \right] \Psi = 0.$$

Therefore

$$\Delta E(\mathcal{Z}, \lambda) = \mathcal{Z}^2 \Delta E(1, \lambda Z^{-4}).$$

## ■ Solution 9.2

```
01    # eigval: Solve 2 by 2 eigenvalue problem for 1st order energies
02    # and correct linear combinations of 3s and 3d0 states
03
04    read `zrenormdata.m`:
05
06    # Solve the 2 by 2 eigenvalue problem
07    # We return to the original matrix elements
08
09    fW := 3^4:
10    fS := 3^2:
11    N0sq := subs(n=3,l=0,Norm2):
12    N1sq := subs(n=3,l=2,Norm2):
13    W00 := fW * subs(n=3,l=0,W[0,0]):
14    W01 := sqrt(N0sq/N1sq) * fW * subs(n=3,l=2,W[0,-2]):
15    W11 := fW * subs(n=3,l=2,W[0,0]):
16    S00 := fS * subs(n=3,l=0,R[0]):
17    S11 := fS * subs(n=3,l=2,R[0]):
18
19    # Solve quadratic to get the two first order energies
20
21    eq := simplify( (W00 - z * S00) * (W11 - z * S11) - W01^2 ):
22    solutions := [solve(eq,z)]:
23    DE1[0] := solutions[1];
24    DE1[1] := solutions[2];
25    evalf(DE1[0]);
26    evalf(DE1[1]);
27
28    # Now get the coefficients in the two linear combinations
29
30    for r from 0 to 1 do
31        eq1 := (W00 - DE1[r]*S00)*C0[r] + W01 * C1[r];
32        eq2 := C0[r]^2 + C1[r]^2 - 1;
33        solve( eq1,eq2, C0[r],C1[r] );
34        assign(");
35        lprint(` coeffs in reference state `.r);
36        print(C0[r], evalf(C0[r]), C1[r], evalf(C1[r]));
37    od:
38
39    save C0, C1, DE1, `1storder`;
```

## J.10    Solutions to Chapter 10 Exercises

■ **Solution 10.1**

Consider the Schrödinger equation (10.2) and denote the energy by $E(\mathcal{Z}, \lambda)$. Perform the scaling transformation $r = \alpha R$, $\boldsymbol{p} = \boldsymbol{P}/\alpha$ and choose $\alpha = \mathcal{Z}^{-1}$ to obtain

$$\left[\frac{1}{2}P^2 - \frac{1}{R} + \Lambda R^d - \mathcal{Z}^{-2} E(\mathcal{Z}, \mathcal{Z}^{d+2}\Lambda)\right] \Psi(R) = 0,$$

where $\Lambda = \mathcal{Z}^{-a-2}\lambda$. Since this is the Schrödinger equation for $\mathcal{Z} = 1$ we must have

$$\mathcal{Z}^{-2} E(\mathcal{Z}, \mathcal{Z}^{d+2}\Lambda) = E(1, \Lambda).$$

Therefore

$$E(\mathcal{Z}, \lambda) = \mathcal{Z}^2 E(1, \mathcal{Z}^{-d-2}\lambda).$$

Now let

$$E(\mathcal{Z}, \lambda) = \sum_{n=0}^{\infty} E^{(n)}(\mathcal{Z})\lambda^n$$

to obtain

$$\sum_{n=0}^{\infty} E^{(n)}(\mathcal{Z})\lambda^n = \mathcal{Z}^2 \sum_{n=0}^{\infty} E^{(n)}(1) \left(\mathcal{Z}^{-a-2}\lambda\right)^n,$$

from which (10.16) follows.

■ **Solution 10.2**

As in the previous exercise the scaling transformation gives

$$\left[\frac{1}{2}P^2 - \frac{1}{R} - \frac{1}{R}(e^{-\Lambda R} - 1) - \mathcal{Z}^{-2} E(\mathcal{Z}, \mathcal{Z}\Lambda)\right] \Psi(R) = 0,$$

where $\Lambda = \mathcal{Z}^{-1}\lambda$. Since this is the Schrödinger equation for $\mathcal{Z} = 1$ we must have

$$\mathcal{Z}^{-2} E(\mathcal{Z}, \mathcal{Z}\Lambda) = E(1, \Lambda).$$

Therefore

$$E(\mathcal{Z}, \lambda) = \mathcal{Z}^2 E(1, \mathcal{Z}^{-1}\lambda),$$

and we have the series expansion

$$\sum_{n=0}^{\infty} E^{(n)}(\mathcal{Z})\lambda^n = \sum_{n=0}^{\infty} \mathcal{Z}^{2-n} E^{(n)}(1)\lambda^n,$$

from which (10.76) follows.

## ■ Solution 10.3

$$E^{(1)} = \frac{\langle \Psi_0 | W_1 | \Psi_0 \rangle}{\langle \Psi_0 | S | \Psi_0 \rangle} = \frac{-q^2 V_1 \langle \Psi_0 | R | \Psi_0 \rangle}{q^2 \langle \Psi_0 | R | \Psi_0 \rangle} = -V_1,$$

and

$$\begin{aligned}
\langle \Psi_0 | S | \Psi_0 \rangle E^{(2)} &= \langle \Psi_0 | W_2 | \Psi_0 \rangle + \langle \Psi_0 | W_1 - SE^{(1)} | \Psi^{(1)} \rangle \\
&= \langle \Psi_0 | W_2 | \Psi_0 \rangle + \langle \Psi_0 | W_1 G W_1 | \Psi_0 \rangle \\
&\quad - E^{(1)} \langle \Psi_0 | S G W_1 | \Psi_0 \rangle - E^{(1)} \langle \Psi_0 | W_1 G S | \Psi_0 \rangle \\
&\quad + (E^{(1)})^2 \langle \Psi_0 | S G S | \Psi_0 \rangle,
\end{aligned}$$

where

$$\langle \Psi_0 | S | \Psi_0 \rangle = q^3,$$

$$\begin{aligned}
\langle \Psi_0 | W_1 G W_1 | \Psi_0 \rangle &= \sum_{a \neq 0} \frac{1}{(-a)} \langle 0 | W_1 | a \rangle \langle a | W_1 | 0 \rangle \\
&= q^4 V_1^2 \left[ (R_{-1,0})^2 - (R_{1,0})^2 \right],
\end{aligned}$$

$$\begin{aligned}
\langle \Psi_0 | S G W_1 | \Psi_0 \rangle &= \sum_{a \neq 0} \frac{1}{(-a)} \langle 0 | S | a \rangle \langle a | W_1 | 0 \rangle \\
&= -q^4 V_1 \left[ (R_{-1,0})^2 - (R_{1,0})^2 \right] \\
&= \langle \Psi_0 | W_1 G S | \Psi_0 \rangle,
\end{aligned}$$

$$\begin{aligned}
\langle \Psi_0 | S G S | \Psi_0 \rangle &= \sum_{a \neq 0} \frac{1}{(-a)} \langle 0 | S | a \rangle \langle a | S | 0 \rangle \\
&= q^4 \left[ (R_{-1,0})^2 - (R_{1,0})^2 \right],
\end{aligned}$$

$$\begin{aligned}
\langle \Psi_0 | W_2 | \Psi_0 \rangle &= -q^3 V_2 \langle \Psi_0 | R^2 | \Psi_0 \rangle \\
&= -\frac{1}{2} q^3 V_2 \left[ 3q^2 - k(k+1) \right].
\end{aligned}$$

Therefore

$$E^{(2)} = -\frac{1}{2} V_2 \left[ 3q^2 - k(k+1) \right].$$

## J.11   Solutions to Appendix A Exercises

## ■ Solution A.1

To derive (A.8) substitute the basis vectors into (A.6) and use (A.3) to obtain

$$e_i \times (e_j \times e_k) + e_j \times (e_k \times e_i) + e_k \times (e_i \times e_j) = 0,$$

$$\sum_m (\epsilon_{jkm} e_i \times e_m + \epsilon_{kim} e_j \times e_m + \epsilon_{ijm} e_k \times e_m) = 0,$$

$$\sum_{\ell m}(\epsilon_{jkm}\epsilon_{im\ell}e_\ell + \epsilon_{kim}\epsilon_{jm\ell}e_\ell + \epsilon_{ijm}\epsilon_{km\ell}e_\ell) = 0,$$

$$\sum_{\ell}\left(\sum_{m}(\epsilon_{jkm}\epsilon_{im\ell} + \epsilon_{kim}\epsilon_{jm\ell} + \epsilon_{ijm}\epsilon_{km\ell})\right)e_\ell = 0.$$

Since the basis vectors are linearly independent each term in the sum over $\ell$ must be zero.  Finally, using antisymmetry properties such as $\epsilon_{im\ell} = -\epsilon_{i\ell m}$, identity (A.8) is obtained.

To derive (A.9) substitute the basis vectors into (A.7) to obtain

$$(e_i \times e_j)\cdot(e_k \times e_\ell) = \delta_{ik}\delta_{j\ell} - \delta_{i\ell}\delta_{jk}.$$

Now use (A.2) and (A.3) to express the left hand side as

$$\sum_{mn}\epsilon_{ijm}\epsilon_{k\ell n}e_m\cdot e_n = \sum_{m}\epsilon_{ijm}\epsilon_{k\ell m}.$$

To derive (A.10) set $j = \ell$ in (A.9) and sum over $\ell$, using $\sum_\ell \delta_{\ell\ell} = 3$ and $\sum_\ell \delta_{i\ell}\delta_{\ell k} = \delta_{ik}$.  Finally, to derive (A.11) set $k = i$ in (A.10) and sum over $i$.

## J.12   Solutions to Appendix B Exercises

### ■ Solution B.1

The derivation is similar to that of Exercise 1.11. Any $n \times n$ matrix $\mathcal{L}$ can be expressed as

$$\mathcal{L} = \sum_{j,k=1}^{n} \alpha_{jk}E_{jk}.$$

Since $\mathcal{L}^T = -\mathcal{L}$ then $\alpha_{jj} = 0$ and $\alpha_{kj} = -\alpha_{jk}$ so

$$\begin{aligned}
\mathcal{L} &= \sum_{j<k}\alpha_{jk}E_{jk} + \sum_{j>k}\alpha_{jk}E_{jk}\\
&= \sum_{j<k}\alpha_{jk}E_{jk} + \sum_{j<k}\alpha_{kj}E_{kj}\\
&= \sum_{j<k}\alpha_{jk}(E_{jk} - E_{kj}) = \sum_{j<k}\alpha_{jk}\mathcal{L}_{jk}.
\end{aligned}$$

Therefore any antisymmetric matrix $\mathcal{L}$ is a linear combination of the $\mathcal{L}_{jk}$, and as in Exercise 1.11 the $\mathcal{L}_{jk}$ are linearly independent. There are $(n^2 - n)/2$ of these matrices for $1 \leq j < k \leq n$.

### ■ Solution B.2

From (B.50) and (1.6)

$$[\mathcal{L}_{jk}, \mathcal{L}_{\ell m}] = [E_{jk} - E_{kj}, E_{\ell m} - E_{m\ell}]$$

$$\begin{aligned}
&= [E_{jk}, E_{\ell m}] - [E_{jk}, E_{m\ell}] - [E_{kj}, E_{\ell m}] + [E_{kj}, E_{m\ell}] \\
&= \delta_{k\ell} E_{jm} - \delta_{jm} E_{\ell k} - (\delta_{km} E_{j\ell} - \delta_{j\ell} E_{mk}) \\
&\qquad - (\delta_{j\ell} E_{km} - \delta_{km} E_{\ell j}) + (\delta_{jm} E_{k\ell} - \delta_{k\ell} E_{mj}) \\
&= \delta_{k\ell}(E_{jm} - E_{mj}) + \delta_{jm}(E_{k\ell} - E_{\ell k}) \\
&\qquad - \delta_{km}(E_{j\ell} - E_{\ell j}) - \delta_{j\ell}(E_{km} - E_{mk}) \\
&= \delta_{k\ell}\mathcal{L}_{jm} + \delta_{jm}\mathcal{L}_{k\ell} - \delta_{km}\mathcal{L}_{j\ell} - \delta_{j\ell}\mathcal{L}_{km}.
\end{aligned}$$

## ■ Solution B.3

Show that $[\mathcal{C}_1, L_{jk}] = 0$:

$$\begin{aligned}
2[\mathcal{C}_1, L_{\ell m}] &= [L_{jk}L_{jk}, L_{\ell m}] \quad \text{(implied sum over } j,k) \\
&= L_{jk}[L_{jk}, L_{\ell m}] + [L_{jk}, L_{\ell m}]L_{jk} \\
&= iL_{jk}(\delta_{j\ell}L_{km} + \delta_{km}L_{j\ell} - \delta_{jm}L_{k\ell} - \delta_{k\ell}L_{jm}) \\
&\qquad + i(\delta_{j\ell}L_{km} + \delta_{km}L_{j\ell} - \delta_{jm}L_{k\ell} - \delta_{k\ell}L_{jm})L_{jk} \\
&= i(L_{\ell k}L_{km} + L_{jm}L_{j\ell} - L_{mk}L_{k\ell} - L_{j\ell}L_{jm}) \\
&\qquad + i(L_{km}L_{\ell k} + L_{j\ell}L_{jm} - L_{k\ell}L_{mk} - L_{jm}L_{j\ell}) \\
&= +i(L_{k\ell}L_{mk} + L_{jm}L_{j\ell} - L_{km}L_{\ell k} - L_{j\ell}L_{jm}) \\
&\qquad + i(L_{km}L_{\ell k} + L_{j\ell}L_{jm} - L_{k\ell}L_{mk} - L_{jm}L_{j\ell}) \\
&= 0.
\end{aligned}$$

## ■ Solution B.4

This is easily done using the expression (B.55) for the nonzero commutators and the antisymmetry property $L_{jk} = -L_{kj}$. For example,

$$[L_1, L_2] = [L_{23}, L_{31}] = -[L_{32}, L_{31}] = -iL_{21} = iL_3,$$

and similarly for the other commutators of the form $[L_j, L_k]$,

$$\begin{aligned}
[L_1, A_1] &= [L_{23}, L_{14}] = 0, \\
[L_1, A_2] &= [L_{23}, L_{24}] = iL_{34} = iA_3, \\
[L_1, A_3] &= [L_{23}, L_{34}] = -[L_{32}, L_{34}] = -iL_{24} = -iA_2,
\end{aligned}$$

and similarly for commutators of the form $[L_2, A_k]$ and $[L_3, A_k]$,

$$[A_1, A_2] = [L_{14}, L_{24}] = [L_{41}, L_{42}] = iL_{12} = iL_3,$$

and similarly for the other commutators of the form $[A_j, A_k]$.

## ■ Solution B.5

The easiest approach is to simplify the expression for $\mathcal{C}_2$ and use the explicit so(4) commutation relations (5.129) to (5.131) rather than the general commutation relations (B.54):

$$8\mathcal{C}_2 = L_{12}(L_{34} - L_{43}) + L_{13}(-L_{24} + L_{42}) + L_{14}(L_{23} - L_{32})$$

$$
\begin{aligned}
&- L_{21}(L_{34} - L_{43}) - L_{31}(-L_{24} + L_{42}) - L_{41}(L_{23} - L_{32}) \\
&+ L_{23}(L_{14} - L_{41}) - L_{32}(L_{14} - L_{41}) \\
&+ L_{24}(-L_{13} + L_{31}) - L_{42}(-L_{13} + L_{31}) \\
&+ L_{34}(L_{12} - L_{21}) - L_{43}(L_{12} - L_{21}) \\
={}& 2L_{12}(L_{34} - L_{43}) + 2L_{13}(-L_{24} + L_{42}) + 2L_{14}(L_{23} - L_{32}) \\
&+ 2L_{23}(L_{14} - L_{41}) + 2L_{24}(-L_{13} + L_{31}) + 2L_{34}(L_{12} - L_{21}) \\
={}& 4(L_{12}L_{34} - L_{13}L_{24} + L_{14}L_{23} + L_{23}L_{14} - L_{24}L_{13} + L_{34}L_{12}).
\end{aligned}
$$

Now using the correspondence (B.62),

$$
8\mathcal{C}_2 = 4(L_3 A_3 + L_2 A_2 + A_1 L_1 + L_1 A_1 + A_2 L_2 + A_3 L_3).
$$

Therefore

$$
\mathcal{C}_2 = \tfrac{1}{2}(\boldsymbol{L} \cdot \boldsymbol{A} + \boldsymbol{A} \cdot \boldsymbol{L}).
$$

Now evaluate the commutators of $\mathcal{C}_2$ with $L_k$ and $A_k$:

$$
\begin{aligned}
2[\mathcal{C}_2, L_k] &= [L_j A_j + A_j L_j, L_k], \quad \text{(implied sum over } j) \\
&= L_j[A_j, L_k] + [L_j, L_k]A_j + A_j[L_j, L_k] + [A_j, L_k]L_j \\
&= i\epsilon_{jk\ell}(L_j A_\ell + L_\ell A_j + A_j L_\ell + A_\ell L_j) \\
&= 0, \quad \text{from Exercise 2.1.}
\end{aligned}
$$

### ■ Solution B.6

Since matrix multiplication is associative and the identity element is $I_n$ we need only verify closure and the existence of the inverse matrix. Let $A, B \in \mathrm{O}(n)$. Then $A^{-1} = A^T$ and $(AB)^T(AB) = B^T A^T AB = B^T B = I_n$ so $AB \in \mathrm{O}(n)$ and $\mathrm{O}(n)$ is a group. $\mathrm{SO}(n)$ is a subgroup of $\mathrm{O}(n)$ since $\det(AB) = \det(A)\det(B) = 1$ if $A, B \in \mathrm{SO}(n)$.

Let $A \in \mathrm{O}(p,q)$ and define $C = GA^T G$. Then $CA = GA^T GA = G^2 = I_n$ and $AC = AGA^T G = (A^T GA)^T G = G^2 = I_n$ so $C = A^{-1}$ is the inverse of $A$. For closure let $A, B \in \mathrm{O}(p,q)$. Then $(AB)^T G(AB) = B^T A^T GAB = B^T GB = G$ so $AB \in \mathrm{O}(p,q)$. Finally the above property of determinants shows that $\mathrm{SO}(p,q)$ is also a group.

### ■ Solution B.7

Again the derivation is similar to that of Exercise B.1 and Exercise 1.11. Any $n \times n$ matrix $\mathcal{L}$ with the block structure (B.72) can be expressed in terms of the elementary matrices $E_{jk}$ as

$$
\mathcal{L} = \sum_{j,k=1}^{n} \alpha_{jk} E_{jk}
$$

$$= \sum_{j,k=1}^{p} a_{jk} E_{jk} + \sum_{j=1}^{p} \sum_{k=p+1}^{n} c_{jk} E_{jk} + \sum_{j=p+1}^{n} \sum_{k=1}^{p} c_{kj} E_{jk}$$

$$+ \sum_{j,k=p+1}^{n} b_{jk} E_{jk}.$$

Since $a_{jk} = -a_{kj}$ and $b_{jk} = -b_{kj}$ then

$$\mathcal{L} = \sum_{1 \leq j < k \leq n} a_{jk}(E_{jk} - E_{kj}) + \sum_{p+1 \leq j < k \leq n} b_{kj}(E_{kj} - E_{jk})$$

$$+ \sum_{j=1}^{p} \sum_{k=p+1}^{n} c_{jk}(E_{jk} + E_{kj}).$$

The number of terms is $p(p-1)/2 + q(q-1)/2 + pq = n(n-1)/2$ so every $\mathcal{L} \in so(p,q)$ is a linear combination of the matrices (B.73).

■ **Solution B.8**

$$\begin{aligned}
[\mathcal{L}_{jk}, \mathcal{L}_{\ell m}] &= [g_{js} E_{sk} - g_{ks} E_{sj}, g_{\ell t} E_{tm} - g_{mt} E_{t\ell}] \\
&= g_{js} g_{\ell t} [E_{sk}, E_{tm}] - g_{js} g_{mt} [E_{sk}, E_{t\ell}] \\
&\quad - g_{ks} g_{\ell t} [E_{sj}, E_{tm}] + g_{ks} g_{mt} [E_{sj}, E_{t\ell}] \\
&= g_{js} g_{\ell t} (\delta_{kt} E_{sm} - \delta_{sm} E_{tk}) - g_{js} g_{mt} (\delta_{kt} E_{s\ell} - \delta_{s\ell} E_{tk}) \\
&\quad - g_{ks} g_{\ell t} (\delta_{jt} E_{sm} - \delta_{sm} E_{tj}) + g_{ks} g_{mt} (\delta_{jt} E_{s\ell} - \delta_{s\ell} E_{tj}) \\
&= g_{\ell k}(g_{js} E_{sm} - g_{mt} E_{tj}) - g_{jm}(g_{\ell t} E_{tk} - g_{ks} E_{s\ell}) \\
&\quad - g_{mk}(g_{js} E_{s\ell} - g_{\ell t} E_{tj}) + g_{j\ell}(g_{mt} E_{tk} - g_{ks} E_{sm}) \\
&= g_{\ell k} \mathcal{L}_{jm} - g_{jm} \mathcal{L}_{\ell k} - g_{mk} \mathcal{L}_{j\ell} + g_{j\ell} \mathcal{L}_{mk} \\
&= g_{k\ell} \mathcal{L}_{jm} + g_{jm} \mathcal{L}_{k\ell} - g_{km} \mathcal{L}_{j\ell} - g_{j\ell} \mathcal{L}_{km}.
\end{aligned}$$

■ **Solution B.9**

Show that $[Q_2, L_{jk}] = 0$:

$$\begin{aligned}
2[Q_2, L_{jk}] &= g_{ac} g_{bd} [L_{ab} L_{cd}, L_{jk}] \\
&= g_{ac} g_{bd} (L_{ab}[L_{cd}, L_{jk}] + [L_{ab}, L_{jk}] L_{cd}) \\
&= i g_{ac} g_{bd} \{ L_{ab}(g_{cj} L_{dk} + g_{dk} L_{cj} - g_{ck} L_{dj} - g_{dj} L_{ck}) \\
&\quad + (g_{aj} L_{bk} + g_{bk} L_{aj} - g_{ak} L_{bj} - g_{bj} L_{ak}) L_{cd} \}.
\end{aligned}$$

Now use $g_{ac} g_{cj} = \delta_{aj}$, etc., to obtain

$$\begin{aligned}
2[Q_2, L_{jk}] &= i(\delta_{aj} g_{bd} L_{ab} L_{dk} + \delta_{bk} g_{ac} L_{ab} L_{cj} \\
&\quad - \delta_{ak} g_{bd} L_{ab} L_{dj} - \delta_{bj} g_{ac} L_{ab} L_{ck} \\
&\quad + \delta_{cj} g_{bd} L_{bk} L_{cd} + \delta_{dk} g_{ac} L_{aj} L_{cd} \\
&\quad - \delta_{ck} g_{bd} L_{bj} L_{cd} - \delta_{dj} g_{ac} L_{ak} L_{cd})
\end{aligned}$$

$$\begin{aligned}
&= i(g_{bd}L_{jb}L_{dk} + g_{ac}L_{ak}L_{cj} - g_{bd}L_{kb}L_{dj} - g_{ac}L_{aj}L_{ck} \\
&\quad + g_{bd}L_{bk}L_{jd} + g_{ac}L_{aj}L_{cj} - g_{bd}L_{bj}L_{kd} - g_{ac}L_{ak}L_{cj}) \\
&= ig_{bd}(L_{jb}L_{dk} - L_{kb}L_{dj} + L_{bk}L_{jd} - L_{bj}L_{kd}) \\
&\quad + ig_{ac}(L_{ak}L_{cj} - L_{aj}L_{ck} + L_{aj}L_{ck} - L_{ak}L_{cj}) \\
&= 0.
\end{aligned}$$

## ■ Solution B.10

Use the antisymmetry of $\epsilon_{\alpha abcd}$ and $L_{jk}$:

$$\begin{aligned}
8w_1 &= \epsilon_{1abcd}L_{cd}L_{cd}, \quad \text{implied sum} \\
&= 2(L_{23}L_{45} - L_{24}L_{35} + L_{25}L_{34}) \\
&\quad + 2(L_{23}L_{45} + L_{34}L_{25} - L_{35}L_{24}) \\
&\quad + 2(-L_{24}L_{35} + L_{34}L_{25} + L_{45}L_{23}) \\
&\quad + 2(L_{25}L_{34} - L_{35}L_{24} + L_{45}L_{23}) \\
&= L_{23}L_{45} - L_{24}L_{35} + L_{25}L_{34},
\end{aligned}$$

since the operators in each product commute. Therefore

$$w_1 = T_2 L_1 - (\boldsymbol{A} \times \boldsymbol{B})_1.$$

Similarly by replacing indices (2345) with (1345), (1245), (1235) and (1234)

$$\begin{aligned}
w_2 &= -(L_{13}L_{45} - L_{14}L_{35} + L_{15}L_{34}) = T_2 L_2 - (\boldsymbol{A} \times \boldsymbol{B})_2, \\
w_3 &= L_{12}L_{45} - L_{14}L_{25} + L_{15}L_{24} = T_2 L_3 - (\boldsymbol{A} \times \boldsymbol{B})_3, \\
w_4 &= -(L_{12}L_{35} - L_{13}L_{25} + L_{15}L_{23}) = -\boldsymbol{L} \cdot \boldsymbol{B}, \\
w_5 &= L_{12}L_{34} - L_{13}L_{24} + L_{14}L_{23} = \boldsymbol{L} \cdot \boldsymbol{A}.
\end{aligned}$$

Therefore

$$\begin{aligned}
Q_4 &= w_1^2 + w_2^2 + w_3^2 + w_4^2 - w_5^2 \\
&= (T_2 \boldsymbol{L} - \boldsymbol{A} \times \boldsymbol{B})^2 + (\boldsymbol{L} \cdot \boldsymbol{B})^2 - (\boldsymbol{L} \cdot \boldsymbol{A})^2.
\end{aligned}$$

## J.13　Solutions to Appendix C Exercises

## ■ Solution C.1

The Taylor series expansion about $\alpha = 0$ of $f(e^\alpha r)$ considered as a function of $\alpha$ is

$$f(e^\alpha r) = \sum_{n=0}^{\infty} \frac{1}{n!} \frac{d^n}{d\alpha^n} f(e^\alpha r) \bigg|_{\alpha=0} \alpha^n.$$

Letting $y = e^\alpha r$ it follows that

$$\frac{d}{d\alpha} = \frac{dy}{d\alpha} \frac{d}{dy} = y \frac{d}{dy} = y \frac{dr}{dy} \frac{d}{dr} = r \frac{d}{dr},$$

and

$$f(e^\alpha r) = \sum_{n=0}^{\infty} \frac{1}{n!} \left( \alpha r \frac{d}{dr} \right)^n f(r) = e^{\alpha r \frac{d}{dr}} f(r).$$

## ■ Solution C.2

$$\begin{aligned}
\tilde{r} &= e^{ia\alpha T_2} r e^{-ia\alpha T_2} \\
&= \sum_{n=0}^{\infty} \frac{1}{n!} [r, -ia\alpha T_2]^{(n)} = \sum_{n=0}^{\infty} \frac{1}{n!} (-ia\alpha)^n [r, T_2]^{(n)}.
\end{aligned}$$

Now use the realization (4.54) and the basic commutation relation $[r, p_r] = i$ to obtain

$$[r, T_2]^{(1)} = [r, T_2] = \frac{1}{2} [r, rp_r] = \frac{r}{a} [r, p_r] = \frac{i}{a} r.$$

The general result $[r, T_2]^{(n)} = \left( \frac{i}{a} \right)^n r$ follows by induction since

$$[r, T_2]^{(n+1)} = [[r, T_2]^{(n)}, T_2] = \left( \frac{i}{a} \right)^n [r, T_2] = \left( \frac{i}{a} \right)^{n+1} r.$$

Therefore

$$\tilde{r} = \sum_{n=0}^{\infty} \frac{1}{n!} (-ia\alpha)^n \left( \frac{i}{a} \right)^n r = \sum_{n=0}^{\infty} \frac{\alpha^n}{n!} r = e^\alpha r.$$

The result for $p_r$ is proved in a similar fashion using

$$[p_r, T_2] = \frac{1}{a} [p_r, rp_r] = \frac{1}{a} [p_r, r] p_r = -\frac{i}{a} p_r.$$

Finally (C.14) can be obtained as follows.

$$\widetilde{T_3} = e^{ia\alpha T_2} T_3 e^{-ia\alpha T_2} = \sum_{n=0}^{\infty} \frac{1}{n!} (-ia\alpha)^n [T_3, T_2]^{(n)}.$$

Since

$$\begin{aligned}
[T_3, T_2]^{(1)} &= [T_3, T_2] = -iT_1, \\
[T_3, T_2]^{(2)} &= [[T_3, T_2], T_2] = -i[T_1, T_2] = -T_3,
\end{aligned}$$

the general results

$$[T_3, T_2]^{(2n)} = (-1)^n T_3, \quad [T_3, T_2]^{(2n+1)} = (-1)^{n+1} T_1, \quad n \geq 0,$$

easily follow by induction and

$$\begin{aligned}
\widetilde{T_3} &= \sum_{n=0}^{\infty} \frac{(-ia\alpha)^{2n}}{(2n)!} (-1)^n T_3 + \sum_{n=0}^{\infty} \frac{(-ia\alpha)^{2n+1}}{(2n+1)!} (-1)^{n+1} iT_1 \\
&= T_3 \cosh a\alpha - T_1 \sinh a\alpha.
\end{aligned}$$

In a similar manner the transformation of $\boldsymbol{A}$ can be obtained.

# Bibliography


[AB65] Abramowitz, M., and Stegun, I. "Handbook of Mathematical Functions," Dover, New York, 1965.

[AD80] Adams, B.G., Avron, J.E., Čížek, J., Otto, P., Paldus, J., Moats, R.K., and Silverstone, H.J. "Bender-Wu formulas for degenerate eigenvalues," *Phys. Rev. Lett.*, **21**, 1914, (1980).

[AD82] Adams, B.G., Čížek, J., and Paldus, J. "Representation Theory of so(4,2) for the Perturbation Treatment of Hydrogenic-Type Hamiltonians by Algebraic Methods," *Int J. Quantum Chem.*, **21**, 153, (1982).

[AD88a] Adams, B.G. "Application of 2-point Padé approximants to the ground state of the 2-dimensional hydrogen atom in an external magnetic field," *Theor. Chim. Acta*, **73**, 459, (1988).

[AD88b] Adams, B.G., Čížek, J., and Paldus, J. "Lie Algebraic Methods and Their Application to Simple Quantum Systems," *Adv. Quantum Chem.*, **19**, 1, (1988).

[AD92] Adams, B.G. "Unified treatment of high-order perturbation theory for the Stark effect in a two- and three-dimensional hydrogen atom," *Phys. Rev.*, **A46**, 4060, (1992).

[AL69] Alexander, M.A. *Phys. Rev.*, **178**, 34, (1969).

[AL73] Alonzo, M., and Valk, H. "Quantum Mechanics: Principles and Applications," Addison-Wesley, Reading, 1973.

[AL74] Alliluev, S.P., and Malkin, I.A. "Calculations of the Stark effect in hydrogen atoms by using the dynamical symmetry $O(2,2) \times O(2)$," *Sov. Phys. -JETP*, **39**, 627, (1974). Russian original in *Zh. Eksp. Teor. Fiz*, **66**, 1283, (1974).

[AR84] Arteca, G.A., Fernández, F.M., Mesón, A.M. and Castro, E.A. "A Method for Summing Strongly Divergent Perturbation Series: The Zeeman Effect in Hydrogen," *Physica*, **128A**, 253, (1984).

[AR90] Arteca, G.A., Fernández, F.M. and Castro, E.A. "Large Order Perturbation Theory and Summation Methods in Quantum Mechanics," Lecture Notes in Chemistry, Vol. 53, Springer-Verlag, Berlin, 1990.

[AU80] Austin, E.J. "Perturbation theory and Padé approximants for a hydrogen atom in an electric field," *Mol. Phys.*, **40**, 893, (1980).

[AV77]  Avron, J.E., Herbst, I.W., and Simon, B. *Phys. Rev. Lett.*, **62A**, 214, (1977).

[AV79]  Avron, J.E., Adams, B.G., Čížek J., Clay, M., Glasser, L., Otto, P., Paldus, J., and Vrscay, E. "Bender-Wu Formula, the SO(4,2) Dynamical Group, and the Zeeman Effect in Hydrogen," *Phys. Rev. Lett.*, **43**, 691, (1979).

[AV81a]  Avron, J.E. *Ann. Phys. (N.Y.)*, **131**, 73, (1981).

[AV81b]  Avron, J.E., Herbst, I., and Simon, B. *Commun. Math. Phys.*, **79**, 529, (1981).

[AV82]  Avron, J.E. "Bender-Wu Formulas and Classical Trajectories: Higher Dimensions and Degeneracies," *Int. J. Quantum Chem.*, **21**, 119, (1982).

[BA36]  Bargmann, V. *Z. Physik*, **99**, 576, (1936).

[BA65]  Barut, A.O., and Fronsdal, C. "On non-compact groups II. Represnetations of the 2+1 Lorentz group," *Proc. Roy. Soc.*, **A287**, 532, (1965).

[BA70]  Barut, A.O., and Böhm, A. "Reduction of a Class of O(4,2) Representations with Respect to SO(4,1) and SO(3,2)," *J. Math. Phys.*, **11**, 2938, (1970).

[BA71a]  Barut, A.O. "Dynamical Groups and Generalized Symmetries in Quantum Theory," Univ. of Canterbury, Christchurch, New Zealand, 1971.

[BA71b]  Barut, A.O., and Bornzin, G.L. "SO(4,2) Formulation of the Symmetry Breaking in Relativistic Kepler Problems with or without Magnetic Charges," *J. Math. Phys.*, **12**, 841, (1971).

[BA75]  Baker, G. A. Jr., "Essentials of Padé Approximants," Academic Press, New York, 1975.

[BE57]  Bethe, H.A., and Salpeter, E.E. "Quantum Mechanics of One- and Two-electron Atoms," Springer-Verlag, 1957 (also published by Plenum, New York, 1977).

[BE71]  Bender, C.M. and Wu, T.T. *Phys. Rev. Lett.*, **27**, 461, (1971).

[BE73]  Bednár, M. "Algebraic Treatment of Quantum Mechanical Models with Modified Coulomb Potentials," *Ann. Phys. (N.Y.)*, **75**, 305, (1973).

[BE73]  Bender, C.M. and Wu, T.T. *Phys. Rev.*, **D7**, 1620, (1973).

[BE80]  Benassi, L. and Grecchi, V. "Resonances in the Stark effect and strongly asymptotic approximants," *J. Phys. B*, **13**, 911, (1980).

[BI61]  Biedenharn, L.C. "Wigner Coefficients for the $R_4$ Group and Some Applications," *J. Math. Phys.*, **3**, 433, (1961).

[BI81a]  Biedenharn, L.C., and Louck, J.D. "Angular Momentum in Quantum Physics. Theory and Applications," ["Encyclopedia of Mathematics and

Its Applications" (G.-C.Rota, Ser. ed.)], Vol. 8. Addison-Wesley, New York, 1981.

[BI81b] Biedenharn, L.C., and Louck, J.D. "The Racah-Wigner Algebra in Quantum Theory," ["Encyclopedia of Mathematics and Its Applications" (G.-C.Rota, Ser. ed.)], Vol. 9. Addison-Wesley, New York, 1981.

[BO66] Bohm, A. "Dynamical Groups of Simple Nonrelativistic Models," *Nuovo Cimento*, **A43**, 665, (1966).

[BO86] Bohm, A. "Quantum Mechanics: Foundations and Applications," 2nd Ed., Springer-Verlag, New York, 1986.

[BR55] Brueckner, K.A. "Many-Body Problem for Strongly Interacting Particles. II Linked Cluster Expansion," *Phys. Rev.*, **100**, 36, (1955).

[BR68] Brink, D.M., and Satchler, G.R. "Angular Momentum," 2nd Ed., Clarendon Press, Oxford, 1968.

[CH91a] Char, B.W., Geddes, K.O., Gonnet, G.H., Leong, B.L., Monagan, M.B., Watt, S.M. "Maple V Language Reference Manual," Springer-Verlag, New York, 1991.

[CH91b] Char, B.W., Geddes, K.O., Gonnet, G.H., Leong, B.L., Monagan, M.B., Watt, S.M. "Maple V Library Reference Manual," Springer-Verlag, New York, 1991.

[CH91c] Char, B.W., Geddes, K.O., Gonnet, G.H., Leong, B.L., Monagan, M.B., Watt, S.M. "First Leaves: A Tutorial Introduction to Maple," Springer-Verlag, New York, 1991.

[CI77a] Čížek, J., and Paldus, J. "An Algebraic Approach to Bound States of Simple One-Electron Systems," *Int. J. Quantum Chem.*, **12**, 875, (1977).

[CI77b] Čížek, J., and Vrscay, E.R. "On the Use of the SO(4,2) Dynamical Group for the Study of the Ground State of a Hydrogen Atom in a Homogeneous Magnetic Field," Group Theoretical Methods in Physics, Proceedings of the Fifth International Colloquium, Academic Press, New York, 1977, pp 155–160.

[CI82] Čížek, J., and Vrscay, E.R. "Large Order Perturbation Theory in the Context of Atomic and Molecular Physics–Interdisciplinary Aspects," *Int. J. Quantum Chem.*, **21**, 27, (1982).

[CI87] Čížek, J., and Vinette, F. "N-dimensional hydrogen atom in an external spherically symmetric field," *Theor. Chim. Acta*, **72**, 497, (1987).

[CL79] "An Algebraic Approach to the Calculation of the Long-Range Intermolecular Forces of a One-Electron Diatomic Molecular Ion," M. Math. Thesis, University of Waterloo, Ontario, 1979.

[CO84a] Cornwell, J.F. "Group Theory in Physics, Vol 1," Academic Press, London, 1984

[CO84b] Cornwell, J.F. "Group Theory in Physics, Vol 2," Academic Press, London, 1984

[CO67] Coulson, C.A., and Joseph, A. "Self-Adjoint Ladder Operators II," *Rev. Mod. Phys.*, **39**, 838, (1967).

[DA55] Dalgarno, A., and Lewis, J.T. "The exact calculation of long-range forces between atoms by perturbation theory," *Proc. Roy. Soc.*, **A233**, 70, (1955).

[DA56a] Dalgarno, A., and Stewart, A.L. "On the perturbation theory of small disturbances," *Proc. Roy. Soc.*, **A238**, 269, (1956).

[DA56b] Dalgarno, A., and Stewart, A.L. "A perturbation calculation of properties of the $1s\sigma$ and $2p\sigma$ states of $HeH^{2+}$," *Proc. Roy. Soc.*, **A238**, 276, (1956).

[DE63] de-Shalit, A., and Talmi, I. "Nuclear Shell Theory," Academic Press, New York, 1963.

[DI61] Dixmier, T. "Représentations Intégrables du Groupe de De Sitter," *Bull. Soc. Math. France*, **89**, 9, (1961).

[DU60] Dupont-Bourdelet, F., Tillieu, J., and Guy, J. "Sur le calcul des énergies perturbées d'ordre quelconque," *J. Phys. Radium*, **21**, 776, (1960).

[ED57] Edmonds, A.R. "Angular Momentum in Quantum Mechanics," Princeton University Press, 1957.

[EP62] Epstein, J.H. and Epstein, S.T. *Amer. J. Phys.*, **30**, 266, (1962).

[FE39] Feynman, R.P. *Phys. Rev.*, **56**, 340, (1939).

[FE84] Fernández, F.M. and Castro, E.A. "Perturbation Theory without Wave Function for Multidimensional Systems," *Intern. J. Quantum Chem.*, **26**, 497, (1984).

[FE85] Fernández, F.M. and Castro, E.A. "Perturbation Calculation on the Hydrogen Atom in Electric and Magnetic Fields," *Intern. J. Quantum Chem.*, **28**, 603, (1985).

[FE87a] Fernández, F.M., Ogilvie, J.F. and Tipping, R.H. "Perturbation theory from moment recurrence relations," *J. Phys. A: Math. Gen.*, **20**, 3777, (1987).

[FE87b] Fernández, F.M. and Castro, E.A. "Hypervirial Theorems, Lecture Notes in Chemistry, Vol. 43," Springer, New York, 1987.

[FE92] Fernández, F.M. "Large-order pertubation theory without a wavefunction for the LoSurdo-Stark effect in hydrogenic atoms," *J. Phys. A: Math. Gen.*, **25**, 495, (1992).

[GA76] Galindo, A., and Pascual, P. "Hydrogen Atom in a Strong Magnetic Field," *Nuovo Cimento*, **34B**, 155, (1976).

[GA91] Galindo, A., and Pascual, P. "Quantum Mechanics II," Springer-Verlag, New York, 1991.

[GI74] Gilmore, R. "Lie Groups, Lie Algebras and Some of Their Applications," Wiley, New York, 1974.

[GO80] Goldstein, H. "Classical Mechanics," 2nd Ed., Addison-Wesley, 1980.

[GR78] Graffi, S. and Grecchi, V. "Resonances in Stark Effect and Perturbation Theory," *Commun. Math. Phys.*, **62**, 83, (1978).

[GR79] Grant, M. and Lai, C.S. "Hypervirial Theorems applied to the perturbation theory for screened Coulomb potentials," *Phys. Rev.*, **A20**, 718, (1979).

[GR80] Gradshteyn, I.S. and Ryzhik, M. "Tables of Integrals, Series and Products," Academic Press, New York, 1980.

[GR89] Greiner, W., and Müller, B. "Quantum Mechanics: Symmetries," Springer-Verlag, Berlin, 1989.

[HE37] Hellmann, H. "Einführung in die Quantenchemie," Deuticke, Vienna, 1937.

[HE74] Hehenberger, M., McIntosh, H.V. and Brändas, E. "Weyl's theory applied to the Stark effect in the hydrogen atom," *Phys. Rev.*, **A10**, 1494, (1974).

[HE79] Herbst, I. "Dilatation Analyticity in Constant Electric Field I. The Two Body Problem," *Commun. Math. Phys.*, **64**, 279, (1979).

[HI60] Hirschfelder, J.O. "Classical and Quantum Mechanical Hypervirial Theorems," *J. Chem. Phys.*, **33**, 1462, (1960).

[HU61] Huby, R. "Formulae for Non-degenerate Rayleigh Schrödinger Perturbation Theory in any order," *Proc. Phys. Soc. (London)*, **78**, 529, (1961).

[HU72] Humphreys, J.E. "Introduction to Lie Algebras and Representation Theory," Springer-Verlag, New York, 1972

[JA90] Janke, W. "Large-order perturbation theory of the Zeeman effect in hydrogen from a four-dimensional anisotropic anharmonic oscillator," *Phys. Rev. A*, **41**, 6071, (1990).

[JO67] Joseph, A. "Self-Adjoint Ladder Operators I," *Rev. Mod. Phys.*, **39**, 829, (1967).

[JO80] Johnson, B.R. "On a connection between radial Schrödinger equations for different power law potentials," *J. Math. Phys.*, **21**, 2640, (1980).

[KA89] Kahaner, D., Moler, C. and Nash, S. "Numerical Methods and Software," Prentice-Hall, New Jersey, 1989

[KI65] Kihlberg, A., and Ström, S. "On the unitary irreducible representations of the (1+4) de Sitter group," *Arkiv För Fysik*, **31**, 491, (1965).

[KI78] Killingbeck, J. "Perturbation Theory Without Wavefunctions," *Phys. Lett.*, **65A**, 87, (1978).

[LA77] Landau, L.D., and Lifshitz, E.M. "Quantum Mechanics (Non-relativistic Theory," 3rd Ed. Pergamon Press, Oxford, 1977.

[LA81] Lai, C.S. "Calculation of the Stark effect in Hydrogen atoms by using the hypervirial relations," *Phys. Lett.*, **83A**, 322, (1981).

[LI65] Lipkin, H.J. "Lie Groups for Pedestrians," North-Holland, Amsterdam, 1965.

[LO65] Löwdin, P.-O. "Studies in Perturbation Theory. IX. Connection Between Various Approaches in the Recent Development–Evaluation of Upper Bounds to Energy Eigenvalues in Schrödinger's Perturbation Theory," *J. Math. Phys.*, **6**, 1341, (1965).

[MA85] Marc, G. and McMillan, W.G. *Adv. Chem. Phys.*, **58**, 209, (1985).

[MC92] McRae, S.M. and Vrscay, E.R. "Canonical Perturbation Expansions to Large Order from Classical Hypervirial and Hellmann-Feynman Theorems," *J. Math. Phys*, , to appear Sept., (1992).

[ME70] Merzbacher, E. "Quantum Mechanics," 2nd Ed., John Wiley, New York, 1970.

[ME70a] Messiah, A. "Quantum Mechanics," Vol. I. North-Holland, Amsterdam, 1970.

[ME70b] Messiah, A. "Quantum Mechanics," Vol. II. North-Holland, Amsterdam, 1970.

[MI72] Miller, W. Jr. "Symmetry Groups and their Applications," Academic Press, New York, 1972.

[NA64] Naimark, M.A. "Linear Representations of the Lorentz Group," Pergamon Press, New York, 1964.

[NE50] Newton, T.D. *Ann. Math.*, **51**, 730, (1950).

[PA26] Pauli, W. "Über das Wasserstoffspektrum vom Standpunkt der neuen Quantenmechanik," *Z. Physik*, **36**, 336, (1926). English translation "On the Hydrogen Spectrum from the Standpoint of the new Quantum Mechanics," in "Sources of Quantum Mechanics," (B.L. van der Waerden, ed.), Dover, (page 387), 1968.

[PA76] Paldus, J. *In* "Theoretical Chemistry: Advances and Perspectives," (H. Eyring and D. Henderson, eds.), Vol 2, p. 131. Academic Press, New York, 1976.

[PO61] Powell, J.L. and Craseman, B. "Quantum Mechanics," Addison-Wesley, New York, 1961.

[PR80] Privman, V. "New Method of perturbation theory calculation of the Stark effect for the ground state of hydrogen," *Phys. Rev. A*, **22**, 1833, (1980).

[RE82] Reinhardt, W.P. "Padé Summations for the Real and Imaginary Parts of Atomic Stark Eigenvalues," *Intern J. Quantum Chem.*, **21**, 133, (1982).

[RO59] Rotenberg, M., Bivens, R., Metropolis, N. and Wooten Jr. J.K. "The 3-j and 6-j Symbols," Technology Press, M.I.T., Cambridge, MA, 1959

[RO84] Rösner, W., Wunner, G., Herold, H., and Ruder, H. "Hydrogen atoms in arbitrary magnetic fields: I. Energy levels and wavefunctions," *J. Phys. B*, **17**, 29, (1984).

[SA71] Saletan, E.J., and Cromer, A.H. "Theoretical Mechanics," Wiley, New York, 1971.

[SA86] Sattinger, D.H., and Weaver, O.L. "Lie Groups and Algebras with Applications to Physics, Geometry and Mechanics," Springer-Verlag, New York, 1986.

[SC26] Schrödinger, E. *Ann. Physik*, **79**, 361, (1926).

[SC59] Schwartz, C. "Calculations in Schrödinger Perturbation Theory," *Ann. Phys.*, **2**, 156, (1959).

[SC65] Schwinger, J. "On Angular Momentum" in "Quantum Theory of Angular Momentum," L.C. Biedenharn and H. van Dam, Eds, Academic Press, New York, 1965.

[SC68] Schiff, L. "Quantum Mechanics," 3rd Ed. McGraw-Hill, New York, 1968.

[SI70] Silverstone, H.J. and Holloway, T.T. "Explicit Formulas for the Nth-Order Wavefunction and Energy in Nondegenerate Rayleigh-Schrödinger Perturbation Theory," *J. Chem. Phys.*, **52**, 1472, (1970).

[SI78] Silverstone, H.J. "Perturbation theory of the Stark effect in hydrogen to arbitrarily high order," *Phys. Rev.*, **A18**, 1853, (1978).

[SI79] Silverstone, H.J., Adams, B.G., Čížek J. and Otto, P. "Stark Effect i Hydrogen: Dispersion Relation, Asymptotic Formulas, and Calculation of Ionsization Rate vis High-Order Perturbation Theory," *Phys. Rev. Lett.*, **43**, 1498, (1979).

[SI81] Silverstone, H.J., and Moats, R.K. "Practical recursive solutions of degenerate Rayleigh-Schrödinger perturbation theory and applications to high-order calculations of the Zeeman effect in hydrogen," *Phys. Rev.*, **A23**, 1645, (1981).

[SI82] Simon, B. "Large Orders and Summability of Eigenvalue Perturbation Theory: A Mathematical Overview," *Intern. J. Quantum Chem.*, **21**, 3, (1982).

[SL60] Slater, J.C. "Quantum Theory of Atomic Structure," Vol. II. McGraw-Hill, New York, 1960.

[ST65] Ström, S. "Construction of representations of the inhomogeneous Lorentz group by means of contarction of representations of the (1+4) de Sitter Group," *Arkiv För Fysik*, **30**, 455, (1965).

[SW72] Swenson, R.J. and Danforth, S.H. "Hypervirial and Hellmann-Feynman Theorems Applied to Anharmonic Operators," *J. Chem. Phys.*, **57**, 1734, (1972).

[TH41] Thomas, L.H. *Ann. Math.*, **42**, 113, (1941).

[VA68] van der Waerden, B.L. (Ed.) "Sources of Quantum Mechanics," Dover, 1968.

[VI88] Vinette, F. and Čížek J. "Perturbation energy expansion using hypervirial theorem and symbolic computation for the N-dimensional hydrogen atom in an external spherically symmetric field," *Computer Phys. Comm.*, **52**, 35, (1988).

[VR77] Vrscay, E.R. "The Use of the SO(4,2) Dynamical Group for the Study of the Ground State of a Hydrogen Atom in a Homogeneous Magnetic Field," M. Math. Thesis, University of Waterloo, Ontario, 1977.

[VR85] Vrscay, E.R. "Algebraic methods, Bender-Wu formulas, and continued fractions at large order for charmonium," *Phys. Rev.*, **A31**, 2054, (1985).

[VR86] Vrscay, E.R. "Hydrogen atom with a Yukawa potential. Perturbation theory and continued fractions–Padé approximation at large order," *Phys. Rev.*, **A33**, 1433, (1986).

[YA91] Yang, X.L., Guo, S.H., Chan, F.T., Wong, K.W. and Ching, W.Y. "Analytic solution of a two-dimensional hydrogen atom. I. Nonrelativistic theory," *Phys. Rev.*, **A43**, 1186, (1991).

[WY74] Wybourne, B.G. "Classical Groups for Physicists," Wiley, New York (1974).

# Index of Symbols Used

$\alpha$      Fine structure constant (see Section 4.12), parabolic scaling factor (see Section 4.14).

$\boldsymbol{A}$      Scaled Laplace-Runge-Lenz vector (Chapter 5.5).

$A_1, A_2, A_3$      Components of $\boldsymbol{A}$.

$A^2$      $A^2 = A_1^2 + A_2^2 + A_3^2$.

$\mathcal{C}_1, \mathcal{C}_2$      Casimir operators for so(4) (see Chapter 5).

$C_a^{(j)}$      Perturbed wavefunction coefficients (see Chapter 7.4.4) and perturbation coefficients for $\langle r^j \rangle$ (see Chapter 10.6).

$\delta_{jk}$      The Kronecker delta symbol defined so that $\delta_{jk} = 0$ if $j \neq k$ and 1 if $j = k$.

$D$      Dimensionality of Euclidean space (see Chapter 4.7).

$D_a^{(j)}$      Renormalized perturbed wavefunction coefficients (see Chapter 8.2.1).

$\Delta E$      Energy correction to unperturbed value: $E = E_0 + \Delta E$.

$E$      Energy eigenvalue.

$E^{(j)}$      $j$ th energy correction to the energy $E$ (see Chapter 7).

$\epsilon_{jk\ell}$      The Levi-Civita symbol (see Appendix A).

$f^{(j)}$      Perturbed parabolic separation constant coefficients (see Chapter 8.4).

$\gamma$      so(2,1) scaling factor $q/\mathcal{Z}$ (see Chapter 4).

$\boldsymbol{\Gamma}$      One of the vector operators of so(4,2) with components $\Gamma_1$, $\Gamma_2$ and $\Gamma_3$ (see Chapter 6.3).

$G$      Resolvent operator (see Chapter 7).

$\boldsymbol{J}$      General angular momentum vector.

$J_1, J_2, J_3$      Components of $\boldsymbol{J}$.

$J^2$      $J^2 = J_1^2 + J_2^2 + J_3^2$, the Casimir operator for su(2) with eigenvalues $j(j+1)$ where $j = 0, 1/2, 1, 3/2, \ldots$.

$J_-, J_+$      $J_- = J_1 - iJ_2$, $J_+ = J_1 + iJ_2$.

| | |
|---|---|
| $J_{-1}, J_0, J_1$ | Spherical components of $\boldsymbol{J}$ (see Chapter 3). |
| $k$ | so(2,1) Casimir operator $T^2$ has eigenvalues $\tau = k(k+1)$ (see Chapter 4.3). |
| $L_n^{(\alpha)}$ | Generalized Laguerre polynomial (see Chapter 4.11). |
| $\boldsymbol{L}$ | Orbital angular momentum vector (see Chapter 3). |
| $L_1, L_2, L_3$ | Components of $\boldsymbol{L}$. |
| $L_-, L_+$ | $L_- = L_1 - iL_2,\ L_+ = L_1 + iL_2$. |
| $L^2$ | $L^2 = L_1^2 + L_2^2 + L_3^2$, the Casimir operator for so(3) with eigenvalues $\ell(\ell+1)$ where $\ell = 0, 1, 2, \ldots$. |
| $\ell$ | Orbital angular momentum quantum number. |
| $_1F_1$ | Confluent hypergeometric function (see Chapter 4.11). |
| $\lambda, \Lambda$ | Perturbation parameters. |
| $m$ | Magnetic quantum number (eigenvalue of $L_3$). |
| $N_a$ | renormalization coefficient (see Chapter 8.2). |
| $n$ | Principal quantum number for a hydrogenic atom or harmonic oscillator (see Chapter 4). |
| $n_1, n_2$ | Parabolic quantum numbers for hydrogenic atom (see Chapter 4 and Chapter 8). |
| $\boldsymbol{p}$ | Linear momentum vector in 3-dim space with components $p_1 = p_x$, $p_2 = p_y$ and $p_3 = p_z$. In coordinate realization in atomic units $\boldsymbol{p} = -i\nabla$ (see Chapter 4). |
| $p_r$ | Radial momentum (see Chapter 4.4, Chapter 4.7 and Chapter 10.5). |
| $\boldsymbol{P}$ | Scaled linear momentum vector in 3-dim space with components $P_1 = P_X$, $P_2 == P_Y$ and $P_3 = P_Z$. In coordinate realization in atomic units $\boldsymbol{P} = -i\nabla$ (see Chapter 4). |
| $p^2$ | $p_1^2 + p_2^2 + p_3^2$. In coordinate realization in atomic units $p^2 = -\nabla^2$. |
| $P^2$ | As for $p^2$ but in scaled (model space) coordinate system. |
| $P_R$ | Scaled radial momentum (see Chapter 4.4). |
| $\Phi, \phi, \Psi, \psi$ | Wavefunctions. |
| $q$ | $T_3$ eigenvalue (see Chapter 4.3). |
| $Q$ | $Q = n_1 - n_2$ (see Chapter 8.4 and Chapter 8.5). |
| $q_1, q_2$ | Alternate parabolic quantum numbers. $q_i = k + 1 + n_i$, where $k = (|m| - 1)/2$ (see Chapter 4.14). |
| $\boldsymbol{r}$ | Position vector in 3-dim space with components $x_1 = x$, $y_1 = y$ and $z_1 = z$. |

| | |
|---|---|
| $r$ | $\|\boldsymbol{r}\| = \sqrt{x^2 + y^2 + z^2}$. |
| $\boldsymbol{R}$ | Scaled position vector in 3-dim space with components $X_1 = X$, $Y_1 = Y$ and $Z_1 = Z$. |
| $R$ | $\|\boldsymbol{R}\| = \sqrt{X^2 + Y^2 + Z^2}$ (see Chapter 4). |
| $S$ | $S = \gamma^2 R$. |
| so(2,1) | Lie algebra of SO(2,1) with generators $T_1$, $T_2$ and $T_3$ (see Chapter 3 and Appendix B). |
| so(3) | Lie algebra of the special orthogonal group SO(3) (Rotation group in 3 dimensions) with generators $L_1$, $L_2$ and $L_3$ (see Chapter 3 and Appendix B). |
| so(4) | Lie algebra of the special orthogonal group SO(4), rotation group in 4 dimensions (see Chapter 5 and Appendix B). |
| so(4,2) | Lie algebra of the Lie Group SO(4,2) with 15 generators $\boldsymbol{L}$, $\boldsymbol{A}$, $\boldsymbol{B}$, $\boldsymbol{\Gamma}$, $T_1$, $T_2$ and $T_3$ (see Chapter 6 and Appendix B). |
| su(2) | Lie algebra of the special unitary group SU(2) with generators $J_1$, $J_2$ and $J_3$. |
| $T_1, T_2, T_3$ | Generators of the so(2,1) Lie algebra. |
| $T^2$ | $T^2 = T_3^2 - T_1^2 - T_2^2$, the Casimir operator for so(2,1) with eigenvalues $\tau = k(k+1)$ (see Chapter 4). |
| $\tau$ | Eigenvalue of the so(2,1) Casimir operator $T^2$. |
| $\vartheta, \varphi$ | $x = r \sin\vartheta \cos\varphi$, $y = r \sin\vartheta \sin\varphi$, $z = r \cos\varphi$ (spherical coordinates). |
| $\xi, \eta$ | Parabolic coordinates (see Chapter 4.14). |
| $x, y, z$ | Physical cartesian coordinates in 3-dim space. |
| $x_1, x_2, x_3$ | $x_1 = x$, $x_2 = y$ and $x_3 = z$ |
| $X, Y, Z$ | Scaled cartesian coordinates in 3-dim model space (see Chapter 4). |
| $X_1, X_2, X_3$ | $X_1 = X$, $X_2 = Y$ and $X_3 = Z$ |
| $Y_{\ell m}(\vartheta, \varphi)$ | Spherical harmonic functions. |
| $W$ | Scaled perturbation (see Chapter 7.3). |
| $\mathcal{Z}$ | Nuclear charge in atomic units ($\hbar = e = m = 1$). |

# Index


accidental degeneracy, 99

Casimir operator, 13
charmonium, 211, 212
    $E^{(1)}$ to $E^{(4)}$, 220
    algebraic RSPT, 212
    energy tables, 313
    iterative formulas, 216
    low order calculations, 213
    resolvent operator, 213
    symbolic calculations, 214
    symbolic programs, 216
commutator, 2
    identities, 12, 17, 18
confluent hypergeometric function, 64
conventional RSPT, 120
    high order iteration, 123
    iteration formula, 122
    low order formulas, 122, 136
    resolvent operator, 121
    Zeeman effect, 129

Dalgarno-Lewis method, *see* Stark effect
Dirac-Coulomb equation, 68

general linear Lie algebra, 3
general radial equations, 71
$gl(n, \mathcal{F})$, 3, 12

harmonic oscillator
    3-dim case, 53
    D-dim case, 56
harmonium, 211, 212
    $E^{(1)}$ to $E^{(4)}$, 220
    algebraic RSPT, 212
    energy tables, 331

    iterative formulas, 216
    low order calculations, 213
    resolvent operator, 213
    symbolic calculations, 214
    symbolic programs, 216
Hellmann-Feynman theorem, 234
hermiticity
    $so(2,1)$ generators, 58
HVHF method, 233
    hydrogenic equations, 236
    hydrogenic systems, 234
    power potentials, 236
    screened Coulomb case, 240
hydrogenic atom
    3-dim case, 50
    D-dim case, 56
    parabolic coordinates, 72
hypervirial theorem, 234

index of symbols used, 443

Klein-Gordon equation, 66

Laguerre polynomials, 63
    hypergeometric form, 64
    Rodrigues' formula, 63
Laplace-Runge-Lenz vector, *see* LRL vector
Levi-Civita symbol, 245
    identities, 245
    properties, 245
Lie algebra, 247
    closure, 6
    closure property, 3
    commutator, 2, 12
    definition, 2
    general linear, 3
    irreducible rep, 10

matrix representation, 4
merging, 6, 105
real, 5
realization, 5
reducible rep, 10, 11
representation, 4
so($n$), 253
so(2,1), 258
so(3), 255
so(4), 256
so(4,1), 259
so(4,2), 259
so(p,q), 256
structure constants, 3
vector cross product, 12
Lie group, 247
  definition, 247, 248
  one-parameter subgroups, 248
  rotation group SO(2), 250
  rotation group SO(3), 250
  SO($n$), 253
  SO(p,q), 256
  special unitary group SU(2), 252
linear algebra
  definition, 1
LRL vector, 82, 94
  classical case, 94, 103
  classical definition, 94
  commutation relations, 98, 103
  identities, 96, 97, 103
  modified definition, 98
  quantum case, 95
  quantum definition, 96
  scaled version, 100
  tilted version, 264

Maple programs
  *alliluev*, 162
  *charmgs*, 221
  *charm*, 219
  *dalgarno*, 136
  *eigval*, 425
  *genstarkdata*, 141
  *genzeemandata*, 184

*gsstark*, 145
*gszeeman*, 186
*harmgs*, 222
*harm*, 219
*hvhfcharm*, 239
*hvhfharm*, 239
*hvhfpower*, 238
*hvhfscr*, 241
*moments*, 208
*reinhardt*, 154
*renorm*, 143
*rpower*, 216
*rzmatrix*, 267
*s1test2d*, 176
*s1test*, 172
*scrtest*, 242
*silverstone*, 168
*stest*, 169
*sumit*, 155
*symcharmgs*, 226
*symharmgs*, 226
*sympower*, 223
*symstark*, 149
*symzeeman*, 189
*test3-0*, 202
*test3-1*, 203
*u1test2d*, 176
*u1test*, 173
*unravel1*, 172
*unravel*, 170
*utest*, 171
*yukawa*, 231
*zee3s3d*, 197
*zrenorm*, 184
modified algebraic RSPT, 129
  high order formulas, 131, 132
  low order formulas, 132
  resolvent operator, 130, 134
  symmetric formula, 131–133, 136
modified LRL vector, 98
  identities, 98

orbital angular momentum, 7
  3-dim case, 54

D-dim case, 78, 79

parabolic coordinates, 72
Pauli spin matrices, 8, 13, 15, 252
perturbation matrix elements, 265
 $R$, 274
 $R(R^2 - Z^2)$, 279
 $RZ$, 275, 278
 $R^d$, 3-dim case, 276
 $R^d$, D-dim case, 266, 280
 $Z$, 277
 $Z^2$, 277
 3-dim case, 265
 symbolic calculation, 267
 symbolic formulas, 269
perturbation theory, 119
 continuum difficulties, 124
 conventional RSPT, 120
 HVHF method, 233
 modified algebraic RSPT, 129
 power series method, 228
power series method, 228

radial momentum, 78
 3-dim case, 46
 D-dim case, 55
real Lie algebra, 5
relativistic systems, 66

scalar operator, 30
 matrix elements, 30
scaled LRL vector, 100
screened Coulomb potentials, 226
 $E^{(1)}$ to $E^{(6)}$, 242
 algebraic RSPT, 227
 energy tables, 355
 power series method, 228
 Schrödinger equation, 227
 Yukawa potential, 227
so(2,1), 5, 258
 $R$ matrix elements, 113
 Casimir operator, 258
 commutation relations, 42, 83
 coordinate realizations, 45
 ladder operators, 43

 matrix representation, 45
 parabolic realization, 76
 realization, 48
 $T_3$ eigenfunctions, see $T_3$ eigenfunctions
 $T_3$ eigenvalues, see $T_3$ eigenvalues
 tilting transformation, 261
 unirreps, 44
so(3), 4, 7, 255
 boson realization, 40
 Casimir operator, 13, 26, 255
 commutation relations, 26
 ladder operators, 26
 matrix representation, 29
 natural realization, 7
 scalar operator, 30
 tensor operator, 37
 unirreps, 28
 vector operator, 30
so(3,1)
 Casimir operators, 87
so(4), 81, 256
 accidental degeneracy, 99
 as coupling problem, 89
 Bohr energy formula, 99
 Casimir operators, 87, 90, 102, 256
 commutation relations, 83, 101
 hydrogenic realizations, 97
 matrix representation, 88, 102
 Pauli realization, 82, 98
 scaled realization, 82, 100
so(4,1), 105, 259
 Casimir operators, 107, 259
 commutation relations, 106
 identities, 107, 117
 representation theory, 107, 108
 scaled realization, 106
so(4,2), 105, 259
 $R$ matrix elements, 113
 $Z$ matrix elements, 114
 Casimir operators, 110, 260
 commutation relations, 110

matrix representation, 111, 112
scaled realization, 108, 109
Z matrix elements, 277
so(n), 253
commutation relations, 254, 255
so(p,q), 256
commutation relations, 258
spherically symmetric systems, 211
charmonium, 211
harmonium, 211
HVHF method, 233
screened Coulomb potentials, 226
Stark effect, 137
$E^{(j)}$ for parabolic states, 166
$X^2$ matrix elements, 161
$\Psi^{(1)}$ to $\Psi^{(6)}$, 148
2-dim case, 174
asymptotic formula, 153
continuum difficulties, 124
conventional iteration scheme, 167
conventional RSPT, 160, 161
Dalgarno-Lewis method, 125, 136
degenerate program, 197
direct summation, 152
energy tables, 281
ground state program, 144
hand calculations, 134
high order formalism, 138
high order general case, 166
high order special cases, 172
iteration scheme, 140
matrix elements, 141
Padé summation, 153
parabolic coordinates, 158
parabolic state energy corrections, 171
scaled $T_3$ equation, 128
scaled hamiltonian, 127
Schrödinger equation, 138
separation constants, 161, 164, 170

series summation, 152
summation program, 155
symmetric formula, 140
symmetric program, 149
su(2), 4, 8
boson realization, 14
fermion realization, 9
isospin realization, 9
natural realization, 8
Pauli spin matrices, 8
su(3), 5, 14
Casimir operator, 14
su(n), 13

$T_3$ eigenfunctions
explicit form, 60
$T_3$ eigenvalues, 45, 49, 51, 54, 57, 67, 70, 77
tensor operator
definition, 37
spherical components, 37
tilting transformation, 261
hydrogenic hamiltonian, 263
operator transformations, 262
tilted LRL vector, 264

vector operator, 30
commutation relations, 31
identities, 31
matrix elements, 33, 36, 38
selection rules, 32

Wigner-Eckart theorem, 37

Yukawa potential, 227
$E^{(1)}$ to $E^{(6)}$, 243
energy series, 233
energy tables, 355

Zeeman effect, 179
$3s-3d_0$ degenerate level, 195
$\Psi^{(1)}$ to $\Psi^{(4)}$, 188
asymptotic formula, 191
conventional RSPT, 129
doubly-degenerate RSPT, 191, 192

energy tables, 295
ground state program, 186
hand calculations, 180
high order formalism, 182
iteration scheme, 183
matrix elements, 184
method of moments, 206
Schrödinger equation, 180
series summation, 191
symmetric program, 189